VOLCANIC SUCCESSIONS

MODERN AND ANCIENT

TITLES OF RELATED INTEREST

Boninites
A. J. Crawford (ed.)

Carbonatites
K. Bell (ed.)

Cathodoluminescence of geological materials
D. J. Marshall

Chemical fundamentals of geology
R. Gill

Crystal structures and cation sites of the rock-forming minerals
J. R. Smyth & D. L. Bish

The dark side of the Earth
R. Muir Wood

Deep marine environments
K. Pickering *et al.*

Deformation processes in minerals, ceramics and rocks
D. J. Barber & P. G. Meredith (eds)

Experiments in physical sedimentology
J. R. L. Allen

Igneous petrogenesis
B. M. Wilson

Image interpretation in geology
S. Drury

The interpretation of igneous rocks
K. G. Cox *et al.*

Introduction to X-ray spectrometry
K. L. Williams

Komatiites
N. Arndt & E. Nisbet (eds)

Mathematics in geology
J. Ferguson

Perspectives on a dynamic Earth
T. R. Paton

Petrology of the igneous rocks
F. Hatch *et al.*

Petrology of the metamorphic rocks
R. Mason

Planetary landscapes
R. Greeley

A practical approach to sedimentology
R. Lindholm

A practical introduction to optical mineralogy
C. D. Gribble & A. J. Hall

Principles of physical sedimentology
J. R. L. Allen

Rutley's elements of mineralogy
C. D. Gribble

Sedimentary structures
J. D. Collinson & D. B. Thompson

Sedimentology: process and product
M. R. Leeder

Simulating the Earth
J. Holloway & B. Wood

Statistical methods in geology
R. F. Cheeney

The young Earth
E. G. Nisbet

VOLCANIC SUCCESSIONS

MODERN AND ANCIENT

A geological approach to processes, products and successions

R. A. F. CAS
Department of Earth Sciences, Monash University

J. V. WRIGHT
Consultant, Sheffield, England

London
UNWIN HYMAN
Boston   Sydney   Wellington

© R. A. F. Cas and J. V. Wright, 1988

This book is copyright under the Berne Convention. No reproduction without permission. All rights reserved.

Published by the Academic Division of
Unwin Hyman Ltd
15/17 Broadwick Street, London W1V 1FP, UK

Unwin Hyman Inc.,
8 Winchester Place, Winchester, Mass. 01890, USA

Allen & Unwin (Australia) Ltd,
8 Napier Street, North Sydney, NSW 2060, Australia

Allen & Unwin (New Zealand) Ltd in association with the
Port Nicholson Press Ltd,
60 Cambridge Terrace, Wellington, New Zealand

First published in 1987
Second impression 1988

British Library Cataloguing in Publication Data

Cas, R.A.F.
    Volcanic successions: modern and ancient:
a geological approach to processes,
products and successions.
1. Rocks, Igneous
I. Title   II. Wright, J.V.
552'.2   QE461
ISBN 0-04-552021-6
ISBN 0-04-552022-4 Pbk

Library of Congress Cataloging in Publication Data

Cas, R.A.F. (Ray A. F.)
    Volcanic successions, modern and ancient.
Bibliography: p. 487
Includes index.
1. Volcanic ash, tuff, etc. 2. Volcanoes.
3. Volcanism. I. Wright, J. V. (John V.) II. Title.
QE461.C378 1987   551.2'1   86-17409
ISBN 0-04-552021-6 (alk. paper)
ISBN 0-04-552022-4 (pbk.: alk. paper)

Typeset in 10 on 12 point Plantin Light by Columns of Reading
and printed in Great Britain by The Oxford University Press, Oxford.

To our families

PREFACE

The idea for this book came into being between 1981 and 1982 when J.V.W. came to Monash University to take up a Monash Postdoctoral Fellowship. During this period a short course on facies analysis in modern and ancient successions was put together, integrating J.V.W.'s extensive volcanological experience in numerous modern volcanic terrains with R.A.F.C.'s extensive sedimentological and volcanological experience in older volcanic and associated sedimentary successions in the Palaeozoic and Precambrian of Australia. The enthusiastic response from the participants to the first short course, taught in May 1982, and to subsequent annual re-runs, encouraged us to develop the short course notes into this book.

The idea for both the short course and the book arose because we felt that there was no single source available that comprehensively attempted to address the problems of analysing, interpreting and understanding the complexity of processes, products and stratigraphy in volcanic terrains. Until 15 years ago, volcanic successions received attention primarily from igneous petrologists with principal interests in geochemistry, mineralogy and magma genesis. Although a number of books covering many aspects of physical volcanology have appeared since then, none has fully treated the subject by trying to integrate approaches from both modern and ancient volcanic successions, and from volcanological and sedimentological perspectives. One of our aims in the book is to provide geologists with a sound basis for making their own well founded interpretations. For that reason we cover not only concepts about processes, and the nature of the products, but also methods and approaches that may be useful in analysing both modern and ancient successions. Most importantly, we treat the diversity of products in volcanic terrains as facies, and we use the method of facies analysis and interpretation as a means of constructing facies models for different volcanic settings. These models will, we hope, be useful as norms for comparison for workers in ancient terrains. The only publication which overlaps with this one to any extent is the excellent book *Pyroclastic rocks* by Dick Fisher and Hans Schmincke.

Many people, organisations and institutions have directly or indirectly contributed to or made the production of this book possible. Foremost we acknowledge our PhD supervisors, George P. L. Walker (J.V.W.) and Gil Jones (R.A.F.C.) for their enlightened and stimulating supervision, and their continued interest thereafter. If anyone in the field of physical volcanology warrants special mention as a source of inspiration through a never-ending succession of outstanding contributions, it is George Walker. No other volcanologist has given so much to the science and its students. Thank you George. Financial support for our

ix

research and other visits to volcanic regions has come from: Commonwealth Postgraduate Award, Macquarie University, Monash University, ARGS and Otago University William Evans Visiting Fellowship (R.A.F.C.), and NERC, Lindeman Trust Fellowship, University of California Santa Barbara, University of Puerto Rico, American Philosophical Society and Monash Postdoctoral Fellowship (J.V.W.). We would also like to acknowledge other colleagues who for some years have co-operated, listened, criticised and encouraged us: Rod Allen, Brian Clough, Keith Corbett, Arthur Day, Warren Edney, Dick Fisher, Chuck Landis, Jocelyn McPhie, John Roobol, Steve Self, Alan Smith, Steve Sparks, Colin Wilson and John Wolff.

This book was written in two years, imposing great personal stresses on our families in the process. In particular, Sue Cas gets special mention not only for tolerating it all, but for her constructive suggestions on style and expression when proofreading the entire manuscript. The book could not have been written in this time without the considerable financial backing resulting from the short courses. In particular, we thank the organisations (especially Aberfoyle Exploration, Broken Hill Proprietary, British Petroleum Minerals, Electrolytic Zinc, Esso Minerals, Gold Fields Exploration, Shell Minerals, Western Mining Corporation, and Zinc Corporation) and individuals who have supported the course and therefore made the book possible. Our extreme gratitude goes to many people who have assisted, always willingly, with the logistics and mechanics. In particular, Warren Edney and, in the earlier stages, Arthur Day managed and co-ordinated a large number of people in all the facets of producing the final manuscript for the publishers, including typing, drafting, photography, copyright releases and proofreading. Without the constant help of Warren and Arthur we would still be labouring five years hence. Warren was ably assisted by Paul Dielemans, whose versatility proved invaluable. We cannot thank the following enough: Pam Hermansen, Monica Leicester and Robyn Sheehan for their impeccable typing skills and patience; Jenny Purdy and Draga Gelt for the excellent drafting; Steve Morton and Bruce Fuhrer for the skilful photography and the patience that all good photographers have; Tim Watson and Barbara Sandys for financial management of the resources needed to produce the manuscript; Bretan Clifford and Stuart Bull for assistance with proofreading; and Val Muscutt of BP London for keeping the mail going between two co-authors trying to write a book from opposite sides of the globe. We also sincerely thank staff and students of the Department of Earth Sciences at Monash University for their patience, interest and encouragement, and for providing the friendly and stimulating atmosphere in which an idea was transformed into reality. In particular, we thank Bruce Hobbs, Mark Bloom, Mike Etheridge, Larry Frakes, Dave Gray, Bob Gregory, Ian Nicholls, Pat Rich and Vic Wall, for making a great department. R.A.F.C. also wishes to thank Professors C. Carron (Fribourg University, Switzerland) and D. Coombs (Otago University, New Zealand) and their departments for making facilities available whilst on study leave, to make final amendments.

Although we take responsibility for the content of the book, various colleagues kindly read parts of the manuscript and offered many useful suggestions. For this we are extremely grateful to Rod Allen, Keith Corbett, Arthur Day, Warren Edney, Ian Nicholls, Steve Self, Colin Wilson and John Wolff.

We also gratefully acknowledge the very constructive comments of Steve Sparks, Peter Sutcliffe and Peter Francis in reviewing the manuscript for the publisher at various stages. We also thank Pete Kokelaar, Steve Self, Steve Sparks and Colin Wilson for providing preprints of manuscripts before publication.

Thanks also go to the editorial and production staff of Allen & Unwin for their punctual and friendly assistance. In particular, Roger Jones and Geoffrey Palmer and their staff are thanked for their extreme efficiency and patience.

Ray Cas
John V. Wright

CONTENTS

LIST OF TABLES

Plate 1 Succession of pyroclastic fall, flow and surge deposits of the Quaternary Okataina Volcanic Centre, New Zealand. Complex geometries and stratigraphic relationships result from the erosion of the oldest (9 000 years BP) Pukerimu pyroclastics (bottom right) and mantling and infilling of the irregular topography by 'younger' fall, flow and surge deposits, of the Mamaku (7 500 years BP), Whakatane (5000 years BP), Rotokawau (4000 years BP; dark basaltic fall layer near top of succession) and Kaharoa (900 years BP) eruptive intervals.

CHAPTER ONE

An introduction to facies analysis in volcanic terrains

Initial statement

Volcanic terrains consist of a greater variety of rock types than any other surface environment on Earth. They include lavas, deposits of explosive pyroclastic eruptions, primary volcanic autoclastic deposits and deposits resulting from the very significant spectrum of sedimentary processes that operate in volcanic terrains. Until the 1960s the amount of detailed and systematic work on the physical processes producing this diversity of rock types was subordinate to studies on the chemistry, mineralogy and petrogenesis of the volcanics. The growing need to understand better the processes operating and the peculiar depositional environments of volcanic terrains, in conjunction with major advances in the field of sedimentology, have led to a major growth in research and understanding of the physical processes.

Studies in *both* modern and ancient volcanic terrains have contributed to this growth in knowledge. The approach to describing, documenting and interpreting the rock types of volcanic terrains has benefited much from the equivalent approach in sedimentology. In particular, the facies concept is proposed as a useful means of documenting and interpreting the characteristics of rock units. The essence of facies analysis is the identification of distinctive characteristics that lend themselves to the interpretation of their origins, depositional processes and environments of deposition. In this chapter we introduce the facies concept, and consider the essential parameters useful in the description and interpretation of facies.

1.1 Introduction

Volcanic terrains are host to a greater diversity of rock types than is any other surface environment. However, until the 1960s the principal emphasis in volcanological research was on mineralogy, geochemistry and magma genesis. Since then, there has been an increasing awareness of the need to understand better the nature of the rock types, the physical processes responsible for their formation, and their significance in terms of depositional setting. This awareness has been stimulated by a diversity of needs in a number of areas, including eruption monitoring and prediction, hazard evaluation, exploration for the resources associated with volcanic terrains, geological survey mapping, academic research and petrological studies. In this book our aim is to provide a comprehensive account of the enormous range of rock types in volcanic terrains, their characteristics, associations and modes of formation, and their depositional and tectonic setting. Thus, we hope to provide readers with the sound geological approach necessary to make their own meaningful interpretations of volcanic successions.

Although we have tried to provide as comprehensive and up-to-date a summary of the subject, its concepts and its literature as time and space permitted, the book should not be considered to be a treatise on volcanology. We have addressed those topics and principles which we considered to be most important to the general aims of the book and, although all developments in the subject have not been treated at the research level, we have referenced the pertinent literature to enable readers to follow up specific topics. Wherever relevant, we have drawn on our experiences and research in both modern and ancient successions and, although this inevitably introduces biases, we have tried to balance this by constant reference to the volcanological literature.

Early subdivisions of the rock types in volcanic terrains into lavas and explosively erupted pyroclastics are now known to be oversimplified, and can be expanded into a fourfold subdivision of lavas, pyroclastics, autoclastic deposits and redeposited volcanic sediments or epiclastics. Lavas are now known to be diverse in character. Lavas have variable geometry, morphology, internal structure, mobility and flow behaviour (Ch. 4), which can be attributed to the variable physical and chemical properties of magmas (Ch. 2). Whereas in the past it was assumed that any fragmental rock in volcanic successions had an explosive, i.e. pyroclastic, origin, it is now generally appreciated that fragmental rocks in volcanic terrains can have diverse origins (Ch. 3). Pyroclastic rocks can themselves be subdivided into pyroclastic fall, flow and surge deposits (Chs 5–9). Autoclastic rocks are non-explosive in origin, originating from the quench-shattering of magma on contact with water or by brecciating during lava flow (Ch. 3). It is now also appreciated that an enormous range of normal erosional, epiclastic sedimentary processes and deposits are important elements of volcanic terrains (Chs 3 & 10–12).

All fragmental volcanic rocks, irrespective of origin, can be described by the non-genetic term 'volcaniclastic'. In this book we devote more attention to volcaniclastic rocks and their origins than to coherent lavas, even though in some volcanic settings lavas are more significant volumetrically. Our emphasis on volcaniclastic deposits arises from their potentially greater importance for interpreting the palaeoenvironment in which volcanism occurred, and because volcaniclastic rocks are more significant volumetrically in the rock record than are lavas. This is in spite of the fact that in modern environments pillow and massive lavas of the basaltic oceanic crust are the most significant volcanic rock type. However, their preservation potential in the rock record is small because they are destined to be subducted at the oceanic trenches as part of the global plate tectonic system.

The great diversity of rock types and processes in volcanic terrains makes the recognition of the origins of rock types in volcanic successions difficult. However, in ancient volcanic successions, recognition of original rock types may be made even more complicated by the effects of deformation, metamorphism and alteration (Ch. 14).

It has also become apparent that stratigraphic relationships in volcanic terrains may be complex

(Ch. 14), and that an understanding of the likely stratigraphic relationships and successions is dependent on an awareness of the different character of different volcanic centres and their stratigraphies (Chs 13 & 14) and of the tectonic settings in which volcanism occurs (Ch. 15).

Attempts to make sense of the diversity of rocks, processes, stratigraphic models and depositional settings of volcanic successions have been aided by major advances in the field of sedimentology. In both volcanology and sedimentology a systematic approach to describing, documenting and interpreting the character of, and relationships between, rock types is necessary. Success is dependent on an awareness of the possible diversity and complexity and on a sound understanding of basic physical and sedimentological principles.

In this book we hope to provide a comprehensive account of the volcanological and sedimentological concepts and principles that can be used in interpreting the complexities of both modern and ancient volcanic terrains. In this chapter we now consider the approach we think is needed to describe and document the characteristics of rock units, and also the basic principles that determine these characteristics. This is a necessary prelude to the discussion of the origins of particular deposits, because successful interpretation of the origins is dependent on making the correct observations in the first instance. Some of the descriptive characteristics of deposits discussed in this chapter are re-emphasised in Chapter 14 as a preliminary step to developing general facies models for a range of volcanic settings. In the remaining chapters we address in detail aspects considered to be relevant to a full understanding of volcanic rocks, their modes of formation and depositional and tectonic setting.

1.2 The facies concept

Different rock types are distinguished because they are texturally or mineralogically different in hand specimen or in thin section. In outcrop they may also be distinguished by their general physical appearance; for example, the presence or absence of some type of depositional structure such as layering, cross-stratification or grainsize grading. Alternatively, perhaps two or three rock types that are regularly interbedded and contain distinctive internal depositional structures may have a unique appearance that distinguishes them from other intervals or associations of rock types. The term 'facies' is used for such distinctive intervals or associations of rocks in outcrop. The facies approach is a convenient way of identifying, describing and interpreting distinctive intervals and/or associations of rock(s) which recur many times in a stratigraphic succession. Although the concept is most commonly applied in sedimentology (see Reading 1978, Selley 1978, R. G. Walker 1984), it is also applicable in volcanic successions, and is even used by metamorphic petrologists to distinguish different metamorphic grades based on significant marker minerals or associations of minerals.

A facies is therefore a body or interval of rock or sediment which has a unique definable character that distinguishes it from other facies, or intervals of rock or sediment. The definable character may be compositional or textural, or may be based on the sedimentary structures or fossils present. A facies is the product of a unique set of conditions in the depositional environment. These conditions may be physical, chemical or biological in origin, and may include such factors as the topography and bathymetry; the mechanisms and rates of material release, transport and deposition; the climate and weather; the nature of the source materials (both chemically and physically); the prevailing chemical condition; and the floral–faunal influences.

A facies can be defined at any scale. At a regional level stratigraphic units such as groups, formations or members are effectively facies because they have an overall lithological character that distinguishes them from other groups, formations or members. At a more local scale, facies may be defined at the scale of an outcrop by an interval of several or more beds which is basically uniform, or even by individual beds, or by both. The degree of detail used in subdividing a stratigraphic succession into facies will largely be controlled by the aims of the study, the information available and the level of understanding that is sought.

Even though associated facies may be different, they may still be genetically related as parts of the same depositional or eruptive event. For example, a single ignimbrite may contain several facies (Chs 7 & 8). An understanding of the spatial and age relationships between facies is therefore important, and success in the interpretation of facies is dependent on an awareness of the possible complications and of *genetically significant associations of facies*.

1.3 Description of facies

The genetic origins (i.e. mode of formation) of a facies may not always be obvious. Initially it is therefore better to avoid genetic facies names (e.g. ignimbrite, agglomerate), which are highly interpretive, until the origins have been clearly thought out. Descriptive terms such as 'rhyolitic, matrix-supported breccia facies' are preferable initially (also see Ch. 12). Such a facies may be an ignimbrite (or part of one), a hydrothermal explosion breccia or a mud flow deposit, to name but three possibilities. To evaluate which possibility is most likely requires careful examination. This should involve description and consideration of the facies properties, derived where possible from a combination of outcrop, hand specimen and thin section observations. On-the-spot application of genetic names may, more often than not, lead to erroneous interpretations.

Few approaches to facies description and analysis have been so systematic and logical as that proposed by Selley (1978). Selley nominated five facies descriptors:

geometry
lithology
sedimentary structures
palaeocurrents or sediment movement patterns
fossils

1.3.1 GEOMETRY

Geometry describes the three-dimensional form or shape (including thickness) of a facies and of its component strata. The preserved geometry of a facies is controlled by:

pre-depositional relief on the depositional surface,
volume of material deposited and the way the
     topography accommodates that volume,
physical properties of the transporting and depositional
     agent,
post-depositional erosion and
subsequent deformation

Pre-depositional relief

The relief on the depositional surface is controlled by the balance between erosion and deposition. Erosion will predominate where slopes are high and the relief is significantly above the base level towards which erosion is working (e.g. sea level, lake level, ocean floor). Erosion will produce negative changes in relief producing valley, gully and canyon, and ridge and plateau-like morphology. Most depositional units deposited in such a terrain will be confined within topographic lows, but some, such as air-fall deposits, may drape over irregular topography (Plate 1). Where the influence of erosion becomes subordinate to deposition, depositional processes will smooth out topographic differences, so in most instances will produce more tabular geometries for deposits (Plate 1). Relief may also be affected by contemporaneous tectonic activity and the emplacement of units with very positive relief, usually due to high bulk viscosity and internal strength (e.g. viscous lavas, debris flows, rock avalanches; see below).

Volume deposited and accommodation by topography

If the volume of material is low compared with the topographic relief into which it is deposited, then this volume will be entirely contained by the topographic depression (Plate 1, Fig. 1.1a). If the volume is large compared with the size of topographic depressions, then the deposit will overspill the topographic low and produce major variations in thickness (e.g. lavas, debris or pyroclastic flow deposits which infill a valley and spill over onto the confining ridge interfluves as a thin veneer; Fig. 1.1b).

(a) Valley accommodation

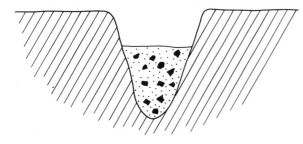

(b) Valley overspill

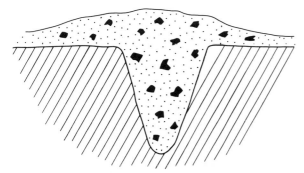

Figure 1.1 Possible relationships between the volume of topographic lows such as valleys, canyons, etc., and the volume of material emplaced within the topographic low during a depositional or eruptive event. (a) A small-volume debris flow is entirely accommodated by a larger deep valley, producing a thick laterally discontinuous deposit. (b) The volume of the debris-flow deposit is too large to be accommodated by the valley. The deposit has an extremely variable thickness and a wider extent due to overspill of the valley margins.

Physical properties of the transporting and depositional agent

The principal consideration here is the rheological properties of the transporting agent (Ch. 2). For example, low viscosity depositional agents will, topography permitting, spread their load as a broad, thin sheet (e.g. turbidity currents, pyroclastic surges, basaltic lavas). High viscosity materials or those with high strength (e.g. rhyolitic lavas, see Plate 2, Ch. 2) will produce a mound-like depositional unit with very significant positive relief, which markedly changes the topography on the depositional surface.

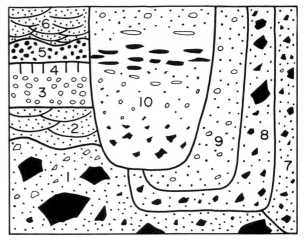

1 lahar – debris flow deposit
2 fluvial sediments
3 plinian pumice fall deposit
4 soil
5 scoria fall deposit
6 fluvial sediments

7 lahar filling in box-canyon cut into 1–6
8 block and ash deposit
9 non-welded ignimbrite
10 welded ignimbrite

Figure 1.2 Schematic representation of the possible effects of erosion on preserved depositional geometry and on facies relationships, which are highly irregular and abrupt, and make correlation and depositional sequence very difficult to determine.

Post-depositional erosion

The majority of volcanic terrains have relatively high slopes which are subject to severe degradation by the agents of erosion: gravity, water, ice and wind. As a consequence, many facies units emplaced during active volcanism, whether they be lavas, pyroclastic flow or fall deposits, and re-deposited volcaniclastic sediment intervals originating from epiclastic processes can have their original depositional geometry dramatically modified (Fig. 1.2), or even have the record of their emplacement or eruption, or both, completely removed.

Deformation

Deformation may have a marked effect on the preserved geometry of a facies. The effects may range from simple block faulting to extreme strain. The latter may be especially significant where hydrothermal alteration has weakened the rock (e.g. S. F. Cox 1981).

The preceding discussion suggests that geometry

by itself may not be a useful diagnostic characteristic of a particular facies type. Nevertheless, associations of facies constituting thick intervals may produce well defined geometries, particularly where influenced by normal sedimentary processes. In this regard, volcanic terrains will be influenced throughout their active life by surface sedimentary processes, and will contain sedimentary intervals with normal facies characteristics but consisting essentially of volcanically derived detritus.

1.3.2 LITHOLOGY

Lithology has three aspects:

physical constituents,
composition and
texture.

In non-volcanic sedimentary successions, all of these aspects of lithology can be very important in elucidating sediment sources and genesis, and in reconstructing the nature of the depositional environment (Selley 1982, Leeder 1982). For example the presence of certain physical components (e.g. shelly fragments, oolites, distinctive lithic fragments) can be diagnostic of depositional conditions and settings. Similarly, the presence of distinctive compositional grain-types (e.g. glauconite, phosphorite) or sediment-types (e.g. radiolarian cherts, evaporites) may be significant enough to establish the depositional setting and conditions. Textural features may also be revealing (e.g. well-rounded beach or aeolian sands, mud-supported debris-flow deposits). Similarly, primary volcanic facies may contain distinctive lithological aspects (shards, accretionary lapilli) that contribute to the understanding of their genesis and depositional setting. These aspects are as important in ancient successions as in modern ones, but may be more difficult to identify and quantify in ancient successions because of the effects of lithification, and perhaps metamorphic and structural overprinting.

With regard to lithology, we will now consider the features listed above that are relevant to the understanding of volcanic successions.

Physical constituents of volcanic successions

Volcanic successions contain varying proportions of lava flows and fragmental or clastic rocks. The principal physical constituents of lavas are crystals or phenocrysts, smaller microscopic crystals called microlites, uncrystallised magma or volcanic glass (which forms the groundmass), vesicles (gas bubbles), xenoliths and xenocrysts. Xenoliths and xenocrysts are, respectively, foreign rocks and crystals incorporated from country rock (e.g. wall rock to the magma chamber or conduit) or from another crystallising magma. All of these components can be present in varying combinations and proportions.

The clastic facies in volcanic successions consist of fragmental aggregates of magmatic clasts, foreign lithic clasts and crystals. The magmatic clasts may vary in vesicularity from dense lava fragments to vesiculated pumice and scoria. They may be glassy or variably crystallised, and of varying grainsize, ranging from large blocks many metres in diameter to micrometre-sized grains. Breakage of fragments may have occurred either during primary volcanic eruption (Ch. 3) or post-eruptively by surface processes (Ch. 10). At any time after formation, primary volcanic rocks may be eroded and the clastic components redeposited by normal surface processes.

The term 'volcaniclastic' is a non-genetic term for any fragmental aggregate of volcanic parentage, irrespective of origin. 'Pyroclastic' refers only to those aggregates formed by explosive volcanic activity *and* deposited by transport processes resulting directly from this activity. Other types of volcaniclastic rock include those fragmented or deposited, or both, by **epiclastic** processes, which are normal surface processes involving weathering and erosion (Ch. 10). 'Epiclastic' is therefore used here to describe deposits or rocks that were produced by normal surface fragmentation processes (weathering, physical abrasion, gravitational collapse) or were finally deposited by normal surface processes (traction, suspension, mass flow; Ch. 10), irrespective of their fragmentation mode, or both. The usage of the term 'epiclastic' thus goes beyond the more traditional provenance sense.

Therefore, epiclastic deposits or rocks may contain fragments with a proven primary volcanic mode of fragmentation (e.g. glass shards, pumice; Ch. 3) which have been transported and redeposited by normal surface processes (e.g. by mud-flows, river transport, turbidity currents, etc.) a long distance from the initial eruption point. We cannot stress enough, therefore, that caution is needed before deciding that fragmental aggregates or rocks with a primary volcanic fragmentation origin have also had a pyroclastic rather than epiclastic transportation mode. *This must be proven rather than assumed.*

Composition

Composition refers to the geochemical, mineralogical and petrological character of a volcanic rock, irrespective of whether it is a lava, pyroclastic or redeposited volcaniclastic. The final composition of a rock may be the end-result of a complex history of processes causing chemical and physical change. These processes include pre-eruptive magmatic processes, both chemical and dynamic (Ch. 2), and co-eruptive and post-eruptive processes that physically separate or fractionate physical constituents (e.g. glassy ash from crystals and lithics; Ch. 11). Hydrothermal activity and, in older volcanic successions, weathering, diagenesis and metamorphism, may have further altered the chemistry of volcanic rocks. In Chapter 2 we briefly consider the effects of magma composition on eruptive behaviour, and also approaches to classifying magmas and volcanic rocks according to chemical and mineralogical composition.

Texture

The term 'texture' encompasses the physical characteristics of the components of a deposit or rock, and also its overall characteristics or bulk properties. The textural properties of an aggregate are a reflection of inherited characteristics from the source, of the mode of fragmentation and of characteristics developed during or after transport and deposition. Any one of these influences may produce a distinctive textural character. Aspects of texture that will be considered here in terms of their process significance or environmental significance are grainsize, rounding, sorting, shape and fabric.

Grainsize and the grainsize characteristics of an aggregate are one of the first characteristics seen in an outcrop. The preserved grainsize of a fragmental aggregate is a reflection of the minimum grainsize available at the source point, the type and efficiency of fragmentation, the competency of the transporting and depositing medium to carry that grainsize, and the degree of physical abrasion during transportion and deposition. These factors apply for both pyroclastic and epiclastic aggregates. For lavas, the size of the phenocrysts reflects physicochemical conditions in the magma chamber and during the ascent of the magma. Factors which influence crystal size include cooling rate, melt composition and structure, nucleation kinetics of each mineral type and sorting processes such as crystal settling.

For fragmental aggregates, whether pyroclastic or epiclastic in origin (*which has to be evaluated in each instance*), grainsize is therefore *not* a reflection of proximity to source or eruption point. For example, huge boulders, metres in diameter, can be transported tens of kilometres from source or eruption point by pyroclastic flows, debris flows, rock avalanches or glaciers. None of these transporting agents needs to produce any significant signs of abrasion or rounding. Equally significant is that very fine grainsizes are possible for ashes and pyroclastic flow deposits near vent if the explosive fragmentation during eruption has been very efficient (e.g. during some hydrovolcanic eruptions, Ch. 3).

Although actual grainsize has no specific value in palaeovolcanological and palaeogeographic reconstructions by itself, the use of overall grainsize population parameters has major application when dealing with modern unconsolidated volcanic successions. Through sieving, the detailed grainsize characteristics of an unconsolidated aggregate can be determined (App. I), and from this information and its graphical representation, statistical grainsize parameters can be calculated. In modern successions the uses of this approach include distinguishing and classifying different types of pyroclastic deposits (Chs 5 & 6).

Although such approaches have added much to understanding volcanological processes in modern terrains, they are not usually applicable to lithified, consolidated successions because it is not possible to disaggregate the rock into all of its original grains, preserving their shape and size. For these successions, only qualitative estimates of grainsize and grainsize parameters are practicable. One usually has to rely on the field outcrop facies characteristics in order to determine the genesis and the palaeovolcanological and palaeogeographic significance (see App. II).

Sorting is a reflection of the degree to which the transporting agent has been capable of separating grains of different hydraulic properties and depositing together grains that are hydraulically equivalent. The hydraulic behaviour of a particle is a measure of the way in which the particle responds when acted on by a fluid, whether the fluid be water, wind, mud or volcanic gas. Factors that affect the hydraulic behaviour of particles include their density, weight and shape. In normal epiclastic or terrigeneous sediments most grains, being mineral or rock fragments, have approximately equal densities and are generally equidimensional. As a result, currents acting on such a sediment population sort grains according to weight, as reflected by grainsize. In such situations it is not uncommon to talk of size sorting. *However, once sediment populations of differing shape, density and weight are mixed, well developed size sorting becomes impossible* even though the populations may be hydraulically well sorted. For example, a beach sediment consisting of rounded quartz grains and blade-like shell fragments is likely to be poorly sorted by size, but will be hydraulically well sorted. In volcanic settings, not only are there major variations in the shape, but also in the densities of the components (e.g. crystals, shards, pumice, lithics). As a result, volcaniclastic aggregates, whether they be pyroclastics or reworked and/or redeposited volcaniclastics, are likely to be poorly sorted according to size, but may be well sorted hydraulically. It is well known that sorting for pyroclastic deposits is poorer than for non-volcanic epiclastic equivalents of the same overall grainsize class. The short duration of many pyroclastic transportation processes also reduces the importance of hydraulic sorting.

Shape is an assessment of the three-dimensional form of a grain. For non-volcanic epiclastic sediments the shape is largely inherited from the morphology of the grain in the source, and can be affected by crystallisation shape, cleavage (mineral and tectonic) and layering, whether it be sedimentary, igneous or metamorphic. For pyroclastic aggregates the mode of fragmentation may also impart distinctive shape-morphology properties (Ch. 3).

Rounding is the degree to which sharp corners and edges have been abraded during transportation or deposition, or both. Generally, rounding is better in sediments that have been subjected to constant energy levels during reworking. However, rounding can also be produced by pyroclastic processes (Chs 3, 8 & 12). (Most sedimentological texts refer to 'roundness' properties rather than 'rounding' properties.)

Fabric is a consideration of the relationships between, and arrangement or packing of grains in, an aggregate. Depositional fabric is clearly a reflection of the transporting mode and depositional conditions, and is more fully discussed in Chapter 12 and Appendix II.

1.3.3 SEDIMENTARY STRUCTURES

Sedimentary structures are probably the most important analytical tool in facies analysis. They are produced before deposition (e.g. erosional features), during deposition (e.g. current generated structures) and after deposition (e.g. soft-sediment deformation, bioturbation) of sedimentary aggregates, and can be referred to as being pre-, syn- and post-depositional in timing, respectively. They are extremely important because they, together with textural aspects, most immediately reflect the depositional conditions, and the modes of transport and deposition. If the structures are produced by fluid flow, then they are especially important because they reflect the fluid dynamics of the host environment and its transportational and depositional agents.

As discussed in Chapter 10, particles of mineral or rock can be transported in particulate fashion (i.e. one by one) or by mass-movement (i.e. bulk aggregates of particles moved instantaneously as one). Particulate movement of granular sediment (coarser than clay) produces an assemblage of tractional structures (cross-stratification, dunes, ripples, etc.). Mass-movement processes frequently deposit a massive, structureless aggregate, although low sediment concentration, low viscosity mass flows or the trailing tails of mass flows may also produce tractional sedimentary structures. Tractional sedimentary structures are therefore not exclusively associated with processes involving particulate sediment transport, or with environments that are 'shallow-water' in aspect. Each case has to be evaluated on its merits.

Pyroclastic processes also involve particulate and mass-movement of clastic aggregates. The grain types and shapes are different as, frequently, are the transporting media and their fluid dynamic properties compared with those of epiclastic process regimes. Such differences should produce distinctive differences in the types of structures and textures produced, and these will be highlighted in Chapters 5–9 inclusive.

1.3.4 SEDIMENT MOVEMENT PATTERNS

The directions of current flow or sediment transport directions can be measured where asymmetrical structures such as ripples, dunes, angle of repose cross-stratification and imbrication, and sole structures, such as flutes, can be used to determine local directions of movement of sediment or palaeocurrent flow. Over a larger area, numerous readings can be used to reconstruct the palaeogeography, and to trace palaeogeographic changes as they have influenced current flow and sediment transport pathways. Furthermore, distinctive regional palaeocurrent patterns develop in certain sedimentary environments (e.g. radial patterns for alluvial and submarine fans and deltas, bimodal for nearshore marine settings, etc.) (Selley 1978).

Flow direction indicators in primary volcanic facies can be used in the same way as structures in epiclastic sediments. However, the structures will be different, as discussed above. For example, Waters (1960) recognised that downstream trailing pipe vesicles in lavas, inclined foreset beds produced by lava deltas and quench-fragmented lavas

flowing into water (e.g. Fig. 4.16) can be used to assess lava flow directions for basaltic lavas. Cummings (1964) described eddies in flow banding that develop on the downstream side of inclusions in rhyolitic lavas. For pyroclastic flows, flow directions have been variously determined using pumice clast alignment and imbrication (Elston & Smith 1970, Kamata & Mimura 1983), alignment of logs (Froggatt *et al.* 1981) and, more commonly, by contouring average maximum lithic clast sizes (e.g. Kuno *et al.* 1964, G. P. L. Walker 1981a; App. I). For pyroclastic surges, structures such as dunes, low-angle cross-stratification, and chute and pool structures (Fisher & Waters 1970; Ch. 7) are useful if present.

Dispersal directions for epiclastic successions are largely topographically controlled. This may also be the case for pyroclastic successions, but the flow mechanisms may be so energetic that they may largely ignore and surmount topographic highs. Nevertheless, flow directions will mainly be radial from the vent, and they may be useful in palaeogeographic reconstruction of the volcanic centre (Chs 13 & 14).

1.3.5 FOSSILS

The use of both body and trace fossils as palaeoenvironmental indicators is essentially the same for both volcanic and non-volcanic successions. However, the most critical thing is to establish whether the fossils are *in situ* or have been transported and redeposited, especially in marine successions, where downslope redeposition is common. Even redeposited fossil remains may be useful in indicating the nearby environmental conditions if the ecological affinities of the organisms are known.

1.4 Facies analysis and interpretation – the importance of associations of facies

Having carefully documented and described individual facies, it is also necessary to look at the association of facies and the relationships between them to evaluate to what degree they are genetically related. Different facies may be the products of one event (e.g. the diverse facies of ignimbrite-forming eruptions, Chs 5–9). Assessment of the spatial and age relationships is therefore important before general models of the deposits and sequence of deposits of specific events can be formulated. On a larger scale, particular types of volcanic centres and settings may consistently produce similar associations of facies which can then be used to formulate general facies and stratigraphic models for those settings (Chs 13 & 14).

1.5 Summary

Since there is such a diversity of rock types and processes in volcanic terrains, the interpretation of their origins has to be addressed with care. In the first instance, this is dependent on careful documentation and description, for which the approach to facies description and analysis used in sedimentology is useful. Having described and interpreted individual facies, consideration needs to be given to the spatial and age relationships between them (Fig. 1.2), and to associations of facies to assess to what degree spatially related facies are genetically related. By doing this, models of the deposits produced by particular events can be developed, and clearer pictures of the depositional processes and environments emerge.

1.6 Further reading

Standard texts that contain a good coverage of sedimentological principles are Blatt *et al.* (1980), G. M. Friedman and Sanders (1978), Leeder (1982) and Selley (1982). For detailed discussion of the facies concept see Reading (1978) and R. G. Walker (1984). For comprehensive discussion of sedimentary structures and their formation see Allen (1982), Collinson and Thompson (1982), Conybeare and Crook (1968) and Potter and Pettijohn (1963). For further, more detailed discussion of the characteristics of volcanic terrains, the processes that operate, their products and their interpretation, read on!

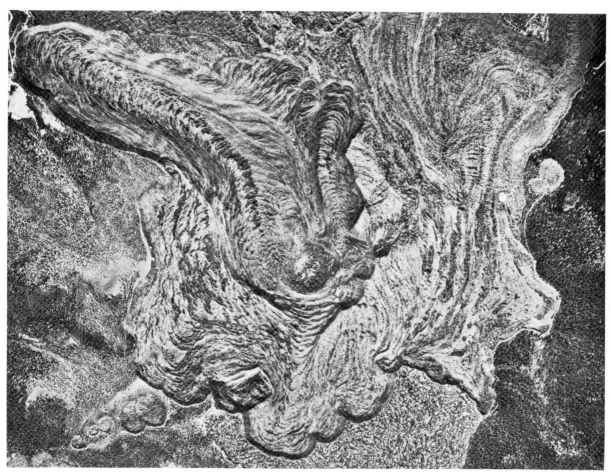

Plate 2 Vertical aerial perspective of Big Glass Mountain, an obsidian flow complex in the Medicine Lake Highlands, east of Mount Shasta, USA. Flows have high viscosity with steep flow-fronts and irregular tops dominated by concentric ridges called ogives. (After Greeley 1977a.)

CHAPTER TWO

Some properties of magmas relevant to their physical behaviour

Initial statement

Magmas may erupt as coherent lavas and then flow coherently or fragment during flow, or they may erupt explosively to form a range of pyroclastic products. At the time of eruption the volcanic products may range in character from pure magmatic liquid to essentially solid. If the erupted material flows, either as a coherent mass or as a particulate mass flow, then the original character of the erupted material will control the form and mobility of the resultant deposit. An understanding of why magmas erupt coherently or explosively and how they behave subsequently requires a brief review of some of the properties of magmas.

In this chapter we briefly introduce the compositional variability of magmas and their classification, and then look at specific properties that are relevant to their rheological behaviour. Rheology is the study of the deformational behaviour of materials. Factors that affect the rheological behaviour of magmas include their temperature, density, viscosity, yield strength and the mechanical or tensile strength. The viscosity of magmas is controlled by many variables, including pressure, temperature, chemical composition (especially volatile and silica contents), crystal content, and bubble content. Finally, we examine the effects of some of these variables on fluid flow states in coherent lavas and particulate debris flows. From the Reynolds Number criterion for fluid turbulence, it is seen that most lavas and debris flows will flow in a laminar fashion because of their high viscosities. However, where the viscosities are *relatively* low (e.g. a very hot, or peralkaline lava, or a fluid-rich debris flow), or where flow velocities are high, then parts, or nearly the whole body of lavas and debris flows, may flow turbulently.

2.1 Magmas – an introduction to their diversity and character

Magmas are molten or partially molten rock materials. They are chemically complex, multi-component silicate systems which have varying compositions, temperatures, crystal contents and volatile contents, and therefore varying rheological properties (McBirney & Murase 1984). These properties have an important bearing on the mode of eruption.

Magmas can have widely different histories. They may be generated within the Earth's crust or upper mantle. They may then crystallise at depth as an intrusive body (to form plutonic or subvolcanic rocks), or be erupted at the Earth's surface to form volcanic rock. The erupted products may vary from pure liquids to essentially pure solids. Magmas which erupt may undergo considerable changes during their rise to the surface. They may, for instance, reside for some time in high level subvolcanic magma chambers where crystallisation may occur along the margins of the chamber, or removal of crystals (phenocrysts) from the melt may occur through settling. Petrological and chemical changes will result from such fractional crystallisation. However, it is now becoming increasingly apparent that some magmas can have much more complex histories than this. The new awareness that many magma chambers may be periodically replenished, periodically tapped and continually fractionated, and that they are *open systems*, questions some conventional geochemical interpretations based on closed and 'static' systems (see O'Hara & Matthews 1981). Replenishment can give rise to the mixing of magmas or the development of compositional zonation. Many volcanic products show evidence of this. Sometimes this mixing is thought to trigger eruptions (Ch. 3). Understanding the fluid dynamic behaviour of magmas in chambers is a rapidly developing field (see J. S. Turner & Gustafson 1981, Huppert & Sparks 1984).

Many factors control the compositional and mineralogical characteristics of an erupted magma. These include the nature of the subsurface source rocks where melting occurs, the earlier history of that source in terms of previous thermal, metamorphic and melting events, the degree of partial melting of the source rocks, the degree of crystallisation in the magma, the extent of segregation of magma from crystals, the amount of contamination by wall rock and the degree of magma mixing before eruption. Discussion of all of these is beyond the scope of this book, but as a result of these factors volcanic rocks can have a diversity of chemical and mineralogical compositions, and physical characteristics (the reader is referred to Hargraves (1980) for a more comprehensive discussion).

Two important topics that we must now consider are:

classification
magmatic associations

2.1.1 CLASSIFICATION

The classification of igneous rocks (and hence the magmas they represent) can be approached in two ways – one based on the chemistry of the magma or rock, and the other on observable modal mineralogy. No one scheme can be regarded as ideal for all purposes, and the approach used will be governed by the desired purpose of making the classification.

For a discussion of magma properties we can adopt a simple chemical classification. The most abundant chemical component in most igneous rocks is SiO_2, which can range from <35 to nearly 80 wt%. Volcanic rocks can be initially divided into:

(a) High silica types (>63% SiO_2), which can be called *silicic* or *acidic*.
(b) Intermediate types (52–63% SiO_2).
(c) Low silica types (<52 to >45% SiO_2) which can be called *basic*.
(d) Magmas or volcanic rocks with <45% SiO_2 are called *ultrabasic*.

Several rock types occur within each of these silica classes. These can be distinguished on the basis of their variable alkali contents. A complementary way of subdividing igneous rocks chemically

Table 2.1 A simple chemical classification for the common volcanic rock types.

	SiO$_2$ (wt%)	Al_2O_3 saturation classes			
		Peraluminous*	Metaluminous†	Subaluminous‡	Peralkaline§
acid	>68	rhyolite or obsidian $\longrightarrow$			pantellerite, comendite
	63–68	rhyodacite $\longrightarrow$			
		dacite $\longrightarrow$			
			latite	trachyte	
intermediate	57–63		andesite	$\downarrow$	phonolite $\uparrow$
	52–57		mugearite		
			tholeiitic basalt		
			hawaiite		
basic	45–52		alkali basalt		
			basanite		
ultrabasic	<45		nephelinite, leucitite		

* Molecular Al_2O_3 > (CaO + Na$_2$O + K$_2$O).
† Molecular Al_2O_3 < (CaO + Na$_2$O + K$_2$O) and Al_2O_3 > (Na$_2$O + K$_2$O).
‡ Molecular Al_2O_3 ~ (Na$_2$O + K$_2$O).
§ Molecular Al_2O_3 < (Na$_2$O + K$_2$O).
Note: Basaltic rocks cover a wide compositional range and can be further subdivided. For more-comprehensive chemical classification schemes, see Yoder and Tilley (1962), Green and Ringwood (1967) and Irvine and Baragar (1971).

(especially acid to intermediate rocks) is to evaluate the relative abundances of molecular Al_2O_3 to total alkalis and calcium (Na$_2$O + K$_2$O + CaO), i.e. the degree of alumina saturation (Shand 1947, K. G. Cox *et al.* 1979). The common volcanic rocks are categorised in Table 2.1 using this approach.

Variation in silica and alkali contents is reflected in the mineralogy, particularly the feldspars and feldspathoids. Acid and intermediate rocks such as dacites and andesites are dominated by plagioclase feldspars, whereas rhyodacites and latites have subequal proportions of potassium feldspars and plagioclase, and rhyolite and trachyte are dominated by potassium feldspars. Pantellerites and comendites are alkali-rich (peralkaline) sodic rhyolites which, compared with more aluminous rhyolites, develop Na-rich feldspars with Na-amphibole or pyroxene. The acidic and more silicic intermediate rocks (>57% SiO$_2$) are usually silica-oversaturated, and the acid rocks in particular will contain free quartz crystals or grains, i.e. modal quartz. Where magmas are undersaturated in SiO$_2$, feldspathoid minerals (e.g. analcite, nepheline, leucite) may occur at the expense of some feldspar.

Examples include phonolites, which have potassium-rich alkali feldspar and minor feldspathoids, basanites with plagioclase, alkali feldspar and feldspathoids, and nephelinites and leucitites, which are feldspar-free ultrabasic rocks. Within the range of basalts, tholeiites may be slightly oversaturated, while alkali basalts tend to be slightly undersaturated.

A comment should be made here on the distinction between the terms 'basic and ultrabasic' and 'mafic and ultramafic'. The former terms are used to describe igneous rocks with low SiO$_2$ contents, the latter are used for rocks with high modal ferromagnesian mineral contents. Similarly 'silicic' and 'acidic' refer to high SiO$_2$ contents, whereas 'felsic' and 'salic' are used for igneous rocks with high modal contents of light coloured minerals (quartz, feldspars). Some mafic rocks can be ultrabasic (e.g. nephelinites) and ultramafic rocks can be basic (e.g. pyroxenites), but generally most basic rocks are mafic and acidic rocks are felsic or salic.

For studies based on field observations of volcanic rocks, an entirely chemical approach to classification is not practical. It also ignores useful mineralogical and textural information. A chemical scheme is also of limited value where rocks have undergone alteration due to hydrothermal or fumarolic activity, weathering or, in some ancient terrains, regional metamorphism. In many cases a more tangible means of classifying rocks based on mineralogy is therefore required. Even in chemically altered rocks, the primary mineralogy and textures can still be identified in many cases. Mineral types and

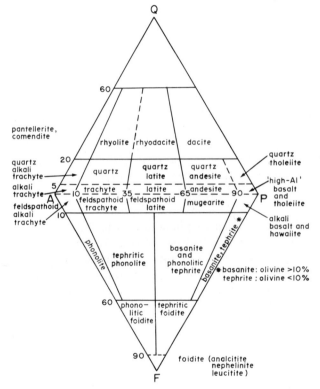

Figure 2.1 Names and modal (mineral volume percentage) compositions of volcanic rocks recommended by the IUGS Subcommission on the Systematics of Igneous Rocks (after Streckeisen 1979) with slight amendments. Minerals at the corners of the QAPF diagram are: Q = silica minerals (usually quartz); A = alkali feldspar (including albite); P = plagioclase; F = feldspathoids (e.g. analcite, nepheline, leucite). Rock names are determined by ignoring mafic mineral contents and recalculating quartz, plagioclase, alkali feldspar and feldspathoid contents to 100%. The sum of mafic mineral modes (M) is used to distinguish basalt ($M =$ 35–90) from andesite ($M =$ 0–35), and basanite (>10% olivine) from tephrite (<10% olivine). Mafic minerals include olivine, pyroxenes, amphiboles, and micas. Most mafic volcanic rocks plot along the P and F sides of the QAPF diagram.

abundances not only reflect magma chemistry, but their textural relations give additional information about the eruptive and cooling history of volcanic rocks. (Useful atlases of igneous rocks and their textures include MacKenzie *et al.* (1982) and Moorhouse (1970), and detailed interpretations of textures can be found in Hatch *et al.* (1972) and K. G. Cox *et al.* (1979).)

Rock names have traditionally been given according to mineral content, and many classification schemes have evolved. The most recent attempt at a standard scheme based on mineralogy is that presented by Streckeisen (1979) shown in Figure 2.1. Here, volcanic rock types are allocated fields on the QAPF diagram, and are classified according to the relative modal abundances of their felsic minerals. This presents problems in distinguishing the members of the mafic–ultramafic rock spectrum. These are distinguished by the abundance of mafic minerals (Fig. 2.1).

For many volcanic rocks modal mineral contents cannot be accurately determined because of the microcrystalline or glassy texture of the groundmass. Where phenocrysts are the only recognisable minerals, and if rock names are based on these alone, then the prefix 'pheno-' should be added. Thus, a rock containing plagioclase and quartz phenocrysts in an undetermined groundmass would be a 'pheno-dacite'. Where plagioclase is the sole felsic phenocryst, the rock is a 'pheno-andesite' or 'pheno-basalt', even though a *complete* modal or chemical analysis could show it to be a dacite. On the other hand, rocks with abundant plagioclase together with clinopyroxene phenocrysts may either be andesite or tholeiitic basalt, depending on the *total* proportion of mafic minerals present. Some very glassy rocks (80–100% glass) are given special names such as **obsidian** or pitchstone, for glass of rhyolitic composition.

In ancient volcanic terrains metamorphosed or metasomatised volcanic rocks in which feldspars are sericitised or albitised and mafic minerals replaced by chlorite, epidote, serpentine or talc, etc., may be given the prefix 'meta-' whenever original textures and mineralogy can be determined. Some special names which have been used for rocks in ancient terrains include **spilite** (albite–

Table 2.2 Some measured temperatures of erupting magmas.

Volcano	Composition	Temperature (°C)	Source
Kilauea, Hawaii 1952–63*†	tholeiitic basalt	1050–1190	MacDonald (1972)
Mt Etna*† 1970–75	hawaiite	1050–1125	Archambault and Tanguy (1976)
Paricutin* 1944	basaltic andesite	943–1057	Bullard (1947)
Santá Maria* 1940	dacite	725	MacDonald (1972)
Mt St Helens† 1980	dacite	850	J. D. Friedman et al. (1981)

Measured by *optical pyrometer and †thermocouple placed into lava. Those from Kilauea are largely optical measurements, and those from Etna are largely thermocouple ones.

chlorite rock) for a meta-basalt, and **keratophyre** and **quartz keratophyre** (albite, quartz and minor chlorite, epidote and iron oxides) for meta-andesite or dacite. However, the rock names in Figure 2.1 should be retained where possible, regardless of alteration state or geological age.

2.1.2 MAGMATIC ASSOCIATIONS

Volcanic rocks can be grouped into various 'associations', 'series', 'suites' or 'clans' based on petrological and chemical distinctions. Examples include the tholeiitic, alkaline and calc-alkaline associations. These may have restricted spatial distributions and be restricted to specific volcano-tectonic settings. For many years petrologists and geochemists have attempted to relate the petrogenesis of modern and ancient volcanics to their tectonic setting using, in particular, their trace element and isotope chemistry. We discuss relationships between volcanism and tectonic setting in Chapter 15.

For a detailed consideration of the petrological characteristics of volcanic rocks, see igneous petrology texts such as Carmichael et al. (1974), K. G. Cox et al. (1979) and Hughes (1983).

2.2 Temperature

Magma temperatures may be estimated in a variety of ways. Direct measurements on lavas may be made, either using a thermocouple probe inserted into a lava flow or lake, or by means of an optical pyrometer (which is especially useful for measuring the temperature of lava fountains). A large amount of temperature information is available for Hawaiian lavas and some of this is summarised by MacDonald (1972). Hawaiian tholeiitic basalts approach the surface between about 1050°C and 1200°C (Table 2.2). For silicic magmas there are fewer data available because such eruptions have not frequently been observed this century. An optically determined temperature for the 1940 dacite dome of Santá Maria and a thermocouple measurement of the Mt St Helens 1980 dacite dome indicate substantially lower eruption temperatures than for basaltic lavas (Table 2.2). Estimates of the typical eruption temperatures of the major magma types are given in Table 2.3.

Table 2.3 Summary of estimates of typical eruption temperatures for volcanic rocks.

Rock type	Temperature (°C)
rhyolite	700–900
dacite	800–1100
andesite	950–1200
basalt	1000–1200

Laboratory experiments may also be used to estimate magma temperatures. Almost all volcanic rocks contain crystals in varying amounts. Therefore, by experimentally determining the liquidus temperature (the temperature above which no crystals are stable) and the solidus temperature (the temperature below which there is no liquid phase) the maximum and minimum temperature limits for the existence of a silicate liquid are found. An uncertainty is introduced by H_2O content, which can strongly affect the liquidus and solidus temperatures. For rocks with a few suspended phenocrysts the liquidus temperature may be a good approximation of the *eruption temperature*. *Crystallisation temperatures* can be estimated by *mineral geothermometry*, which makes use of temperature-dependent compositional relationships between coexisting minerals assumed to be in mutual equilibrium. Frequently used geothermometers include magnetite–ilmenite mineral pairs (Spencer & Lindsley 1981), olivine–ilmenite (D. J. Anderson & Lindsley 1981) and two pyroxenes (B. J. Wood & Banno 1973). However, it must be borne in mind that laboratory estimates do not necessarily reflect *emplacement* temperatures. For example, lava flows may have erupted at higher temperatures than those indicated by mineral pairs which may have crystallised after eruption, and the final emplacement temperature of hot pyroclastic flows may be governed more by the extent of mixing with air and country rock than by initial magma temperature.

2.3 Density

There have been few measurements of the densities of igneous rocks at elevated temperatures, despite the obvious importance of such data in understanding magmatic systems. The results obtained for four volcanic rocks by Murase and McBirney (1973) are given in Figure 2.2. Densities are markedly different for the different compositional types but, as expected, all show a decrease in density with increasing temperature. Density is also dependent on pressure, increasing in proportion to the confining pressure (Stolper & Walker 1980, Kushiro 1980).

Bottinga and Weill (1970) and Nelson and Carmichael (1979) considered the partial molar volumes of rock-forming oxide components in silicate liquids as a function of temperature. The

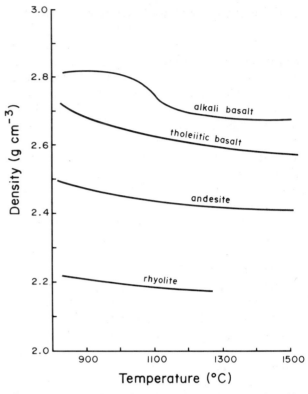

Figure 2.2 Densities of some molten volcanic rocks with varying temperature at atmospheric pressure (after Murase & McBirney 1973).

density of any magmatic liquid can thus be estimated from its chemical composition using the empirical methods of Bottinga and Weill (1970) and Richet *et al.* (1982).

The effect of density on the fluid dynamical behaviour of magmas is an important petrological variable affecting their chemical characteristics. Recent exciting work has modelled the mixing of dense, ultrabasic magmas with lighter, more fractionated basaltic magmas in mid-ocean ridge magma chambers (Huppert & Sparks 1980a, b), and of heavier, wet basic magma with more silicic magma within high-level chambers typical of stratovolcanoes (Huppert *et al.* 1982a, J. S. Turner *et al.* 1983). The reader is referred to these papers, as well as the review paper of Huppert and Sparks (1984), and McBirney (1980), for an insight into this type of study and the petrological implications.

2.4 Viscosity and yield strength

In lay terms, *viscosity* is a measure of the consistency of a substance. For our purposes it is a reflection of the internal resistance to flow by a substance when a shear stress is applied. In pure fluids this resistance to flow is essentially caused by molecular or ionic cohesion. In magmas it is complicated by the presence of solids (crystals) and gas bubbles (McBirney & Murase 1984). Furthermore, the processes of uprise and pressure release, crystallisation, cooling and degassing ensure that the viscosity of all magmas changes during their history. Consideration of the viscosity of magmas is important because it affects the mobility and form of unfragmented, coherently erupted lavas (Ch. 4), and because it may affect the rate of vesiculation (Ch. 3), a significant factor at a time when explosive fragmentation and eruption are imminent. These applications will be discussed further in subsequent chapters. The relevance of viscosity to fluid flow states is discussed in Section 2.5.6.

To define viscosity quantitatively, we first need to consider fluid rheologies. Some fluids, such as

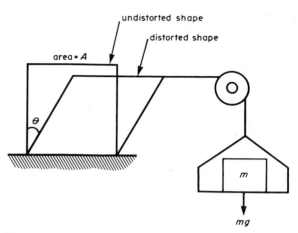

Figure 2.4 Distortion of a cube of foam rubber by an applied stress, *mg*. Strain is a measure of the degree of deformation, measured here as the angle θ. The rate of strain is the rate at which the foam cube deforms, and can be represented by dθ/d+. (After Allen 1970a.)

air and water, will flow (i.e. deform) when an infinitesimally low shear stress is applied, and these are called *Newtonian* fluids. For Newtonian fluids the relationship between the shear stress and strain rate (rate of deformation) is linear (Fig. 2.3). Only some very fluidal, high temperature magmas with low concentrations of crystals behave in Newtonian fashion. *Non-Newtonian* substances are those in which the relationship between shear stress and strain rate is variable (Fig. 2.3; called *pseudo-plastic* substances), or in which a *yield strength* must be exceeded, after which the relationship between shear stress and strain rate may be linear or non-linear (Fig. 2.3). Substances for which an initial yield strength must be exceeded and after which the relationship between shear stress and strain rate is linear are called *Bingham* substances (Fig. 2.3).

The viscosity of a substance can be quantitatively defined as the ratio of the shear stress to rate of strain. For a foam cube for example (Fig. 2.4), if a shear stress σ is applied, the viscosity of the foam, η, is given by

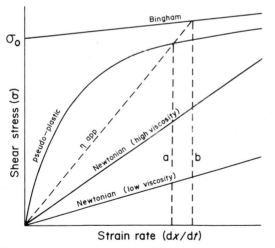

Figure 2.3 Flow curves for a Bingham, a pseudo-plastic, and two Newtonian substances. η_{app} is the apparent viscosity of the pseudo-plastic substance at strain rate *a*, and of the Bingham substance at strain rate *b*: σ_0 is the yield strength of the Bingham substance. Stress and strain rate are explained in text and Figure 2.4. (After Wolff & Wright 1981.)

$$\eta = \sigma \left/ \frac{d\theta}{dt} \right. \qquad (2.1)$$

for which the unit is the Pa s (= 1 dyn s cm^{-2} ≡ 10 poise).

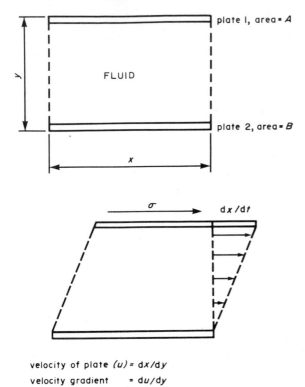

velocity of plate $(u) = dx/dy$

velocity gradient $= du/dy$

Figure 2.5 Diagrammatic representation of parameters used to define viscosity in a liquid. Application of a shear stress σ to the upper plate confining a liquid induces a velocity gradient du/dy.

For fluids it is not practical to measure $d\theta/dt$. A more practical parameter to measure is the vertical velocity gradient, du/dy, induced by applying a shear stress, σ ($= F/A$), to the upper plate of two plates of known equal area (A) which confine a fluid (Fig. 2.5). The viscosity of the fluid can be written as

$$\eta = \sigma \left/ \frac{du}{dy} \right. \tag{2.2}$$

This equation is valid for pure Newtonian fluids. Equation 2.2 can also be written as

$$\sigma = \sigma_0 + \eta \left(\frac{du}{dy} \right)^n \tag{2.3}$$

where σ is the total shear stress and σ_0 is the stress required to initiate flow (= the yield strength of a Bingham substance). For pure Newtonian sub-

stances $\sigma_0 = 0$ and $n = 1$ (Eqn 2.2); for pseudo-plastic materials $\sigma_0 = 0$ and $n < 1$; and for Bingham substances σ_0 (= yield strength) has a finite value and $n = 1$. Shear stress and the yield strength of the material are usually expressed in Newtons per square metre ($1\ \mathrm{N\ m^{-2}} = 10^{-1}$ dyn $\mathrm{cm^{-2}}$); strain rate is expressed in reciprocal seconds ($\mathrm{s^{-1}}$).

There are very few estimates of the viscosities of magmas. MacDonald (1972) summarised a number of measurements from lava flows. However, nearly all of these assume Newtonian rheology. In most of these cases, and in subsequent work, lava viscosities have been calculated from the Jeffreys equation:

$$\eta = \frac{g\varrho \sin \alpha d^2}{nV} \tag{2.4}$$

where η is the viscosity, g the acceleration due to gravity, ϱ the density, α the slope angle of the terrain, d the thickness of flow, $n = 3$ for broad flows or 4 for narrow flows and V is the velocity of flow.

More recent field and laboratory measurements have indicated that at sub-liquidus temperatures, lavas and common igneous melts generally have non-Newtonian rheologies (Robson 1967, Shaw *et al.* 1968, Shaw 1969, Murase & McBirney 1973, Pinkerton & Sparks 1978, McBirney & Noyes 1979, McBirney & Murase 1984). This behaviour is due to the presence of dispersed crystals and gas bubbles, and possibly due to the development of molecular structural units in a silicate melt. At above-liquidus (supra-liquidus) temperatures Newtonian rheology is applicable.

By assuming Newtonian behaviour, lower apparent viscosities (Fig. 2.3) are calculated by the Jeffreys equation for faster-moving flows. For a non-Newtonian lava, which is say pseudo-plastic, its apparent viscosity decreases with increasing strain rate (Fig. 2.3). Hence, a fast-moving lava will appear less viscous or more fluid than when moving more slowly.

The most accurate published field determinations of lava viscosity are given by Shaw *et al.* (1968) and Pinkerton and Sparks (1978). Shaw *et al.* (1968) used a rotating shear vane to measure the

Table 2.4   Results of field measurements of physical properties of basaltic lavas.

	Makaopuhi lava lake, Hawaii 1968	Etna 1975
composition	tholeiitic basalt	hawaiite
temperature (°C)	1130–1135	1086
phenocryst content (vol%)	25–35	45
yield strength (N m^{-2}) (=10^{-1} dyn cm^{-2})	70–120	370
Bingham viscosity (Pa s)	6.5–7.5 × 10^2	9.4 × 10^3

Data from the Makaopuhi lava lake are from Shaw *et al.* (1968) and Etna 1975 from Pinkerton and Sparks (1978). Compared with the Hawaiian lava, the Etna lava was at a lower temperature and had a higher phenocryst content, which would be responsible for its higher yield strength and plastic viscosity.

viscosity in the tholeiite of the 1968 Makaopuhi lava lake, Hawaii. Although Pinkerton and Sparks (1978) used a variety of methods to measure the rheological properties of small lava flows erupted on Mt Etna in 1975, the results in Table 2.4 are only from a specially developed penetrometer. Results from both Makaopuhi and Etna indicated that the lavas behaved in a pseudo-plastic manner, but could be approximated closely to a Bingham model with a definite yield strength (Table 2.4). *Bingham* or *plastic viscosities* of the lavas have been determined (Eqn 2.3), and these are also shown in Table 2.4. There are no field measurements of Bingham viscosities for more-felsic or salic lavas.

Viscosities obtained experimentally for five rocks of varying compositions at varying temperatures from Murase and McBirney (1973) are given in Figure 2.6. Sub-liquidus viscosities are only apparent viscosities (Fig. 2.3).

Empirical methods have also been developed for calculating the viscosities of silicate melts from their chemical composition and temperature (Bottinga & Weill 1972, Shaw 1972). According to McBirney and Murase (1984), the validity of these estimates is only established at supraliquidus temperatures, i.e. for crystal-free magmas. However, viscosity contrasts between different compositions can be represented in a qualitative way.

Viscosity and yield strength are important in controlling not only fluidity and lava mobility, but also the resultant geometry and morphology of lavas (Ch. 4).

2.5 Factors controlling viscosity in magmas

Not all of the factors that control viscosity in magmas are well understood, nor is the way in which they all interact. The principal factors that contribute to viscosity that have been studied include:

pressure
temperature
volatile content, especially dissolved water
  content
chemical composition
crystal content
bubble content

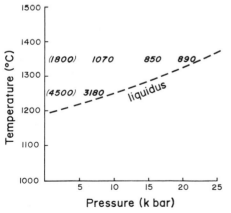

Figure 2.6 Viscosity of an andesite melt with varying temperature and pressure. Numbers indicate the viscosity of the melt in poises; those in parentheses represent the viscosity of the same melt at 1 atm pressure calculated by the method of Bottinga and Weill (1972). (After Kushiro *et al.* 1976.)

2.5.1 PRESSURE

In a series of experiments carried out in a piston cylinder apparatus at supra-liquidus temperatures Kushiro (1976, 1978, 1980) and Kushiro *et al.* (1976) showed that in natural and synthetic melts, the viscosity becomes lower with increasing pressure (Fig. 2.6), especially at high pressures. Significant specific findings of these experiments were the following.

(a) As pressure increases at constant temperature, the rate at which the viscosity becomes lower in basaltic magma is less than the rate in andesitic magma (the viscosity of basaltic magma is initially significantly lower anyway).

(b) The viscosity of an andesitic melt with 4 wt% H_2O is lower by a factor of 20 than in its anhydrous equivalent at the same temperature and pressure.

(c) The viscosity of a $NaAlSi_2O_6$ (jadeite) melt decreased by an order of magnitude more than a melt of $Na_2Si_3O_7$ did with increasing pressure, which Kushiro relates to a change in the co-ordination number of Al from four to six, implying that the melt structure is an important influence on viscosity. (Note: although the viscosity became lower, the density increased with an increase in pressure.)

2.5.2 TEMPERATURE

Viscosity is very dependent on the temperature of the magma (Fig. 2.7). Both field and experimental data show that the viscosity of all magmas increases significantly on cooling (H. Williams & McBirney 1979), partly due to crystallisation. However, at equivalent temperatures and pressures different magmas have different viscosities, suggesting that compositional aspects are also important in determining their viscosities.

2.5.3 VOLATILE CONTENT

Dissolved water content has a marked effect on the viscosity of magmas (Fig. 2.8; H. Williams & McBirney 1979, McBirney & Murase 1970). At

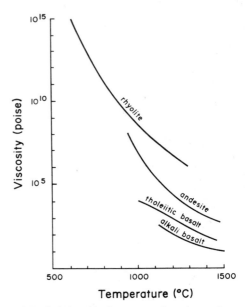

Figure 2.7 Relationship between viscosity and temperature for some volcanic rocks. The rhyolite was glassy or liquid through the entire temperature range. The rocks are the same as in Figure 2.2. (After Murase & McBirney 1973.)

fixed temperatures the viscosity of a particular magma becomes lower with increasing water content, especially for more silicic magmas (Shaw 1963, 1972). The solubility of water in magmas is controlled by temperature, pressure, the presence of other volatiles (H. Williams & McBirney 1979) and the presence of exchangeable cations (those not in tetrahedral co-ordination sites; Burnham 1979, Mysen *et al.* 1982). The solubility of water in magma increases with decreasing temperature and increasing pressure, and decreases with increasing abundance of other volatiles. Burnham (1979) indicated that the effect of water on a silicate melt is to depolymerise the melt by breaking Si–O–Si bridges. A H^+ ion exchanges with cations not in tetrahedral co-ordination sites (e.g. Na^+), hydrolysing one of the tetrahedral co-ordinating oxygens to OH^-. Stolper (1982) showed that *dissolved* water exists in both hydroxl (OH^-) and molecular (H_2O) forms in silicate glass, and by inference in silicate melts. At low total water contents (<3 wt%), the rate at which hydroxyl ion concentration increases and melt viscosity decreases are both high. However, at higher water contents,

(a)

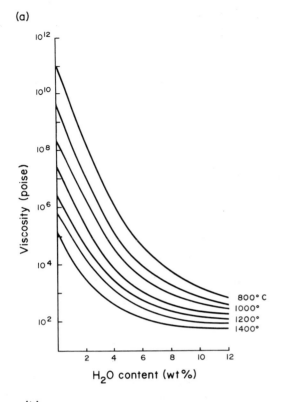

(b)

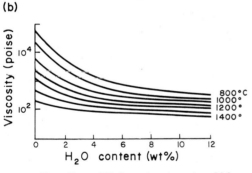

Figure 2.8 The effect of H_2O on the viscosity of (a) granitic and (b) basaltic melts at varying temperatures. (After Murase 1962.)

basis, it has been shown experimentally that water is more soluble in silicic melts than in mafic ones (Burnham 1979). However, on an equimolal basis the solubility of water is the same in all magma types. Furthermore, because water solubility increases with decreasing temperatures (H. Williams & McBirney 1979), most silicic magmas, because of their lower eruption temperatures, could contain more water than mafic ones do, if adequate water is available in the subsurface source area. Other factors probably also influence the actual water content of magmas, e.g. H_2O is geochemically incompatible with most silicate systems, and is therefore concentrated in more-evolved magmas.

Pressure is an important control on the solubility of water in a magma (H. Williams & McBirney 1979; Ch. 3). As a magma rises in the subsurface and the confining pressure decreases, water will begin to exsolve from the magma, and crystallisation occurs. The effect of this is to increase the viscosity and the strength of the magma. If the exsolved volatile content is low, then this increased viscosity may be sufficient to stop the magma from disrupting explosively, as discussed in Chapter 3.

However, notwithstanding all of the foregoing, it is significant that the effect of dissolved water in lowering viscosity is greater for silicic magmas than for basic ones (Fig. 2.8) because there are more Si–O bonds to break, so an erupting silicic magma with a low water content (e.g. one that has degassed) will be more viscous than a basic one with an equivalent weight percentage water content at the same temperature. This implies that some factor other than pressure, temperature and dissolved water content affect the viscosity of magma; namely, the magma composition. '

The exact effects of other volatiles is poorly known, being dependent on their solubilities and abundances. Chlorine and fluorine have a marked effect on magma rheology. Peralkaline rocks have high Cl and F contents which are thought to considerably reduce viscosities and yield strengths of magmas of these compositions (Schmincke 1974, Wolff & Wright 1981, 1982). The supraliquidus viscosities of pantellerites are typically two orders of magnitude below those of calc-alkaline rhyolites (Wolff & Wright 1981). Carbon

as water is added, the rate of increase of hydroxyl ion concentration falls, as does the rate of decrease of melt viscosity, but the concentration of molecular water increases significantly (Stolper 1982). The effect of dissolved water on magma viscosity is therefore due more to the concentration of network-breaking hydroxyl ions than to the total dissolved water content. On a weight percentage

dioxide has a low solubility at low pressures, but its solubility increases markedly in the presence of H_2O (Mysen 1977, Burnham 1979). However, CO_2 increases polymerisation, and therefore viscosity, in melts by forming CO_3^{2-} complexes (Eggler & Rosenhauer 1978, Mysen *et al.* 1982).

2.5.4 CHEMICAL COMPOSITION

The overall composition of a magma affects its viscosity in a complex fashion. The elements in a magma can be divided into *network formers* and non-network formers. Si^{4+} and to a lesser extent Al^{3+} and Fe^{3+} are the principal cation network formers. Silica content is important in contributing to the viscosity of magmas, because Si–O bonds are the strongest cation–anion bonds in a magma, mineral or rock. Even well above its liquidus temperature, a magma has a well defined structure (Burnham 1979, Hess 1980, Mysen *et al.* 1982), and its strength and shear resistance can be attributed to intermolecular bonds, and particularly Si–O bonds. Al–O bonds are also important in this regard, since they are also much stronger than other cation–oxygen bonds, though not as strong as Si–O bonds. O^{2-} is both a network and non-network former. In the former role, its principal function is to form cation–oxygen tetrahedra with Si^{4+}, Al^{3+} and Fe^{3+}. Mysen *et al.* (1982) suggest that silicon–oxygen tetrahedra represent the basic building block to a range of network structural units, these being SiO_4^{4-} (monomers), $Si_2O_7^{6-}$ (dimers), $Si_2O_6^{4-}$ (chains), $Si_2O_5^{2-}$ (sheets) and three-dimensional units. Overall this sequence of units corresponds to increasing degrees of polymerisation and viscosity. The type of network structural unit present depends on the ratio of non-bridging oxygens to silicon (NBO : Si) and the types of non-tetrahedrally co-ordinated cations present, which are called *network modifiers* (Mysen *et al.* 1982).

If the NBO : Si ratio decreases, magmas will be more polymerised. Similarly, the higher the field strength of the network modifying cations, the more polymerised the melt will be at a given NBO : Si ratio (Mysen *et al.* 1982). Peralkaline magmas with relatively high Na^+ and K^+ ion concentrations will be of relatively low viscosity because of the effect of Na^+ and K^+ in lowering the degree of melt polymerisation. Similarly, basic magmas (higher overall NBO : Si and more network modifiers) will have lower viscosities than acidic ones will at the same temperature and volatile contents.

Some minor components can have opposing effects. For example, TiO_2 reduces silica activity and the degree of polymerisation, whereas P_2O_5 increases silica activity and the degree of polymerisation (Ryerson & Hess 1980, Mysen *et al.* 1982).

2.5.5 CRYSTAL CONTENT

The effect of crystals suspended in a magma is to increase the effective or bulk viscosity of the magma (discussed further in Ch. 11). The effective viscosity can be estimated from the Einstein–Roscoe equation (McBirney & Murase 1984):

$$\eta = \eta_0(1 - R\phi)^{-2.5} \qquad (2.5)$$

in which η is the effective viscosity of a liquid with a volume fraction of ϕ suspended solids, η_0 is the viscosity of the liquid alone and R is a constant whose best estimated value for lavas is 1.67 (Marsh 1981, McBirney & Murase 1984). However, results calculated using Equation 2.5 are frequently at variance with values of lava viscosities measured in the field, due to the larger sizes and higher concentrations of crystals than those for which Equation 2.5 was designed (McBirney & Murase 1984). To overcome this McBirney & Murase (1984) designed a computer program for calculating effective viscosities in crystal-bearing magmas.

2.5.6 BUBBLE CONTENT

The effect of bubbles on the bulk viscosity of the magma can be variable, depending on the degree of vesiculation, the size and distribution of bubbles, and the viscosity of the magma interstitial to the bubbles (see Sparks 1978a). As described above, dissolved water contributes significantly to lowering magma viscosity. When the water exsolves, the magma viscosity begins to increase. However, the exsolved phases, are very low viscosity fluids, which may affect the overall *bulk* viscosity. In low viscosity magmas such as basalt, exsolution of volatiles may have relatively little effect on bulk viscosity because the low viscosity is largely due to the effects of temperature and composition. The presence of abundant fluid bubbles, may enhance the already low viscosity. In more acidic magmas however, the viscosity of the magma is initially high, and may be significantly affected by exsolution. Vesiculated rhyolite, for example, will have a very high bulk viscosity, irrespective of the degree of vesiculation, unless it is peralkaline in character. By contrast, the mechanical strength of the rhyolite may be low due to the high vesicle content and overall physical heterogeneity, especially if bubble walls are thin. Hence in spite of the overall high bulk viscosity, such a vesiculated rhyolite, may be very susceptible to mechanical, explosive fragmentation, as discussed in Chapter 3.

2.6 Strength

Volcanic products, like all substances, can deform in a ductile or brittle manner when subjected to a stress. Given the great diversity of physical states of erupting volcanic materials (liquids to solids), the deformation takes many forms, including highly plastic as reflected by lava flow, to essentially brittle, as reflected by explosive disruption or fragmentation of a rhyolite dome. Given the wide range of physical states, volcanic materials therefore have very diverse mechanical strengths. The mechanical strength of rock decreases rapidly as the degree of partial melting increases (van der Molen & Paterson 1979, Shaw 1980), the converse also being true. Therefore as a magma crystallises, its strength increases (in an uncertain relationship – McBirney & Murase 1984) as the proportion of crystals increases. Consideration of mechanical strength is most significant in the context of explosive fragmentation, because the greater the mechanical strength of the erupting material, the greater the tendency to resist explosive disruption. In this regard, fragmentation can be effected by both tensile stresses and shear stresses. The former

occurs for example during explosive growth of bubbles when the gas pressure exceeds the tensile strength and surface tension force of the magma. The latter can occur, for example, when vesiculated magma is crushed by a velocity-induced shear stress that exceeds the shear strength of the erupting mass, when magma discharge rates are high. The mechanics of fragmentation will be discussed further in Chapter 3. Units of strength are the same as stress or force units (dyn cm^{-2}, Pa, etc.).

2.7 Fluid flow character

The physical properties of magmas, and of aggregates of pyroclastic and epiclastic debris, will control to a large extent the nature of resultant flows of lava and debris away from the source. Viscosity, or bulk viscosity (for heterogeneous aggregates), is especially important in controlling not only flow mobility and form, but also the fluid flow state. There are essentially two fluid flow states:

laminar flow
turbulent flow

In *laminar flow*, fluid streamlines are smooth and parallel, and no mixing of streamlines occurs. The fluid is therefore free of eddies and vortices. In *turbulent flow*, streamlines are highly irregular and are dominated by eddying, so high degrees of

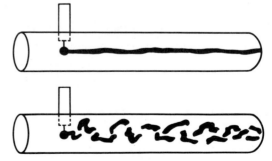

Figure 2.9 Reynolds' experiment, in which a dye streak is injected into a steady flow of water through a glass tube to define (a) laminar flow and (b) at increased flow velocity above a critical value, turbulent flow in which eddies disperse the dye and eventually colour the whole flow.

mixing of the fluid occur. In 1883, the English physicist Reynolds first defined the criteria distinguishing the two flow states in experiments carried out by passing Newtonian fluids through pipes (Fig. 2.9). The defining parameter for Newtonian fluids, the Reynolds Number (Re), is given by

$$Re = UD\varrho/\eta \qquad (2.6)$$

where U is the average velocity, D the pipe diameter, ϱ the density and η the viscosity; Re is dimensionless. For open channel flow (e.g. rivers, lavas), D is replaced by R, where R is the hydraulic radius = A (cross-sectional area) /P (wetted channel perimeter), so that

$$Re = UR\varrho/\eta \qquad (2.7)$$

For flows in both pipes and channels, the transition between laminar and turbulent flow lies between Re values of 500 and 2000.

For non-Newtonian Bingham substances, such as debris flows and lavas, the Reynolds Number criterion for turbulence is inadequate because of the high strength and viscosity of the substance (Hampton 1972, Hiscott & Middleton 1979). For such substances the criterion for turbulence is

$$Re \geqslant 1000 \, B \qquad (2.8)$$

where B is the Bingham Number, which is given by

$$B = \tau_c D/\eta U \qquad (2.9)$$

where τ_c is the strength of the substance and the other parameters are as defined in Equation 2.6. Equations 2.6 and 2.9 may be combined to produce a criterion for turbulence in Bingham substances, known as the Hampton Number (Hiscott & Middleton 1979):

$$\varrho U^2/\tau_c \geqslant 1000 \qquad (2.10)$$

The critical point about the Reynolds Number is that it is inversely proportional to the viscosity, and directly proportional to the velocity. Because of this, lavas may move by laminar flow or turbulently, but generally, because of their relatively high viscosities, *most lavas flow in laminar fashion*. Only low viscosity, highly fluidal magmas experience turbulent flow, and then usually only where relatively high terrain slopes cause acceleration to relatively high velocities (but see Section 4.12 on Archaean komatiite lavas). High viscosity lavas move at low velocities, even on steep slopes, because of their high internal yield strength. Flow banding, which is characteristic of rhyolites (Fig. 2.10), is a reflection of laminar flow. It is usually preserved only in very viscous, highly siliceous

Figure 2.10 Flow-banding in spherulitically devitrified obsidian dome rhyolite. Holocene Okataina Complex, North Island, New Zealand.

Figure 2.11 Flow-aligned plagioclase phenocrysts in the Permian Bombo Latite, Kiama, New South Wales, Australia. These aligned phenocrysts reflect laminar flow paths.

(a) Longitudinal section

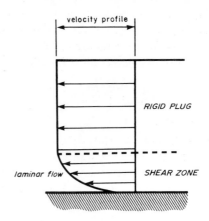

(b) Plan view

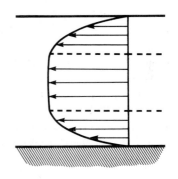

(c) Cross-sectional view

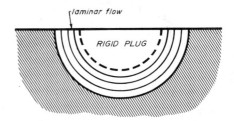

Figure 2.12 Velocity profiles for flow of a Bingham substance in a subaerial setting. Length of arrows is proportional to fluid velocity. (After A. M. Johnson 1970.)

flows. In less viscous flows, laminar flow may be reflected by alignment of platy or elongate phenocrysts (Fig. 2.11).

A. M. Johnson (1970) suggested that lavas (Ch. 4) may have rheological similarities to cohesive debris flows (Ch. 10), and should flow similarly. Many debris flows and lavas approximate Bingham substances, and both frequently develop interior channels flanked by non-moving 'dead zone' margins, which are frequently levée-like in form and so act to contain the more active part of the flow. The formation of levées and steep margins in debris flows and lavas is attributed to their cohesive character and the existence of a yield strength during flow (A. M. Johnson 1970, Hulme 1974; Ch. 4 for lava levées). Observations of natural debris flows, laboratory experiments and theoretical modelling have shown that the flow character of many debris flows cannot be described simply as laminar or turbulent. As a consequence of a finite yield strength, the bulk of the flow moves as an essentially internally inert 'plug' upon a zone of laminar flow, which is a zone of very high shear between the coherent plug and the channel walls (A. M. Johnson 1970; Figs 2.12 & 13). However, flows can be turbulent given high velocities and low bulk viscosities. Where flows move within valleys incised into bedrock, the

Figure 2.13 Cross-sectional view of a basaltic volcaniclastic debris flow of late Eocene–early Oligocene age, Waiareka–Deborah volcanics, Bridge Point, South Island, New Zealand. The lower stratified succession into which the debris flow has eroded consists of primary and redeposited pyroclastics. The cross-sectional view is similar to that in Figure 2.12c, with a semicircular channel-form, a lower laminar-flow layered zone and an inner massive plug core in which large clasts have been concentrated.

bedrock constitutes the channel walls, but where flow occurs on an unconfined surface, the lateral edges of the flow solidify and become the *de facto* channel walls. More will be said about lava flow types and resultant characteristics in Chapter 4, about pyroclastic flows and their deposits in Chapters 5 and 7–9, and about epiclastic mass flows in Chapter 10.

2.8 Further reading

An excellent summary of the properties of magmas that affect their rheology is given by McBirney & Murase (1984). Other physical properties of magmas have been adequately reviewed by Murase and McBirney (1973), H. Williams and McBirney (1979) and Hargraves (1980). A good review of the rheological variables of magmas (including lunar magmas) is given by Whitford-Stark (1982) and detailed review of the factors controlling the structure of silicate melts is given by Hess (1980) and Mysen *et al.* (1982). The peculiar properties of Archaean komatiite lavas are briefly considered in Chapter 4, Section 4.12.

Plate 3 The township of Vestmannaeyjar on the island of Heimaey, 1973, half buried in tephra fall-out from the eruption column. A basaltic lava from the new active volcano flows to the left into the sea and fragments. The older 5000–6000-year-old volcano. Helgafell, stands watchfully in the background to the right. The stark contrast is produced by a recent snowfall which has melted and been buried around the volcano and lava flow. Note houses that are partially buried beneath snow covered tephra deposits. (After S. Jonasson in Gunnarsson 1973.)

CHAPTER THREE

Volcaniclastic deposits: fragmentation and general characteristics

Initial statement

Volcanic successions can consist of both coherent lavas and a range of volcaniclastic deposits. Lavas, their characteristics and the parameters which control their characteristics, are discussed in Chapter 4. In this chapter we consider how volcaniclastic deposits (particularly those formed by primary volcanic processes) are formed, and some of their general characteristics. The use of the non-genetic term 'volcaniclastic', opens the way to explore the range of origins of all fragmental volcanic rocks. Although some have explosive pyroclastic origins, many do not. Misinterpretation of fragmental volcanic deposits as explosive pyroclastic rocks is commonplace, and usually results from a lack of understanding of the types of fragmentation processes and the characteristics of

the deposits from each process, and on whether the deposits have been redeposited from the site of fragmentation. Nomenclature of volcaniclastic rocks is then further discussed in Chapter 12.

3.1 Introduction

The fragments in volcaniclastic rocks can be produced both by primary volcanic processes (those that are essentially contemporaneous with eruption) and by secondary surface processes (weathering, erosion, mass-wastage). Both of these groups of processes can produce generally similar textural types (e.g. breccias, sand-sized aggregates, mud-sized aggregates). To facilitate discussion of specific differences and a logical treatment of the processes and products, it is now appropriate to introduce the specific modes of formation of volcaniclastic deposits, these being

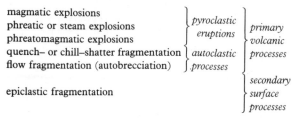

The adjective 'pyroclastic' is applied here to describe *explosive* eruptions, irrespective of their origin, and the products that are fragmented and deposited contemporaneously with such eruptions. Autoclastic processes are non-explosive, and consist of quench-fragmentation and flow fragmentation. Reworked or redeposited (or both) pyroclastic and autoclastic material becomes epiclastic upon reworking or redeposition.

3.2 Fragmentation due to magmatic explosions

Dissolved magmatic volatiles affect not only the viscosity (Ch. 2) and the freezing and melting temperatures of magmas, but also the nature of volcanic eruptions, in terms of whether they might be explosive or non-explosive. The dominant volatile component associated with most volcanic activity is water. Carbon dioxide is also an important early exsolving volatile for basalts (D. W. Macpherson 1984). The solubility of volatiles in magma is controlled, at least in part, by confining pressure (Ch. 2), and their solubility decreases as the magma rises to shallower crustal levels with lower confining pressures. At a certain depth the solubility will decrease sufficiently for carbon dioxide and water to begin exsolving from the magma and become separate fluid phases. The depth at which this occurs depends on the magma type, the actual volatile content and the vapour pressure of the dissolved water and carbon dioxide relative to the confining pressure (e.g. Burnham 1972; Fig. 3.1). Exsolution will commence when the vapour pressure equals the confining pressure. The higher the magmatic volatile content, the

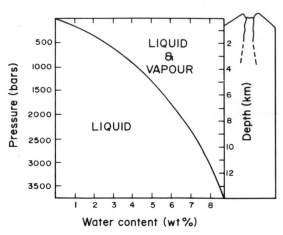

Figure 3.1 Solubility of water in andesitic magma as a function of confining pressure, which is related to crustal depth. (After Burnham 1972.)

higher the vapour pressure exerted, and the greater the depth at which exsolution will begin. D. W. Macpherson (1984) suggested that, in submarine basalts at least, carbon dioxide will exsolve and form vesicles before water does.

The confining pressure at any point in the subsurface is a function of the local stress field, which can be defined by three orthogonally orientated stress components, called the maximum principal stress (σ_1), the intermediate principal stress (σ_2) and the minimum principal stress (σ_3) (Hobbs *et al.* 1976). At relatively shallow crustal levels, the pressure regime is not simply hydrostatic, so the confining pressure is not just the vertical lithostatic load (= σ_1). $\sigma_1 \neq \sigma_2 \neq \sigma_3$ and the confining pressure is best defined as the average of the three (= ($\sigma_1 + \sigma_2 + \sigma_3$)/3; M. A. Etheridge *pers. comm.*).

For rhyolitic magmas L. Wilson *et al.* (1980) suggested that the depth of exsolution of water can be determined using the relationship

$$\varrho_{cr}gh_e = \left(\frac{n}{s}\right)^2 - P_{s'} \qquad (3.1)$$

where ϱ_{cr} is the crustal density (kg m^{-3}), g is the acceleration due to gravity, h_e the depth of exsolution (km), n the total weight fraction of exsolved volatile water, $P_{s'}$ the pressure at the surface (bars; 1 bar $= 10^5$ Pa) and s is a constant ($= 0.0013$) found from the solubility of water relationship $n_d = sP$, where n_d is the weight fraction of water dissolved at pressure P.

If the exsolution of the volatile component is not induced by crystallisation, it is called *first boiling* or decompressional exsolution–vesiculation. Its immediate effect is to increase the viscosity and strength of the magma, because in the case of water it no longer contributes to lowering the viscosity. First boiling should be reflected by the presence of *vesicles*, however small. This exsolved phase has a vapour pressure, which has the potential to explosively disrupt the magma if it continues to increase. The vapour pressure is essentially dependent on the initial volatile content of the magma, and on its temperature. In basic magmas both carbon dioxide and water are relevant in this regard; in more-acidic magmas water is more important because of its higher abundance.

Crystallisation has the effect of concentrating the dissolved volatile components in the remaining liquid magma, and this will result in higher vapour pressures. The latent heat released through crystallisation will also help to maintain high temperatures and keep vapour pressures high. These effects could lead to boiling if the vapour pressure becomes equal to the confining pressure.

This type of boiling is called *second boiling* (F. J. Turner & Verhoogen 1960, Burnham 1972) or crystallisation-induced exsolution–vesiculation, and is a very significant event. Vesiculation in a magma could be produced by both first and second boiling. Under certain circumstances, second boiling alone, caused by crystallisation, could initiate exsolution and vesiculation. Burnham (1972) has calculated that an andesitic magma with no more than 2.8 wt% water will become saturated with water at about 2 km below the Earth's surface ($\equiv 500$ bars; Fig. 3.2). If the magma rises further, cools or cools and crystallises, then the water will separate as a high temperature liquid or steam (or both). The exsolved phase causes an enormous increase in pressure within the magma chamber (Burnham 1972) and the potential increase in volume of the system during this crystallisation and boiling can be up to 53%.

Once a magma has evolved to this stage, there are two distinct situations in which explosive fragmentation of the magma due to exsolution of volatiles can occur:

a sealed near-surface magma chamber or conduit
an open vent, erupting vesiculating magma

3.2.1 EXPLOSIVE FRAGMENTATION FROM A SEALED, NEAR-SURFACE MAGMA CHAMBER OR CONDUIT

If the magma chamber pressure equals or exceeds the minimum principal stress in the country rock *and* the tensile strength of the country rock, then the roof of the magma chamber and the volcanic edifice will fail, possibly in a major explosive event. In the example that is cited above for an andesitic magma with 2.8 wt% water, Burnham (1972) calculated that as crystallisation and cooling proceed, the internal pressure of the magma chamber should theoretically increase until at the point of full crystallisation (700°C) the internal pressure will have reached 13×10^7 Pa (or 1300 bars). This would be far in excess of the confining pressure of 5×10^7 Pa (or 500 bars) and the typical tensile strength of the country rock ($\sim 1.5 \times 10^7$ Pa or 150 bars). Therefore, before this stage, fracture and failure of the roof

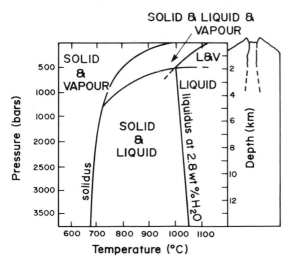

Figure 3.2 Solid–liquid–vapour fields for an andesitic magma with 2.8 wt% H_2O. (After Burnham 1972.)

of the chamber, and probably explosive disruption of the magma and the volcanic edifice, will have occurred. Such an event could happen at any time in the history of a volcano when the magma chamber or conduit is sealed (i.e. the vent is closed or blocked off) as long as a volatile-rich magma is being supplied to the chamber or conduit. For a sealed magma chamber or conduit which is also overlain by a body of water (ocean, lake), the hydrostatic pressure of the water column, added to the lithostatic load, also contributes to the confining pressure constraining exsolution and vesiculation.

3.2.2 EXPLOSIVE FRAGMENTATION OF A VESICULATING MAGMA ERUPTING FROM AN OPEN VENT

There are two separate situations to consider here:

subaerial vents and
subaqueous vents.

Subaerial vents

Studies of the way in which explosive expansion of volatiles fragments a magma in an open vent have been made by Verhoogen (1951), McBirney (1963, 1973), McBirney and Murase (1970) and Bennett (1974). Other studies have contributed to a better understanding in indirect ways. However, the most significant study is that of Sparks (1978a), in which there is both a critical evaluation of the previous studies and an attempt to evaluate quantitatively all of the factors involved in bubble formation and growth in magmas. The discussion here is a summary of some of Sparks' excellent review. Another useful discussion is given in H. Williams and McBirney (1979, their Ch. 4).

Consideration of nucleation theory is beyond the scope of this discussion (readers are referred to Sparks 1978a), but if a magma becomes even slightly supersaturated with volatiles, then nucleation of bubbles will occur. Observations of natural pumices and scoria show that vesicles are rarely less than 5 μm in diameter, suggesting that this may be a minimum stable nucleation size. The growth of a fluid bubble thereafter is controlled by (a) the diffusion of volatiles dissolved in the magma into the bubbles and (b) by the rate at which the confining pressure falls as the bubble or the magma, or both, rise. In the case of an open vent, the confining pressure is essentially the *magmastatic pressure* of the magma column. Sparks (1978a) calls the lowering of the magmastatic pressure during bubble or bubble and magma rise *decompression*. Growth rates due to diffusion are controlled by the composition, solubility, concentration and the degree of supersaturation of the volatiles. Decompressional growth of bubbles is controlled by the rise velocity of the magma, the rate at which the magma is disrupted and removed at the free surface in the vent, and by the rise of the bubbles within the magma body (Sparks 1978a). In practice, both diffusion and decompression may operate together.

Other factors which affect bubble growth are properties of the magma; density, surface tension, viscosity, and the solubility and diffusion coefficient of the gas in the magma. Some relevant equations relating these factors for H_2O follow.

$$-\log_{10} D = \log_{10} \eta + 5.82 - 4100/T \qquad (3.2)$$

where D is the diffusion coefficient, η is viscosity and T is the temperature in Kelvin (Sparks 1978a). This equation was derived from the experimental data of Scholze and Mulfinger (1959) using artificial silicate melts at 1 atm pressure ($\sim$101 kPa) and at temperatures between 1000 and 1400°C. Sparks (1978a) comments that the relationship should be used with care because of the particular compositions, viscosities and temperatures used. Nevertheless, it implies that diffusion is dependent on viscosity, and that diffusion rates in basic magmas should be higher than in silicic melts. The estimates of Sparks suggest that for basaltic magmas at 1100–1300°C, $D = 10^{-5}$–10^{-6} cm^2 s^{-1}, whereas with rhyolitic melts at 750–850°C, $D = 10^{-7}$–10^{-8} cm^2 s^{-1} for H_2O.

Burnham (1983) expresses the relationship between diffusivity and viscosity, slightly differently:

$$\log_{10}D = 2 \log_{10}T - 1.33 \log_{10}\eta - 5.74 \qquad (3.3)$$

Vesicle size is related to diffusivity by

$$R = 2\beta(Dt)^{\frac{1}{2}} \qquad (3.4)$$

where R is the radius, β the growth rate constant, D the diffusion coefficient and t is time. The values for D cited for basaltic and rhyolitic magmas suggest that bubbles in basalt should grow to a radius 10–50 times larger than in rhyolite in a fixed time (Sparks 1978a).

The pressure of the fluid in the growing bubble, P_b is given by

$$P_b = P_m + 2\sigma'/r + \varrho_m(r\ddot{R} + \tfrac{3}{2}\dot{R}^2) + 4\eta\dot{R}/r \quad (3.5)$$

where P_m is the magmastatic pressure, σ' the surface tension, r the radius, $\dot{R}$ the growth rate, $\ddot{R}$ the acceleration rate, ϱ_m the magma density and η the dynamic viscosity (equation from Rosner & Epstein 1972, Sparks 1978a). Equation 3.5 is extremely significant in understanding the constraints on bubble growth, because all the forces constraining and resisting bubble growth are represented:

(a) the magmastatic pressure of the magma column ($= \varrho_m g h$; g = acceleration due to gravity, h = height of magma column);
(b) the surface tension force of the magma in the bubble walls ($= 2\sigma'/r$);
(c) the inertial force of the magma in the bubble walls ($= \varrho_m(r\ddot{R} + \tfrac{3}{2}\dot{R}^2)$) and
(d) the viscous force of the magma in the bubble walls ($4\eta\dot{R}/r$).

Bubbles cannot continue to grow infinitely, because they do not grow in isolation from each other. Bennett (1974) suggested that bubbles growing near to each other cease growing because high-viscosity magma liquid has to be forced through intricate channels between bubbles which are trying to expand. The rapid exsolution of volatiles causes a rapid increase in the viscous resistance to growth of bubbles, and increases the tensile strength of the magma (McBirney 1973, Sparks 1978a, L. Wilson *et al.* 1980). In addition, the bubbles will not burst because there is no significant pressure gradient across bubble walls (Fig. 3.3). Under these circumstances, volatiles will continue to diffuse from the magma to the bubble, and the fluid pressure will continue to rise until equilibrium is reached between the fluid pressure in the bubble

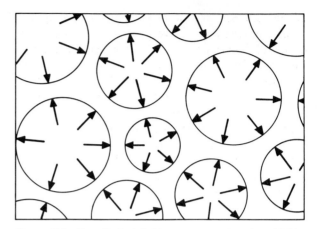

Figure 3.3 Hypothetical fluid pressure vectors in a highly vesiculated magma. No appreciable pressure gradient exists across bubble walls. Bubbles cease to grow when the fluid pressure in the bubbles equals the vapour pressure of the residual volatile fraction within the (volatile saturated) magma of the bubble walls.

and the vapour pressure of the volatile still dissolved in the magma (Sparks 1978a).

If the frothed magma does not fragment when bubble growth has ceased, how and where does it fragment? According to Sparks (1978a), the magma disrupts or fragments explosively at its free surface in the vent, because it is there that a high pressure gradient exists between the vesiculating magma and the atmosphere. The sequence is shown schematically in Figure 3.4. Gas pressures at the fragmentation surface are controlled by the proportion exsolved and their type (Fig. 3.5). According to L. Wilson *et al.* (1980), magma fragments when the void fraction in the magma is approximately 0.77 (based on examination of modern pumice fragments). Bennett (1974) and L. Wilson *et al.* (1980) also suggest that as the' vesiculating and fragmenting magma accelerates upwards through the conduit to speeds ranging from subsonic to supersonic, the shear stress produced at the high velocities and, most importantly, the rate at which this stress and the strain rate increase may be sufficient to overcome the short-term tensile strength of the magma in the bubble walls, leading to disruption. The sequence shown in Figure 3.4 implies that the disruption surface should migrate downwards. However, counter-

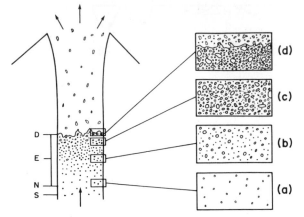

Figure 3.4 Gas bubble nucleation, growth and disruption sequence in a magma column in an open vent. S is the level of gas saturation, N is the level of bubble nucleation, E is the interval of bubble growth and exsolution, D is the level of magma disruption. Degrees of progressive exsolution and bubble growth stages are represented schematically in (a)–(d): (a) at early stages nucleation occurs and bubbles grow uninhibited; (b) growth continues and new nuclei are created; the larger bubbles begin to interfere with one another's growth; (c) the magma froth is saturated with bubbles and growth rates are retarded and eventually cease; (d) the fragmentation surface engulfs the froth and propagates down by the bursting of bubbles. (After Sparks 1978a.)

acting this is the rise velocity of the bubbles and of the magma, the latter being, in part, produced by volume increases associated with continued exsolution and expansion of volatiles, and by upward supply of magma from deeper levels.

Many other aspects of explosive magmatic eruptions could be discussed, given space. Many of these are discussed by L. Wilson *et al.* (1978, 1980) and L. Wilson (1980a), including controls on the depth of the disruption surface in the conduit, eruption rates, eruption velocities and heights of eruption columns, and the interplay between magma properties and conduit geometry in controlling these. Reference is made to these studies in Chapters 5 and 6. By way of example, though, eruption rates in the order of 1.1×10^6 $m^3 \ s^{-1}$ are feasible, and eruption velocities can be as high as $600 \ m \ s^{-1}$, producing eruption columns up to 45–55 km maximum height. Eruption intensity, eruption rate and eruption velocity are controlled by volatile contents (L. Wilson *et al.* 1978, 1980).

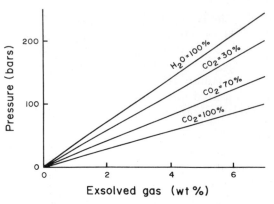

Figure 3.5 Gas pressure exerted by H_2O/CO_2 mixtures at the point of explosive disruption of vesiculated magma as a function of the total weight percentage of exsolved gases. (After L. Wilson 1980a.)

Subaqueous vents

For subaqueous vents, the same physical constraints control whether explosive expansion of volatiles will occur. Two other complications exist, however: first, the effects of the hydrostatic pressure in constraining explosive expansion and, secondly, the physical interaction between hot magma and cold water.

The pressure gradient in a body of pure water is 1 bar per 10 m. The *critical point* (pressure) of *pure water* is 216 bars, which is the pressure above which there is no distinction between vapour and liquid (= fluid), and at which volume decrease with increasing pressure is slight (Fig. 3.6). Below this confining pressure, the volume rises exponentially with decreasing pressure, and rapid boiling can occur. In other words, at confining pressures of 216 bars or more, H_2O bubbles will not expand significantly (McBirney, 1963), and certainly not rapidly enough to overcome the short-term tensile strength of the magma (which is necessary for explosive disruption to occur), even if their internal pressure were to exceed the confining pressure. These pressures would have to be unreasonably high, given present estimates of the vapour pressure exerted by volatiles in magmas (Sparks 1978a, L. Wilson *et al.* 1980). A pressure of 216 bars corresponds to a water depth of 2160 m, so it is implausible that explosive fragmentation of a magma erupting subaqueously in pure water could occur at this depth or greater (McBirney 1963). Explosive fragmentation could occur at shallower depths if the original magmatic water content were high enough (McBirney 1963; Fig. 3.7).

However, the effect of a solute such as salt is to raise the pressure of the critical point (Sourirajan & Kennedy 1962, G. Green *pers. comm.*; Fig. 3.8), the critical pressure increasing with increasing NaCl content. In sea water the concentration of NaCl is about 3.5%, and this raises the critical pressure of sea water to about 315 bars, corresponding to water depths of about or slightly less than 3150 m, given the slightly greater density of sea water. Therefore, for subaqueous eruptions in sea water explosive fragmentation is not possible below this depth and, as McBirney (1963) has pointed out, it is more likely to occur at much shallower depths (<1 km). However, it should be noted that vesiculation can occur at very much

(a) Basalt

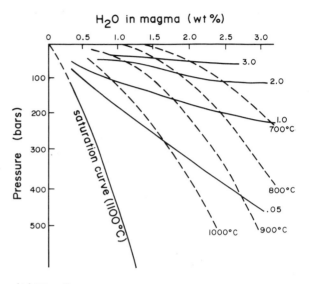

(b) Rhyolite

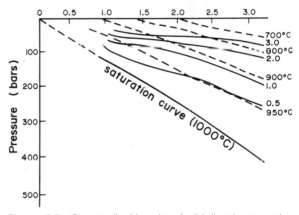

Figure 3.7 Gas to liquid ratios (solid lines) at varying magma temperatures (broken lines) for magmas with varying magmatic water content at different pressures. (a) Basalt, (b) rhyolite. (After McBirney 1963.)

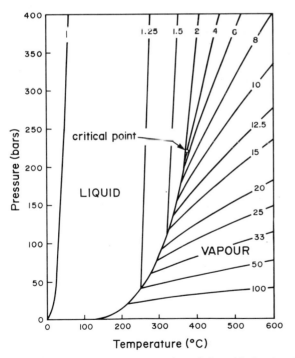

Figure 3.6 Graphical representation of the critical point of pure water (after Kokelaar 1982). Numbered curves are the *P–T* paths of equal volume change. Note that beyond the critical point these curves no longer converge to define a liquid/vapour phase boundary.

greater depths than those corresponding to the critical pressure, but the growth rate of such vesicles will be insignificant. It is only below the critical point that the growth rates can be rapid enough to cause explosive expansion and disruption.

The foregoing discussion of the constraints on subaqueous explosive eruptions is a representation of the 'conventional wisdom', as first discussed in

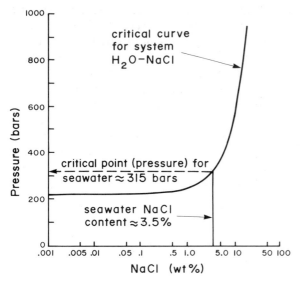

Figure 3.8 Critical point for the system H_2O–$NaCl$. For sea water with an approximate $NaCl$ content of 3.5 wt%, the pressure at the critical point is about 315 bars. (After Sourirajan & Kennedy 1962.)

detail by McBirney (1963). Recently, however, Burnham (1983) has proposed that highly explosive eruptions could even be possible at depths of 10 km. His analysis concerns the rhyolitic tuffs that are hosts to the Kuroko ores in Japan, and which are supposed to have erupted at water depths of about 4 km (Ohmoto 1978, Ohmoto & Skinner 1983, Burnham 1983). Burnham argues that very rapid exsolution (seconds or less) of even 1 wt% or less of magmatic H_2O from rhyolitic magma during second boiling (crystallisation) should be capable of releasing enough energy to cause failure of the country rock around the magma chamber. Sudden decompression of the magma associated with the failure of the country rock should produce still further exsolution which, according to Burnham, should release further energy – sufficient to cause explosive eruption of the magma at great depths on the sea floor.

There is little doubt that exsolution of volatiles or vesiculation, due to either first or second boiling, could release enough energy and increase the fluid pressure of the magma chamber sufficiently for it to cause failure of the country rock. Abundant vesicular submarine lavas are testimony to this. However, Burnham's analysis does not convincingly demonstrate that sufficient vesicularity, or explosive expansion of vesicles, is produced to cause the magma to erupt *explosively* at such depths. Burnham suggests that the vesicularity produced in his analysis should be about 17% at 4 km to 37% at 3 km, which coincides with the vesicularity range measured in volcanic debris associated with the Kuroko ores (20–30%; Burnham 1983). However, there are insufficient data on vesicle sizes. Also, based on a comparison with explosive subaerial pyroclastic eruptions of rhyolitic magma, with documented vesicularity of 70–80% (Sparks 1978a), it seems that the calculated and cited vesicularity values for the Kuroko rocks are inadequate to allow explosive eruption to take place. There is no doubt that vesiculation of magma can take place under the conditions outlined by Burnham (1983), and that magma fluid pressures will be high enough to allow eruption of vesiculated magma as lava, but it has *not* been effectively demonstrated that the pressure of the H_2O phase *in the vesicles* will be sufficiently in excess of the ambient hydrostatic pressure to allow the vesicles *to expand explosively* against this hydrostatic pressure, which at depths of 4 km is in excess of the critical point of water.

On the same count, although boiling of the sea water occurs when it comes into contact with the lavas in the vent, the confining hydrostatic pressure is too great to allow explosive expansion of the boiling bubbles in the sea water at such depths (Section 3.4). However, fragmentation could occur due to quench-fragmentation (Section 3.6).

3.3 Magma mixing as a means of triggering explosive eruptions

The foregoing section on magmatically controlled explosions sets out the principles involved in, and the constraints imposed on, explosive eruptions that are driven purely by magmatic volatiles. The cases outlined all focus on a simple magma rising in the subsurface as a closed system, and undergoing exsolution of its own volatiles, which then drive the explosive eruption. However, recent developments in studies of magma genesis and fluid behaviour have shown that many magmas are not simple closed systems, and that their chemistry and eruptive history are controlled by the mixing of separately formed or separately evolved magma batches (Sparks *et al.* 1977, O'Hara & Matthews 1981, Huppert *et al.* 1982a).

The realisation that eruption history and the triggering of explosive eruptions may be controlled by magma mixing stems from the recognition of, and the frequent occurrence worldwide of mixed pumice fragments in largely silicic pumice fall and ignimbrite deposits (Sparks *et al.* 1977; also see Eichelberger 1980; e.g. Fig. 3.22, Chs 5–7). Mixed pumice is pumice that contains a streaky, fluidal layering of salic and more-mafic components. The mechanics of magma mixing are adequately discussed in Huppert *et al.* (1982a), and the references therein, so will not be discussed further here. However, the way in which magma mixing could initiate eruption, especially explosive eruption, will be briefly reviewed here.

The basic essence of magma mixing in triggering eruptions is that two magmas at different temperatures and with different volatile contents come into contact in a subsurface magma chamber. Many mixed pumices in rhyolitic fall or ignimbrite deposits contain mixed rhyolite–basaltic or andesitic phases, suggesting that basic magma has come up into a chamber of rhyolite magma. The actual eruption may be initiated under one of several different circumstances (Sparks *et al.* 1977, Huppert *et al.* 1982a, J. S. Turner *et al.* 1983):

(a) First, the addition of a volume of magma to a chamber may cause the total magma chamber fluid pressure to exceed the minimum principal stress and the tensile strength of the country rock, leading to fracturing of the roof of the chamber and release of the magma. If the combined volatile content of the magmas is high enough, then the release of confining pressure may precipitate increased exsolution and bubble growth to produce explosive eruption. Where volatile content is low, perhaps due to previous degassing of the magma during earlier eruptions, lava eruption, with little or no explosive activity, may occur.

(b) Where a rhyolitic magma with a moderate volatile content is injected by a basic magma from below, the superheating of the lowest part of the rhyolitic magma will induce convective uprise of this magma. Also, such superheating of the rhyolite will by itself (to a certain extent) increase the volatile vapour pressure in the magma, even without decompression, and may cause further exsolution of the remaining volatiles. As it rises, it decompresses, volatiles exsolve and the gas pressure and the total fluid pressure in the chamber may rise sufficiently to cause explosive eruption.

(c) The rising injecting basic magma may contain relatively high volatile contents, which exsolve during uprise and are transferred by convection, diffusion and mixing into a low volatile rhyolitic magma, leading again to fluid pressure build-up and explosive eruption.

(d) As a hot basic magma comes into contact with a colder rhyolitic magma, transfer of heat from the basic to the rhyolitic magma will cause rapid cooling of the basic magma, which may cause crystallisation and so lead to exsolution in the residual basic fluid, again building up the total chamber fluid pressure.

The circumstances causing eruption can involve any or several of the foregoing. Varying degrees of mixing of the magmas may be produced by turbulent transfer of heat, and by density changes in the magma(s) caused by crystallisation and exsolution of volatiles. However, much of the above discussion is speculative at present, and has been little tested on a case-history basis. Mixed pumice eruptions may also occur without the triggering effect of injection of a more basic magma into a more silicic chamber. For instance, withdrawal of magma from a stable, zoned chamber (e.g. rhyolite to andesite, or phonolite to latite) could result in the mixing of the two layers (Blake 1981a, b).

3.4 Phreatic or steam explosions and phreatomagmatic eruptions

The interaction between hot magma and water produces what can collectively be called **hydrovolcanic** activity, and produces a range of volcaniclastic products. The water could be a substantial subsurface groundwater reservoir or a surface body of water (e.g. crater lake, caldera or non-caldera lake, sea water). The immediate effect is to cause superheating, boiling, volatolisation, build-up of gas pressure of the external water and, confining pressure constraints permitting, explosive expansion of the gas produced. By this, the heat energy of the magma is transferred into the mechanical energy driving explosion. The intensity of the activity is controlled by the water : magma mass ratio and the amount of superheating of the water (Sheridan & Wohletz 1981, 1983, Wohletz 1983, Wohletz & Sheridan 1983, Colgate & Sigurgeirsson 1973). *Fragmentation probably occurs as a function of both quenching and explosive activity.* The quenching occurs as a consequence of the contact between hot magma and 'colder' bodies of water, steam, debris and rock.

The ratio of water mass to magma mass controls the type of resultant activity (Fig. 3.9). If the

water : magma mass ratio is low (less than about 0.2, Fig. 3.9) then the external water mass contributes very little to the fragmentation of the magma. In such a case, explosive eruption of the magma can be driven only by magmatic volatiles (magmatic explosions), given the constraints discussed above, or the explosive activity is confined to the eruption of steam, and little or no solid ejecta. Steam explosions are called **phreatic** explosions and significant proportions of country rock fragments may be erupted by these. Where the interaction between external water and magma produces significant explosive eruption of magmatic ejecta driven by both external and magmatic volatiles, the explosions are **phreatomagmatic**. Such eruptions are thought to occur where the water : magma mass ratios are equal to or slightly greater than 0.3 (Fig. 3.9; Sheridan & Wohletz 1981, 1983, Wohletz 1983, Wohletz & Sheridan 1983), and are caused by major increases in the degree of superheating and energy transfer. As a result, the degree of magma fragmentation is likely to be very high (Fig. 3.9; e.g. Self & Sparks 1978) and explosive activity will be very intense. Large volumes of pyroclastic detritus can be formed, and significant eruption columns or plumes may be produced. However L. Wilson *et al.* (1978) point out that because much of the thermal energy of the magma is used in superheating the external water, the mechanical efficiency is less than that of a magmatic eruption. Eruption columns for hydrovolcanic eruptions are therefore likely to be smaller than for magmatic explosive eruptions with equivalent eruption rates. Nevertheless, hydrovolcanic eruptions can be equally as devastating as magmatic ones. According to Sheridan and Wohletz (1981), where water : magma mass ratios are greater than 0.3 the level of superheating and energy transfer is less efficient, and the intensity of resultant explosions will be less. Not only does the relative degree of interaction between water and magma control the degree of fragmentation and explosiveness (Fig. 3.9), but it will also control the physical character of the transporting agent and the field facies characteristics that are produced (Chs 5–9; Sheridan & Wohletz 1983, Wohletz 1983, Wohletz & Sheridan 1979).

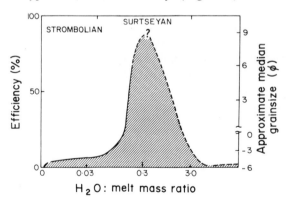

Figure 3.9 Efficiency of transfer of thermal energy to mechanical energy resulting from the interaction of hot magma with cold water as a function of the mass ratio of magma to water. The curve also shows approximate median grainsize of melt fragments. The maximum on the curve corresponds to maximum efficiency of energy transfer and therefore explosiveness (water : magma mass ratio ~0.3). For ratios less than 0.1, fragmentation of magma by external water will be minimal. (After Wohletz 1983.)

Where the interaction between magma, or quenched magmatic debris, or both, and water is explosive and self-sustaining, it is called a fuel–coolant interaction (Colgate & Sigurgeirsson 1973, Peckover *et al.* 1973, Wohletz 1983, Sheridan & Wohletz 1983). The explosive energy results from the sudden transfer of heat from the magma to the water producing instantaneous vaporisation and explosive expansion of the water. The interaction between the water and the magma, and the mode of fragmentation can both occur in several ways, and have been studied experimentally. The processes can be viewed as a cycle consisting of distinctive stages, and have been summarised by Wohletz (1983), based on the models of Buchanan (1974), Board *et al.* (1974) and Corradini (1981).

Stage 1. The initial contact between magma and water produces a vapour film between the magma and water.

Stage 2. The vapour film expands, but at its outer margin, at the contact with the water, condensation of the vapour can occur, and the film may collapse in places. Expansion and collapse can occur several times, until the energy associated with collapse causes fragmentation of the magma. Film collapse and fragmentation can occur in several discrete ways, including the penetration of linear jets of water into the magma; impact of a volume of collapsing water with sufficient intensity to create a stress wave capable of caving in the magma surface; direct water–magma interaction leading to quenching, vaporisation and explosive expansion; and the development of an unstable fluid interface associated with the transmission of a shock wave.

Stage 3. The mixing of the collapsed vapour film with the magma increases the surface area of the interface between the two.

Stage 4. Stage 3 facilitates rapid heat transfer as water encloses melt fragments.

Stage 5. Formation of a new vapour film as water is suddenly vaporised by superheating, which then leads to a reversion to stage 2.

Several situations exist in which hydrovolcanic eruptions can occur, but the constraints controlling explosive hydrovolcanic eruptions are essentially the same as for magmatic explosive eruptions:

- a magma rising in the subsurface, without an open vent or conduit, interacting with ground water

- a magma in an open vent leading to a body of water (lake, ocean)

- a lava flowing into a body of water or over water-saturated sediment

- a pyroclastic flow moving into a body of water or over water-saturated sediment

- magma rising to heat subsurface hydrothermal fluids already at temperatures near boiling point

3.4.1 INTERACTION WITH GROUND WATER

In this situation a magma encounters a ground water reservoir. The gas pressure produced by superheating of the ground water, and the resultant expansion of the ground water as it changes phase from water to steam must equal or exceed the minimum principal stress in the country rock and the tensile strength of the country rock before failure of the overburden and explosive eruption can occur. Kokelaar (1982) discussed the intrusion of hot magmas into water-saturated unconsolidated sediments, emphasising that, even if explosive activity does not occur, the sediments could be extensively mixed as the interstitial fluid is heated, expands, begins to convect, perhaps even being locally vaporised, and fluidises and turbulently mixes the sediment. Quench-fragmentation of the magma may also occur to varying degrees as it is intruded into cold, water-saturated sediments (Fig. 3.10). Dynamic mixing of the quenched debris and the sediment could occur if the pore water boils and fluidises the clastic pile, producing deposits called peperites (see Section 3.6). Intrusive interaction is also capable of producing pillow-like pods of magma (Fig. 3.11) and spectacular soft-sediment deformation structures (e.g. Kokelaar 1982), including load cast-like features.

Figure 3.10 Basalt (light colour) intruding water-saturated, unconsolidated sediments, producing quench shattering of the magma, and fluidisation-driven turbulent mixing of the sediment and water as it is boiled, leading to destruction of original sedimentary structures. Bunga beds, Late Devonian Boyd Volcanic Complex, Bunga Head, New South Wales.

Figure 3.11 Pillow-like pods of magma intruding water-saturated, unconsolidated sediments. Pods tend to be less regular than proper pillows in shape and size, and may frequently also have quench-shattered fragmented margins. Pods may at times also be entirely isolated in the sediment, linked to the intrusive body by a feeder canal only. (a) Oligocene Angahook Basalt Member intruding Point Addis Limestone, Airey's Inlet, Victoria, Australia (photograph by A. Day). A small inclusion of limestone can be seen near base of basalt pod. (b) Eocene–Oligocene Waiareka–Deborah volcanics intruding the Totara Limestone, Totara Terraces, North Island, New Zealand. (c) Blow Hole Latite Member, Permian Gerringong Volcanics, Kiama, New South Wales, Australia.

3.4.2 INTERACTION WITH SURFACE WATER

The eruption of magma from a subaqueous vent can produce several discrete effects, which can occur in various combinations and in varying intensities. In its simplest form this interaction may be non-explosive, leading to outpouring of submarine lavas such as pillow lavas. In this situation there will be significant transfer of heat from the lava to the water, but the absence of explosions and quench-fragmentation suggests that there is a thin film of steam at the interface between the hot lava and the water which acts as an insulator (J. G. Moore 1975). The second type of effect involves quench-fragmentation due to the chilling effect of the cold water on the lava (see Section 3.6). Granulation may be localised to the margins of the lava directly in contact with the water, or may be more pervasive, leading to a thick pile of hyaloclastite (quench-fragmented debris). Pervasive quench-fragmentation will also lead to penetration of water into the body of the lava and, because of the increased surface area of the granulated lava rubble, higher degrees of heat transfer take place with the consequent increased potential for explosive activity.

The explosive activity stems not from magmatic volatiles (although they may participate to varying degrees), but from the boiling of ambient water and its explosive expansion. This ambient water may be situated in an open vent, and will also occur as interstitial water in a pile of hyaloclastite of the fragmental pile around the vent of a subaqueous volcano. If the interaction between hot magma, quenched debris and water is explosive, prolonged and self-sustaining, fuel–coolant interaction occurs as discussed above (Colgate & Sigurgeirsson 1973, Peckover et al. 1973, Kokelaar 1983). For maximum explosive efficiency and transfer of thermal energy into mechanical energy, water : magma mass ratios of 0.25–0.3 are required (Colgate & Sigurgeirsson 1973, Sheridan & Wohletz 1981, 1983, Wohletz & Sheridan 1979).

Kokelaar (1983) has argued that much of the *apparent* explosive activity associated with surtseyan-type marine volcanoes is not so much due to fuel–coolant type interactions, but expansion of water which is part of a fluidised slurry of tephra, hyaloclastite and water in the vent. As the hot magma passes upwards through the slurry, the two are mixed. The rapid expansion of the water in mixed pockets of slurry causes jetting of fragments of magma, slurry and water upwards into the atmosphere.

The scenario developed here relates to shallow vents. Explosive activity at greater depths is subject to the constraints of the confining hydrostatic pressure of the water column, given that the exploding agent is superheated sea water. The gas pressure in bubbles must exceed the hydrostatic pressure of the overlying water body before explosive expansion can occur. As before, if the hydrostatic pressure is greater than 216 bars for a freshwater lake ($\equiv$ water depth of 2160 m) or 315 bars in sea water ($\equiv$ water depth of about 3150 m), then explosive eruption cannot occur. In shallower marine water, shallow lakes or crater/caldera lakes, catastrophic hydrovolcanic explosions are possible because of the lower hydrostatic pressures and the potentially large volumes of water available.

In many cases it is very difficult to assess whether the first or second of these situations occurred, or to what degree magmatic volatiles were important. Hydrovolcanic explosions involving ground water or surface water, or both, are considered to have been important in the explosive events of Surtsey (1963) (Kokelaar 1983), Taal (1965) and Capelinhos (1957), all three of which are marine volcanoes, and Vesuvius (1906), Kilauea (1790, 1924) and Mt St Helens (1980) (Sheridan & Wohletz 1981), as well as the phreatic eruptions of Ruapehu (North Island, New Zealand) in the last decade (Nairn et al. 1979).

3.4.3 LAVA FLOWING INTO WATER OR OVER WATER-SATURATED SEDIMENT

The flow of hot lava into a body of water or over water-saturated sediments could produce local explosive activity as trapped water is superheated and expands rapidly (e.g. Fig. 3.12). The basalt flow entering Heimaey harbour in Iceland during the

Figure 3.12 (a) Explosion breccia produced by interaction of superheated steam produced by the flow of a latite lava over unconsolidated near-shore marine sediments. Quench-shattering is also evidenced by the close fit of many fragments. Permian Bombo Latite Member, Gerringong Volcanics, Bombo, New South Wales, Australia. (b) Archaean subaqueous flow-top breccia from Teutonic Bore, Western Australia. Breccia comprises pillow fragments with *in situ* hyaloclastite (chill-fragmented) margins set in a matrix of granular hyaloclastite.

1973 eruption (Plate 3) produced local hydro-volcanic explosions (Colgate & Sigurgeirsson 1973). The fragmented lavas described by Schmincke (1967a) from the Columbia River Basalt were considered to have resulted from the flow of lava onto, and injection of lava into, water-bearing sediments. In this case the fragmentation of the lava may have been due to a combination of local phreatomagmatic explosions as well as quench-fragmentation (Section 3.6). The resultant deposits are mixtures of fragments and pre-existing uncon-solidated wet tuffaceous lacustrine sediments which

Schmincke called peperites (see Section 3.6). As in the case of magmas intruding unconsolidated sedi-ments (see above), spectacular soft-sediment de-formation structures, lava loadcasts, and lava pil-lows can result from the flow of lava over wet, unconsolidated sediments. Explosive activity re-sulting from the flow of lavas into water has, in some cases, produced secondary cones or craters called littoral cones and pseudocraters (Fisher 1968, Thorarinsson 1953; Ch. 13). These are rootless vents because they are unattached to a subsurface chamber.

3.4.4 PYROCLASTIC FLOWS MOVING INTO WATER OR OVER WATER-SATURATED SEDIMENT

G. P. L. Walker (1979) has interpreted the 50 000-year-old ash falls of the Rotoehu Ash in New Zealand (Nairn 1972) resulting from immense explosions in marine littoral environments pro-duced when the coeval Rotoiti ignimbrite flowed into the Bay of Plenty. Trapped, superheated sea water developed a high pressure and ultimately caused the explosions that dispersed the Rotoehu Ash (G. P. L. Walker 1979). Although there is debate about the position of the source vents for the Rotoehu Ash (I. A. Nairn *pers. comm.*), Walker's mechanism is appealing. Such littoral explosions would also generate rootless vents, and their formation has previously been discussed by J. V. Wright and Coward (1977) for Palaeozoic welded ignimbrites in northern Snowdonia, North Wales (Ch. 9). The 18 May pyroclastic flow deposits of Mt St Helens contain numerous phreatic explosion craters resulting from the passage of the flows over water-saturated debris and ground within the de-vastated watershed of Spirit Lake and the Toutle River (Rowley *et al.* 1981, Christiansen & Peterson 1981). The superheating of the ground water produced numerous phreatic explosions, although no major eruption columns are known to have developed.

3.4.5 MAGMA RISING INTO A HYDRO-THERMAL SYSTEM

Nairn (1979) discussed a special type of hydro-volcanic explosive eruption involving hydrothermal fluids in considering the 1886 Rotomahana–Waimangu eruption in New Zealand. Although it appears that the interaction of basaltic magma with water-saturated lake sediments and lake waters may have been the source of some of the explosive activity, as evidenced by the mixture of basaltic fragments and lake sediment in the eruption deposits, the Rotomahana area had a long history of high-level hydrothermal activity before the 1886 eruption. Nairn (1979) proposed that if the hydrothermal fluids at depth were already near to boiling point when hot basaltic magma was injected from below, which is likely, then the extra thermal input from the basalt would have been sufficient to boil the hydrothermal fluids. Flashing of these fluids to steam and induced steam pressures greater than the lithostatic load would have produced explosive activity. Lowering of overburden pressure by initial explosions would produce instantaneous boiling in the highly pressured, high temperature, subsurface hydrothermal reservoir system, so producing significant, wholly hydrothermal explosions. Because of the high temperature of subsurface hydrothermal fluids, the efficiency of energy transfer from thermal to mechanical energy would be greater in these circumstances than in the previously discussed situations involving contact between hot magma and cold surface water or ground water.

Although some hydrothermal explosions are fuelled by rising magma, others such as those associated with the Kawerau geothermal field in New Zealand (Nairn & Wiradiradja 1980) are not. These explosions were apparently due to a reduction of fluid pressure in the hydrothermal system, leading to the flashing to steam of the superheated water. Products include a breccia of hydrothermally altered ignimbrite, rhyolite and tuff in a silty hydrothermal clay matrix.

3.5 An introduction to the products of pyroclastic eruptions

The principal aim of this section is to introduce the components of pyroclastic deposits and their characteristics. No attempt will be made to discuss the transportational and depositional modes, the aggregate properties and field characteristics of different pyroclastic deposits, which will be dealt with in Chapters 5–10 and 12, or the nomenclature of volcaniclastic aggregates (Ch. 12).

However, it should be pointed out that the mechanics of fragmentation by pyroclastic processes, as outlined above, are not the only processes by which volcanic fragments are formed during eruption. Fragments also form in the vent through physical collision and abrasion, and the term 'milling' is sometimes used to describe such abrasive processes. Once ejected from vent, large fragments may further fragment on impact with the Earth's surface.

Pyroclastic deposits are composed of **pyroclasts**, which is a loose term for any fragment released in a volcanic explosion or eruption. Pyroclasts can have a wide range of sizes, irrespective of their origin. Fragments greater than 64 mm in diameter are called *blocks* or *bombs*, those between 64 mm and 2 mm in diameter are called *lapilli*, and those less than 2 mm in diameter are called *ash* (see Table 12.5). The classification of pyroclastic deposits is further treated in Chapter 12. **Tephra** is a collective term for all pyroclastic deposits, including the deposits of pyroclastic flows, surges and falls (Self & Sparks 1981) and are discussed in Chapter 5. Three principal kinds of pyroclasts or components are found in pyroclastic deposits:

juvenile fragments
crystals
lithic fragments

3.5.1 JUVENILE FRAGMENTS

Juvenile fragments represent samples of the erupting magma. They may therefore be partially crystallised, or uncrystallised, depending on the pre-eruptive history of the magma. On eruption they rapidly chill to partially crystallised or uncrystallised glassy fragments. Because different magma compositions have different densities, viscosities and fluidity, fragments from different magmas will develop varying morphologies during explosive ejection from the vent. Other factors controlling the morphology of particles include the degree of vesiculation of the magma, whether the style of fragmentation is magmatic or phreatomagmatic and the mode of transport (Chs 5–9). These aspects have all been reviewed for ash-sized (<2 mm) material by Heiken (1972, 1974), Wohletz (1983) and Sheridan and Marshall (1983), and the principal points will be reviewed here.

Basaltic pyroclasts can have very varied morphologies because of the often very fluidal nature of basaltic magmas. Around vent, coherent lumps of magma may be so hot and fluidal as to re-amalgamate into lava, which flows into or away from the vent. Other such lumps may retain identity and accumulate as aggregates of poorly defined **spatter fragments** called **agglutinates** (Fig. 3.13, Ch. 6). Large basaltic pyroclasts may also chill sufficiently during flight to retain the shapes developed in flight. Shaped bombs include various

Figure 3.14 Basaltic bombs, some with included xenoliths. (a) Mt Leura, Victoria, Australia. Note flanged fusiform and spindle shapes attained during flight. (b) Large bomb and included xenolith from Mt Elephant, Victoria, Australia, showing internal flow and vesicle development.

fusiform, finned and ribbon forms (Fig. 3.14; see MacDonald (1972) for more photographs). Sometimes vesiculation occurs or continues to occur after eruption, producing bombs with cracked or 'breadcrusted' chilled glassy skins and expanded, vesiculated interiors (Fig. 3.15).

Magmatically fragmented juvenile clasts usually reflect their explosive origin by virtue of a high vesicle (gas bubble) content (Fig. 3.16), unless the magma is so fluidal that the fragment flows plastically on landing. Magmatic explosive eruptions of basalt and basaltic andesite are dominated by lapilli-sized vesiculated fragments, called **scoria** (Fig. 3.16). This is now used as a general term for all dark-coloured vesiculated fragments encompassing basic through intermediate compositions. Scoria can show a wide range in vesicularity, and

Figure 3.13 Basaltic spatter. Forming a spatter rampart on the crater rim of Mt Napier, Victoria, Australia. Note how individual fluidal clots have moulded into each other.

Figure 3.15 Breadcrust surface on a rhyolitic bomb whose outer chilled surface crust has been cracked open due to vesiculation and expansion of the interior of the clast after chilling of its surface. From Mt Tarawera, North Island, New Zealand.

hence density. The larger fragments tend to have a ropy or stringy surface texture. The morphology of lapilli and ash-sized particles derived from scoria eruptions can be very varied. Those formed during very fluid basaltic eruptions producing a large proportion of lava spray often have smooth glassy surfaces moulded by surface tension (Fig. 3.17). Such fragments are termed **achneliths** (G. P. L. Walker & Croasdale 1972), and would include the drop-like shapes called Pele's tears and, in an extreme form, the long threads or clusters of

threads called Pele's hair. More viscous eruptions produce fragments with more-ragged shapes. Phreatomagmatically fragmented juvenile clasts are frequently more blocky, less vesicular and therefore less cuspate than the pyroclasts of magmatic explosions (Heiken 1972, 1974, Self & Sparks 1978, Self *et al.* 1980, Wohletz 1983; Fig. 3.18), unless the magma had vesiculated before phreatomagmatic eruption. In such cases, distinction between magmatic and phreatomagmatic deposits may not be simple. This is considered further below, in discussing shards.

Phreatomagmatic basaltic pyroclasts are often erupted as a fawn-coloured to brown glass called sideromelane, which readily alters by hydration and oxidation from yellowish-brown to reddish-brown palagonite. Palagonite formation occurs readily through normal weathering (Hay & Iijima 1968, MacDonald 1972; Ch. 14).

For further reading about the morphology of basic pyroclasts, see MacDonald (1972), Heiken (1972, 1974, 1978) and G. P. L. Walker and Croasdale (1972).

Pumice (Fig. 3.19) is the common product of explosive magmatic eruptions involving viscous silicic to intermediate magmas, including phonolites and trachytes. It is usually light in colour, highly vesicular and, when formed from porphyritic magmas, contains crystals. Generally, in the literature the term is applied to the larger grainsizes, and the terms *pumice block* or *bomb*, *pumice lapilli* and

Figure 3.16 Basaltic scoria. Note the highly vesicular and delicate angular form of the scoria fragments. Some larger fragments may have a ropy or stringy surface texture. (a) Large clast about 10 cm across from Tower Hill, Victoria, Australia. (b) Coarse scoria from Mt Napier, Victoria, Australia.

Figure 3.17 Smooth-surfaced achneliths composed of chilled basaltic lava spray. 1959 Kilauea Iki eruption, Hawaii.

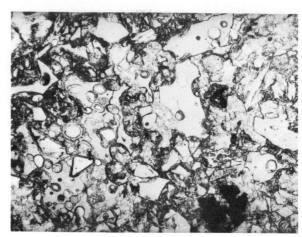

Figure 3.18 Photomicrograph showing juvenile basaltic clasts of slightly vesicular brown glass, produced by phreatomagmatic fragmentation, Quaternary Cape Bridgewater volcano, Victoria, Australia. Clast shapes suggest that vesicles did not play a primary role during fragmentation. Area shown is about 5 mm across.

ash should be used when appropriate (see 'Lithological classification' in Ch. 12). Pumice clasts generally have densities of less than or about $1.0 \mathrm{~g~cm}^{-3}$, and hence may float if deposited in water (see Fig. 10.21). Slightly denser juvenile fragments can be distinguished, as for example in Figure 3.20, by terms such as semi-vesicular andesite. Pumice clasts from within a single pumiceous pyroclastic deposit will show a range in density, and this is controlled by two factors. First, there is variation in the degree of vesiculation of clasts; this is illustrated by Figure 3.21a, which shows the range in density of pumice clasts within a limited size range. Secondly, there is variation of density with grainsize of pumice clasts; average density of pumice clasts increases with decreasing grainsize (Fig. 3.21b) as the volume of unbreached vesicles relative to total clast volume decreases. Also, pumice need not be of uniform composition. Mixed pumice, consisting of streaky, often fluidal layering of (say) rhyolitic and more-basic components (Fig. 3.22) is widespread in pumice fall deposits and ignimbrites. Pumice fragments may be equant, elongate or platy. They may also be angular or rounded, the rounding being due to abrasion in the vent, eruption column or during pyroclastic flow (see Ch. 7).

Figure 3.19 Vesicular pumice from the Recent Taupo plinian pumice fall deposits, High Level Road, east of Rotorua, New Zealand.

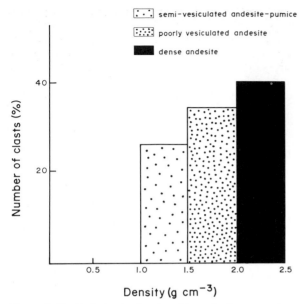

Figure 3.20 Variation in density of juvenile fragments in a single pyroclastic deposit. The example is a block and ash-flow deposit from Mt Pelée, Martinique. (After J. V. Wright *et al.* 1980.)

Shard, glass shard and ash shard are terms often used to describe the angular ash-sized glass particles which result from the magmatic explosive fragmentation of pumice vesicle walls (Fig. 3.23). We restrict 'shard' to glassy (vitric), usually microscopic magmatic fragments, and we do not use the term to describe fragments of broken crystals (or lithics), unlike as has sometimes been done by workers studying ancient volcaniclastic rocks. Morphologically glass shards can have various forms. Those which are commonly illustrated in texts have a variety of Y and cuspate shapes, and these are characteristic of explosive silicic magmatic eruptions (Fig. 3.23a). If deposited in a hot enough state and under a sufficient overburden load pressure, shards may deform plastically and weld (Chs 7 & 8; Fig. 3.23b).

Heiken (1972, 1974), carrying out electron microscopic studies, was the first to draw attention systematically to the fact that many phreatomagmatic glass shards were more blocky and less vesicular than traditional cuspate, magmatically formed shards (Fig. 3.24). However, more recently Wohletz (1983), also using the scanning electron

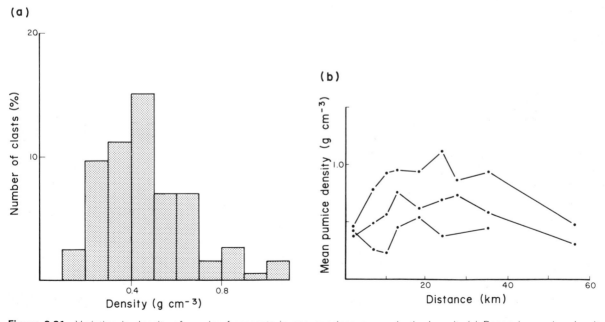

Figure 3.21 Variation in density of pumice fragments in one pumiceous pyroclastic deposit. (a) Range in pumice density measured in clasts >32 mm at vent. (b) Variation in average density of pumice clasts with grainsize and distance from the vent. The example is the pumice-fall deposit erupted from Askja volcano, Iceland in 1875. (After Sparks *et al.* 1981.)

Figure 3.22 Mixed pumice, consisting of bands of grey (intermediate) and white (silicic) layers from the Waimihia pumice, High Level Road, east of Rotorua, New Zealand.

microscope to study natural and experimentally formed ashes, has documented a great diversity in the morphologies of phreatomagmatic pyroclasts. Wohletz's *principal* glassy pyroclast types are the following (Fig. 3.25):

Type 1. Blocky shards with curviplanar surfaces and low vesicularity. These are still the most abundant shard types both in basic and in acidic hydrovolcanic ashes.

Type 2. Vesicular pyroclasts with irregular shapes and smooth fluid form surfaces. These are known only from basaltic ashes, and are common on Surtsey. They suggest significant vesiculation of the magma prior to explosive interaction with external water.

Type 3. Shards finer than 63 μm. These are known only from basaltic ashes, and are moss-like in appearance, having convoluted shapes and highly irregular surfaces formed by several or many globular masses attaching together.

Type 4. Spherical or drop-like shards, only known from the fine fraction of basaltic ashes. They have smoothly curved surfaces, and are usually attached to blocky grains or agglutinated as botryoidal surface encrustations.

Type 5. Plate-like or crescent-shaped shards, with at least one curved face that is demonstrably part of a burst vesicle wall. These are derived from vesiculating magmas that interacted with external water, and illustrate the difficulty of sometimes distinguishing magmatic and phreatomagmatic ashes.

The shape of the shards produced during hydro-volcanic explosions depends on the viscosity, sur-

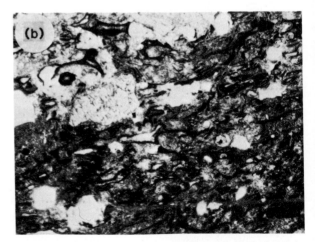

Figure 3.23 Photomicrographs of glass shards with typical cuspate and branched forms. (a) Non-welded redeposited shards in the marine, Tertiary Dali Ash, Rhodes Island, Greece. Area shown is about 1.7 mm across. (b) Welded shards from the Pleistocene Kaingaroa Ignimbrite, North Island, New Zealand. Area shown is about 2.5 mm across.

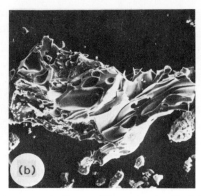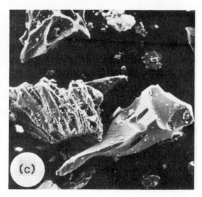

Figure 3.24 SEM photomicrographs of glass shards (from >63 μm size fraction). (a) and (b) Pumice-flow deposit, Mt Pelée. (c) Lower Bandelier ignimbrite, New Mexico.

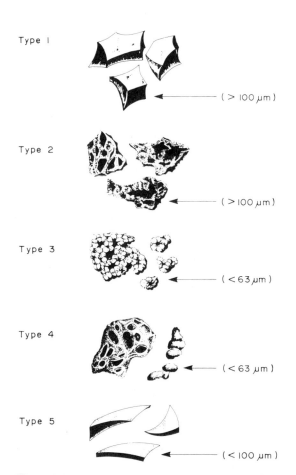

Type 1 (> 100 μm)

Type 2 (> 100 μm)

Type 3 (< 63 μm)

Type 4 (< 63 μm)

Type 5 (< 100 μm)

Figure 3.25 Wohletz's five principal morphological types of phreatomagmatic ashes. Typical size indicated in parentheses.

face tension and yield strength of the melt, and the rate of heat energy release and the degree of vesiculation of the magma prior to explosive inter-action with water. Fragmentation may be of a brittle, ductile or viscous nature (Wohletz 1983). Brittle deformation occurs during quenching and solidification, producing blocky fragments of Wohletz's Type 1. More-ductile behaviour would produce more-irregular elongate fragments (Type 2), perhaps where the rate of quenching is not so high, allowing deformation to occur after fragmen-tation, as particles and vapour mix turbulently.

Types 3 and 4 clasts results from more-fluidal interaction between magma and water or water vapour. The moss-like morphologies are probably due to the viscous effects of the melt, whereas the drop-like particles result from the surface tension effects in the magma (Wohletz 1983).

The traditional distinction between cuspate mag-matic shards and more blocky phreatomagmatic shards seems to be somewhat simplistic. In practice a complete spectrum of phreatomagmatic shard types probably exists. The moderately to highly vesiculated shards from Surtsey, one of the classical phreatomagmatic type examples, is testimony to this. There, as in similar Surtseyan Tertiary suc-cessions exposed on the Otago coast of the South Island of New Zealand near Oamaru (Cas *et al.*, in prep.), sea water probably acted to catalyse explos-ive fragmentation of an already vesiculated basaltic magma that was in a state of incipient magmatic explosive disruption. This problem is further dis-cussed in Section 6.8.3.

3.5.2 CRYSTALS

Free crystals and angular fragments of crystals are released during the explosive disruption and breakage of porphyritic magmas and juvenile fragments, and form a discrete juvenile component from the vesiculated or non-vesiculated magmatic fragments in the pyroclastic deposit. Selvedges of glassy groundmass may be attached. Crystals can be present in both magmatic and phreatomagmatic pyroclastic deposits. Some non-juvenile crystals may also be derived by fragmentation of accessory and accidental lithics (see below), and these are called **xenocrysts**.

3.5.3 LITHIC FRAGMENTS

The term 'lithics' generally describes the dense components in a pyroclastic deposit. Lithics may be subdivided into non-vesiculated juvenile magmatic fragments (**cognate lithics**), country rock that has been explosively ejected during eruption (**accessory lithics**) and clasts picked up locally by pyroclastic flows and surges (**accidental lithics**). 'Cognate' therefore refers to fragments that have solidified from the erupting magma. Accessory and accidental lithics are also called **xenoliths**. In many cases it may be very difficult to separate accessory from

accidental lithics. Upper mantle derived peridotite xenoliths, which are common in alkaline basaltic scoria deposits, crystalline xenoliths in some andesitic pyroclastic deposits and intensely hydrothermally altered lithics are easily identified examples of accessory lithics. Distinctive accidental lithic clasts may sometimes be correlated with specific local sources. Lithics are usually angular, but in some circumstances may be rounded due to in-vent abrasion during eruption. However, angular lithic clasts may also be produced by quench-fragmentation, autobrecciation (see below) and epiclastic processes (Ch. 10).

3.6 Quench– or chill–shatter fragmentation

The sudden contact between a hot coherent body of magma and cold water or water-saturated sediment causes rapid heat loss from the magma in the contact zone. This sets up thermal stresses (which are essentially tensile in character) as quenching and chilling, cooling and associated contraction of the magma occurs. This can occur where magma is erupted subaqueously (Pichler 1965, de Rosen-Spence *et al.* 1980), subglacially (J. G. Jones 1969, Furnes *et al.* 1980), where lava flows into water (Fuller 1931, Waters 1960, Moore & Fiske 1969, J. G. Jones 1969) or over water-saturated sediments (Waters 1960, Schmincke 1967a), and where magma is intruded into water-saturated sediments or country rock (Pichler 1965, Kokelaar 1982). Aggregates of quench-fragmented debris have been called hyaloclastites by Rittmann (1962), Pichler (1965) and Honnorez and Kirst (1975).

The products of quench or chill shattering are granulated glassy fragments of varying sizes, frequently splintery to blocky in shape, and frequently with sharp edges and corners, and planar margins (Figs 3.10, 12 & 26; also see Section 4.11). In thin section they consist of angular fragments of glass, often having perlitic cracks. Honnorez and Kirst (1975) suggested that basaltic hyaloclastites can be distinguished from phreatomagmatically fragmented debris on the basis of morphometric parameters. Fragments in which more than 20% of

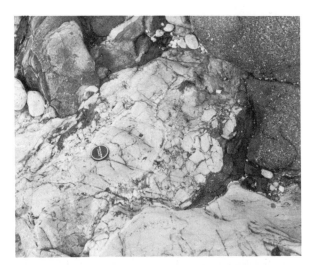

Figure 3.26 Quench-fragmented basalt at the margin of a basaltic dyke, Upper Devonian Boyd Volcanic Complex, New South Wales, Australia.

the perimeter is planar, rather than concave or convex, are likely to be of quench origin (Honnorez & Kirst 1975). According to Honnorez and Kirst (1975), plots of the percentage of the grain margins that are planar, against the numbers of corners and inflection points, are more discriminating than the previous method of assessing the percentage of the perimeter that is planar. This approach has some weaknesses, in that it assumes that all quenched debris are derived from magmas which are non-vesicular at the time of quenching, and that all vesiculated debris with convex or concave margin segments and corners (remnants of vesicle walls) are phreatomagmatically fragmented. Neither assumption is necessarily true, and there may be many situations where partially vesiculated magma is quenched, or where quenching and magmatic or phreatomagmatic disruption occur simultaneously. This could be the case with some of the specimens used by Honnorez and Kirst (e.g. Fig. 3.12).

Pichler (1965) points out that for viscous, silicic magmas the processes of quench-fragmentation and autobrecciation (see Section 3.7) may also be closely interrelated and contemporaneous. For shallow-water settings, hydrovolcanic explosions may also occur simultaneously. However, below maximum depths of about 3150 m (sea water) only quench-fragmentation or flow brecciation can account for fragmentation of magmas, because the hydrostatic pressure of the overlying water body prevents explosions.

However, not all lavas that erupted subaqueously or by flowing into water are fragmented (e.g. basaltic pillow lavas). In these situations, lava must be insulated from the cold water body by a boundary layer. The boundary layer may be a quench-fragmented carapace of granulated glass, which insulates the inner part of the lava body from quenching (Pichler 1965), or a superheated layer of water vapour (H. Williams & McBirney 1979, Kokelaar 1982), or a composite of both.

Where quenching results from the interaction of either an intrusion or a lava flow with wet, unconsolidated sediment, the quenched debris and the wet sediment may be dynamically mixed as the pore water is superheated, boils, fluidises the aggregate and turbulently mixes it (Kokelaar 1982).

To such rocks the term 'peperite' is given, derived from the Italian 'peperino', suggesting a spotty, pepper and salt-like texture. The formation of peperites could also involve significant phreatic and phreatomagmatic explosive activity, and peperitic textures could even be formed when pyroclasts from local phreatomagmatic and phreatic explosions are ejected into the air and fall into unconsolidated, water-saturated sediments.

The interpretation of peperites, particularly at the type locality at Gergovia volcano in the Auvergne region of France, has been debated for many years (see J. G. Jones (1967a) for a critical review). Originally the term 'peperite' was introduced by Scrope in several of his publications (e.g. Scrope 1858) for basaltic breccias with a carbonate clastic matrix, resulting from explosive eruption and air-fall deposition of basaltic pyroclasts into a lake in which carbonate sedimentation was occurring, a view strongly supported by J. G. Jones (1967a). Michel-Levy (1890) and Michel (e.g. 1948) considered the peperites to have formed from intrusion of basalt into the carbonate sediment, presumably accompanied by quenching or explosive activity, or both, together with mixing or redeposition, or both.

Schmincke (1967a) has interpreted peperites from the Ellensburg Formation of Washington as resulting from the flow of basaltic lavas over water-saturated former lake sediments and partial intrusion of parts of the base of the lava into the sediments, resulting in quench-fragmentation, local steam explosions, and mixing of the basalt and sediment to produce deposits with peperitic texture.

It is therefore apparent that the term 'peperite' should be used with care, and not in a genetic sense, because of the possible diverse origins of deposits with peperitic texture (Ch. 12).

3.7 Flow fragmentation (autobrecciation) and its products

If a viscous congealed lava continues to flow, or if the congealed viscous crust of a lava is moved by continued flow of the inside of the lava, this congealed lava becomes stressed, deforms and may

Figure 3.27 (a) Autobrecciated Quaternary andesite, Pinnacle Ridge, Mt Ruapehu, North Island, New Zealand. Outcrop is about 3.5 m high. (b) Autobrecciated flow-banded Lower Devonian rhyolite, Snowy River Volcanics, Buchan, Victoria, Australia.

stretch plastically and break into slabs or blocks (e.g. aa lavas, Ch. 4), or it may fracture in a brittle fashion if the viscosity and strain rate are high enough, to produce smooth faced blocks and blocky lavas (MacDonald 1972; Fig. 3.27). The blocks may be free to tumble, or they may be sintered together, or be enclosed in non-fragmented magma.

3.8 Epiclastic fragmentation

In most volcanic terrains, huge volumes of volcaniclastic debris are also produced and transported by processes which are not of primary volcanic origin – i.e. they are not vent-related processes. Such processes are collectively called **epiclastic processes** (epi = upon, referring to processes at the Earth's surface) and include gravitational collapse or mass-

wastage, chemical and physical weathering, and transportation processes involving wind or water. More details of epiclastic processes and products in volcanic terrains will be discussed in Chapter 10. Their influences and relative significance should never be underestimated.

Epiclastic processes are capable of producing fragments of all sizes, and of transporting them, as well as fragments produced by primary volcanic processes (this chapter), well away from source. Therefore, a volcanic composition by itself does not mean that the deposit is pyroclastic or near-vent, no matter what the size of the fragments or the angularity of the fragments (Chs 1, 10 & 12).

3.9 Further reading

Much has been written about the fragmentation modes of different volcaniclastic deposits, but no other comprehensive review such as that presented here is known to us. For more details on fragmentation modes and the characteristics of the resultant deposits, the reader is referred to the specific references cited within this chapter. MacDonald (1972) and Fisher and Schmincke (1984) contain useful illustrations and discussions on products and processes in various places in their respective books.

Plate 4 Dark, flat-lying, Miocene flood basalt lava flows of the Columbia River Plateau, Oregon, USA. Stratigraphically these belong to the Picture Gorge Basalt which unconformably overlies terrestrial sediments (pale coloured rocks) and the John Day ignimbrite (dark layer to top of bluff at bottom right).

CHAPTER FOUR

Lava flows

Initial statement

Lavas are flows of coherent magma that are erupted at the Earth's surface during effusive volcanic activity which is essentially non-explosive, or, for some basaltic lavas, that are fed by lava fire fountains. Lava flows show great variations in size and shape, and in their surface and internal features. In this chapter we describe these variations and discuss the different controls on lava morphology, such as effusion rate, physical properties (particularly viscosity and yield strength) and environmental factors. Many of the differences between lava flows can be treated by initially dividing them into low viscosity (low silica) and high viscosity (high silica) types. The characteristics of the flows are thereafter controlled by the nature of the environment into which they are extruded, or the nature of the environment into which they then flow, or both.

4.1 Introduction

The principal requirement for the effusive eruption of magma as coherent lava is that the eruption not be explosive or that it be a relatively small fire fountain of lava. This requires that the exsolved volatile content (Ch. 3) of the magma chamber immediately before eruption, and of the magma during eruption, be sufficiently low to prevent the build-up of a gas pressure which could cause fully developed explosive fragmentation (Ch. 3). With the possible exception of some basic and ultrabasic magmas, most subaerially erupted magmas could potentially contain sufficient magmatic water to cause explosive eruptions, given the initial availability of sufficient water in the magma source regions. Therefore, for most volcanoes, and excepting basaltic volcanoes, the volume of volcaniclastics erupted far exceeds the volume of coherent lavas. However, basaltic volcanic centres are usually exceptions to this, because basic magmas usually contain lower magmatic volatile contents.

For coherent magmas to be erupted from magma sources with high volatile contents the magma has to degas to prevent build-up of magmatic gas pressure. This can be done by the following methods.

(a) Direct escape of exsolving volatiles, either gradually through the vent, hydrothermal springs and fumaroles or, more rapidly, by episodic hydrothermal (or steam) explosions (Ch. 3).

(b) A previous or contemporaneous phase of explosive activity during which most of the volatiles of the magma and magma chamber have been removed. This suggests that lavas could be a terminal event of many explosive eruptions and, indeed, this seems to be the case (e.g. the 1980 Mt St Helens eruption).

The exsolution and degassing of volatiles from a magma will increase its viscosity and yield strength, which will affect the mobility, flow distance and thickness of any lava flow (Ch. 2). However, compared with more-silicic lavas, most basic lavas are fluidal and mobile, irrespective of their water content. This is because basic magmas are less polymerised than silicic magmas, since they contain fewer intermolecular Si–O bonds or bridges (Burnham 1979; Ch. 2). In general, the volume and length of lava flows decrease as the content of SiO_2 increases. Thus, basalt flows are usually much more voluminous and longer than rhyolite flows.

4.2 Size and form of subaerial lava flows

The largest known lava flows are **flood lavas**. These include both continental **flood basalts** and those of mid-ocean ridges (e.g. Iceland). Rarer Archaean ultramafic flows (**komatiites**; Section 4.12) and extensive phonolite and trachyte flows that have been described from the East African Rift can also be considered as flood lavas. Such lavas are erupted from large fissures, which are laterally continuous fractures that can be called line sources (see below).

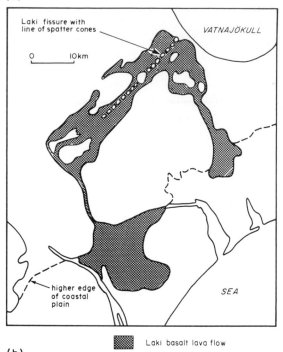

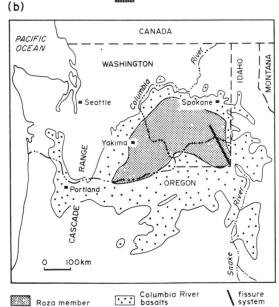

Figure 4.1 (a) The 1783 Laki basalt lava flow in southern Iceland erupted from 115 craters along the 25 km fissure (after Thorarinsson 1969). (b) The Columbia River flood basalt plateau with the distribution of the Roza Member and its fissure vent system (see also Fig. 13.7) (after Swanson et al. 1975).

The largest historic lava of this type is the 1783 Laki basalt flow (Fig. 4.1a) which travelled about 40 km and has a volume of 12.3 km³ (Thorarinsson 1969). Very much larger flood basalts over 200 km long are known in the geological record. In the mid-Miocene Columbia River Plateau of the western USA, flows of the Roza Member (Fig. 4.1b) travelled 300 km and its volume exceeded 1500 km³ (Swanson et al. 1975). This and other lavas in the Columbia River Plateau (e.g. Swanson & Wright 1981) are volumetrically three to five times larger than many of the world's largest stratovolcanoes (Ch. 13). For a general description of flood basalt 'volcanoes' and provinces, see also Chapter 13.

The largest lava unit that has been described is, indeed, that of the Roza Member in the Columbia River Plateau. The Roza Member actually consists of a small number of lava flows and, typically, stratigraphic sections show one or two thick flows, but sometimes three or more. Also, a number of thinner flows (or flow units; see Section 4.3) occur near vent. Cumulatively these flows originally covered a minimum area of 40 000 km² (Fig. 4.1b). All of the flows were probably erupted over a period of a few hundred years or less, but individual flows are thought to have been erupted in a matter of days and are themselves (except the thin near-vent flows) still extremely large. The largest individual lava flows must have volumes up to 700 km³ (Swanson & Wright 1981).

The average thickness of single lava flows in measured sections of the Columbia River Plateau basalts is between 15 and 35 m (Waters 1961). Comparable lava flow thicknesses are found in the flood basalts of the Deccan Traps in Central India (Subbarao & Sukheswala 1981); Choubey (1973) found average thicknesses of between 15 and 30 m. In the flood basalts of eastern Iceland, G. P. L. Walker (1963) found that the average flow thickness was 17 m.

Other basaltic lavas can be considered to be erupted from central volcanoes, point-source vents or small restricted fissures. These lava flows are much less voluminous and cover much smaller

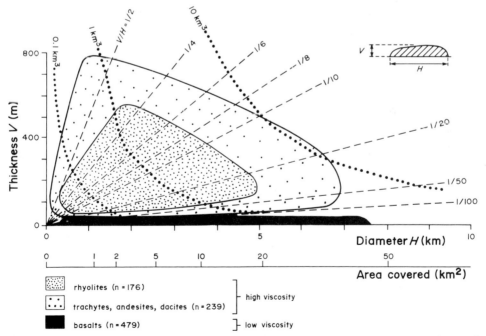

Figure 4.2 Dimensions of lavas of different compositions. The two scales along the x-axis give the area covered by the extrusion and the diameter of a circle having this area. The broken lines give the aspect ratio V/H. The dotted lines give the volumes of circular disc-like bodies of the dimensions shown as a rough guide to the volumes of the lava extrusions. (After G. P. L. Walker 1973a.)

areas. For example, since 1535, Mt Etna in Italy has erupted <4 km^3 of lava; the largest single flow was 0.5 km^3 (Wadge *et al.* 1975). On Mauna Loa historic flank eruptions of basalt along the south-west and north-east rifts range up to 0.5 km^3 in volume (Lipman 1980).

More-silicic lavas are usually small volume flows, and are not as extensive as basaltic ones. Flow distances are in most cases less than a few kilo-metres, although intermediate lavas (especially) can flow in excess of 25 km. Rhyolitic lavas generally do not travel further than 3–4 km and do not exceed about 1 km^3. A notable exception is the Chao dacite flow from Chile, which travelled 12 km and has a volume of 24 km^3 (Guest & Sanchez 1969). Andesites, dacites and rhyolites can be up to a few hundred metres thick.

The extensive phonolites and trachytes in East Africa (King 1970) are, however, very different, and compositional factors (alkaline and peralkaline affinities) suggest that these lavas may have been erupted with viscosities close to those of basalts.

A parameter frequently used in the description of lava geometry is **aspect ratio**. This can be defined as the ratio of the average thickness (V) to the horizontal extent (H), i.e. V/H, where H is taken as the diameter of a circle with a surface area equal to that of the lava. In practice, thickness and surface area are measured with a grid superimposed on topographic maps of the lava. G. P. L. Walker (1973a) plotted aspect ratios for a large number of lavas. He distinguished low viscosity, low aspect ratio lavas (mainly basalts) from high viscosity, high aspect ratio lavas (trachytes, andesites, dacites and rhyolites; Fig. 4.2). Note that the convention we have adopted is to refer to a high aspect ratio lava as one being thick in relation to its area, and a low aspect ratio one as being relatively thin, which is the opposite sense to that used by G. P. L. Walker (1973a).

4.3 Factors affecting the morphology of subaerial lavas

The shape of lava flows can be related to three main controls:

effusion rate
physical properties
slope

However, in reality the controls of lava shapes are likely to be many, and they are complexly inter-related. For instance, effusion or lava discharge rate is itself dependent on a large number of factors: vent shape and dimensions, viscosity, yield strength and magma pressure gradient within the volcano.

4.3.1 EFFUSION RATE

G. P. L. Walker (1973a) considered effusion rate to be the most important controlling factor on lava form. He showed that the distance travelled by lava flows was proportional to the effusion rate. It was thought that this was due to the effect of cooling. Lavas erupted at higher rates would travel further before cooling would lower their viscosity and inhibit movement.

The effusion rate observed in lava flows of more-basic compositions ranges from <0.5 to 5000 m^3 s^{-1} (Table 4.1). G. P. L. Walker (1971) suggested that basaltic lavas extruded at high effusion rates form far-reaching flows composed of a single flow unit, termed **simple lavas** (Fig. 4.3), whereas those

Table 4.1 Effusion rates of some basic lava flows.

Eruption	Average volumetric effusion rate (m^3 s^{-1})
Laki 1783	5×10^3
Etna 1865–1975 ($n = 17$)	15–45
Etna 1975	0.3–0.5
Mauna Loa 1851–1950 ($n = 10$)	100
Askja 1961	33 (800)*
Paricutin 1943–52†	0.7

* This higher effusion rate is for the first eight hours of the eruption; the eruption went on to last for five weeks.
† The lava evolved from basaltic-andesite to andesite during the course of its eruption.

(a) Two simple lava flows

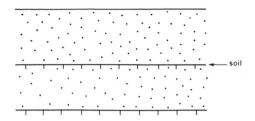

(b) Compound lava flow

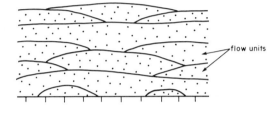

(c)

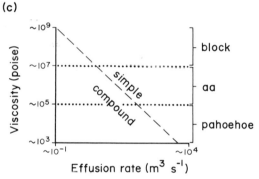

Figure 4.3 Simple and compound lava flows. (After G. P. L. Walker 1971.)

thousands of flow units) and how new boccas (small openings) feeding new flow units commonly formed at the fronts of mature flow units which had otherwise ceased to flow.

Large single flow unit flood basalts found in the geological record are believed to be erupted at very high effusion rates. Order of magnitude calculations by Swanson *et al.* (1975) suggest that the effusion rate along the Roza fissure vent system was 1 km^3 d^{-1} km^{-1} (d = day) for one flood basalt flow. This calculation uses an individual lava flow volume of 700 km^3 (approximately half the total volume of the Roza Member) for each of what appears to be two main lava flows (see Section 4.2), an eruption duration of seven days and a length of 100 km for the fissure vent system. This leads to an effusion rate of 1×10^6 m^3 s^{-1} for the whole vent system or 1×10^4 m^3 s^{-1} km^{-1}. These estimates of discharge rate for the Roza flood basalt flows are comparable with those estimated for highly explosive ignimbrite-forming eruptions (10^4–10^7 m^3 s^{-1}, Ch. 8).

A survey of historic more silicic, higher viscosity lavas (Table 4.2) shows that average effusion rates are between 0.05 and 11.6 m^3 s^{-1}, generally a few orders of magnitude lower than those for basic lavas. More-viscous extrusions might also be expected to form compound lavas at low extrusion rates. For example, the Santiaguito dacite dome, Santá Maria volcano, Guatemala, is a compound

Table 4.2 Effusion rates of some andesitic and dacitic lavas (after Newhall and Melson 1983).

Eruption	Average volumetric effusion rate (m^3 s^{-1})
Santorini, Greece 1886–70	0.7
Santiaguito, Santá Maria volcano 1922–present	0.4
Mt Lamington 1951–6	5.8
Bezymianny 1955–present	1.8
Colima 1975–6	0.05
Augustine 1976	11.6
Mt St Helens 1980–present	0.5
Usu 1910 (Meiji–Shinzan cryptodome*)	3.5
Usu 1943–5 (Showa–Shinzan cryptodome)	1.2
Usu 1977–present (Usu–Shinzan cryptodome	0.6

* Cryptodomes are explained in Section 4.8.

extruded at low effusion rates produce flows composed of small flow units which pile up close to the vent and produce **compound lavas**. The Laki 1783 flow in Iceland was extruded at a relatively high effusion rate (Table 4.1), and is composed of only a few flow units, so it can be considered a simple lava. The historic lava flows of Mt Etna have been erupted at much lower effusion rates (Table 4.1), and form compound lava flows or flow-fields (Wadge 1978, Pinkerton & Sparks 1976). Wadge also demonstrated that the maximum distance attained by these lava flows increases linearly with increasing effusion rate for flows greater than 1 km long. Pinkerton and Sparks (1976) describe the formation of the 1975 flow-field (composed of many

silicic lava which began in 1922 and continues to grow to date; it now consists of at least 14 recognisable flow units (Rose 1972a).

4.3.2 PHYSICAL PROPERTIES

Hulme (1974), who modelled lavas as Bingham substances (Ch. 2), indicated that the principal factor governing their shape was their non-New-tonian rheology. His theoretical analysis and experiments with kaolin suspensions, which are close to Bingham substances in rheology, showed that aspect ratio was mainly dependent on yield stress.

For a Bingham substance to flow downhill, it must form a layer deep enough for the shear stress at the base to exceed the yield strength. Close to the lateral margins the depth is not great enough for downhill flow to occur, and dead zones of stationary fluid form levées along margins. The depth and width of a flow, and the width of each dead zone, are related to five independent initial parameters: effusion rate (F), the slope (α) and three properties of the fluid – Bingham viscosity (η), yield strength (σ_0) and specific weight $(g\varrho$, where g is the acceleration due to gravity and ϱ is the density of the fluid). The critical depth (d_c) which must be exceeded for any flow to occur is given by

$$d_c = \sigma_0/g\varrho\alpha \qquad (4.1)$$

For lavas with higher yield strengths, d_c is therefore larger, and the thickness of the lava flow is greater.

The aspect ratio of a lava flow can be predicted from

$$\text{aspect ratio} = \sigma_0/(F\eta)^{0.25} (g\varrho)^{0.75} \quad (4.2)$$

Aspect ratio therefore depends mainly on yield strength. Equation 4.2 predicts that lavas with low yield strengths, such as basalts, give rise to flows with lower aspect ratios, and more-silicic lavas with higher yield strengths will occur as higher aspect ratio flows, which is in general agreement with field observations (Fig. 4.2).

From Equation 4.2, aspect ratio would seem to be insensitive to changes in effusion rate, but in reality this is more complicated because of the effect of temperature variations. A change in effusion rate will lead to a change in the temperature within

flowing lava at any given point, and this will have a marked effect on the yield strength. An increased effusion rate would increase the temperature, and therefore reduce the yield strength, at any particular point in the lava.

4.3.3 SLOPE

Flow width varies inversely with ground slope (Hulme 1974). However, the effect of slope on lava length has been shown to be small compared with other factors (G. P. L. Walker 1973a).

4.4 Eruption of subaerial basaltic lavas

Basaltic lavas are erupted from either fissures or central (also called point source) vents. Fissure systems that feed large flood basalts may be very large (e.g. >100 km in length, Fig. 4.1b). Central vents are typical of larger basaltic volcanoes, scoria cones and other types of smaller basaltic volcanoes. However, these smaller centres are commonly associated with fissures, and even on the large volcanoes fissures may control flank eruptions. Many eruptions of basaltic lava may begin along a large length of a fissure, but activity quickly localises to a few point sources or 'nodes' (L. Wilson & Head 1981, Delaney & Pollard 1982). Even for the large flood basalts this also seems to be true (Ch. 13).

Basaltic lavas can issue from vents as:

(a) coherent flows from small boccas (openings), or from the overspill or breaching of a lava lake ponded in a crater or
(b) fire fountains of lava that reconstitute around the vent and then flow away.

Many eruptions of basalt lava flows begin with a phase of fire-fountaining of gas-rich magma, succeeded by the extrusion of coherent flows of relatively gas-poor magma. There would also be periods when lava is issuing as coherent flows and fountains at the same time, either from the same vent or separately along a fissure. Flows formed from agglutinated lava spatter are associated with spatter cones and spatter ramparts (Fig. 4.1a, Chs 6

& 13). Lavas in which obvious spatter fragments are observed can be called **clastogenic lavas**; fragments will be flattened, stretched and deformed as in some welded tuffs, and in many ways they form by an analogous mechanism to welded air-fall tuffs (Ch. 6).

Sparks and Pinkerton (1978) suggested that the loss of volatiles during lava fountaining has an important effect on the rheology of the lava. Degassing of the lava leads to considerable undercooling, rapid growth of quench crystallites, a rapid increase in the viscosity and the development of a high yield strength. Thus, highly gas-charged magmas giving rise to intense lava fountaining are likely to generate higher viscosity basaltic flows with higher yield strengths. Magmas with lower initial gas contents should therefore form more-fluidal lavas from less-vigorous fire fountains or lava lakes. The lavas erupted in 1961 at Askja in Iceland changed from higher viscosity and higher yield strength aa to pahoehoe flows (Section 4.5.1) later in the eruption as the intensity of fire-fountaining waned (Sparks & Pinkerton 1978).

4.5 Features of subaerial basaltic lava flows

Many of the features of basalt lava flows have been well documented from studies in Hawaii, and we refer the reader to the descriptions and illustrations in MacDonald (1967, 1972). Basaltic lava flows contain a large array of surface features, but the preservation potential of many of these in the geological record is very limited. We shall split our description of some of the features of subaerial basaltic lavas into the following:

  pahoehoe and aa lavas
  flood basalts
  plains basalts

4.5.1 PAHOEHOE AND AA LAVAS

These are the Hawaiian names given to the two main types of basaltic lava flow that have been distinguished (Figs 4.4–6). **Pahoehoe** lavas are

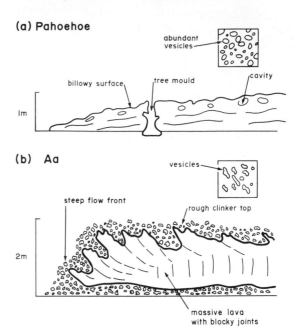

Figure 4.4 Longitudinal sections through the two main types of subaerial basaltic lava flow. (After Lockwood & Lipman 1980.)

characterised by smooth, billowy and sometimes ropy and 'toe' surfaces. In contrast, **aa lavas** have exceedingly rough spinose and fragmented surfaces. These are both end-member types with all transitions between them; **slabby** and **block lavas** resemble aa, but are less spinose, with fragments that are more regular in form. Pahoehoe and aa commonly form in the same lava flow. Pahoehoe may change downslope to aa, but the opposite has never been observed.

The early character of most lavas erupted on Hawaii are almost always pahoehoe. *Pahoehoe* is generally a very fluid, fast flowing lava but it can also form from viscous magma at low effusion rates. Generally small, highly mobile flows advance as a coherent unit with a smooth rolling motion. Larger, less mobile flows advance by protrusion of bulbous 'toes' of lava. On Hawaii, pahoehoe is common on smooth, gentle slopes (see below), and tends to form rather thin flows (often less than 1–2 m; Figs 4.4 & 5a). Internally, pahoehoe lavas are characterised by large numbers of smooth, regular spheroidal vesicles. Many flows contain more than 20%

Figure 4.5 (facing page and above)   Pahoehoe lava flows. (a) Succession of five thin flow units exposed within the crater rim of Mt Hamilton, Victoria, Australia. These flow units have non-vesicular bases with narrow oxidised margins, which grade into highly vesicular upper and middle portions. White inclusions within the base of some flow units are locally derived vein quartz xenoliths. (b) Smooth, billowy pahoehoe surface of the 1975 flow in Kilauea caldera, Hawaii. (c) Shelly pahoehoe, Mauna Iki. (d) Crust of shelly pahoehoe, Mauna Iki. (e) and (f) Ropy pahoehoe, Mauna Ulu 1969–74 flows (near Mauna Ulu). (g) Pahoehoe toes in a Mauna Ulu 1969–74 flow fed by lava tubes down the Hilina fault system (about 8 km from the vent). Figure (circled) indicates scale. (h) Section through a pahoehoe toe buried within a compound lava, Mt Eccles, Victoria, Australia. (i) Weathered ropy pahoehoe surface on the 5000–6000-year-old Harman Valley flow, Wallacedale, Victoria, Australia.

vesicles, though it is not uncommon to find parts of flows with 50% vesicles.

Swanson (1973) described several different types of pahoehoe lava flow formed during the 1969–71 activity from Mauna Ulu in Hawaii. A very vesicular, cavernous type, called shelly pahoehoe (Figs 4.5c & d), formed when gas-charged lava welled out of a fissure with little or no accompanying fountaining. A relatively smooth and denser type formed from the fall-out of fire fountains >300 m in height. The third type, characterised by over-lapping, denser (<20% vesicles), pahoehoe toes and lobes (Fig. 4.5g), formed when largely degassed magma issued from tubes several kilometres from the vent.

The ropy type of pahoehoe (Figs 4.5e, f & i), although perhaps the most distinctive, is actually quantitatively limited in extent (MacDonald 1972). The ropes consist of a regular train of corrugations a few centimetres in height, their long axes being perpendicular to or convex to the local direction in which the flow is moving. Fink and Fletcher (1978) have done a structural analysis of these features. They can be interpreted as folds which develop at the surface of a fluid whose viscosity decreases with depth. The braided appearance and more-complex structures found in many pahoehoe flows can be explained by the superposition of two or more episodes of folding.

These pahoehoe surface features generally have a low preservation potential in the geological record (e.g. Fig. 4.5i). Ropy surfaces and toes may be preserved, especially if quickly covered over by another lava flow or flow unit (e.g. Fig. 4.5h). If found, convex trains of pahoehoe ropes can be used as palaeoflow direction indicators, although caution is required in determining flow direction based on only one or two occurrences, since some pahoehoe ropes may be a response to local eddies on the flow surface (MacDonald 1972).

When a thickened crust forms on a flow, lava tubes commonly form internally (Fig 4.7). Lava tubes are almost exclusively restricted to pahoehoe flows. They can range in size from less than 1 m in diameter to large caves >30 m wide and 15 m high, and can form large distributary networks which can carry lava below the nearly stationary lava surface for distances of many kilometres. Some of the best examples that have been described are from the Quaternary basaltic provinces in Australia (e.g. Ollier & Brown 1965, Ollier 1969, Atkinson *et al.* 1975; Fig. 4.7), and Atkinson *et al.* (1975) report a system of lava tubes which may have extended for more than 100 km in north Queensland. Tubes may later collapse to produce large open channels and depressions on the surface of older flows (Figs 4.7b & c).

Lava tubes are important because they inhibit radiative heat losses from the surface of a flow, and enable the flow to travel long distances. Tube-fed

Figure 4.7 (a) Cave formed by lava tube on Mauna Iki. (b) Collapsed lava tube on Mauna Iki. (c) Cave exposed by collapse of lava tunnel roof, Byaduk Caves, Victoria, Australia.

pahoehoe flows can achieve lengths much greater than aa flows of equivalent effusion rate.

Peterson and Swanson (1974) observed lava tubes forming during the 1970–1 activity of Mauna Ulu, Hawaii. They were observed to form by:

(a) gradual roofing-over of a lava stream from its levées by the accretion of lava spatter along the edges and

(b) cooling of a lava surface to produce a crust, beginning at the levées and growing inward and downstream.

Ollier and Brown (1965) previously suggested that thick flows would develop shear planes, and that only the hottest, thickest layers would keep flowing, leaving voids or tubes. However, Peterson and Swanson (1974) found no evidence for this in Hawaii.

Other surface features that occur on pahoehoe flows are **hornitos, pressure ridges** and **tumuli** (lava blisters; Fig. 4.8). Hornitos are small, rootless spatter cones up to several metres high formed by explosions due to, for instance, trapped ground water. Pressure ridges are elongate uplifts of the lava crust, occurring subparallel to the flow direction at flow margins, but perpendicular in central portions. They are thought to be due to upward pressure from still-liquid lava flowing beneath the solidifying surface. Tumuli are small mounds or dome-like blisters up to 20 m or more in diameter on the crust of a lava flow, again caused by pressure from underflowing lava, or pressure associated with volatilisation of groundwater.

Aa flows are generally thicker (from 2 to 3 m, up to about 20 m) than pahoehoe flows and advance much more slowly. The jagged flow-front (Fig. 4.4b) creeps forward and steepens until a section becomes unstable and breaks off. Collapse is

◀**Figure 4.6**   Aa lava flows on Hawaii. (a) 1868 lava flow on Mauna Loa. (b) Detail of fragmented clinker top to the 1868 Mauna Loa flow. (c) Spinose top of pre-Missionary flow from Mauna Loa.

Figure 4.8 (a) Pressure ridge in a thick columnar jointed cooling crust, Wallacedale, Victoria, Australia. Uplifted columns have separated along the axis of the pressure ridge to produce radial V-shaped fractures. Tilted vesicle layers in foreground are parallel to the former flow surface. (b) Tumulus with large radial fractures formed on the 1919 pahoehoe lava flow in Kilauea caldera, Hawaii. (c) Tumulus in columnar jointed flow surface, Wallacedale, Victoria, Australia.

repeated as the flow slowly advances in caterpillar-track fashion over an autobrecciated layer of fragmented lava. Internally, aa lava is characterised by irregular elongate vesicles that are drawn out in response to internal flow, and a stratification consisting of a solid massive lava body sandwiched between layers of fragmented clinker that may be welded together (Figs 4.4b & 6).

The transition from pahoehoe to aa is generally regarded to result from the increase in viscosity caused by cooling, gas loss and greater crystallinity with time. Peterson and Tilling (1980) made a detailed study of the transition, which occurs at some critical point in the relationship between viscosity and rate of shear strain. If the viscosity is low, then the transition only occurs if there is a high rate of shear; for example, as caused by flow over a steep slope. If viscosity is high, only a low rate of shear is required. At the transition, stiff clots form in parts of the flowing lava where the shear rate is greatest, and remaining fluid adheres to these. Also, fragments of solidified pahoehoe crust are incorporated into the flow, and masses and frag-

ments of aggregate gradually complete the transition to aa.

However, aa lavas also form at vent. When lavas have a moderate to high viscosity, pahoehoe lavas will only form at low effusion rates, whereas aa will form when effusion rates have exceeded a critical value (this was 2×10^{-3} m^3 s^{-1} on Etna in 1975: Sparks & Pinkerton 1978). On the other hand, lavas with low initial viscosities will form pahoehoe even at high effusion rates (there is no limiting effusion rate).

Both pahoehoe and aa lavas form levées. In a study of levées formed by lavas of Mt Etna, Sparks *et al.* (1976) found four principal types of levées (Fig. 4.9). **Initial levées** are formed because of the yield strength of the lava, as indicated by the studies of Hulme (1974) (see Section 4.3.2 above). These form in both pahoehoe and aa flows. **Accretionary levées** were observed near boccas, and consisted of piles of clinker accreted to smooth pahoehoe lava channels. The clinkers weld themselves together to form a steep, solid levée. In flows where aa has developed fully, the flow front

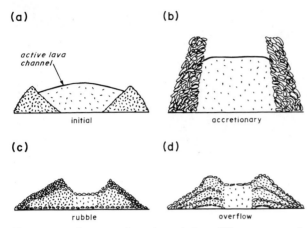

Figure 4.9 Cross sections through four different types of lava levée observed on Mt Etna. Heavier stipple is massive lava; sparsely stippled areas represent flowing lava. (After Sparks *et al.* 1976.)

advances and the sides also expand by avalanching of aa debris. These **rubble levées** are at angles of repose of 30–35°. The fourth type, **overflow levées** form when lava repeatedly floods over existing rubble levées. Most levées on Mt Etna are hybrids of two or more of the four types. Thus, although Hulme's (1974) theory of levée formation was confirmed by these observations, accretionary, rubble and overflow levées nucleate and modify initial levées.

4.5.2 FLOOD BASALTS

Flood basalts form extensive sheets of lava with very low aspect ratios (Plate 4, Fig. 4.10a). Compositionally, these lavas are dominantly tholeiitic (e.g. Swanson & Wright 1981), although commonly they can be alkali basalts, e.g. in the Ethiopian province (Mohr 1983; Ch. 13) and in the Deccan Traps of India (Krishnamurthy & Udas 1981). They are pahoehoe flows, and sometimes ropy surface features are preserved. Many of the larger flows of this type must have ponded as vast lava lakes, taking years to tens of years to solidify, as indicated by the well developed massive **columnar jointing** (Fig. 4.10).

Cooling is accompanied by contraction, and takes place from the cooling surfaces (principally the top and bottom of the flow) inwards. The

tensional stresses set up during contraction may produce regular joint sets perpendicular to the cooling surfaces, and usually vertical to sub-vertical in orientation. Well defined intersecting joint sets may produce regular polygonal columns. The joint faces (and columns) propagate inwards from the cooling surfaces as the 'cooling front' advances inwards. This progressive propagation may be reflected by complementary sub-horizontal joints within columns, or by a segmentation pattern on the vertical joints, reflecting successive propagation stages (Fig. 4.10b). Columnar jointing can exhibit a two- or three-tiered arrangement (Spry 1962, MacDonald 1967, 1972; Figs 4.10b & d). The bottom consists of thick, usually well formed columns normal to the base of the flow. Above this **colonnade**, a layer of thinner, less regular, often chaotic columns essentially normal to the flow top, but highly irregular in structure, is found. This layer is called the **entablature**. There may be an upper colonnade above this. Two-tiered columnar jointing is common in the Columbia River basalts (e.g. Swanson & Wright 1981).

Recently, Kantha (1980, 1981) proposed that columnar jointing in basalts results from a fluid dynamic process operating in the lava during cooling. Double-diffusive convective processes, due to temperature and chemical differences between the top and bottom of a stagnant melt, are thought to drive columnar 'finger' motions in the melt. When solidification eventually occurs, contraction cracks would have preferred propagation paths along the boundaries of adjacent 'basalt fingers', giving rise to columnar jointing. Similar 'salt fingers' can be produced experimentally, and also occur in nature in the oceans. Kantha (1980, 1981) pointed out the striking similarities of basalt columns to these. Although Kantha's ideas are very interesting, not all columnar jointing can be attributed to 'finger motions'. Welded tuffs, for example, often display very well developed columnar jointing (Ch. 8) which cannot be explained by this process.

Lava tubes, lava channels, and other large scale flow features are generally lacking in flood basalts. This may be because they did not form, or because they were destroyed by later movements within the

Figure 4.10 (a) Flood basalt lava flow of the Picture Gorge Basalt in the Columbia River basalt plateau, Oregon. (b) Icelandic flood basalt with lower columnar jointed colonnade and upper entablature. Note horizontal segmentation pattern on vertical joints (see text). (c) Top of columnar jointed flood basalt lava in Iceland showing polygonal form of columns. (d) Two-tiered columnar jointing, Campaspe River, Victoria, Australia. (e) Large uniform columns in a thick, massive flow, Organ Pipes, Victoria, Australia.

ponded lakes of lava, perhaps by convective circulations. Large, circular down-sag structures have been described, which may result from magma withdrawal.

Palaeoflow direction in flood basalts can be determined if spiracles or pipe vesicles are present. These are concentrations of vesicles in small, curved pipe-like structures found at the base of flows. They form when bubbles of steam from heated ground water rise into the lava, and are then stretched in the direction of flow as it continues to move (Waters 1960, MacDonald 1967).

4.5.3 PLAINS BASALTS

Although these are large basaltic flows, Greeley (1977b, 1982) grouped them as a separate type from flood basalts. Plains basalts have characteristics of both flood basalts and the smaller shield-building pahoehoe lavas, such as those in Hawaii that were discussed above (and in Ch. 13). They are typified by the Snake River Plain, in the western USA (Greeley 1977b, 1982). Lavas have been erupted from central vents to produce low coalescing shields, or from fissures to produce sheets. Lavas are compound, and flow units are up to about 10 m thick. Lava tubes and lava channels are an important means of flow propagation.

4.6 Submarine basaltic lavas

The formation of pillows or pillow lava (Figs 4.11 & 12) would generally be regarded the most distinctive feature of basaltic lavas erupted under water. From studies of the present ocean floor and of ancient successions, submarine pillow lavas are also known to be intimately associated with massive or sheet flows (Fig. 4.11).

There has been considerable debate about the formation of pillow lavas (e.g. J. G. Jones 1968, J. G. Moore 1975, Vuagnat 1975, de Wit & Stern 1978). In many two-dimensional outcrops, most pillows appear to be discrete entities, although careful observation commonly reveals some interconnected pillows. However, good three-dimensional observational data show that many apparently

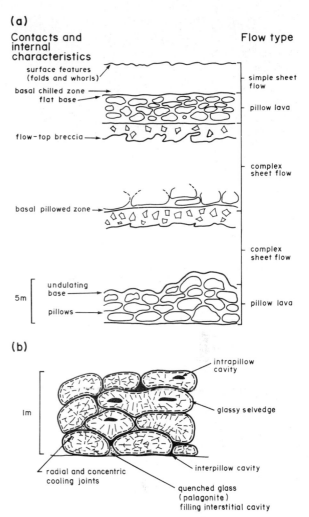

Figure 4.11 (a) Succession of submarine pillow lavas and sheet flows (after Hargreaves & Ayres 1979). (b) Detail of pillows. In cross section pillows can vary from 10 cm spheres to large forms several metres across. They are usually at least several tens of centimetres in diameter.

discrete pillows represent cross sections through interconnected lava tubes (Fig. 4.13). Although erupted subaerially, submarine observations of the March–May 1971 Mauna Ulu flows (J. G. Moore *et al.* 1973) indicated that pillows formed and the lava advanced by the budding of subaqueous lava tubes. This process is therefore quite analogous to the digital advance of subaerial pahoehoe lava and the formation of pahoehoe toes, as first suggested by Lewis (1914).

Figure 4.12 (a) Pillows with well developed radial cracks and thin quenched margins, Boatmans Harbour, Oamaru, South Island, New Zealand. Inter-pillow spaces are filled with pelagic and skeletal carbonate sediment. (b) Steeply dipping (tectonic) pillow lavas in the Franciscan Formation, California. Way up is from right to left. (c) Tropically weathered pillows in the Rio Orcovis Formation, Puerto Rico.

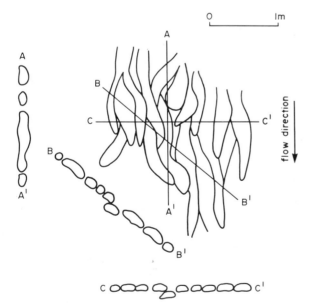

Figure 4.13 Plan view of, and three orientated cross sections through, pillow lavas. (After Hargreaves & Ayres 1979.)

Palaeoflow directions in pillow lavas can be determined by measuring the direction of budding from re-entrant selvedges (Fig. 4.14). The shapes of pillows also allow the determination of younging directions in ancient deposits (e.g. Fig. 4.12b).

Massive flows of basaltic lava have frequently been encountered during sea-floor drilling, and have been described as sheet flows from the Galapagos rift valley (Ballard *et al*. 1979). These may have a variety of surface features, including folds and whorls like subaerial pahoehoe, or they may be flat or broken. The transition from pillowed to massive morphology, within a single flow or between flows (Fig. 4.11), could reflect an increased discharge rate. Ballard *et al*. (1979) interpreted sheet flows as analogous to modern subaerial unchannelled pahoehoe flows erupted at high discharge rates, and pillow basalts as analogous to tube-fed pahoehoe lavas erupted at much lower discharge rates.

Submarine basaltic lavas are erupted either along fissures at mid-ocean ridges or from central vents at seamounts (Ch. 13). Mid-ocean ridge (MOR) volcanic activity produces quiet effusion of pillow and sheet lava flows (Bonatti 1967). There is little physical interaction between lava and sea water, and this is generally restricted to the formation of thin glassy crusts. There may be minor quench shattering and autobrecciation, or collapse pits with breccias in sheet flows (Ballard *et al.* 1979), but there is generally little fragmental volcaniclastic material produced (Fig. 4.11). Seamounts have been observed to have both pillow and sheet flows at their summits (Lonsdale & Batiza 1980). They also have extensive amounts of hyaloclastite (Fig. 4.15). These may form debris flows (hyaloclastite stone streams of Lonsdale & Batiza (1980)) down

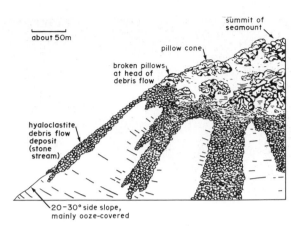

Figure 4.15 Sketch of summit area of a seamount near the East Pacific Rise. (After Lonsdale & Batiza 1980.)

(a)

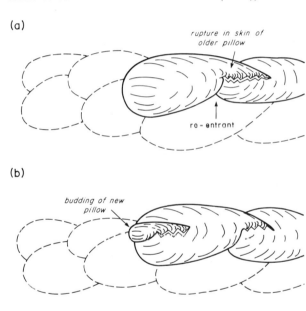

(b)

(c)

Figure 4.14 Cross section showing the development of re-entrant selvedges by budding of a new pillow. (After Hargreaves & Ayres 1979.)

the sides of seamounts. Some of the largest may have debouched into the ocean basin and account for the thick sequences of this type of deposit in the ocean crust found in off-axis drilling of mid-ocean ridges (Schmincke *et al.* 1979). Lonsdale and Batiza (1980) traced debris flows back into broken pillows and pillow lavas at the seamount summit, and suggested that they were formed, in part, by hydrovolcanic explosions. Quench-fragmentation and simple gravitational collapse are probably also important processes.

4.7 Subaerial basaltic lavas flowing into water

When basaltic lavas flow from land into water (e.g. a lake, the sea or glacial meltwater ponds formed during the eruption of intraglacial volcanoes), **lava deltas** are often built out from the shore (J. G. Jones & Nelson 1970, J. G. Moore *et al.* 1973). In general, such deltas consist of a lower part of palagonitised hyaloclastite breccias and pillow lavas characterised by steep foreset beds (up to 40°) which have been termed **flow-foot breccias** (J. G. Jones & Nelson 1970; Fig. 4.16). These are capped with near flat-lying massive lavas. A passage zone between these marks the approximate water level at that time. J. G. Jones and Nelson (1970) showed how relative movements of water level and a volcanic pile or terrain over a period could be

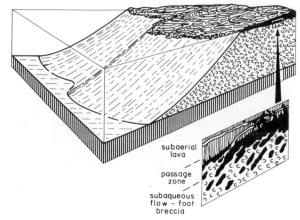

Figure 4.16 Form and structure of a basaltic lava which has brecciated on flowing into water. Thickness of breccia unit as depicted is of the order of 100 m. (After J. G. Jones & Nelson 1970.)

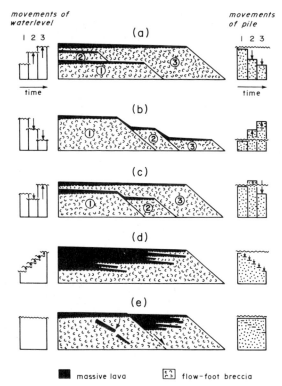

Figure 4.17 Structural relationships in successions of basalt flows which have flowed into water during periods of vertical movement of water level or of the volcanic pile. Relationships are valid for vertical scales ranging from 1 cm = 10 m to 1 cm = 100 m. (After J. G. Jones & Nelson 1970.)

deciphered from such successions (Fig. 4.17). Furnes and Sturt (1976) described how, in macro-tidal environments (tidal ranges of several metres), the rising tide would overstep advancing lava flows, producing complex relationships between different lithologies. Hyaloclastite breccias interfingering and alternating with massive lava bodies continuously build up to high tide level. The whole succession is then covered by massive subaerial lava.

If a lava has a high viscosity and high yield strength, it may interact differently with water, and may flow underwater maintaining continuity. The June 1969 flow from Mauna Ulu entered the sea as a narrow flow of aa, and travelled underwater for several hundred metres without a lava delta being constructed (J. G. Moore *et al.* 1973). In contrast, the lava flows erupted in March–May 1971 from Mauna Ulu, which were lower viscosity pahoehoe flows, constructed a lava delta (J. G. Moore *et al.* 1973), including a significant pillow lava component, when they flowed into the sea.

Other features associated with the flow of basaltic lava into water are **pseudocraters** or **littoral cones** (Thorarinsson 1953, Fisher 1968; Chs 3 & 13). These are rootless vents with small craters, formed by the explosive release of steam trapped at the base of a lava flow.

4.8 Subaerial andesitic and dacitic lavas

High aspect ratio andesitic (including some basaltic andesites) and dacitic lava flows are common on stratovolcanoes (Ch. 13). On these volcanoes they can be, and often are, associated with the eruption of pyroclastic flows (Ch. 13). Eruptions of these types of lavas have been common this century and in historic times (Table 4.2).

Lavas of these compositions typically occur as small-volume, short block flows (sometimes with well developed levées), and as domes (Fig. 4.18); an exception already mentioned is the large Chao dacite flow in northern Chile. Some andesite and dacite lavas form spectacular spires and pitons with very high aspect (Figs 4.18d & e). These lavas must have been extremely viscous and have had very

Figure 4.18 Andesitic and dacitic lavas. (a) Andesitic block lava erupted high on Colima volcano, Mexico in 1975. Note the well developed levées. (b) Dacitic block lava on Néa Kaiméni, Santorini. (c) The 1981 Mt St Helens dacite dome (after M. & K. Kraft in Christiansen & Peterson 1981). (d) The 1902 dacitic spine of Mt Pelée (after Bullard 1976). (e) Gros Piton (dacitic), St Lucia, West Indies.

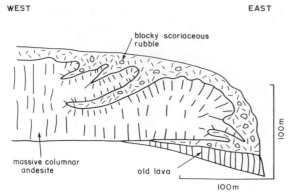

Figure 4.19 Section showing flow-front of pre-historic andesite lava exposed in the north crater wall of Soufrière, St Vincent. (After Sigurdsson 1981.)

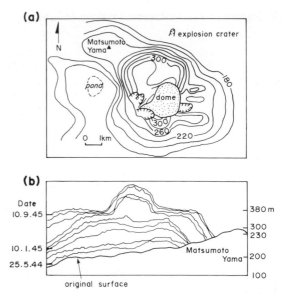

Figure 4.20 (a) Contour map of the 1945 Showa-Shinzan cryptodome of Usu volcano, Hokkaido, Japan. (b) Profiles showing growth of the cryptodome. (After Minakami *et al.* 1951.)

high yield strengths; some even have striated and gouged margins, showing that they were nearly solid when extruded (e.g. Fig. 4.18d). These lavas generally have very high crystal contents.

Andesite and dacite lavas have steep flow-fronts with screes of autobrecciated lava. Sigurdsson (1981) described the flow front of a short andesite lava on Soufrière volcano, St Vincent. This consists of columnar jointed lava lobes surrounded by a thick and irregular layer of blocky and scoriaceous lava rubble (Fig. 4.19), and forms about one-third of the 900 m flow. Expansion of the flow-front as the lava moved required the outward bulldozing of the block and scoria rubble, which is tens of metres thick. The lava lobes in the flow-front may have originated by injection of lava into the collar of rubble.

Internally, behind their flow-fronts, andesitic and dacitic lavas are usually massive, with columnar or blocky jointing. Andesitic lava flows sometimes have a well developed, often flat-lying, sheeted structure with aligned tabular and platy phenocrysts. This **flow foliation** is generally attributed to shear partings developed during laminar flow. In ancient rocks this can sometimes be confused with textures developed in densely welded tuffs (Ch. 8). (Crystals in the lavas should have euhedral regular shapes, and not be broken and fragmented as in the pyroclastic rocks.) Dacite lavas often have a steep flow layering which may be flow folded.

Andesites and dacites are also commonly em-

placed as **cryptodomes**. A cryptodome is a dome-like uplift of the surface rocks in a volcanic area, seemingly caused by a near-surface intrusion. Some of the best-documented occurrences of cryptodomes are those described from Usu volcano, Japan (Minakami *et al.* 1951; Table 4.2). This volcano has an historic record of major ground surface changes. During 1910, 1943–5 (Fig. 4.20) and 1977–8, areas of up to 1 km diameter were uplifted 150–200 m by the intrusion of lava at a shallow depth. The precise thickness of sediment overlying the 1943–5 cryptodome is uncertain, although in the final stages of its growth lava could be seen glowing through large cracks in a thin mantling layer 3–10 m thick. In this case the lava was a hypersthene dacite. Other recorded cryptodomes include the Roche's lava on Montserrat, West Indies (Rea 1974), which has locally uplifted tuffs and fossiliferous limestones on the flanks of the volcano. A similar cryptodome, Brimstone Hill on St Kitts in the Lesser Antilles (P. E. Baker 1969), has dragged up on its flanks patches of Plio-Pleistocene limestone which are now dipping outwards at about 45°.

4.9 Eruption of subaerial rhyolite lava flows

As far as we are aware, there has only been one observed historic eruption of rhyolite lava. This was during the 1953–7 activity which formed the Tuluman Islands, two new islands in the St Andrew Strait, northern Papua New Guinea (M. A. Reynolds & Best 1976, M. A. Reynolds *et al.* 1980). The final phase of the eruption, beginning in November 1956, produced subaerial lava flows. Earlier phases were characterised by dominantly submarine activity, and produced fields of floating, vesicular lava blocks. Many rhyolite lavas and domes often occur in arcuate distributions about central calderas or volcanic depressions, as seen, for example, in the Taupo Zone of New Zealand, the Valles and Long Valley calderas and Mono Craters in the western USA and La Primavera volcano in

Mexico (Fig. 4.21a). R. L. Smith and Bailey (1968) suggested that extrusion of rhyolite lavas commonly follows resurgence of magma after climactic ignimbrite eruptions which result in caldera subsidence (Ch. 8). In many of these situations it seems that the lavas have been extruded around the ring fault or fracture on which caldera collapse took place. At Mono Craters an arcuate line of rhyolite lavas is thought to represent activity over part of an actively developing ring fracture system around the foundering roof of a large crustal magma body in a pre-caldera stage of evolution (Hermance 1983, Rundle & Eichelberger 1983). In some examples (e.g. La Primavera, Fig. 4.21a) the caldera may become filled with a lake, and these post-caldera rhyolites are emplaced in association with lacustrine sediments of the caldera. However, the geology of rhyolitic volcanic centres will be expanded upon later, in Chapter 13. The Tuluman Islands also

(a) La Primavera volcano

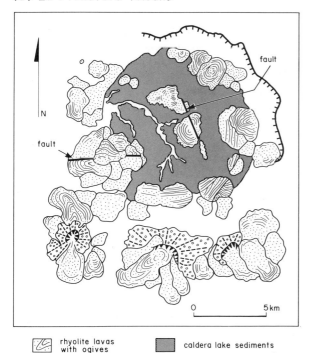

(b) Southern Lipari

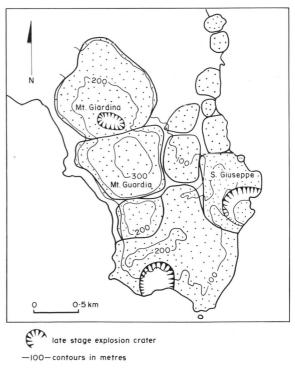

Figure 4.21 (a) Map of the distribution and surface features of the rhyolite lava flows of La Primavera volcano, Mexico. Some of the linear features are faults; curved features are ogives (see Plate 2 and text) (after Clough 1981). (b) Map of the rhyolite domes of southern Lipari, Aeolian Islands (after Richardson 1978).

seem to occur on an arcuate line of rhyolite lavas, but here M. A. Reynolds *et al.* (1980) have speculated that these lie above a ring fracture developing above a mantle hot spot.

In other areas, rhyolite lavas do not seem to be obviously associated with a caldera, e.g. the spectacular concentration of rhyolite domes in the southern part of Lipari in the Aeolian Islands, Italy (Fig. 4.21b), and in Papua New Guinea on the D'Entrecastreaux Islands and at Talasea in New Britain (I. E. M. Smith 1976, Smith & Johnson 1981, Lowder & Carmichael 1970).

Many rhyolitic lavas are associated with pyroclastic deposits, each lava being almost invariably associated with preceding phases of explosive pyroclastic activity. The style of explosive activity can vary from mainly phreatic eruptions, producing rings of lithic breccias surrounding the lava, e.g. Panum Crater (one of the Mono Craters), to highly explosive plinian and ignimbrite-forming eruptions. During such explosive phases a pumice cone or tuff ring can be built-up around the vent (e.g. Fig. 4.22e). Even while the rhyolite lava is growing, explosive eruptions may continue, and evidence for

Figure 4.22 Rhyolitic lavas. (a) Mt Guardia dome, Lipari (photograph by S. Hall). (b) Cerro El Chato dome, La Primavera volcano. (c) Cerro El Colli mesa lava, the youngest of the La Primavera lavas (after Clough 1981). (d) Mesa El Majahuate mesa lava, La Primavera volcano. (e) Coulée of La Primavera volcano which has flowed to left from a vent at the summit of the pumice cone seen to right (after Clough 1981).

this would be unusually large amounts of obsidian ejecta amongst the pyroclastic deposits. Also, visible craters may be present on some of the rhyolite domes, and this can be seen in some of the domes of Lipari (Fig. 4.21b). Formation of cratered domes is attributed to the late stage build-up of gas below the viscous magma, which is released in the form of an explosion with no fresh magma effusion.

Most rhyolite lavas would seem to reach the surface through a circular conduit, which presents a much smaller cooling surface to the country rocks than a fissure vent. However, some of the dome lavas of La Primavera are elongate, and surface ridges are parallel to the caldera ring-fault (Fig. 4.21a), suggesting that rhyolite lavas may also be extruded along fissures (Clough 1981). The Circle Creek rhyolite in Nevada is thought to have been erupted through a large fissure system (Coates 1968). This is a rhyolite lava flow covering 130 km^2, with multiple vents aligned on linear trends. These trends are thought to represent fissures which closed to a series of subcylindrical vents as lava was extruded, in a similar manner to that observed in basaltic fissure eruptions. It was suggested that the motive force for this extrusion was the weight of fissured crust that downsagged into a magma chamber, thus forming a sag-basin as opposed to a caldera. In eastern Iceland, Gibson and Walker (1963) have traced composite or mixed rhyolite and basalt lavas to composite dykes which would be feeder fissures. The Tarawera Volcanic Complex of New Zealand (Cole 1970) consists of a cluster of rhyolite domes and associated pyroclastics, along a NNE trend which reflects a crustal fissure–fracture zone within the Okataina Volcanic Centre.

4.10 Features of subaerial rhyolite lava flows

Our description of the features of subaerial rhyolite lava flows can be subdivided into the following:

shape
lithology
surface features
growth and internal structure

4.10.1 SHAPE

Rhyolite lavas (Fig. 4.22) can be subdivided according to their shape into:

(a) *domes* (or tholoids), which are circular in plan with a small surface area (Figs 4.22a & b),
(b) *mesa lavas*, which are lavas with an approximately circular plan forming biscuit-shaped bodies (Figs 4.22c & d) and
(c) *coulées*, which are lavas which form when flow is asymmetric and concentrated to one side of the vent producing an extrusion which is elongate in plan (Fig. 4.22e).

Although these terms most commonly apply to rhyolitic lavas, they can also be used to describe the form of some dacite, and even andesite, lavas. These three lava types develop in response to the varying controlling factors discussed previously. Rhyolite lavas have a wide range in thickness from <50 m to >500 m (Fig. 4.2). However, the average thickness is probably about 100 m (in descriptions in the literature and Fig. 4.2, rhyolite domes tend to be over represented compared with the thinner coulées, because domes tend to survive much longer as topographic features).

Some of the thinnest rhyolite lavas known are aphyric, and have been ascribed unusually low viscosities. For example, the early rhyolites of Long Valley caldera (Bailey *et al.* 1976) contain some flow units only 50 m thick which have flowed up to 6 km. The aphyric condition may suggest an extremely high magma temperature at the time of eruption as the cause of the increased fluidity. Other thin rhyolite lavas recorded are the acid–basic mixed lavas of Iceland (Gibson & Walker 1963), having an average thickness of 60 m. These lavas may have had reduced yield strengths and viscosity due to superheating on contact with basic magma.

One of the factors which could determine the shape of rhyolite lava flows is the presence of a confining crater built by earlier pyroclastic explosions. This is probably true for smaller rhyolite domes, but larger ones may exceed the critical crater volume and flow away from the crater area laterally. Many domes, on the other hand, do not appear to be associated with an earlier construc-

tional crater (preceding pyroclastic material being more widely dispersed from the vent), but it is possible that such a feature could have been completely submerged beneath the succeeding dome.

Rhyolite lavas also commonly form cryptodomes,

which are generally termed '*intrusive rhyolites*' by workers in ancient successions, based on demonstrable intrusive contacts. The Devonian Boyd Volcanic Complex in Eastern Australia shows excellent examples of such intrusive rhyolite lavas. Some of the rhyolite domes of La Primavera

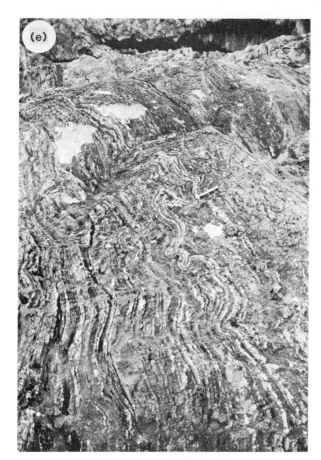

Figure 4.23 (facing page and above)   Lithologies of subaerial rhyolite lava flows. (a) Flow-banded obsidian from a glass flow, Newberry Crater, Oregon. (b) Platy jointed obsidian dome, Okataina Complex, New Zealand. (c) Interbanded obsidian and spherulitic layers, Rocche Rosse flow, northern Lipari. (d) Flow-folded obsidian (now partly perlitised), La Primavera volcano. (e) Flow-folded Upper Devonian rhyolite lava at Tathra, New South Wales, Australia (photograph by S. Ralser). (f) Stony rhyolite lava with basaltic inclusions, southern Lipari (photograph by S. Hall).

volcano which are found in contact with caldera lake sediments (Fig. 4.21a) are thought to have formed as crytodomes (Clough *et al.* 1981, 1982; Ch. 13). Caldera lake sediments are locally folded and faulted, and invariably dip away from the rhyolite lavas.

4.10.2 LITHOLOGY

In rhyolite lava flows a variety of lithologies and textural features can be found: **obsidian**, layers containing **spherulites**, **pumiceous** layers, horizons of **stony rhyolite** (lithic rhyolite), and in lavas where hydration of obsidian has occurred, **perlite**.

Black, vitreous obsidian sometimes occurs as thick foliated layers, often interbanded or as lenses, within layers of the other lithologies (Fig. 4.23). This layering, or flow foliation, is frequently folded. Obsidian usually forms a chilled glassy carapace around rhyolite lavas, commonly about 10 m thick over the top and around the flow front, with a thinner layer along the base (Fig. 4.24). The cores of many lavas usually consist of stony rhyolite. Some of the thinner 'obsidian flows' and 'glass flows' may be obsidian throughout their interiors.

Spherulites are radiating aggregates of alkali feldspar, with or without cristobalite and tridymite,

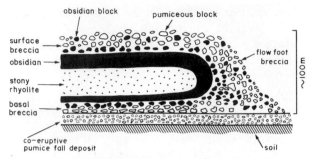

Figure 4.24 Schematic section showing distribution of lithologies in a rhyolite lava flow.

which are commonly found in the glassy carapace (Fig. 4.25a; Ch. 14). They commonly have diameters of 0.1–2 cm, but can be much larger and occasionally grow up to nearly 10 cm. They often comprise specific flow layers (Fig. 4.23c). However, they are usually superimposed on flow structures, and the flow foliation can be traced through, and is not deflected by the spherulites, showing that crystallisation took place after flowage of the lava had nearly ceased. Factors governing the development of spherulites are discussed by Lofgren (1971a; Ch. 14). Higher water contents in some layers could promote growth rates of spherulite fibres locally.

Some spherulitic growths are, in fact, **litho-physae**, which are radiating aggregates of fibrous crystals which have formed around an expanding vesicle (Fig. 4.25b). These vesicles have formed in a melt while it was still capable of flowing. Crusted and broken lithophysae often testify to later flowage.

More-vesiculated pumiceous layers may occur interbanded with obsidian and spherulitic layers. Many rhyolite lavas are capped by blocks of pumice or more pumiceous lava (Fig. 4.24). In older flows these are unlikely to be preserved. Pumiceous breccias formed during flow can also be found at their bases (Fig. 4.24), as well as co-eruptive pumiceous pyroclastic deposits.

The principal lithological component of most rhyolite lavas, especially domes, is foliated stony rhyolite (Fig. 4.23f). This is formed by post-eruption crystallisation of the melt to a finely crystalline rock. This may occur during emplacement, as well as during subsequent cooling. With

many young rhyolite lavas, however, little of this rock is seen because erosion will have had insufficient time to cut through and expose the interiors.

Bands and lenses of light grey perlite are formed by the hydration of obsidian. Obsidian adsorbs water from the atmosphere, forming an hydrated layer which thickens with time as the water diffuses into the glass. From measurements of the thickness of the hydration rind on artifacts collected from archaeological sites and experimental studies (I. Friedman & Long 1976), it is known that the square of the rind thickness is approximately proportional to time, and varies from less than 0.5 $(\mu m)^2$ per 1000 years to as much as 30 $(\mu m)^2$ per 1000 years. This variation is partly due to

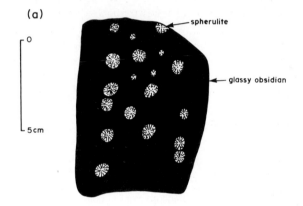

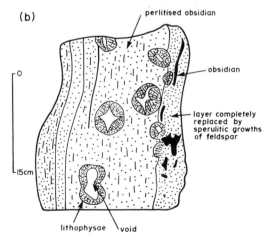

Figure 4.25 (a) White spherulites in black, glassy obsidian; this attractive rock is sometimes called 'snowflake' obsidian. (b) Lithophysae in perlitised flow banded obsidian. Both examples from Mexico.

differences in composition of the obsidian (trachytic obsidian hydrates much more rapidly than rhyolitic), and partly to differences in climate.

In older rhyolites the obsidian is divided into small rounded kernels of structureless obsidian about 1 cm across, surrounded by concentric cracks in light grey friable perlite (this can be seen in the flow-folded obsidian in Fig. 4.23d). The cracks (called *perlitic cracks*) which subdivide the glass are probably created by thermal stresses set up during cooling of the lava. Water then diffuses from these cracks into the bordering glass, and an hydration front advances on the residual kernel of non-hydrated obsidian. The front is visible in thin section because the refractive index of the hydrated glass is lower than that of the non-hydrated glass.

It is also becoming increasingly apparent that many rhyolite lavas contain a very small, but significant, proportion of **basic inclusions**. Basaltic inclusions are found in several of the lavas of La Primavera volcano, in some of the rhyolites in southern Lipari (Fig. 4.23d) and in some from the Late Devonian Boyd Volcanic Complex, southeastern Australia. These basic inclusions exhibit varying degrees of original fluidity. Some show embayed and crenulate margins, indicating that the inclusion was still fluid while the host obsidian was liquid. In other examples the basic inclusions show angular contacts, showing that they had cooled sufficiently to solidify while the lower melting temperature obsidian was still capable of flowing. The inclusions within La Primavera lavas are very similar to the basaltic-andesite lavas which have erupted around the volcano. Although these basic inclusions, with their fluidal characteristics, form only a volumetrically small proportion of their host lavas, their presence indicates that these are mixed lavas resulting from the coexistence of rhyolitic and basaltic magma (Ch. 3). This bimodal association is characteristic of rhyolite volcanoes (Ch. 13).

4.10.3 SURFACE FEATURES

The upper surfaces of young rhyolite lavas are typically blocky and rough. They have curved, concentric ridges called **ogives** (Fig. 4.21a & Plate 2). These are concave in the upflow direction, and

have spacings of tens of metres. Ridges on rhyolite lavas have been variously interpreted as ramp structures (MacDonald 1972), folds on the surface of the lavas (Loney 1968, Fink 1980a) or as 'squeeze-up' extrusions through cracks during stretching of the flow surface.

Loney (1968) described the Southern Coulée of Mono Craters, in which ogives there were open anticlines within the flow foliation caused by longitudinal compression of the lava stream due to external resistance in their advance, and the analogy was made between these features and the compression waves in glaciers. Fink (1980a) suggested that some surface ridges have a similar origin to pahoehoe ropes, and presented results of folding analysis to show them to be compatible. The wavelength and amplitude of the folds are dependent on the temperature gradient, the contrast

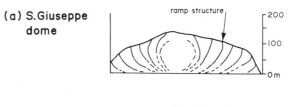

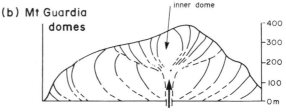

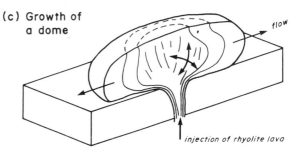

Figure 4.26   (a) and (b) Cross sections through two of the rhyolite domes in southern Lipari (located on Fig. 4.21b). Horizontal scale same as vertical. Mt Guardia is also pictured in Figure 4.22. (c) 3-dimensional view of the growth of a rhyolite dome (after Richardson 1978).

Figure 4.27 Ramp structure in the Rocche Rosse obsidian coulée, northern Lipari (photograph by S. Hall).

between the surface and interior viscosities and the ratio of the compressive stress (due to flow) to the gravitational stress (due to the weight of lava). Fink's analysis also allows calculation of the minimum viscosity of many flows, for which such data are unavailable.

The predominant flow foliation orientation on the surface of La Primavera rhyolite lavas (Fig. 4.21a) is subvertical (Clough 1981). The same is true of the rhyolite lavas of Lipari (Figs 4.26–28). This suggests that in these cases the ogives are an outer manifestation of an internal ramp structure

(see below) rather than surface folds. Fink's analysis can only apply to a limited period during the emplacement history of a rhyolite lava flow. Once the surface is cold, it will behave in a brittle manner, and fracturing and ramping could then be important and superimposed on these earlier features.

Other surface features of rhyolite lavas include pumice diapirs. Fink (1980b, 1983) described in detail pumiceous diapirs in the Big and Little Glass Mountain obsidian flows, northern California (Plate 2). These rose from the basal pumiceous

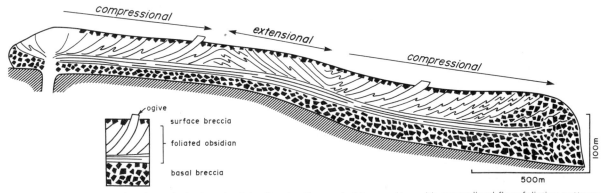

Figure 4.28 Cross section through the length of the Rocche Rosse obsidian coulée, with generalised flow foliation patterns. (After Hall 1978.)

layer of the flows, due to the gravity instability caused by the density inversion inherent in the flow. Dome spacings of between 50 and 70 m were measured, and elongation and surface folding of the diapirs indicates they emerged before the flows stopped moving. Crevasses may be common, and these form by thermal contraction during cooling of the surface and by radial expansion near the outer margin of the advancing flow (Fink 1980b).

At their flow-fronts rhyolite lavas are steep, and are typically terminated by talus aprons of brecciated lava and pumice blocks. These are built partly by autobrecciation of the lava and, like some andesite lavas, advance in bulldozer fashion. Later gravitational collapse of the unstable steep flow front would also be an important process in the building of such aprons.

4.10.4 GROWTH AND INTERNAL STRUCTURE

Flow foliation in rhyolite lavas consists of interbanded and foliated layers of varying crystallinity (obsidian and stony rhyolite), spherulite content and vesicularity. Different lithological layers are thought to have been batches of physically heterogeneous lava with attendant variations in water content, crystal content and, perhaps, temperature. The prominent foliations are then generated during stretching, shearing and attenuation during flow. Recently, Nelson (1981) suggested that lithological differences in rhyolite flow banding could result from thermal feedback (and temperature increases) in layers, due to shear stresses in the moving lava. Local temperature increases would have the effect of reducing local water solubility, increasing diffusion rates, and increasing nucleation rates and growth rates of gas bubbles, thus causing highly vesicular bands to parallel shear planes.

Flow directions and flow history of rhyolite lavas can be determined by structural analysis of the orientation of foliations, folds, vesicles, tabular and platy crystals, and stretched-out volcanic glass. Late-stage movements of cooled lava are indicated by slickensides and tension gashes. However, there are only a few studies of this kind documented in the literature (Christiansen & Lipman 1966, Loney 1968, Benson & Kittleman 1968, Fink 1980a & b, 1983, Huppert et al. 1982b).

Early ideas suggested that rhyolite domes were viscous extrusions from a central vent and were **endogenous**, that is, formed by addition of lava from within the lava body. Ideally, such bodies should have a concentric structure, and this was produced experimentally by Reyer (in H. Williams 1932) by squeezing a viscous substance through a narrow aperture. However, in most cases, the internal flow-foliations of the lava, while concentric at the margins when viewed in plan, are either vertical or steeply inclined in the core and dip inward at low angles at the basal margin. This is the so-called **ramp structure** (Fig. 4.26). As lava is repeatedly injected into the growing dome, the ramp structures move outwards, both radially and tangentially (Fig. 4.27). With time, each fraction of lava becomes progressively attenuated, due to stretching and shearing, especially in the basal part of the dome. Slickensides found on ramp structures indicate movement directions on the foliation surfaces, which must have cooled to be solid enough to allow such structures to form.

Rhyolite coulées often contain very well developed flow foliations. Again, these often seem to develop ramp structures, steep at the top of the flow, curving down to be asymptotic against the base in the upflow direction (Fig. 4.27). Once the width and height of the coulée is established, most of the movement is concentrated in a thin zone of shearing at the base. The normal situation will be for rhyolite flows to be compressive, as lava builds up behind the brecciated flow-front which retards the advancing flow. By analogy with the analysis of similar structures and the stress distributions found in glaciers, slip or shear planes will be inclined back towards the vent, and it is these that are the ramp structures (Hall 1978). If the flow became extensional (such as just in front of a sudden increase in slope) shear planes would have the opposite dip. Hall (1978) has described these features in the Rocche Rosse coulée in northern Lipari (Fig. 4.27).

4.11 Subaqueous silicic lavas

When silicic lavas are erupted into shallow water, ice- or water-bearing sediments, then domes and small lava bodies surrounded by quenched and fragmented volcaniclastic debris form (Fig. 4.29). Pichler (1965) extended the term 'hyaloclastite' to cover the quenched and fragmented silicic rocks formed in subaquatic environments (Section 3.6). However, descriptions of *acid hyaloclastites* are not common in the literature (Pichler 1965, de Rosen-Spence *et al.* 1980, Furnes *et al.* 1980).

Pichler (1965) described shallow-water submarine silicic lavas and hyaloclastites from the Island of Ponza, Italy (Fig. 4.29). They are rhyolitic to quartz-latitic in composition. The lava bodies occur as dykes and stock-like masses. The outer parts of these are glassy and extensively brecciated and jointed, with radial prismatic cooling joints. Hyaloclastite formation may be a continuous process, with fresh magma being intruded into the water-bearing hyaloclastites, and chilling of the glassy selvedge, which then bursts into fragments with the influx of new magma. This process would be continued until the flow of lava had ceased. Pichler thought that there was a close connection between the formation of hyaloclastites and autobrecciation of the quenched surface of the lava. The silicic hyaloclastites mostly occur as unstratified, loose glassy sand-sized fragments containing angular juvenile lava blocks, often vesiculated. (Previously, these were generally referred to as volcanic conglomerates and tuffs, pumiceous agglomerates or lava breccias.) The hyaloclastites on Ponza are associated with calcareous shallow marine sediments, and marine fossils have been

reported within the hyaloclastites. Also, stratified units of hyaloclastite material are found and attributed to sedimentary redistribution, while some explosive activity at the surface could also have formed pyroclastic deposits.

De Rosen-Spence *et al.* (1980) described subaqueous silicic lavas in the Rouyn–Noranda mining area, Canada. The lavas form domes, lobes and tabular bodies that are somewhat larger than those described by Pichler (1965). The largest lava is 10 km long and 400 m thick. Hyaloclastites are associated with small tongues and lobes of lava. Deposits resembling flow-foot breccias are found, and these are believed to form by avalanching at the flow front as the lavas advanced. Volcaniclastic turbidites also occur suggesting a *relatively* deep-water setting, although no specific depth is indicated.

Subglacially erupted rhyolite lavas and hyaloclastites have been documented from Iceland (Furnes *et al.* 1980, de Rosen-Spence *et al.* 1980) and, as previously mentioned, rhyolite lavas thought to be erupted into a caldera lake or water-bearing sediments occur in La Primavera volcano (Clough *et al.* 1981). These two examples are discussed further in Chapter 13. Also, as pointed out at the beginning of Section 4.9, the only observed eruption of rhyolite lava, at the Tuluman Islands, was mainly submarine. Some of the coarse volcaniclastic deposits illustrated in photographs (M. A. Reynolds *et al.* 1980) may contain a high proportion of quench-fragmented debris.

Large volume, deep-water submarine silicic lavas seem to be very different in their characteristics, judging by an example from Australia (Cas 1978a), namely the poorly vesicular dacite–andesite and rhyodacite porphyries of the Merrions Tuff. Two of these porphyries form thick, regionally extensive units (>1200 km^2) in the mid-Palaeozoic Hill End Trough of New South Wales (Fig. 4.30). The sediment sequence in this trough consists of thick mass-flow deposits and turbidites (5 km thick) deposited in a deep-water environment. The porphyries do not show cross-cutting relationships to enclosing sediments, and there is no evidence for them being post-sedimentation sills – indeed, there is evidence that the porphyries were exposed to

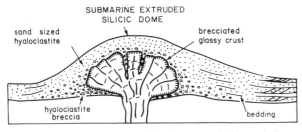

Figure 4.29 Geology of a silicic lava extruded into shallow water. Thickness of dome depicted is of the order of 200 m. (After Pichler 1965.)

(a)

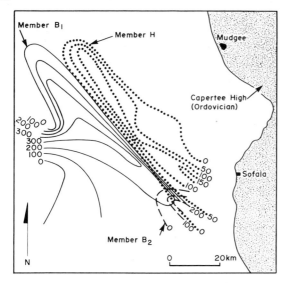

(b)

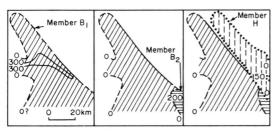

Figure 4.30 (a) Isopach maps of the three extensive silicic lavas within the Merrions Tuff, Australia. These are constructed relative to the unfolded configuration of the formation, isopachs in metres. (b) Inferred sequence of emplacement and spatial relationships of the lavas. Not shown are the deep-water marine sediment horizons separating each unit. (After Cas 1978a.)

erosion accompanying the emplacement of sediment overlying one of the units, and their presence has influenced later patterns of sedimentation. Texturally, the porphyries show no evidence of vitriclastic texture, broken crystals, pumice or foreign lithic fragments, which may suggest they were deep-water welded ash flows or ignimbrites (see Chs 8 & 9). In places the units have fragmented margins, thought to be partly quenched and partly autobrecciated.

Aspect ratios of the two larger units are very low,

and flow volumes are two orders of magnitude larger than for most silicic lavas. Cas (1978a) suggested that, although subaerial silicic flows are demonstrably viscous and immobile, subaqueously erupted and emplaced ones may behave fluidly and be highly mobile. This would be because of the inability of volatiles to escape under the high hydrostatic pressure in a deep-water environment (Ch. 3). Such an explanation is needed to account for the wide areal extent and voluminous nature of the silicic submarine Merrions Tuff lavas. A consequence of this is that the products of deep-water silicic volcanism may be markedly different from those of subaerial or shallow-water silicic volcanism. However, we know of no similar descriptions in the literature. Some accounts in earlier volcanic literature have described extensive submarine rhyolite lavas from North Wales, but these units are now known to be welded ignimbrites.

Investigation of modern oceans has not found large silicic flows, although some oceanic islands are known to have erupted small volumes. This suggests that there may also be some tectonic control on the nature of this type of volcanism in the past, and we will discuss large ensialic submarine basins in Chapter 15.

4.12 Komatiites – peculiarities of the Archaean

The discussion about the physical properties of magmas in Chapters 1 and 2, and of the features of lavas and their eruptions discussed in this chapter so far, have been relevant to 'normal' or penecontemporary magmas and lavas of the Phanerozoic, and even the bulk of the Proterozoic. However, during the Archaean there was a unique, but common group of magmas and lavas called komatiites. Their unique character appears to have been not just compositional, but also in terms of their physical properties and behaviour as lavas.

Komatiites are ultrabasic volcanic rocks – either lavas, tuffs, hyaloclastites or autobreccias (Arndt & Nisbet 1982). Compositionally, komatiites are unique because they are extremely high magnesian rocks, with MgO contents greater than 18% (Arndt

Figure 4.31 Spinifex texture characteristic of komatiitic lavas. (After Donaldson 1982.)

& Nisbet 1982). This is still much higher than normal basalts (<10% MgO), and there is an intermediate group of rocks, called komatiitic basalts, with MgO contents between 18% and 10%.

Mineralogically, the rocks are ultramafic, the only crystalline phases in nearly all cases being olivine, pyroxene and chromite, which have usually been pseudomorphed by metamorphic assemblages. Texturally, komatiites are distinguished by the presence of needle-shaped crystals or crystallites of olivine which criss-cross each other and/or are arranged in radiating sheaths or clusters known as **spinifex texture** (Fig. 4.31). This texture gives many komatiites a distinctive appearance in hand specimen or thin section. Komatiitic basalts can have pyroxene spinifex textures, rather than olivine spinifex textures. Spinifex textures are commonly accepted as resulting from the extremely rapid cooling (also called quenching) of the lavas.

Komatiites are thought to have been erupted at temperatures between 1400 and 1700°C, and viscosities have been estimated at 10–100 poises, compared with recent basalts with temperatures of 1200°C and viscosities of >5000 poises (Huppert *et al.* 1984; also see Tables 2.2–4). Huppert *et al.* (1984) calculated that the Reynolds Numbers for komatiitic magmas being discharged at varying velocities from fissures of varying width, may have been in the range 5.2×10^3 to 1.0×10^6, well above the critical boundary of 2000 distinguishing laminar flow from turbulent flow conditions (Ch. 2). Komatiites therefore probably flowed turbulently, and their rheology, unlike most lavas, would not have been like a Bingham substance, but more likely pseudoplastic or even Newtonian in character (Ch. 2). For a fissure only 10 m wide and 100 m long, magma discharge rates may have been >10 000 $m^3 s^{-1}$ (cf. Table 4.1).

Fluid turbulence in lavas facilitates convective heat loss, and this was probably responsible for the abnormally high cooling rates in komatiites, and the consequence of this, spinifex texture (Huppert *et al.* 1984).

The extremely high temperatures of komatiites may have allowed them to melt and thermally erode the substrate over which they flowed, producing lava channels incised into, and bounded by, the substrate (Huppert *et al.* 1984).

Komatiite lavas have thicknesses ranging from <1 m to >10 m. They frequently form compound lava units (Fig. 4.3), and single cooling units consisting of multiple flows that essentially cooled as one (cf. ignimbrites, Ch. 8). Many normal features of basaltic lavas have been described in komatiites: pillows (rare), polygonal cooling joint sets, vesicles, and brecciated margins of both hyaloclastite and autobreccia. Komatiites therefore appear to have flowed in subaqueous, and probably in subaerial environments as well.

From the studies of Pyke *et al.* (1973) and Arndt *et al.* (1977), a general facies model highlighting the internal textural zonation of komatiitic flows (Fig. 4.32) has been widely accepted and used as a datum for comparison in other studies of komatiitic rocks (e.g. Arndt & Nisbet 1982).

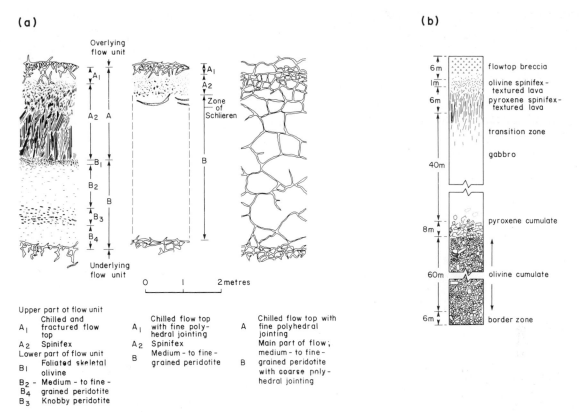

Figure 4.32 Facies models for (a) komatiite lavas and (b) komatiitic basalt lavas, highlighting the internal textural zonation. (After Arndt *et al.* 1977.)

4.13 Further reading

For a quantitative analysis of the eruption of basaltic lava, the most rigorous study is that by L. Wilson and Head (1981); we shall refer to this work again in Chapter 6. Some of the best descriptions and interpretative analyses of submarine lavas come from detailed examination of Archaean successions in Canada. We recommend to the reader the excellent studies by Dimroth *et al.* (1978) and Hargreaves and Ayres (1979) for submarine basaltic lavas, de Rosen-Spence *et al.* (1980) for submarine silicic lavas and Cousineau and Dimroth (1982) for an andesitic submarine pile. Fink (1983) is the best published account documenting the structure and emplacement of a rhyolite flow, and is a good guide to the type of structural mapping needed to study these lavas. The geology of Archaean komatiite lavas is summarised in Arndt and Nisbet (1982) and Nesbitt *et al.* (1982); however, the main approach has been the study of their geochemistry and petrogenesis.

Plate 5 Pit dug to study a section through a young scoria-fall deposit erupted from Sunset Crater in about AD 1065, San Francisco Volcanic Field, Arizona, USA. Most pyroclastic deposits when freshly erupted are loose fragmental aggregates — so use a spade and leave your hammer at home!

CHAPTER FIVE

Three types of pyroclastic deposits and their eruption: an introduction

Initial statement

Pyroclastic deposits form directly from the fragmentation of magma and rock by explosive volcanic activity. They can be grouped into three genetic types according to their mode of transport and deposition:

> falls
> flows
> surges

In this introductory chapter on pyroclastic rocks we set out the differences between these three basic types of deposit. We also describe the eruption style and the deposits of different kinds of pyroclastic falls, flows and surges, based on studies of Quaternary volcanoes. Recent work on pyroclastic deposits from modern volcanic successions has largely concentrated on their genesis, and here we relate the deposits to the physical processes controlling their formation, transportation and deposition. Accretionary lapilli receive special mention, since these can be important indicators of certain types of eruption and process, and for distinguishing pyroclastic deposits from other volcaniclastic sediments.

5.1 Introduction

Three basic types of pyroclastic deposit have been distinguished in the literature:

> pyroclastic fall deposits
> pyroclastic flow deposits
> pyroclastic surge deposits

These types can all be formed by any of the

pyroclastic explosive eruption mechanisms introduced in Chapter 3 (magmatic, phreatomagmatic and phreatic). The essential characteristics of the main pyroclastic deposit types are initially summarised here, before their more detailed description and discussion in the later sections below, and in subsequent chapters (Chs 6–9). The components found in pyroclastic deposits have been described in Chapter 3. The approach and methods used to study and analyse modern pyroclastic deposits are described and discussed in Appendix I. Particularly relevant to this chapter is the discussion on grainsize distribution in Appendix I.

5.1.1 PYROCLASTIC FALL DEPOSITS: DEFINITION

A fall deposit is formed after material has been explosively ejected from a vent, producing an eruption column, which is a buoyant plume of tephra and gas rising high into the atmosphere. The geometry and size of a deposit reflects the eruption column height, and the velocity and direction of

Figure 5.2 Several separate pyroclastic fall deposits forming the bedded sequence mantling erosional topography. Hills and valleys are cut into older massive pumiceous pyroclastic flow deposits of the Oruanui ignimbrite (20 000 years BP), near Lake Taupo, New Zealand.

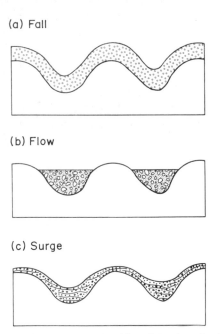

(a) Fall

(b) Flow

(c) Surge

Figure 5.1 Geometric relationships of the three basic types of pyroclastic deposit overlying the same topography. (After J. V. Wright *et al.* 1980.)

atmospheric winds (G. P. L. Walker 1973b, L. Wilson *et al.* 1978). As the plume expands, pyroclasts fall back to Earth, under the influence of gravity, at varying distances downwind from the source, depending on their size and density (or **terminal fall velocity**; Ch. 6) so forming *eruption plume derived fall deposits*. The largest fragments will be explosively ejected on ballistic trajectories, and these are unaffected by the wind and are called *ballistic clasts*. Other fine-grained pyroclastic fall deposits are generated in part from ash elutriated out of the top of moving pyroclastic flows forming *ash-cloud derived fall deposits*; examples of this type of pyroclastic fall deposit can be more voluminous and may be further dispersed than those of ash from eruption columns (Section 5.2).

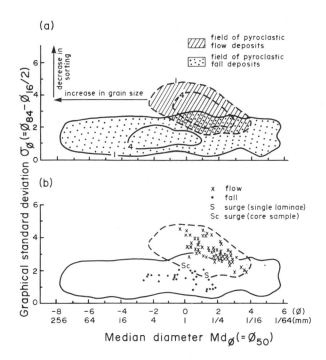

Figure 5.3 (a) Md_ϕ/σ_ϕ plot often used to show grainsize characteristics of unconsolidated modern pyroclastic deposits. The grainsize distribution of a sample is first determined by mechanical analysis (App. I). Cumulative curves of the distribution are then drawn on arithmetic probability paper and the two Inman (1952) parameters of median diameter and graphical standard deviation, which is a measure of sorting, are derived. Solid lines labelled 1% and 4% are contours for the field of pyroclastic fall deposits and within these 99% and 96%, respectively, of sieve analyses of fall deposits occur (based on over 1300 analyses) (after G. P. L. Walker 1971). Broken lines are similar contours for the field of pyroclastic flow analyses (based on about 800 analyses) (after G. P. L. Walker 1971, G. P. L. Walker *et al.* 1980). (b) Example of an Md_ϕ/σ_ϕ plot. All of the samples are from the products of one large Mexican eruption. This produced pumiceous pyroclastic fall, surge and flow deposits. The pyroclastic flow deposits are called the Rio Caliente ignimbrite (after J. V. Wright 1981).

Figure 5.4 Two pyroclastic fall deposits. (a) Scoria fall deposit on Santorini. Note planar stratification and the degraded top of the deposit which is a palaeosol. Rule is 30 cm long. (b) Close-up, showing good sorting (for a pyroclastic deposit) of a pumice-fall deposit from the Lower Bandelier Tuff, New Mexico (Ch. 6).

Fall deposits show **mantle bedding**; that is, they locally maintain a uniform thickness while draping all but the steepest topography (Figs 5.1 & 2). Although pyroclastic deposits are generally poorly sorted, fall deposits are relatively well sorted (σ_ϕ values are normally $\leqslant 2.0$, Figs 5.3 & 4) because of aeolian fractionation during transport. Sometimes they show planar internal stratification or lamination (which has been called shower bedding; Fig. 5.4a) due to variations in eruption column behaviour, but never cross-stratification or bedforms showing erosion or truncation of the underlying layers. Near to the vent, some air-fall deposits are welded, or pass into agglutinated spatter (Ch. 3). Carbonised wood is generally lacking but, when found, is usually restricted to near-vent deposits.

5.1.2 PYROCLASTIC FLOW DEPOSITS: DEFINITION

These are the deposits left by surface flows of pyroclastic debris which travel as a high particle concentration gas–solid dispersion. They are gravity controlled, hot and, in some instances, may be partly fluidised (Ch. 7). As a general rule, deposits are topographically controlled, filling valleys and depressions (Figs 5.1, 5 & 6). However, certain *'violent'* pumiceous pyroclastic flows emplaced at extremely high velocities are known to form a topography mantling pyroclastic flow facies. We will discuss this special facies in Chapter 7.

Internally, pyroclastic flow deposits are generally massive and poorly sorted ($\sigma_\phi \geqslant 2.0$), but sometimes show grading of larger clasts known as coarse-tail grading (Fig. 5.6). Poor sorting in flow deposits is attributed to high particle concentration, and not to turbulence, with the dominant flow mechanisms probably being laminar or plug flow, or both (Ch. 7). The superposition of a number of flow units (each flow unit being regarded as the deposit of a single pyroclastic flow) can give the appearance of internal stratification (e.g. Fig. 5.16a, below); however, a diffuse layering is occasionally observed within individual flow units, and is due to internal shearing during transport. Pyroclastic flow deposits sometimes contain *'fossil fumarole pipes'* or **gas segregation pipes** (e.g. Fig. 5.15c, below), from

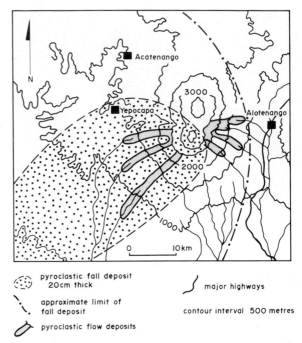

legend:
- ⬡ pyroclastic fall deposit 20cm thick
- ＼ approximate limit of fall deposit
- pyroclastic flow deposits
- ∫ major highways
- contour interval 500 metres

Figure 5.5 Distribution of pyroclastic flow deposits from the 1974 eruption of Fuego volcano, Guatemala. The pyroclastic flow deposits fill canyons and valleys on the lower slopes of the volcano. Their distribution contrasts with the pyroclastic fall deposits produced in the same eruption. (After D. K. Davies *et al.* 1978a, Rose *et al.* 1978.)

which the fine-ash fraction has been lost by gas streaming through the moving pyroclastic flow, or after the flow came to rest (C. J. N. Wilson 1980; Ch. 7). Such gas streaming produces pipes and other pods enriched in heavier crystals, lithics or larger vesicular fragments, which are important features that distinguish these primary pyroclastic mass-flow deposits of pyroclastic debris from epiclastic flows of volcanic material.

Pyroclastic flows are emplaced at high temperatures (Table 5.1). Evidence for a high emplacement temperature is also very important in distinguishing pyroclastic flow deposits from epiclastic debris flow deposits. This evidence would include the presence of:

(a) carbonised wood,
(b) pink coloration due to thermal oxidation of iron, or dark coloration due to crystallisation of finely-disseminated microlites of magnetite

Figure 5.6 Pyroclastic flow deposits. (a) Filling a valley on the lower slopes of Fuego after the 1974 eruption (Fig. 5.5). Note the poor sorting and the large lava blocks showing overall reverse grading. The new channel cut through these deposits is approximately 40 m deep, and this was incised in two wet seasons (after Vessell & Davies 1981). (b) Pumiceous pyroclastic flow deposit (ignimbrite) ponded over a steeply dipping pumice-fall deposit. The pumice-fall deposit mantles former topography and is internally stratified This is called the Granadilla pumice, and the pyroclastic flow was erupted later in the same eruption on Tenerife. A younger, thin pumice-fall deposit overlies the flat top of the pyroclastic flow deposit and this is capped by a palaeosol (photograph by J. A. Wolff). (c) and (d) Thick deposit of a single pumice flow which choked a large valley cut into older volcanics at Micoud, St Lucia, Lesser Antilles. Contact is to the right of scale figures.

Table 5.1 Some measured emplacement temperatures of pyroclastic flow deposits.

Deposit	Temperature (°C)	Method	Source
Komagatake 1929 pyroclastic flow deposit	390 (12 days after eruption)	direct measurement	Kozu (1934)
Mt St Helens 1981 pumice-flow deposits	300–750 (near vent 750–850)	direct measurement	Banks and Hoblitt (1981)
Vesuvius AD 79 ignimbrite	~400	palaeomagnetic & infra-red spectrum of carbonised wood	D. V. Kent et al. (1981)
Upper Bandelier ignimbrite	550–800	welding experiments	R. L. Smith and Bailey (1966), Ch. 8
Prehistoric Mt St Helens block- and ash-flow deposit	550–600	palaeomagnetic	Hoblitt and Kellogg (1979)

(or other iron or manganese oxide minerals), which may be oxidised to haematite, producing the pink colour,

(c)   a zone(s) of welded tuff and

(d)   a thermal remanent magnetism (TRM; Hoblitt & Kellogg 1979).

Carbonised wood is common in pyroclastic flows erupted from volcanoes in tropical or wooded temperate areas, but is absent or scarce in those erupted from volcanoes in dry climates.

Although the term 'ignimbrite' is widely used for the deposit of any pyroclastic flow, we reserve it for the deposits of pumiceous pyroclastic flows (see Section 5.4.2 and Ch. 7).

5.1.3   PYROCLASTIC SURGE DEPOSITS: DEFINITION

A surge transports pyroclasts along the surface as an expanded, turbulent, low particle concentration gas–solid dispersion. Deposits mantle topography but are also topographically controlled, and they tend to accumulate, or are thickest in depressions (Fig. 5.1). Characteristically, they show unidirectional sedimentary bedforms: low angle cross-stratification, dune-forms, climbing dune-forms, pinch and swell structures, and chute and pool structures have all been described. Deposits are often enriched in denser lithics and crystals. Individual laminae are generally well sorted, but core samples incorporating a number of laminae can be poorly sorted (Fig. 5.3). They can contain small gas segregation pipes, produced by gases escaping from preceding flow deposits, and carbonised wood.

Of course, surges are a type of flow, but the term pyroclastic flow has traditionally been associated with the high concentration flows, and it is appropriate to classify the fundamentally different types of deposits produced by flows and surges separately, even though there may essentially be a spectrum (see Ch. 7 for further discussion on the distinction, and the debate surrounding this).

5.2   Eruptions producing pyroclastic falls

Upward transport of pyroclasts high into the Earth's atmosphere may occur in two ways:

in eruption columns accompanying explosive eruptions
in ash clouds accompanying pyroclastic flows

5.2.1   EXPLOSIVE ERUPTION COLUMNS

The eruption columns produced by explosive eruptions may take many forms (Figs 5.7 & 8), and their energetics and dynamics have been discussed recently by Settle (1978), L. Wilson et al. (1978), Sparks and L. Wilson (1982) and Sparks (1986). The height reached by an eruption column, together with the atmospheric wind velocity profile (which may vary with height, e.g. Fig. 13.28), controls the dispersal of pyroclasts (Fig. 5.8). Observed eruption columns have attained heights between 2 and 45 km (Table 5.2; L. Wilson et al. 1978). Plume height is a function of vent radius, gas exit velocity, gas content of the eruption products and the efficiency of conversion of thermal energy during the entrainment of cool atmospheric air (L. Wilson et al. 1978). In all highly explosive eruptions, the thermal energy released is completely dominant over the initial kinetic energy released from decompression and expansion of the gas phase. The style of explosive activity is also important in controlling the character of the eruption column. Discrete instantaneous explosions produce transient plumes, whereas prolonged release of fragmented magma in a steady state eruption forms a long-term, maintained plume. If discrete explosions occur in rapid succession (within seconds to a few minutes) a maintained plume may also form.

Eruption columns can be divided into three parts (Sparks & L. Wilson 1976, Sparks 1986):

(a)   an initial gas thrust part, due to rapid decompression of the gas phase,

(b)   an upper convection plume which is driven by the release of thermal energy from juvenile particles. In this region buoyancy is dominant and the top is defined by the level of neutral

Figure 5.7 Two explosive eruption columns. (a) The 'big umbrella' or 'mushroom' above Lassen Peak, northern California, in the eruption of 22 May 1915. Eyewitness reports indicate that the height of the column was of the order of 10–15 km (photograph courtesy of Loomis Museum Association). (b) Above Mt St Helens on 22 July 1980. This shows the convective column mushrooming out, and a cloud of ash rising from a pyroclastic flow which moved towards the photographer. Column height is about 15 km (after J. W. Vallance in Christiansen & Peterson 1981).

buoyancy, H_B, where the column bulk density equals that of the surrounding atmosphere and

(c) an umbrella region (also called a downwind plume), where the column spreads radially or downwind, or both, to form an umbrella cloud. The umbrella cloud extends from height H_B to height H_T, the level to which the column continues to rise due to its momentum (Sparks 1986).

The height of the initial gas thrust phase varies with the style of the activity (Ch. 6). In most eruption columns, the lower gas thrust part makes up less than 10% of the total column height (L. Wilson *et al.* 1978). For discrete explosions (strombolian and vulcanian eruptions; Ch. 6) this ranges from a few tens of metres to a few hundred metres

(E. Blackburn *et al.* 1976, Self *et al.* 1979). For maintained eruption columns the range is from a few hundred metres to a few kilometres in some eruptions (1.5–4.5 km for initial gas velocities of 400–600 m s^{-1}; L. Wilson 1976, Sparks & L. Wilson 1976). Rapid deceleration of the gas thrust phase occurs between these heights, above which particles are incorporated into an eruption column driven by convection. A maintained convecting eruption column could reach heights of greater than 40 km during some large explosive (plinian) eruptions (L. Wilson *et al.* 1978). A convecting plume will rise until it reaches a level in the atmosphere (H_B) with the same density, and then it will mushroom, spreading radially or laterally, or both, downwind (Figs 5.8 & 9). In eruption columns that form from discrete explosions, convective recovery only takes columns to heights of a few kilometres,

(a)

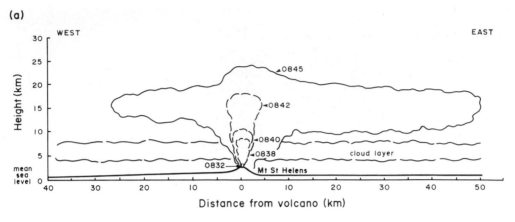

(b)

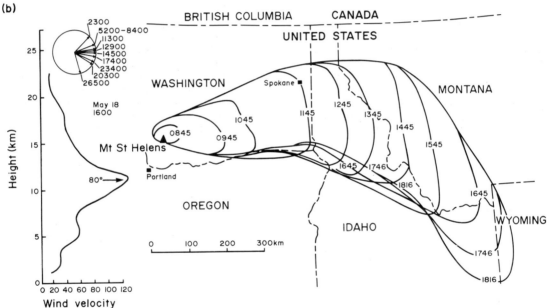

unless explosions occur in quick succession, in which case a maintained plume forms.

L. Wilson *et al.* (1978) and Settle (1978) have independently shown that the maximum height of an eruption column (H_T) is proportional to the fourth root of the rate of release of energy, and hence the fourth root of the mass eruption rate. For maintained eruption columns the height can be predicted from

$$H_T = 8.2\dot{Q}^{1/4} \qquad (5.1)$$

(after Morton *et al.* 1956, L. Wilson *et al.* 1978),

where H_T is the height of the column in metres and $\dot{Q}$ is the steady rate of release of energy in watts. $\dot{Q}$ is related to the eruption conditions at vent by:

$$\dot{Q} = \beta v \pi r^2 s(\theta - \theta_a)F \qquad (5.2)$$

in which β, v, s and θ are, respectively, the bulk density, velocity, specific heat and temperature of the erupting fluid, θ_a is the temperature to which the eruption products ultimately cool ($\sim$270 K in most cases), r is the vent radius and F is an efficiency factor (discussed below). The bulk density, β, is related to the density of the magmatic gas,

(c)

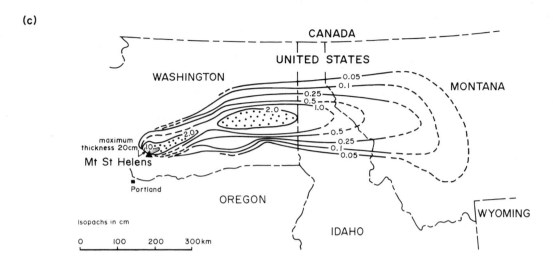

Figure 5.8 Development of the eruption column, downwind plume and dispersal of pyroclasts in the 18 May 1980 eruption of Mt St Helens. (a) East–west profile schematically showing early vertical growth and lateral expansion of the plume. (b) Isochron map showing maximum downwind extent of the edge of the plume carried by the fastest-moving wind layer (as observed on satellite photographs). On the left an average wind speed profile measured at Spokane at 16.00 h is given. Circular wind diagram shows average directions to which wind was blowing at different altitudes, and were again measured at Spokane at 16.00 h. (c) Isopachs of the 18 May pyroclastic fall deposit. Note the secondary thickening of the air-fall deposit 300 km downwind; the significance of this will be discussed in Chapter 6. (After Sarna-Wojcicki *et al.* 1981.)

Table 5.2 Some data on observed eruption columns.

Eruption	Average volumetric eruption rate $(m^3 \ s^{-1})$	Plume height (km)	Duration (h)
Hekla 1947	17 000	24	0.5
Hekla 1970	3333	14	2
Soufrière 1902	11–15 000	14.5–16	2.5–3.5
Bezymianny 1956	230 000	34–45	0.5
Fuego 1971	640	10	10
Heimaey 1973	50	2–3	
Ngauruhoe 1974	10	1.5–3.7	14
Santá Maria 1902	120 000	28	18–20
Mt St Helens 18 May 1980	6200	16	9
Soufrière 22 April 1979	12 600	18	0.23

Volumetric eruption rates are given in terms of dense rock equivalent (App. I).
Plume heights are above the top of the volcano, not sea level.
The data on Hekla 1947, refer to the first 30 min of the eruption.
The data on Heimaey refer to the first few weeks of the eruption.
Information is taken largely from L. Wilson *et al.* (1978), with data on Santá Maria (1902) from S. N. Williams and Self (1983), Mt St Helens (1980) from Harris *et al.* (1981) and Sarna-Wojcicki *et al.* (1981), and Soufrière, St Vincent (1979) from Sparks and L. Wilson (1982).

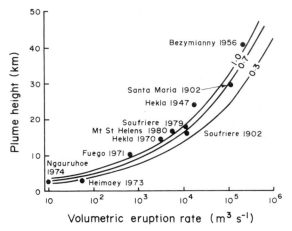

Figure 5.9 Relationship between plume height and volumetric eruption rate. The theoretical curves for F-values of 1.0, 0.7 and 0.3 are discussed in the text. Observed plume heights for ten eruptions are plotted from Table 5.1. (After L. Wilson *et al*. 1978.)

ϱ_g, the density of the pyroclasts, ϱ_m, and the weight fractions of gas and pyroclasts, N and x_m:

$$\frac{1}{\beta} = \frac{x_m}{\varrho_m} + \frac{N}{\varrho_g} \qquad (5.3)$$

If it is assumed that the predominant gas is water and that the erupting fluid is at atmospheric pressure, then for $\theta = 1200$ K, ϱ_g is 0.18 kg m^{-3}. The thermal properties of magma are dominated by the solid phase for gas contents of a few per cent by weight, and the value of s, the specific heat, is taken as 1.1×10^{-3} J kg^{-1} K^{-1}.

The maximum height of the eruption column, H_T, can also be expressed as a function of the volume discharge rate of magma (Sparks 1986; Fig. 5.9):

$$H_T = 5.773(1 + n)^{-3/8}[\sigma \phi s(\theta - \theta_a)]^{\frac{1}{4}} \qquad (5.4)$$

where ϕ is the volume discharge rate in cubic metres
per second, s is the specific heat, θ is the initial temperature of the erupting material, θ_a is the atmospheric temperature at sea level, σ is the magma density and n is the ratio of the vertical gradient of the absolute temperature to the lapse rate.

Figure 5.9 depicts theoretical curves showing the relationships between maintained eruption column

height and volumetric eruption or discharge rate of magma calculated from Equations 5.1–5.3, together with the heights of some observed eruption columns (Table 5.1). The calculations coincide well with recorded column heights. The efficiency factor, F, measures the efficiency of conversion of heat to potential or kinetic energy, and curves with values of $F = 1.0$, 0.7 and 0.3 are used in Figure 5.9. F is mainly controlled by the degree of fragmentation of the magma in the explosive event. Here we are only considering magmatic eruptions, not the special case of explosions generated by magma–water interaction (which will be discussed below). In eruptions which generate a higher proportion of ash-sized ejecta, virtually all of the magmatic heat can be converted to mechanical energy. Many plinian deposits have a substantial proportion of fine-grained particles, and Sparks and L. Wilson (1982) estimated at least 70% efficiency in the conversion of heat in selected plinian columns. On the other hand, strombolian eruptions produce a comparatively much higher proportion of coarse debris (because of a lower degree of fragmentation; Ch. 6), and columns are likely to be much less efficient in the conversion of heat. Consequently, observed eruption columns from this type of activity should fit a theoretical curve with a low F-value, and this seems to be the case for the 1973 Heimaey eruption in Iceland (Fig. 5.9).

The maximum theoretical height expected for a stable maintained eruption plume is about 55 km (L. Wilson *et al*. 1978). This corresponds to an initial gas velocity of 700 m s^{-1} (greater muzzle velocities are unlikely to occur on Earth; McGetchin & Ullrich 1973, L. Wilson 1976), which leads to a volume eruption rate of 1.1×10^6 m^3 s^{-1}.

Equation 5.1 strictly applies to the vertical rise of an eruption column into a still atmosphere with no wind. This should be broadly applicable to most large explosive eruptions, where upward velocities of a particle-rich plume are likely to be much greater over much of its height than the transverse wind velocity. For strong winds and moderate to small eruption columns the effect of wind on column height can be significant, and this is discussed by Settle (1978). A standard atmosphere with a vertical decrease in temperature (environmental lapse rate) of 6.5°C km^{-1} is also used in the calculations of the theoretical curves in Figure 5.9. However, substantial departures from standard atmosphere can occur, and the scatter in the data from observed eruptions in Figure 5.9 may partly reflect variations in vertical atmospheric temperature gradients (L. Wilson *et al.* 1978). These effects have again been illustrated by Settle (1978).

To estimate the rise height of a plume generated from a discrete explosion, another equation must be used:

$$H_T = 1.37Q^{1/4} \qquad (5.5)$$

(Morton *et al.* 1956). This has the same form as Equation 5.1, but Q is the total energy released in joules.

During phreatomagmatic eruptions, a great deal of heat, that in normal magmatic eruptions would be used to drive a convective plume, is used instead in the conversion of water to steam (the heat of vaporisation of water is 580 cal g^{-1} (1 cal $\simeq$ 4.18 J) at atmospheric pressure and 298 K) (L. Wilson *et al.* 1978, Self & Sparks 1978). The thermal energy used in vaporisation can only be recovered by condensation of the steam. Consequently, eruption column heights should be lower in a phreatomagmatic eruption than in a magmatic eruption with the same volumetric rate of discharge. Sparks and L. Wilson (1982) indicated that the effects of steam in controlling column height are probably small except where the mass of steam is comparable with the mass of ash.

5.2.2 ASH CLOUDS ACCOMPANYING PYROCLASTIC FLOWS

During pyroclastic flow-forming eruptions, much of the explosively ejected fragmented magma particles may fail to be included in the resulting pyroclastic flow deposit. Hay (1959) first showed that an enrichment of crystals took place in a small basaltic andesite pyroclastic flow from the 1902 eruption of Soufrière, St Vincent, and he attributed this to the selective loss of vitric ash. Lipman (1967) found a similar enrichment in crystals in a pumiceous rhyolitic ignimbrite erupted from Aso caldera, Japan. Since these studies, G. P. L. Walker

(1972) and Sparks and Walker (1977) have demonstrated that enrichment of crystals is a typical feature of ignimbrites, and must be accounted for by substantial volumetric losses of the vitric component of the original magma which is deposited in associated air-fall ash deposits. Much of this ash is elutriated out of the moving pumice flows by gas streaming through and up, out of the flows. The ash rises above the pumice flows in an upper turbulent ash cloud, which is taken to great heights by huge convective plumes (Fig. 5.7b). Because the ash particles are very fine-grained (nearly all <1 mm), there may be nearly 100% efficiency of conversion of heat to convective energy to drive the plumes.

The types of air-fall ash deposits which result are variously termed *layer 3 deposits* (Sparks *et al.* 1973), *co-ignimbrite ash-fall deposits* (this is our preferred term; Sparks & Walker 1977), and *vitric air-fall ash deposits* (J. V. Wright *et al.* 1980). What is significant here is that these deposits can be very extensive, and can have volumes which are comparable with those of ignimbrites. It is now thought that many of the extensive large ash layers found in deep sea cores are of this type (e.g. Ninkovich *et al.* 1978, Sparks & Huang 1980). We will describe co-ignimbrite ash-fall deposits in more detail in Chapter 8, but will first consider other types of deposits from ash clouds associated with pyroclastic flows (Section 5.6.2).

As a final comment on eruptions producing pyroclastic falls, the high plumes generated, particularly during plinian-type and ignimbrite-forming eruptions, must penetrate the level of the tropopause in the atmosphere (at heights of <6 to 18 km, depending on latitude and season) and contribute fine ash and gaseous species to stratospheric dust veils. Some climatologists have therefore thought that volcanic eruptions might promote periods of climatic cooling. The topic is beyond the scope of this book, but two critical reviews, by Rampino *et al.* (1979) and Self *et al.* (1981), have suggested that volcanic dust veils are only likely to cause short-term (<10 years) very minor temperature fluctuations (in the order of <0.5°C), and are unlikely to trigger ice ages or glaciations, or even minor fluctuations in the 10–100 year range.

Rampino *et al.* (1979) even suggested that it may be climatic variations (leading to stress changes in the Earth's crust) that augment volcanic eruptions rather than vice versa.

5.3 Pyroclastic fall deposits: types and description

The description and interpretation of pyroclastic fall deposits can be approached in a number of ways. The most useful for the volcanologist working on modern pyroclastic deposits has been the quantitative scheme of G. P. L. Walker (1973b), and we will use this as a basis for a detailed description of pyroclastic fall deposits and their explosive mechanisms in Chapter 6. This is a genetic scheme and divides explosive **magmatic** eruptions from open vents into two groups. The first represents a spectrum of increasing dispersal and fragmentation: **hawaiian, strombolian, sub-plinian, plinian** and **ultraplinian.** Phreatomagmatic eruptions constitute the second group, for which two types have been described: **surtseyan** and **phreatoplinian.** These two types have extremely high degrees of fragmentation, and are, respectively, generally basic to intermediate, and acidic in composition although air-fall type and composition cannot be considered in mutually exclusive terms. **Vulcanian** air-fall deposits generated by explosion from closed vents are also defined in the scheme.

However, before discussing this scheme and the resultant deposits (Ch. 6), are there any simpler divisions we can use that still retain some genetic considerations to distinguish modern pyroclastic fall deposits in the field? Three types of pyroclastic fall deposits can be distinguished on broad lithological and genetic grounds:

(a)   scoria-fall deposits,
(b)   pumice-fall deposits,
(c)   ash-fall deposits.

Scoria-fall deposits are composed largely of vesiculated basalt to basaltic andesite magma (Fig. 5.4a). These are the deposits characteristic of hawaiian and strombolian explosive activity (Ch. 6). Near the vent they are associated with lava

spatter cones and scoria cones. They can be very coarse-grained, with the predominant grain size >64 mm, and contain large ballistic bombs, including irregularly shaped bombs and spatter fragments (Ch. 3). Away from the cones, scoria fall deposits are finer-grained and usually <5 m thick.

Pumice-fall deposits (Fig. 5.4b) are composed largely of vesiculated high viscosity magmas (andesite to rhyolite, phonolite and trachyte). They form widely dispersed sheets, and are the sub-plinian, plinian and ultraplinian deposits in Walker's scheme (Ch. 6). Deposits of one eruption are rarely >10 m thick, but very close to the vent deposits as thick as 25 m are known. At vent, the predominant grainsize may be >64 mm, and the deposits contain large lithic and pumice blocks and bombs.

Ash-fall deposits can be formed by a whole spectrum of pyroclastic processes. Phreatomagmatic eruptions characteristically form fine-grained deposits and these often contain accretionary lapilli (Section 5.8). Co-ignimbrite ash-fall deposits can be very extensive examples. They may also contain accretionary lapilli caused by rain flushing, and would be difficult to distinguish from silicic phreatomagmatic (phreatoplinian) ash-fall deposits in the absence of field criteria (Ch. 6). Dense-clast pyroclastic flows may produce equivalent lithic ash-fall deposits. Vulcanian eruptions typically produce ash-fall deposits which may range from dense lithic-rich to scoriaceous types. Close to the vent, these deposits may contain abundant ballistic blocks and bombs. Phreatic eruptions produce lithic ash-fall deposits, and ballistic blocks may be very abundant around the vent. As well as these, pumice and scoria fall deposits have ash-fall deposits as their distal equivalents, and their character depends on downwind aeolian fractionation processes. Air-fall ash deposits range in thickness from <1 mm near vent, to >1 m thick more than 100 km away for co-ignimbrite ash-fall deposits and phreatoplinian deposits.

An alternative non-genetic approach uses lithological descriptions based on dominant grainsize and component types, as shown in Tables 12.5 & 7. For example, in this case most pumice-fall deposits would be pumice lapilli deposits. Most of the coarser near-vent equivalents of the deposits discussed above would then be called volcanic breccias. We will discuss the use of these two terms in Chapter 12.

5.4 Pyroclastic flow-forming eruptions

Pyroclastic flows (Fig. 5.10) are potentially the most destructive of all volcanic phenomena, due to the large distances that some types are capable of travelling and to their high temperature. Serious loss of life has been caused by several small historic pyroclastic flows. Small historic flows have been observed to move up to about 20 km from vent at speeds as high as 60 m s^{-1} (J. G. Moore & Melson 1969, D. K. Davies *et al.* 1978a). However, field studies of older Quaternary deposits suggest that the larger flows (forming ignimbrites) have travelled distances of >100 km from vent, and theoretical analysis based on measurements of the heights of mountains climbed by pyroclastic flows suggests that average speeds of >100 m s^{-1} are common (Ch. 7).

Pyroclastic flows are generated by a number of different mechanisms (Fig. 5.11). From what we understand of observed modern eruptions, these can be split initially into two main types:

lava-dome or lava-flow collapse
eruption column collapse

Figure 5.10 Two pyroclastic flows. (a) Towering ash cloud 4000 m above a pyroclastic flow moving down the Rivière Blanche from Mt Pelée during an eruption in December 1902 (after La Croix 1904). (b) Pumiceous pyroclastic flow erupted on 7 August at Mt St Helens in 1980. This flow travelled at speeds in excess of 30 m s^{-1}. (After P. W. Lipman in Rowley *et al.* 1981.)

(a) Gravitational dome collapse

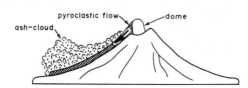

(b) Explosive dome collapse

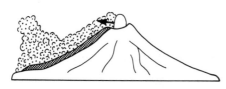

(c) Landslide triggering
explosive collapse
of cryptodome

(d) Discrete explosions
interrupted column
collapse

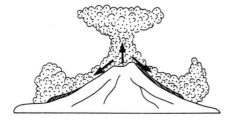

(e) Continuous gas streaming
interrupted column collapse

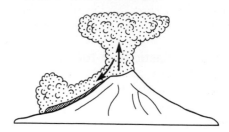

(f) Upwelling at vent
"instantaneous collapse"

(g) Vertical explosion from
dome eruption column collapse

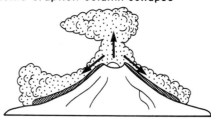

(h) Continuous eruption
column collapse

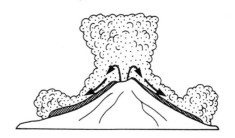

Figure 5.11 Mechanisms generating pyroclastic flows. The pyroclastic flow proper is a high particle concentration underflow. The ash cloud gives rise to other deposits (Fig. 5.13).

5.4.1 LAVA-DOME OR LAVA-FLOW COLLAPSE

This mechanism typically operates on steep-sided andesitic volcanic cones, but also occurs during the eruption of silicic domes not related to major edifices. Fragmental flows of broken lava are generated when an unstable, actively growing lava-dome or lava-flow collapses from the summit or high on the flanks of the volcano. Collapse may be simply gravitational (which is not strictly pyroclastic), or could be an explosively directed blast (Figs 5.11a & b). However, pressure release within a dome due to an initial gravitational collapse could lead to explosive collapse so, in some cases, both processes may have occurred. Explosions could also be triggered by contact of the growing dome with ground water. Such an eruption could therefore be considered to be phreatomagmatic. This also leads to the possibility that phreatic explosions could generate pyroclastic flows containing no juvenile fragments (e.g. Sheridan 1980). It is therefore important to realise that different processes may have occurred at about the same time, and the relative importance of each is, perhaps, difficult to distinguish.

These types of pyroclastic flow we will term block and ash flows, but other terms in use are lava debris flows, hot avalanche deposits (P. W. Francis et al. 1974) and nuées ardentes (see Ch. 12). Block and ash-flows are small-volume pyroclastic flows, and even the deposits of many separate flows or flow units accumulated during the same eruption typically have volumes <1 km^3.

Examples of historic eruptions during which explosive lava-dome or lava-flow collapse was observed are the eruptions of Mt Pelée, Martinique in 1902 and 1929–32 (La Croix 1904, Perret 1937), Merapi, Indonesia in 1942–3 (van Bemmelen 1949), the eruptions of Hibok-Hibok, Philippines (1951) (MacDonald & Alcaraz 1956), Mt Lamington, Papua New Guinea (1951) (G. A. Taylor 1958), and Santiaguito, Guatemala (1973) (Rose et al. 1978). Historic examples where simple gravitational collapse of a dome occurred are the eruptions of Merapi in 1930 and 1942–3 (Neumann van Padang 1933, van Bemmelen 1949) and Santiaguito in 1967 (Stoiber & Rose 1969).

Here we must also ask whether the 1980 eruption of Mt St Helens should be considered to be another example of an explosive dome collapse. The explosive eruption of 18 May was initiated when a giant landslide, triggered by an earthquake, released the confining pressure on a rising dacitic dome (or cryptodome; Ch. 4) which was intruded high into the north flank of the volcano (Christiansen & Peterson 1981). A large rockslide avalanched, and was quickly followed by an explosive directed blast (Fig. 5.11c; also see Fig. 10.6). Explosions were generated by flashing of superheated ground water as well as release of magmatic gases when the dome and its hydrothermal system were exposed and depressurised by the landslide. The avalanche formed a relatively '*hot* and *dry*' volcaniclastic debris flow consisting almost entirely of older volcanic rocks with little juvenile material (<1%; Voight et al. 1981). At the time of emplacement much of the deposit was as hot as 100°C, and it is perhaps debatable whether it should be termed a pyroclastic flow deposit. What to call the deposit of the blast has again been somewhat debatable, but it has been widely regarded as a pyroclastic surge (Section 5.6) and, more recently, as a pyroclastic flow (Section 7.12). Although the eruption was an explosive dome collapse, the eruption and its deposits seem to be more complicated than those generating the block and ash flows that we have described previously. The events at Mt St Helens also triggered a nine hour dacitic plinian eruption with pumice flows forming an ignimbrite (Section 5.5).

There may have been several historic eruptions in which there has been a collapse of a sector of the volcano similar to the one observed at Mt St Helens. The eruption of Bezymianny in 1956 (Gorshkov 1959) produced a directed blast and pyroclastic flows (as well as a very high eruption column dispersing air-fall ash; Fig. 5.9), and is sometimes given as an example of explosive dome collapse (e.g. J. V. Wright et al. 1980). The '*agglomerate flow*' of Gorshkov (1959) may have been a similar volcaniclastic debris flow to the Mt St Helens rockslide avalanche, containing a large proportion of non-juvenile material, judging from the horseshoe-shaped amphitheatre that was left after the eruption. However, these deposits also had a substantial amount of juvenile material and were 'identifiable' as pyroclastic flow deposits. Other eruptions, at Bandai-san, Japan in 1888 (Nakamura 1978), and Sheveluch, Kamchatka in 1964 (Gorshkov & Dubik 1970), seem to have produced similar chaotic deposits, but made entirely or almost entirely from non-juvenile fragments. Therefore, there could be a broad spectrum of deposits produced by collapses of this kind, ranging from recognisable explosively generated block- and ash-flow deposits (or even other types of pyroclastic flow deposit), to volcaniclastic debris flow deposits with no (or very little) juvenile material. Distinguishing such deposits, which lack a high proportion of juvenile fragments, from those formed by epiclastic debris flows is going to be, even in the late Quaternary record, very difficult. Criteria by which to identify such hot, dry volcaniclastic debris-flow deposits have been discussed by Ui (1983) and Siebert (1984).

5.4.2 ERUPTION COLUMN COLLAPSE

In this case, the effective density of a vertical ash-laden eruption column is greater than that of the atmosphere, and gravitational collapse occurs, generating a pyroclastic flow. All of the historic examples of this type have again produced small-volume pyroclastic flow deposits. Many of these were probably formed by interrupted, partial column collapse events. Observations suggest that such small collapses occur when either a short explosion ejects a dense slug of pyroclastic fragments to an altitude of a few hundred metres, part of which then falls back (Fig. 5.11d), or as overloaded parts of a more maintained vertical column produced by continuous gas streaming, collapse (Fig. 5.11e). Both types of collapse event are common during vulcanian activity, but pyroclastic flows are not always generated in such eruptions, and air-fall deposits may be the sole products (Ch. 6). Observed historic eruptions during which this type of pyroclastic flow formed are Mt Lamington (1951) (G. A. Taylor 1958), Mayon, Philippines (1968) (J. G. Moore & Melson 1969), Fuego (1974) (D. K. Davies *et al.* 1978a)

and Ngauruhoe (1975) (Nairn & Self 1978). All produced scoria flows or scoria and ash flows. These types of pyroclastic flows have also been called nuées ardentes and pyroclastic avalanches (Nairn & Self 1978).

Some older eyewitness accounts recorded by Wolf (1878) of the eruption of Cotopaxi, Ecuador, in 1877 suggest that another mechanism should be

Figure 5.12 Development of a pumice flow erupted from Mt St Helens on 22 July 1980. The sequence is not accurately timed, but it begins at 19.01 h (Pacific Daylight Time) and lasts about 45 s. A photograph taken at 19.07 h above Mt St Helens is shown in Figure 5.7b. (Photographs by H. Glicken.)

considered for the eruption of scoria flows. Many of the local people who had observed the eruption described it as 'a pan of rice boiling over'. This suggests that these pyroclastic flows may have originated directly out of the vent, and formed without the collapse of an eruption column, or from a column so dense that it only rose a small height above the vent and instantaneously collapsed and

'bubbled over' (Fig. 5.11f).

It thus seems likely that so-called 'column collapse' that sources pyroclastic flows can take different forms. These include variations ranging from discrete column collapse of high, well maintained columns to partial collapse events from the margins of an unstable, but established, column, to discrete collapse followed by essentially continuous fountaining of pyroclastic debris, to a more passive boiling over, directly out of the vent.

It is now also thought that block and ash flows may be produced by collapse of eruption columns (Fig. 5.11g). Fisher and Heiken (1982) suggested that some of the explosions in the early stages of the Mt Pelée 1902 eruption were vertical rather than directed laterally. Collapse of a vertical column, or a slug of lava debris out of it, generated block and ash flows rather than a directed blast. It was eruptions of this type that occurred on 8 and 20 May 1902, and led to the destruction of St Pierre, and the death of 30 000 people.

The deposits of pumiceous pyroclastic flows are termed **ignimbrite**, and some of these can be very large volume deposits ($>1000 \text{ km}^3$). Few ignimbrites have been erupted this century. Those that have are only small-volume deposits (Ch. 8), and there is little observational information for these. The generally known examples are the Valley of Ten Thousand Smokes ignimbrite erupted from Katmai, Alaska, in 1912 (C. N. Fenner 1920, Curtis 1968), those formed during the eruptions of Komagatake, Japan, in 1929 (Aramaki & Yamasaki 1963) and those from Mt St Helens in 1980. Two notable, and larger, ignimbrite-forming eruptions occurred last century: Krakatau, west of Java, in 1883 (Self *et al.* 1981) and Tambora, also in Indonesia, in 1815 (van Bemmelen 1949, Self *et al.* 1984).

Small-volume pumice flows, like scoria flows, are perhaps in many cases generated by interrupted column collapse. Nobody has yet observed a large-volume ignimbrite-forming eruption, although as early as 1960, R. L. Smith (1960a) suggested that they could be formed by an eruption column collapse mechanism, but on a larger scale. Sparks and L. Wilson (1976) and Sparks *et al.* (1978) presented a theoretical model for the formation of ignimbrites based on the continuous gravitational collapse of a plinian eruption column (Fig. 5.11h). This models helps to explain many features of ignimbrites (Chs 7 & 8), and has since become popular among workers in this field. Continuous collapse of plinian eruption columns from heights of several kilometres could account for the large volume and wide distribution of some ignimbrites.

The model is also appealing because it explains why many ignimbrites are underlain by plinian pyroclastic fall deposits (Fig. 5.6b, Chs 6 & 8).

However, observations of the Mt St Helens 1980 eruption suggest that many of the pumiceous pyroclastic flows, which under our definition form ignimbrite, were not generated by collapse of a high eruption column (Rowley *et al.* 1981), but from low columns. Many pumice flows seemed to spread out from bulbous inflated masses of pyroclasts as they upwelled a short distance above the vent. The sequence of photographs in Figure 5.12 of activity on 22 July illustrate this particularly well, showing the development and instantaneous collapse of a fountain about 500 m high. Descriptions such as a 'pot boiling over' were given (Rowley *et al.* 1981), and there are obvious similarities to the eyewitness descriptions given of the Cotopaxi eruption in 1877. During other periods of activity, partial gravitational collapse of the margins of maintained columns was observed. None of the Mt St Helens pumice flows travelled very far, and all are minor in volume.

These new observations suggest that column collapse as the only mechanism for the generation of ignimbrites may have been overemphasised in recent years, as suggested above. In some instances 'spluttering' or 'frothing' at the vent may be more important. We will develop and expand these ideas on eruption mechanisms of ignimbrites through Chapters 6 and 8.

5.5 Pyroclastic flow deposits: types and description

Most pyroclastic flow deposits are composed of more than one flow unit. Each flow unit is usually regarded as the deposit of a single pyroclastic flow (Fig. 5.13), one of perhaps several or many generated during the course of the same eruption (Sparks *et al.* 1973, Sparks 1976). However, it is certainly possible that as a pyroclastic flow advances it could split into several subflows (R. L. Smith 1960a; and observed at Mt St Helens), each represented in the field by a discrete depositional flow unit. In the field pyroclastic flow units may be

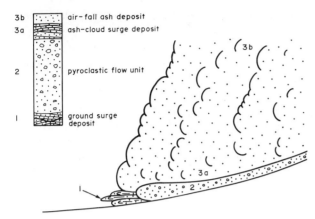

Figure 5.13 Schematic diagram showing the structure and idealised deposits of one pyroclastic flow.

seen to be stacked on top of each other, or be separated by other pyroclastic layers (fall or surge deposits) or reworked epiclastic deposits.

From the foregoing discussion on pyroclastic flow forming eruptions, it appears that three main types of pyroclastic flow deposit are recognised in modern volcanic successions (Figs 5.14–16):

block- and ash-flow deposits
scoria-flow deposits
pumice-flow deposits or ignimbrite

5.5.1 BLOCK- AND ASH-FLOW DEPOSITS

These are topographically controlled, unsorted deposits having an ash matrix and containing large generally non-vesicular, cognate lithic blocks which can exceed 5 m in diameter (Figs 5.14a & 15a). Some of these blocks contain radially arranged cooling joints which show they were emplaced as hot blocks (Fig. 5.15b). Clasts should be all of the same magma type, and therefore the deposits should be, or almost be, monolithologic. Individual flow units are reversely graded in many examples (Figs 5.14a & 15a). They may contain gas segregation pipes (Figs 5.15c & d), although these are not found too commonly in block and ash deposits (Ch. 7), and carbonised wood. Surface manifestations include the presence of levées, steep flow fronts and the presence of large surface blocks, all of which again indicate a high yield strength during

flow. No welded examples are known to us, although Sparks (*pers. comm.*) reports one on the southern flanks of Mt Pelée.

Homogeneous clast composition, hot blocks and gas segregation pipes are the field criteria for distinguishing these pyroclastic flow deposits from types of sedimentary debris deposits such as rock avalanches and debris flows (Ch. 10).

5.5.2 SCORIA-FLOW DEPOSITS

These are topographically controlled, unsorted deposits with variable amounts of basaltic to andesitic ash, vesicular lapilli and scoriaceous ropy surfaced clasts up to 1 m in diameter (Figs 5.14b & 5.15d–f). In some circumstances they may contain large non-vesicular cognate lithic clasts (Fig. 5.15f). Reverse grading of larger clasts within flow units is common, and fine-grained basal layers are sometimes found at the bottom of flow units. Gas segregation pipes and carbonised wood may also be present. The presence of levées, channels and steep flow-fronts indicates a high yield strength during flow. Again, we know of no welded examples.

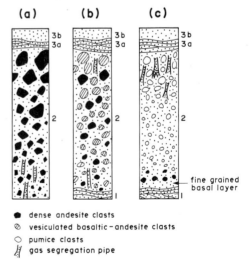

Figure 5.14 Idealised sections of the three main types of pyroclastic flow deposit and associated layers deposited by the mechanisms suggested in Figure 5.13. (a) Block and ash-flow deposit. (b) Scoria-flow deposit. (c) Pumice-flow deposit or ignimbrite.

Figure 5.15 Block- and ash-flow and scoria-flow deposits. (a) Reversely graded block and ash-flow deposit, formed by collapse of a rhyolitic lava flow. This was erupted towards the end of the 700 years BP Kaharoa eruption of the Tarawera volcanic centre, New Zealand. Top of spade handle is base of block and ash-flow deposit, other layers are earlier co-eruptive products. (b) Hot block in block and ash-flow deposit, San Pedro volcano, northern Chile (after P. W. Francis *et al.* 1974). (c) Gas segregation pipes in the 1902 block and ash flow deposits erupted from Mt Pelée (after Fisher & Heiken 1982). (d) Scoria flow deposit erupted from Mt Misery volcano, St Kitts, Lesser Antilles. Note the concentration zones of scoria which seem to be associated with flow unit boundaries and coarser-grained pipes which have been emphasised by rain washing. Arrow points to a carbonised log from which a ^{14}C age of 2860 years BP was obtained (photograph by M. J. Roobol). (e) and (f) The scoria flow deposits (dark) erupted in 1975 at Mt Ngauruhoe, New Zealand. Note thin lobate flow front and dense juvenile fragments with more scoriaceous clasts.

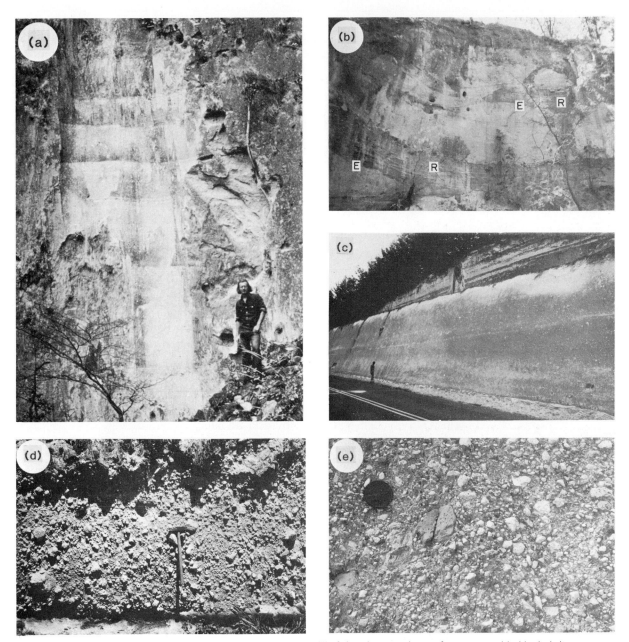

Figure 5.16 Some general features of pumice-flow deposits. All of the photographs are from non-welded ignimbrites or non-welded zones of welded ignimbrites. (a) Stacked thin flow units of the Rio Caliente ignimbrite, Mexico. Flow unit boundaries are picked out by fine-grained basal layers (after J. V. Wright 1981). (b) Flow units of the Rio Caliente ignimbrite interbedded with fluviatile reworked pumice (R) and erosion surfaces (E); arrow points to two flow units filling in small channels cut into the succession. No soils are present, and field evidence elsewhere shows that these erosional events must have been local and short-lived, and occurred during the same eruption. Height of cliff section is about 16 m (after J. V. Wright 1981). (c) Thick, massive flow unit of the Oruanui ignimbrite in New Zealand, which is poorly sorted and texturally very homogeneous throughout the thickness seen (horizontal lines are bulldozer scrapings). (d) Coarse, poorly sorted pumice-flow deposit on St Lucia, Lesser Antilles. (e) Close-up showing poor sorting in an ignimbrite. This is from a flow unit of the Acatlan ignimbrite, Mexico. Dark clasts are lithics.

Figure 5.16 continued (f) 'Fossil fumaroles'; crystal and lithic enriched gas segregation pipes in the Taupo ignimbrite, New Zealand (photograph by C. J. N. Wilson).

5.5.3 PUMICE-FLOW DEPOSITS OR IGNIMBRITES

Ignimbrites are typically poorly sorted, massive deposits containing variable amounts of ash, rounded pumice lapilli and blocks occasionally up to 1 m in diameter (Figs 5.14c & 16). Within flow units, larger pumice fragments can be reversely graded, while lithic clasts can show normal grading. However, ungraded flow units are as common. A fine-grained basal layer is usually found at the bottom of flow units (Fig. 5.16a). The coarser, smaller-volume deposits usually form valley infills, while the larger-volume deposits may form large

ignimbrite sheets that bury all but high topographic features. Sometimes they may show one or more zones of welding (Ch. 8). Their common salmon-pink colour, the presence of carbonised wood and a thermal remanent magnetisation are all ways of distinguishing non-welded ignimbrites from the deposits of pumiceous mud flows. Also, ignimbrites sometimes contain gas segregation pipes (Fig. 5.16f).

5.6 Origins of pyroclastic surges

It is now apparent that dilute, low particle concentration, turbulent, pyroclastic surges can be generated in many different ways. Volcanic base surges, first described from the phreatomagmatic eruptions of Taal volcano, Philippines, in 1965 by J. G. Moore *et al.* (1966) and J. G. Moore (1967), are *only one type* of pyroclastic surge. Pyroclastic surges are known to form in three situations, associated with:

phreatomagmatic and phreatic eruptions
pyroclastic flows
pyroclastic falls

5.6.1 SURGES ASSOCIATED WITH PHREATOMAGMATIC AND PHREATIC ERUPTIONS

These eruptions can generate a **base surge** which is a collar-like, low cloud expanding radially in all directions from the locus of a phreatomagmatic or phreatic explosion and/or by the collapse of the phreatomagmatic or phreatic eruption column (Figs 5.17 & 18). The term 'base surge' was originally applied to the radially outward moving basal clouds observed and photographed in nuclear explosions (Fig. 5.17), to which J. G. Moore (1967) likened similar features observed during the Taal 1965 eruptions, and some other observed historic eruptions.

The eruption of Taal on 28–30 September 1965 took place when water gained access to rising basaltic magma on the southwest side of Volcano Island, Lake Taal (J. G. Moore *et al.* 1966, J. G.

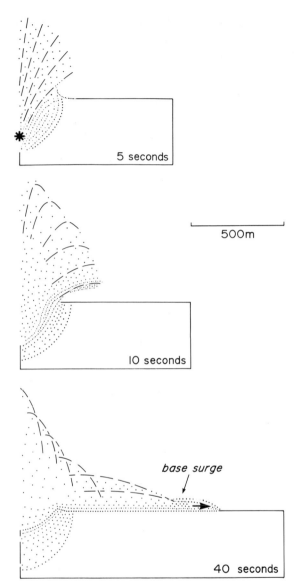

Figure 5.17 Sequential diagram showing formation of a base surge after an underground explosion equivalent to 100 kilotons of chemical explosives. (After J. G. Moore 1967.)

Moore 1967; Fig. 5.19). Explosions produced a series of base surges (Fig. 5.19) which spread out radially with 'hurricane velocity', causing extensive damage and loss of life. These debris-laden clouds obliterated and shattered all trees within 1 km of the explosion centre, and sandblasted objects up to 8 km away. Initially, velocities may have been as high as 100 m s^{-1} (J. G. Moore 1967).

Base surges result from the explosive interaction of magma and water and are probably in many cases 'cold and wet' (Ch. 7). In the entire area affected by base surges from the Taal 1965 eruption, no evidence of charred wood was found on surviving trees or in the deposits. In the zone where ash was plastered on to objects (Fig. 5.19b) the ash must have been mixed with water rather than steam to have been so sticky, and surges would have had temperatures below 100°C (J. G. Moore 1967). However, some phreatomagmatic eruptions have produced hot pyroclastic surges. During the phreatomagmatic eruptions forming the Ukinrek maars, Alaska, in 1977 (e.g. Fig. 5.18c), pyroclastic surges charred tree branches and trunks (Self *et al.* 1980). As discussed in Chapter 3, Sheridan and Wohletz (1981) suggested that there is a natural division between wet and dry base surges, depending on the water : magma mass ratio in phreatomagmatic explosions (Fig. 3.9). With a low water : magma mass ratio '*dry and hot*' base surges may be produced.

Base surges are commonly associated with the formation of small volcanic craters, called variously maars, tuff rings and tuff cones (Ch. 13). These are common features in areas of basaltic volcanism, and without the interaction of ground or surface water or sea water, the basaltic magma would have erupted to form scoria cones and lava flows. There have been a number of eruptions of this type in the 20th century. For descriptions and analysis of this type of activity, the reader is referred to Moore (1967) and Waters and Fisher (1971), who show spectacular photographs of the eruptions of Capelinhos in 1957–8 in the Azores (Figs 5.18a & b) and Taal, Philippines, in 1965–6, and the papers by Kienle *et al.* (1980) and Self *et al.* (1980) describing the formation of the Ukinrek maars, Alaska (Fig. 5.18c). Maars and maar-like constructional landforms can be formed by eruptions of other magma types, including carbonatitic, phonolitic and rhyolitic compositions. For good descriptions of the base-surge deposits associated with prehistoric phonolitic and rhyolitic eruptions of this type see Schmincke *et al.* (1973) and Sheridan and Updike (1975), respectively.

Base surges are also known to be erupted from major volcanoes. They should be common products

Figure 5.18 Phreatomagmatic eruptions producing base surges. (a) At Capelinhos in October 1957. Height of vertical column to top of photograph is about 2200 m. US Air Force photograph (after J. G. Moore 1967). (b) Capelinhos in September 1957. Steam has blown downwind to expose a dense debris-laden central column collapsing to feed a base surge. On the right-hand side of the photograph the surge is moving across the ocean surface (after Waters & Fisher 1971). (c) East Ukinrek maar in 1977. Note chevron shape of base surge cloud (moving to the left) which in this case was directed by lows in the maar rim and shallow valleys (after J. Faro in Kienle et al. 1980).

of andesitic stratovolcanoes with crater lakes, and other volcanoes with caldera lakes. Phreatic and phreatomagmatic eruptions from the crater lake of Ruapehu volcano, New Zealand, have been common this century, and base surges were observed in the eruption of April 1975 (Nairn et al. 1979). The 1979 eruption of Soufrière, St Vincent, which was through a crater lake, also produced base surges (Shepherd & Sigurdsson 1982). The Quill stratovolcano on St Eustatius, also in the Lesser Antilles, has a long history of phreatomagmatic activity, and base-surge deposits form an important part of the pyroclastic succession found in its ring plain. These vary from basaltic andesite to rhyolite in composition, and were produced by a number of eruptions over the past ~30 000 years as the volcano emerged from the sea and grew to its present form (Roobol,

Smith & Wright *unpub. data*). The rhyolitic base-surge deposits form part of a thicker pyroclastic sequence generated during an ignimbrite-forming eruption. Rhyolitic base-surge deposits are also known in association with phreatoplinian phases of the Askja, Iceland, 1875 eruption and the Minoan (1470 BC) eruption of Santorini (Self & Sparks 1978; Ch. 6). Phonolitic base surges also were generated late during the AD 79 eruption of Vesuvius, when large amounts of water from a deep aquifer under the volcano gained access to the magma chamber. The deposits are associated with phreatoplinian air-fall layers (Sheridan et al. 1981; Ch. 6).

(a) Thickness of total ejecta (surge plus fall deposits)

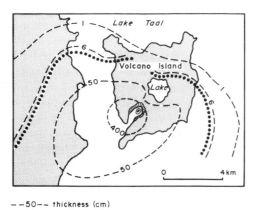

- - 50 - - thickness (cm)

●●●●●●● range of accretionary lapilli

(b) Thickness of base surge deposits

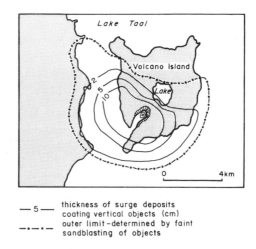

——5—— thickness of surge deposits coating vertical objects (cm)

—·—·— outer limit - determined by faint sandblasting of objects

(c) Maximum clast size in base surge deposits

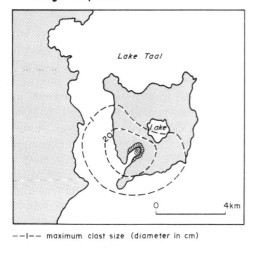

—-1-— maximum clast size (diameter in cm)

(d) Distribution of dune bedforms in base surge deposits

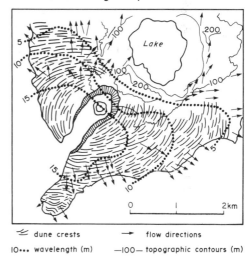

⇜ dune crests   → flow directions

10●●● wavelength (m)   —100— topographic contours (m)

Figure 5.19 General distributional characteristics of the deposits of the 1965 eruption of Taal in the Philippines. Flow directions of major base surge movement in (d) were measured in the field from the sand blasting, tilting and coating of trees and houses. (After J. G. Moore 1967.)

5.6.2 SURGES ASSOCIATED WITH FLOWS

Thin, stratified pumice and ash deposits are often found associated with pyroclastic flow deposits of various kinds. When associated with the bases of flow units, they are called **ground surges**, and when associated with the tops they are called **ash-cloud surges**. These types have different mechanisms of

generation. Compared with base surges they can be considered to be hot and dry.

The term 'ground surge' was coined by Sparks and Walker (1973), but it was used by these authors to mean any type of pyroclastic surge. More recently the term has become used just for those surges found at the bases of pyroclastic flow units, or associated with some fall deposits (Section 5.6.3;

Fisher 1979, J. V. Wright *et al.* 1980). Ground surges are thought to be the same as the 'ash hurricanes' described by G. A. Taylor (1958) from the 1951 Mt Lamington eruption. Taylor observed these to form at the same time as high concentration pyroclastic flows (or his 'ponderous ash flow nuées') directly from the crater without an accompanying vertical eruption column, or from collapsing eruption columns (Fisher 1979).

Ground surges are envisaged as precursors to dense, high concentration pyroclastic flows, preceding their flow-fronts. There are a number of ways in which they can be generated:

(a)   from a directed low concentration blast,
(b)   out of the head of a moving pyroclastic flow or
(c)   by earlier, smaller collapses of the margins of a maintained vertical eruption column.

The concept of a low concentration blast preceding the main part of a pyroclastic flow stems largely from early ideas on understanding the 8 May 1902 eruption of Mt Pelée, which was thought to have been a directed blast. We have already discussed this eruption, and how it is now thought to have generated block and ash flows by collapse of an eruption column. Fisher *et al.* (1980) and Fisher and Heiken (1982) suggested that St Pierre was destroyed by an ash-cloud surge, although G. P. L. Walker and McBroome (1983) suggested that it was by a violent pyroclastic flow (Ch. 7). Several historic block- and ash-flow deposits produced by explosive lava-dome collapse have obvious surge deposits associated with them, but again some of these could be ash-cloud surge deposits. However, Rose *et al.* (1977) described a ground surge produced by explosive collapse directed out of the lava front at Santiaguito in September 1973, and because the surge does not mantle the associated block- and ash-flow deposit, they suggest that it probably preceded it. The initial explosion of Mt St Helens was an obvious directed blast, and its effect on the forest in its path is well known. The deposits from the initial explosion certainly show some characteristics of a surge deposit, as we have alluded to previously, and this is how they have been described by J. G. Moore and Sisson (1981) and Hickson *et al.* (1982). However, the stratigraphy

is more complicated than that of normal ground-surge deposits, and Hoblitt *et al.* (1981) have drawn attention to this. G. P. L. Walker (1983) suggested that the blast was a high concentration pyroclastic flow emplaced at very high velocities, like some violent ignimbrites (Chs 7 & 8). Like the Mt Pelée event, this event and its deposits are the source of much debate. Pumice flows forming ignimbrite did not begin to erupt for another four hours after the initial explosion at Mt St Helens.

Studies by C. J. N. Wilson (1980, 1981, 1984) and C. J. N. Wilson and Walker (1982) suggest that the flow-heads of some pyroclastic flows (especially pumice flows) may ingest large volumes of air, and may be inflated and highly fluidised (Chs 7 & 8). At the front of the moving flow, basal friction will cause an overhang which will act as a funnel for air, in much the same way as a subaqueous mass flow incorporates water (Allen 1971, Simpson 1972). Cold air when heated would rapidly expand, and surges of highly fluidised pyroclasts would be jetted out of the head and upper parts of the flow front (Fig. 5.13; Ch. 7); material ejected at higher positions on the flow-front would contribute to the ash cloud. This could also be another mechanism for generating turbulent, low concentration surges continually advancing in front of some pyroclastic flows. The escaping gas and ash gives the flow-head its 'billowing' or 'sprouting' appearance, as seen, for example, by Perret (1937) in some Mt Pelée pyroclastic flows erupted during 1929–32. This type of jetting of material from the flow-head explains some other facies associated with ignimbrites, and these will be discussed further in Chapter 7.

The third mechanism we can envisage for the generation of ground surges is by repeated minor collapse of a maintained eruption column before major ignimbrite-forming collapse. This could also apply for some ignimbrite-forming eruptions, and Fisher (1979) suggested such a model. Turbulent mixing and intake of cold air at the margins of the eruption column could overload parts of it, and small-scale collapse could generate precursor surges.

More recently, however, G. P. L. Walker *et al.*

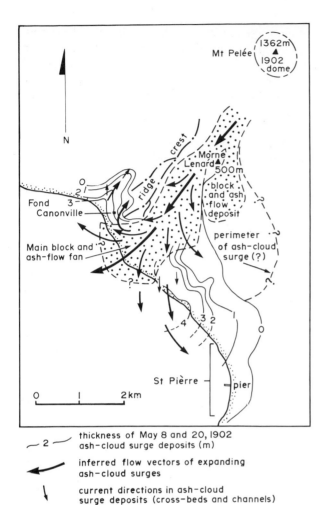

Figure 5.20 Distribution of the block and ash-flow and associated ash-cloud surge deposits from the 8 and 20 May 1902 eruptions of Mt Pelée, and their inferred flow directions. The main block and ash-flow lobe fills the Rivière Blanche. Note how at Fond Canonville ash cloud surges moved around the ridge and then in an opposite direction to the main flow. (After Fisher *et al.* 1980, Fisher & Heiken 1982.)

described one from the Taupo ignimbrite (Ch. 7), and suggested that some other examples of deposits previously called ground surges were deposited by this alternative mechanism. Towards the vent, the quite remarkable ground layer of the Taupo ignimbrite passes into a coarse-grained breccia (it contains blocks >1 m in diameter near the vent), and nearly always lacks internal stratification. On the other hand, ground surges are never as coarse-grained, and have well developed planar stratification or low angle cross-stratification. However, criteria to distinguish the deposits generated by all of these different mechanisms have not been clearly identified.

Ash-cloud surges are turbulent, low density flows generated in the overriding gas and ash cloud as observed above historic pyroclastic flows (Fig. 5.10). The towering ash cloud contains material elutriated from the top of the moving pyroclastic flow, which forms a basal underflow (Figs 5.11 & 13). However, most of the ash rising into the ash cloud is deposited later as a fine-grained ash-fall deposit. In some cases ash-cloud surges could become detached from the moving pyroclastic flow and move independently.

Fisher (1979) discussed the formation of ash-cloud surges in the Upper Bandelier ignimbrite. Fisher *et al.* (1980) and Fisher and Heiken (1982) discussed their formation during the Mt Pelée 1902 eruption. They suggest that block and ash flows were confined to valleys, while fully turbulent, dilute high energy ash-cloud surges moved down the mountain continually expanding outwards (Fig. 5.20). Gravity segregation within individual ash-cloud surges occurred as they expanded, resulting in secondary block and ash underflows with high particle concentrations, which did not travel as far. Fragment-depleted ash-cloud surges are thought to have devastated St Pierre. Burnt wood and other high temperature effects in St Pierre indicate that the flows were hot. The deposits only had a maximum thickness of 1 m in St Pierre, where they are fine-grained, and generally massive, but internal stratification can be found. However, G. P. L. Walker (1983) has questioned the *ash-cloud* interpretation of these deposits, and in some ways has reverted back to older ideas by suggesting

(1981a) and C. J. N. Wilson and Walker (1982) suggested that some crystal- and lithic-rich deposits at the bottoms of some ignimbrite flow units are generated *within* the flow-head. These occupy the same stratigraphic position as the ground surge, but they are not deposited from a separate, dilute low-concentration flow, therefore not by a pyroclastic surge. G. P. L. Walker *et al.* (1981a) suggested that these deposits be called **ground layers**. They

they were high-concentration blasts similar to that at Mt St Helens (Ch. 7). Ash-cloud surges and their deposits were certainly observed to develop at Mt St Helens 1980, and are described by Rowley *et al.* (1981).

5.6.3 SURGES ASSOCIATED WITH FALLS

There is evidence that some pyroclastic surges, associated with magmatically erupted air-fall deposits, are formed by the collapse of an eruption column (or margins of it) without the generation of an accompanying pyroclastic flow. Such surges would again be termed ground surges (Fisher 1979). Roobol and Smith (1976) described pre-historic 'pumice and crystal ground surge deposits' inter-bedded with pumice fall deposits on Mt Pelée, extending up to 2 km away from the vent, and suggested that they formed by gravity collapse of plinian eruption columns. No doubt surges found interbedded with pumice-fall deposits can be generated by other mechanisms. For example, small amounts of external water gaining access to the erupting magma (from surface ground water or a deep aquifer) could generate hot, dry base surges (Sheridan & Wohletz 1981). Sheridan *et al.* (1981) suggested surge deposits interbedded with the early erupted pumice-fall deposit of the AD 79 Vesuvius eruption (the Pompeii pumice of Lirer *et al.* (1973); Ch. 6) were formed in this way; these surges were generated before the major phreatomagmatic activity which produced the wet base surges and phreatoplinian layers mentioned previously.

5.7 Pyroclastic surge deposits: types and descriptions

From the above description, pyroclastic surge deposits can be divided into three types:

  base-surge deposits
  ground-surge deposits
  ash-cloud surge deposits

5.7.1 BASE-SURGE DEPOSITS

Base surges produce stratified, laminated, some-times massive deposits containing juvenile fragments, ranging from vesiculated to non-vesiculated cognate lithic clasts, ash, crystals and occasional accessory lithics. Large ballistic lithics may form bomb sags close to the vent. Surges produced in phreatic eruptions are composed almost totally of accessory lithics, plus perhaps minor amounts of accidental lithics. Juvenile fragments are usually less than 10 cm in diameter, due to the high degree of fragmentation caused by the water–magma interaction. Base surges can accumulate thick deposits (>100 m) around some phreatomagmatic craters (Ch. 13), although they thin rapidly away from the vent. Deposits found in the successions of stratovolcanoes are generally thin (<5 cm to <5 m). Internally, deposits show unidirectional bedforms, and climbing dune-forms can be common (Figs 5.21a, b & 22). Near vent it may be difficult to distinguish planar-bedded surge deposits from

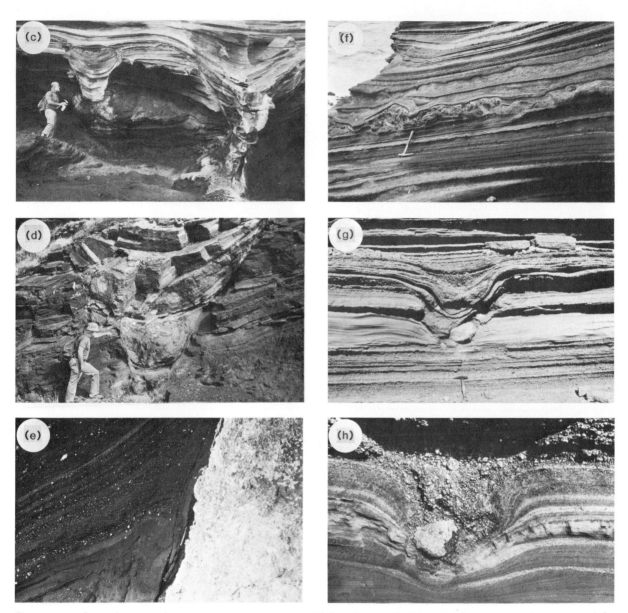

Figure 5.21 Some features of basaltic base-surge deposits. (a) Climbing duneforms, surge flow direction from right to left. Thin planar layers are air-fall deposits. From Lake Purrumbete maar, Western Victoria, Australia. (b)–(f) are from the coastal slopes of Koko crater, Oahu, Hawaii, but most of the deposits in this area are thought to have been erupted from the Hanauma Bay crater complex, 1–5 km to the southwest. (b) Climbing duneforms, surge direction left to right from Hanauma Bay, Hawaii. (c) and (d) U-shaped erosional channels; U-shaped bases and stratigraphy of the deposits suggest that these were fluviatile erosional gullies reshaped and re-emphasised by younger base surges from Hanauma Bay. (For complete story see Fisher 1977.) Note how planar-bedded base surge layers thicken into the channels (cf. Fig. 5.1c). (e) Ash plastered against the almost vertical side of a wall of eroded basement of reef limestone. (f) Planar bedded deposits with penecontemporaneous slumping. (g) and (h) Bomb sags in planar-bedded deposits, Tower Hill, Victoria.

Figure 5.22 Rhyolitic base-surge deposits erupted from the Quill, St Eustatius (<22 240 years BP), Lesser Antilles. The base-surge deposits dominate the stratigraphic interval visible in (a). Here there is a thin lower unit of base-surge deposits (prominent white layer) separated from the thicker upper unit with well developed dunes by a plinian pumice-fall deposit. The more-massive layer above these shown in (b) is a co-eruptive ignimbrite.

Figure 5.23 Some ground surge and ash-cloud surge deposits. (a) and (b) Ground surge deposits at the base of the Upper Bandelier ignimbrite. (c) Ground surge deposit separating two flow units in the Upper Bandelier ignimbrite. The dark (pink) stratified surge deposit is clearly associated with the upper darker flow unit and they were emplaced as one thermal package. (d) Ash-cloud surge deposits between two flow units of the Upper Bandelier ignimbrite. Local field relations and the photograph show that the thinly laminated surge deposits are associated with the pumice-rich top of the lower flow unit (cavernous weathering). (e) Fine-grained 1902 ash-cloud surge deposit in the churchyard at St Pierre. Internal stratification is found in this. (f) The church at St Pierre is thought to have been destroyed by an ash-cloud surge.

Figure 5.24 (above and facing page) Accretionary and cored lapilli. (a)–(f) Associated with different rhyolitic deposits. (a) Within a rhyolitic ignimbrite from the Devonian Snowy River Volcanics, eastern Victoria, Australia. The matrix contains an abundance of flattened and fragmented lapilli. (b) and (c) show exceptional concentrations of accretionary lapilli from phreatoplinian air-fall ashes of the Oruanui eruption, Lake Taupo, deposited in a small crater lake inside the scoria cone, Pukeonake, New Zealand. These have been reworked, as shown by the erosion surface in (c). (d) Lapilli within an ignimbrite about 500 000 years BP in New Zealand (photograph by C. J. N. Wilson). (e) In gas segregation pipes within the Oruanui ignimbrite, New Zealand. (f) From a thick concentration of accretionary lapilli within the body of the Oruanui ignimbrite. (g) Cored lapilli at Koko Crater, Oahu, Hawaii. (h) Cored lapilli in basaltic base surge deposits at Cape Bridgewater volcano, Victoria.

flat-bedded air-fall deposits. Surge deposits usually show some low-angle truncations, and therefore these are key features to look for (criteria to distinguish these two types of deposit in such situations are discussed further in Ch. 7). U-shaped erosional channels have also been described (Figs 5.21c & d) and their formation has been discussed by Fisher (1977).

Base-surge deposits often show evidence of being wet and 'sticky' when deposited. Accretionary lapilli are common (Section 5.8). Deposits may be plastered and stuck to vertical or almost vertical surfaces (Fig. 5.21e), and layers often deform plastically, which can be seen when large bombs impact (Figs 5.21g & h), or there is penecontemporaneous slumping (Fig. 5.21f). Also, **vesiculated tuffs** with entombed gas cavities may be present. Note, however, that vesiculated tuffs are not solely diagnostic of a base-surge origin as indicated by Lorenz (1974), and can be found in phreatomagmatic ash-fall layers, and not necessarily near the vent. They only show that ash was nearly saturated

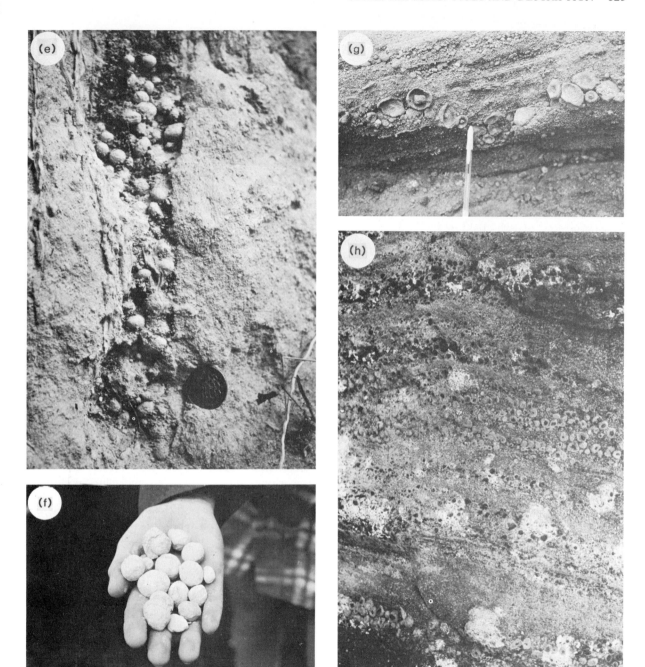

with water at the time of deposition, and that trapped air or steam could not escape. When basaltic in composition, juvenile material is usually, to some extent, hydrated and altered to palagonite (Chs 13 & 14). Such deposits can be lithified, but should not be confused with welded tuffs.

5.7.2 GROUND-SURGE DEPOSITS

Ground surges produce stratified deposits generally less than 1 m thick which are typically recognised at the base of pyroclastic flow units (Figs 5.23a, b & c). The deposits are composed of ash, juvenile vesiculated fragments, crystals and lithics in varying proportions, depending on the constituents present in the eruption column. They are typically enriched in denser components (less well vesiculated juvenile fragments, crystals and lithics) compared with accompanying pyroclastic flow deposits (Sparks 1976). Again, they show unidirectional bedforms; carbonised wood and small gas segregation pipes may be present.

5.7.3 ASH-CLOUD SURGE DEPOSITS

The products of ash-cloud surges are stratified deposits generally less than 1 m thick found at the top of, and as lateral equivalents to pyroclastic flow units (Figs 5.23d & e). They show unidirectional bedforms and pinch and swell structures, and may occur as discrete separated lenses (Fisher 1979). The grainsize and proportions of components depend on the type of the parent pyroclastic flow. One would intuitively expect such deposits to be enriched in vitric particles. However, those associated with the Bandelier Tuffs (Fisher 1979; Fig. 5.23d) are enriched in crystals, and this must be due to further gravity segregation within the ash cloud, as ash-sized particles with a significant proportion of crystals are elutriated out of the parent pumice flow. The ash cloud surges described by Fisher and Heiken (1982) from the 1902 eruption of Mt Pelée (Fig. 5.23e) have very similar component proportions to both their parent and secondary block and ash flows, but this is not surprising because there is little density difference between dense ash-sized juvenile fragments and crystals. Ash-cloud surge deposits again can contain small gas segregation pipes.

5.8 Accretionary lapilli

Accretionary lapilli are lapilli-sized pellets of ash commonly exhibiting a concentric internal structure (J. G. Moore & Peck 1962; Fig. 5.24). They have been described from pyroclastic fall, surge and flow deposits. They are believed to form by the accretion of fine ash around some nucleus, either a water droplet or solid particle. This could occur during rain flushing (J. G. Moore & Peck 1962, G. P. L. Walker 1981a) either of the downward plume from an eruption column or of the accompanying ash cloud of a pyroclastic flow. However, perhaps more frequently, they form in the steam-rich columns of phreatomagmatic and phreatic eruptions (Self & Sparks 1978; the examples shown in Fig. 5.24), perhaps around consensing water droplets. They can then be transported and deposited by fall, base surge or flow processes. Basaltic base-surge deposits often seem to contain the variety called **cored** or **armoured lapilli**, which have a recognisable lithic core and a thick (sometimes 1–2 cm) shell of unstructured ash (Figs 5.24g & h). Perhaps these form in the outward-moving base-surge cloud as solid fragments pick up a coating of sticky wet ash. Accretionary lapilli also form by gases streaming up through pyroclastic flow deposits, and they occur in segregation pipes (G. P. L. Walker 1971; Figs 5.24e & f).

It is important to stress that accretionary lapilli *are not indicative* solely of pyroclastic fall deposits, as often seems to be assumed by workers in ancient successions. They may occur in pyroclastic fall, flow or surge deposits. Indeed, stratified deposits several metres thick, with accretionary lapilli, are more likely to be base-surge deposits. Also, some accretionary lapilli can survive a limited amount of reworking and redeposition, and can therefore be found in epiclastic volcanic sediments (Figs 5.24b & c). Furthermore, they can form *well away from vents* in pyroclastic flows and their trailing ash clouds, as well as in secondary eruption columns generated when pyroclastic flows interact explosively with surface water into which they flow (Chs 8 & 9). They are therefore not indicative of exclusively near-vent depositional settings.

5.9 Further reading

Aspects of the geology of pyroclastic deposits are fully developed and discussed in our later chapters. In Chapter 12 we will present a classification, and in Chapter 12 and Appendix II we consider criteria that may distinguish these rocks in ancient volcanic successions.

Plate 6 Two rhyolitic plinian pumice-fall deposits (white layers) erupted in the past 5000 years from Hekla volcano, Iceland. Top photograph is an outcrop 15 km from the vent and the bottom one shows the same two fall deposits considerably thinner, 50 km from the vent. The pumice falls are interbedded with alluvial sediments and soils. Dark layer just above lower pumice fall in bottom photographed is a distal scoria fall erupted from Katla volcano.

CHAPTER SIX

Modern pyroclastic fall deposits and their eruptions

Initial statement

Pyroclastic fall deposits, their eruptions and the physical controls on their formation are now considered in detail. The classification scheme of G. P. L. Walker (1973b) characterises fall deposits by their dispersal and degree of fragmentation, and this approach is used as a framework for our description. Important volcanological assessments can also be made, because these two parameters are related to the height of an eruption column and the nature of the fragmentation process. In this chapter we also consider some surprising features of the Mt St Helens ash-fall deposits and the implications of these for plinian eruptions. We also focus on the characteristics of distal silicic ash-fall layers. Finally, the properties of welded air-fall tuffs which occur near the vent on a number of volcanoes are

described. Criteria for distinguishing these from welded ignimbrites (Ch. 8) and a thermal facies model for pyroclastic fall deposits are also presented.

6.1 Introduction

The different types of pyroclastic fall deposit are ill-defined in the literature. They are described by a number of terms according to certain styles of explosive activity, named generally after individual volcanoes or volcanic areas where the activity was first observed, or of which the style of eruptions was thought to be characteristic. Examples include *strombolian, hawaiian, vesuvian* and *peléan*. One exception to this is *plinian*, which is named after Pliny the Younger, from his account to the historian

Tacitus of the eruption of Vesuvius in AD 79. This type of eruption and the resulting deposit could also have been called vesuvian. However, in the literature this refers to another style, exemplified by the basaltic eruption of Vesuvius in 1906, which involved a long-sustained gas stream with little ash being released (MacDonald 1972).

This approach to nomenclature has produced many problems. First, the style of eruption can change during the course of one eruption, and certainly during the long-term history of a volcano. For example, during its history Mt Pelée has shown a variety of styles of eruption other than peléan. It has also, for instance, experienced plinian eruptive phases (Roobol & Smith 1976). Secondly, particular eruption styles can occur in places other than the places used to name the particular eruption style. For example, hawaiian style eruptions are not just confined to Hawaii. Lastly, many deposits from historical or relatively recent eruptions have not been studied immediately after their eruption with the aim of correlating particular facies characteristics with *observed* eruption styles. In many cases, particularly with historic eruptions, the details of eruption characteristics have been *inferred* from a

(a)

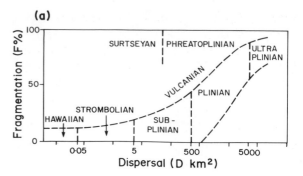

(b)

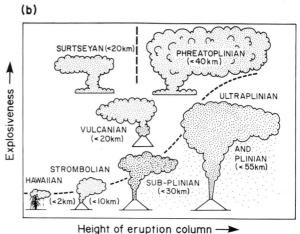

Figure 6.2 (a) *D–F* plot used to characterise different types of pyroclastic fall deposit (after G. P. L. Walker 1973b, and updated in J. V. Wright *et al.* 1980). (b) Cartoon explaining *D–F* plot in terms of eruption column height and 'explosiveness'.

study of the deposits well after the eruption has ended, by people who did not observe the eruption.

However, many of the poorly defined terms are entrenched in the geological literature and it would be naive to assume that they could be abandoned. The only practical solution is to improve the definition of existing terms by more quantitative analysis applied to well preserved young deposits for which good accounts of the eruption are available. From these studies, better descriptions, that can be used as a guide to interpret equivalent deposits in the rock record, may result.

The first serious attempt to describe and classify explosive volcanic eruptions producing pyroclastic falls quantitatively was by G. P. L. Walker (1973b). Walker's approach was based on the characteristics

(a)

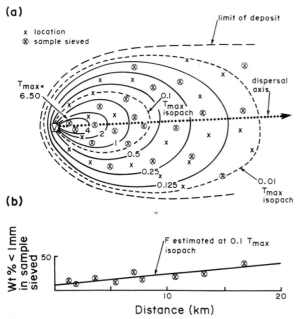

(b)

Figure 6.1 Representation of method used to obtain the two parameters *D* and *F*. See text for explanation.

of the fall deposits examined in the field, and not on the characteristics of the eruptions as was generally the practice previously. This quantitative scheme (Figs 6.1 & 2) relies on accurate mapping of the distribution of fall deposits and detailed granulometric analysis to determine two parameters: dispersal (D) and fragmentation index or degree of fragmentation of the deposit (F). The empirical measure of D used is the area enclosed by the $0.01T_{max}$ isopach (where T_{max} is the maximum thickness of the deposit; Fig. 6.1a). The empirical measure of F chosen is the percentage of a deposit finer than 1 mm at the point on the axis of dispersal where it is crossed by the $0.1T_{max}$ isopach; this can only be determined from the sieve analysis of a sample collected either at this point or, more practically, obtained graphically from sieve analyses of a few samples collected near the dispersal axis (Fig. 6.1b).

G. P. L. Walker (1973b) initially characterised three types of pyroclastic fall deposit on the basis of their values of D and F: *hawaiian–strombolian*, *surtseyan* and *plinian* (Fig. 6.2a). A distinction between strombolian and hawaiian types based on D was proposed, and another distinction, based on F, between normal and violent strombolian, was also proposed. Also, *sub-plinian* was proposed as a new type, intermediate in character between strombolian and plinian. Since Walker's original plot was published, later studies have refined this, and other types of pyroclastic fall deposit have been characterised: *ultraplinian*, *vulcanian* and *phreatoplinian* (Fig. 6.2).

The D–F plot is based on the measurable characteristics of a deposit, but it also reflects some of the essential features of the eruption, even though many changes in observed style of activity may have occurred throughout eruption. For any deposit, this plot is a reflection of not only the eruption column height, since it is this which largely controls D, but also the 'explosiveness' or degree of fragmentation of the magma (Fig. 6.2b). High F-values, for instance, may result from very high intensity eruptions (high volumetric eruption rates) or magma–water interactions. This is therefore a most useful way of making volcanological assessments of, and comparisons between, pyroclastic fall deposits whose eruptions were not observed, and whose original extent is still reasonably intact.

Although the plot of D against F gives a basis for detailing types of pyroclastic fall deposits and their eruptions, it is important to point out here that further research is increasingly revealing a number of its shortcomings. The meaning of F is not as clear as was suggested above. High F-values may not prove to be the result of high degrees of fragmentation, but may also reflect 'wet' eruption plumes in which premature deposition of fines is promoted by rain-flushing. This problem is highlighted in the discussion of distal silicic ash-fall layers (Section 6.9). Also, the fields for phreatomagmatic ash-fall deposits, which are now simply divided into surtseyan and phreatoplinian, are far from satisfactory (Section 6.8).

Before we describe the different types of pyroclastic fall deposit and their eruptions, two parameters that are essential to understanding the deposition and analysis of pyroclastic fall deposits need to be discussed, these being terminal fall velocity and muzzle velocity.

6.2 Terminal fall velocity and muzzle velocity

The distance that individual pyroclastic fragments are transported from the vent is controlled by many interacting factors. The most important are the heights to which particles are taken in the eruption column, the angle of ejection, the wind strength and the terminal fall velocity of the particles.

When an object falls through the air, it accelerates until it reaches a constant or terminal velocity (TV), which is the velocity at which the force of gravity and aerodynamic drag forces are in a state of balance. Particles with smaller terminal fall velocities will travel downwind further for a given eruption column height and wind speed than larger particles with a lower terminal fall velocity. Data on the terminal fall velocities of pyroclasts are given by G. P. L. Walker et al. (1971) and in Appendix I.

G. P. L. Walker (1971) showed that for polycomponent pyroclastic fall deposits it is useful to

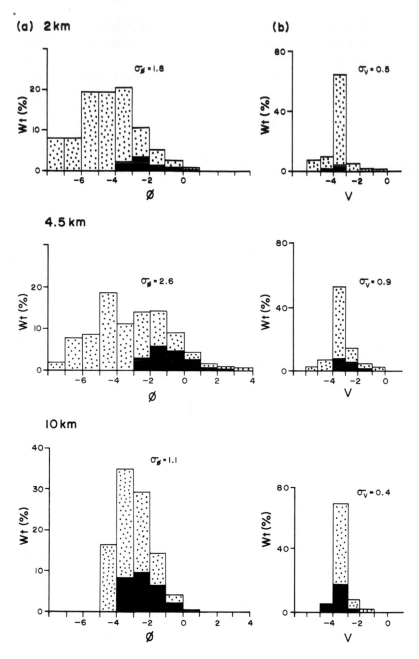

Figure 6.3 Grainsize characteristics of three samples of the Middle Pumice, a pyroclastic fall deposit on Santorini (Fig. 13.30) taken at increasing distances from the probable vent. (a) Histograms of the grainsize distributions. Grainsize distributions of air-fall deposits on a weight percentage basis are a function of the terminal fall velocity of ejecta, which is controlled by both grainsize and density. Less than 3 km from source, samples of the fall deposit contain >90 wt% pumice, and have unimodal histograms and a low σ_ϕ value. The proportion of dense components (lithics and crystals) increases from source. Between 3 and 5 km from source this results in a bimodal grainsize distribution, with a coarse mode due to pumice and a fine secondary mode due to the denser components, and an increase in σ_ϕ. At greater distances (>5 km) a decrease in the proportion of very coarse pumice clasts results in a restricted pumice size range with a mode closely corresponding to that of the dense components. The grainsize distribution is unimodal and sorting improves markedly. (b) Histograms of grainsize in weight percentages plotted against the terminal fall velocity of ejecta; V is defined as $-\log_2 TV$ where TV is the terminal velocity in metres per second. These group together all particles which fall at the same rate in the same class. By doing this, all the grainsize histograms become strongly unimodal.

plot histograms of weight percentages against terminal fall velocity, so grouping together all particles which fall at the same rate. When this is done, grainsize histograms of pyroclastic fall samples become strongly unimodal (Fig. 6.3). Median terminal fall velocity in an air-fall deposit gradually decreases with distance (Figs 6.3 & 4). The slope on

the median terminal fall velocity–distance curve (Fig. 6.4) is controlled by eruption column height and wind speed. For the deposits shown in Figure 6.4, the wind speed was approximately the same, and the slope of the line is therefore a function of eruption column height.

One of the most useful physical parameters in the

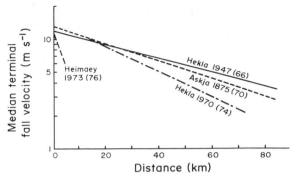

Figure 6.4 Median terminal fall velocity plotted against distance from source for some pyroclastic fall deposits. For each deposit an indication of the windspeed (in km h^{-1}) is given in parentheses. (After Self *et al.* 1974.)

comparison of explosive pyroclastic fall eruptions is the initial gas or **muzzle velocity** at the vent. During observed eruptions this can be determined by measuring the fall times of ballistic blocks and bombs which are unaffected by the wind, or by analysing films of eruptions. In older deposits one can measure the average maximum size of the largest fragments, and if the vent location is known these sizes can be used to estimate the minimum muzzle velocity based on the calculations of L. Wilson (1972). In L. Wilson's (1972) paper the ranges of particles ejected from vent, and the fall times of particles released from an eruption column (or ash cloud), are computed for various values of particle radius, density, launch velocity, launch angle and release height (see App. I).

For any deposit, on a plot of average maximum clast size against distance from vent, a line drawn along the top of the resulting scatter will show the maximum range of fragments of a given size (e.g. Figs 6.15 & 21, below). When maximum lithic or denser juvenile sizes are plotted. This line usually shows a sharp inflection a few kilometres from the vent, and this is thought to reflect the distance range of ballistic fragments (e.g. Figs 6.15 & 21, below). Maximum pumice sizes usually do not show this inflection, because larger pumice bombs tend to break on impact with the ground surface, and owing to their low density even the largest clasts are affected by the wind to some extent. Measurements of the largest lithic fragments are

therefore going to give the most reliable estimates of muzzle velocities. For most practical purposes this is going to involve only lithics much greater than 20 cm in diameter.

6.3 Hawaiian–strombolian

These types of pyroclastic fall deposit are the products of mildly explosive eruptions of basaltic or near-basaltic magmas. Such eruptions eject scoria and relatively fluid lava spatter, and are often accompanied by the simultaneous effusion of lava (Ch. 4; Plate 3). Vents for these eruptions can be fissures or simple conduits. However, observations and theoretical considerations suggest that activity along fissures is quickly localised to a number of points (L. Wilson & Head 1981). This happened, for example, during the Heimaey eruption in 1973 (Thorarinsson *et al.* 1973). Explosive activity builds scoria (cinder) or spatter cones, or both, at the vent, with scoria-fall deposits of limited areal extent and volume being deposited around and downwind of the vent. Scoria cones may be the sites of persistent activity over decades or longer, such as Stromboli (Chouet *et al.* 1974) and Northeast Crater, Mount Etna (McGetchin *et al.* 1974), but more generally they are monogenetic cones (Ch. 13) produced by what can be considered to be single eruptions lasting usually a few weeks to a few months, such as Heimaey in 1973 (Thorarinsson *et al.* 1973, Self *et al.* 1974).

6.3.1 CHARACTERISTICS OF THE DEPOSITS

Deposits of scoria cones often consist of rather poorly bedded, very coarse-grained and sometimes red (oxidised) scoria with metre-sized ballistic bombs and blocks (Figs 6.5–7). Many of the observed beds are not simply air-fall layers, but include mass-flow deposits formed by avalanching and rolling of scoria down unstable slopes as the cone built up. Such beds are laterally discontinuous. Grain flow (Ch. 10) of the loose granular material during downslope movement produces reverse grading (see Fig. 6.10c). A variety of bombs and blocks may be found: large scoriaceous fragments,

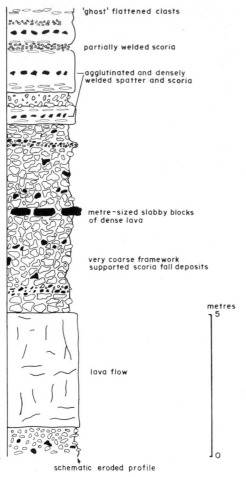

'ghost' flattened clasts

partially welded scoria

agglutinated and densely welded spatter and scoria

metre-sized slabby blocks of dense lava

very coarse framework supported scoria fall deposits

metres
5

lava flow

0

schematic eroded profile

Figure 6.5 Section through uppermost part of scoria deposits at Ohakune craters, near Ruapehu volcano, New Zealand. (After Houghton & Hackett 1984.)

less well vesiculated lava having spindle and cowpat shapes, sometimes bombs with breadcrusted surfaces, and dense lava blocks and slabs. Large accessory lithics of country rock are usually uncommon, but petrologically important mantle-derived nodules may occur as 'cored' lithics with a rind of lava around them. Bomb sags are not a common feature. This is because ballistic bombs land in a thick accumulating bed of coarse, loosely packed, unstratified scoria (cf. surtseyan and base-surge deposits, where bomb sags are common because of the finer grainsizes and the often wet, plastic state of the ash pile). Layers of agglutinated lava spatter and scoria can be conspicuous (Fig.

6.8). Complete welding-together of the clasts may occur, and this is one way in which lavas may be generated (Ch. 4). Rapid accumulation of spatter and scoria is needed to produce such agglutinated and welded layers, and clastogenic lavas (see Section 6.10 on welded air-fall tuffs).

Hawaiian activity produces a much higher proportion of lava spatter at the vent, due to lava fountaining. Consequently, the formation of spatter deposits, spatter cones and ramparts at the vent (Figs 6.8g) and lava flows is likely.

The downwind fall deposits are finer-grained and composed almost entirely of scoria (Figs 6.9 & 10), and are volumetrically small (Table 6.1). Closer to the vent, ballistic bombs will be found and planar stratification defining fall units may be prominent (Figs 6.10 & 5.4a). Deposits usually contain **achneliths**, which are juvenile fragments with smooth, glassy surfaces formed from solidified lava spray (G. P. L. Walker & Croasdale 1972; Section 3.5). These would include the pear-shaped forms called **Pele's tears**, although a wide variety of shapes are possible (see Figure 3.17); the most extreme form would be the filaments of basaltic glass known as **Pele's hair** (Duffield *et al.* 1977). Achneliths are especially common in hawaiian scoria-fall deposits. Eruption column heights and muzzle exit velocities during hawaiian and strombolian activity are low. Consequently, scoria-fall deposits usually have a limited dispersal (D is low) and the fragmentation of magma is low (F is low in Fig. 6.11).

Table 6.1 Volume estimates of the three strombolian scoria fall deposits in Figure 6.9 (excluding volumes of the cones).

Deposit	Volume (km³)	DRE* (km³)
Galiarte	0.02	0.01
Serra Gorda	0.06	0.03
Cone 301	0.02	0.01

* Dense rock equivalent used for these basaltic deposits is 3.0 g cm^{-3}.

6.3.2 MECHANISMS AND DYNAMICS

In hawaiian activity the eruption column is essentially a lava 'fire' fountain formed when jets of disrupting magma are released, almost continuously

Figure 6.6 Cone-building strombolian scoria-fall deposits. (a) Mt Leura, Victoria, Australia and (b) Megalo Vourno, Santorini.

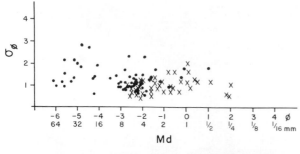

Figure 6.7 Md_ϕ/σ_ϕ plot for some strombolian pyroclastic fall deposits. Solid circles are samples collected from scoria cones, and crosses are from downwind fall deposits. (After G. P. L. Walker & Croasdale 1972, with additions for cone deposits after Houghton & Hackett (1984), and J. V. Wright *unpub. data* from Santorini.)

in some cases, through the vent. Lava fountain heights are generally less than about 200 m (MacDonald 1972), and in such cases magma would be ejected at velocities of a few tens of metres per second (L. Wilson & Head 1981). The predominant products of these lava fountains are large spatter pieces which fall back around the vent area. Poorly developed convective plumes above lava fountains may take the smallest ash-sized particles derived

Figure 6.8 (opposite and above) Agglutinated and welded deposits from scoria cones and a spatter rampart. (a) Red Rock Complex, Victoria, Australia. Non-vesicular, banded zonation represents oxidised margins of welded spatter fragments. Interiors have vesiculated (photograph by R. Allen). (b) Coherent incipiently agglutinated scoria clasts, Mt Leura, Victoria, Australia. (c) The largely quarried strombolian cone at Ohakune, New Zealand, craters with the two agglutinated and densely welded layers shown in Figure 6.5 occurring directly below each of the benches. (d) and (e) Densely welded scoria in the cone at Balos, Santorini. Note the columnar jointing and welding zonation in (d). (f) and (g) Agglutinated lava spatter from part of a spatter rampart at the Sproul in the San Francisco volcanic field, Arizona.

from lava spray up to heights of a few hundred metres, but all coarser fragments will already have fallen out of the column.

The mechanisms and dynamics of strombolian activity have been discussed by E. Blackburn et al. (1976), L. Wilson (1980a) and L. Wilson and Head (1981). Eruptions consist of a series of discrete time transient explosions separated by periods of less than 0.1 s to several hours. Explosions are thought to be generated when one or a number of large gas bubbles (<1 to >10 m in diameter) burst the magma surface (of a lava lake) at vent (E. Blackburn et al. 1976; Fig. 6.12a). These types of explosions can only occur in low-viscosity magmas in which bubbles can rise relatively rapidly and expand. Explosions are driven by the excessive pressure within each bubble. When each one bursts at the surface, it blasts off as pyroclasts the fragmented remains of the magma which formed the upper skin of the bursting bubble (E. Blackburn et al. 1976, L. Wilson 1980a). If there is a pause in activity or, as in the waning stages of an eruption, there is a pause in the activity and a crust has time to form on the magma surface, then this may be ejected during renewed bubble burst events (Fig. 6.12b). This mechanism may account for the slabby lava blocks found in some deposits (Figs 6.5 & 10b).

The pressure in the bursting bubbles is related to their size and the history of their rise through the magma, both of which, in turn, are governed by the physical properties of the magma (Ch. 2). Theoretical analysis (L. Wilson 1980a) and observed activity (Chouet et al. 1974, Self et al. 1974, E. Blackburn et al. 1976) suggest maximum initial gas velocities in these strombolian explosions of 300 m s^{-1}. In their analysis of 15 explosions from film of the Heimaey eruption in 1973, E. Blackburn et al. (1976) found the maximum initial velocity in one burst was 230 m s^{-1}, but the mean was 157 m s^{-1}. Generally, much lower initial velocities (<100 m s^{-1}) were observed in the activity of Stromboli in 1971 and 1975 (Chouet et al. 1974, E. Blackburn et al. 1976). Initial high gas thrust velocities rapidly decrease with height (up to heights of a few tens to one or two hundred metres), above which particles are transported in the upper part of the eruption column driven by convection

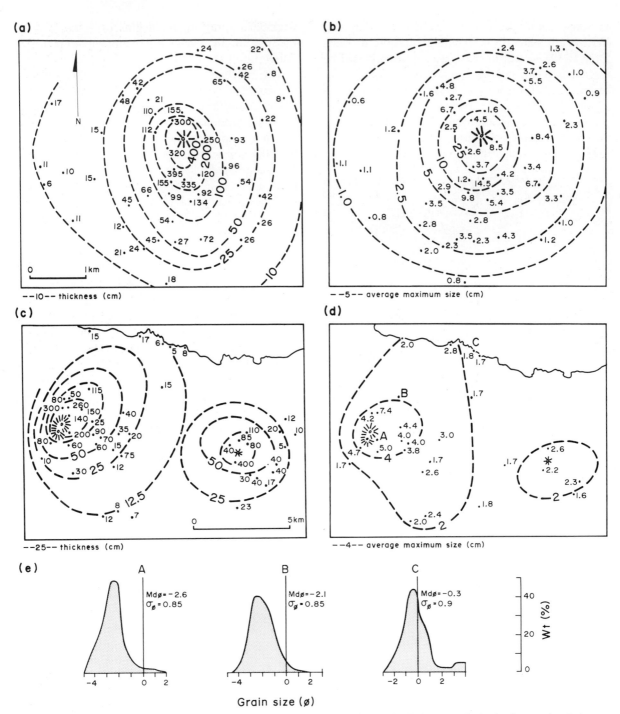

Figure 6.9 Thickness and grainsize characteristics of some strombolian pyroclastic fall deposits in the Azores. Isopleth maps show the average diameter of the three largest scoria clasts. (a) and (b) Scoria-fall deposit from the Galiarte cone, Terceira (after Self 1976). (c) and (d) Scoria-fall deposits from Serra Gorda (west) and Cone 301 (east) on São Miguel. (e) Grainsize distribution curves for the Serra Gorda deposit at the three locations in (d) (after Booth *et al.* 1978). Volumes for the three scoria-fall deposits are given in Table 6.1.

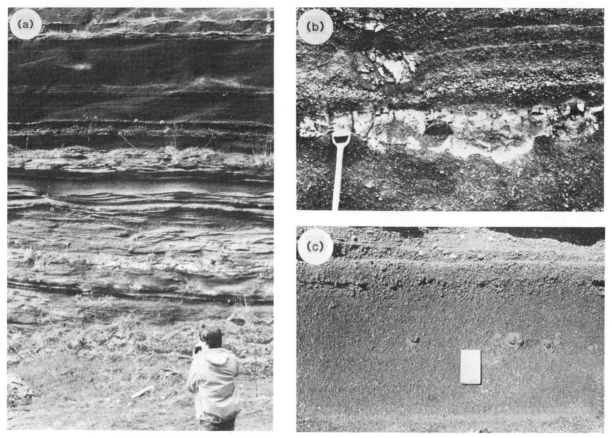

Figure 6.10 (a) Faintly stratified black scoria, Mt Leura, Camperdown, Victoria, Australia. Scoria overlies phreatomagmatic base-surge and fall deposits. (b) Cognate basaltic bomb in scoria, Mt Leura. (c) Slight reverse grading and faint stratification in scoria fall, Tower Hill, Victoria, Australia.

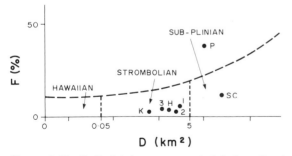

Figure 6.11 $D–F$ plot for some scoria-fall deposits described in the text. 1, 2 and 3 are the downwind deposits for the Galiarte, Serra Gorda and 301 cones. H is Heimaey (1973); K is Kilauea (1959); P is Parícutin; SC is Sunset Crater (see Section 6.5 and Plate 5). (After G. P. L. Walker 1973b, Self 1976, Booth *et al.* 1978, Amos & Self *unpub. data* and sieve data of J. V. Wright on the Kilauea 1959 scoria.)

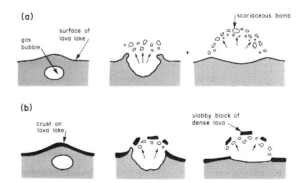

Figure 6.12 Three stages depicting the rise, expansion and bursting of gas bubbles for two contrasting situations during strombolian eruptions. (After L. Wilson 1980a.)

(E. Blackburn *et al.* 1976). If explosions occur in rapid succession (e.g. every 1–2 s), then a maintained eruption column, driven by convection, could reach heights of 5–10 km, as observed in the 1973 Heimaey eruption (see E. Blackburn *et al.* 1976). When explosions occur at longer intervals (e.g. several minutes), the convection cloud remaining after each gas thrust phase may have dissipated before the next explosion, as observed at Stromboli in 1975 (E. Blackburn *et al.* 1976). In this type of activity a fine-grained, well stratified scoria fall deposit of more limited dispersal could be built up.

6.3.3 CLASSIFICATION

The distinction between hawaiian and strombolian pyroclastic fall deposits was only tentatively defined by G. P. L. Walker (1973b) because there were limited data available at that time. There are still very few quantitative data on these deposits, especially those of hawaiian eruptions. Following G. P. L. Walker (1973b), hawaiian basaltic activity is so weakly explosive that any pyroclastic deposit which results, has a D of less than 0.05 km^2, while strombolian activity produces a deposit with a D of more than 0.05 km^2 (Fig. 6.2). This criterion, together with the distinction between the eruption mechanisms we have discussed, should only be considered as a general guide in distinguishing between hawaiian and strombolian fall deposits. Lava fountaining can reach such heights that, although observed activity would be considered typically hawaiian, the resulting deposits would be much more widely dispersed. The 1959 Kilauea Iki lava fountains reached heights of 600 m and the downwind scoria-fall deposit has a D-value of about 0.7 km^2 (Richter *et al.* 1970). By Walker's definition it is strombolian (Fig. 6.11). However, this scoria-fall deposit is composed almost entirely of achneliths or fragments of them (Fig. 3.17), and these should be a very large component of scoria-fall deposits, even of those of wide dispersal, formed by lava fountaining. During strombolian activity, if the intervals between explosions are so long that a maintained column and convective plume cannot form, then the scoria-fall deposit will be more restricted, and may be hawaiian

in its dispersal characteristics. In these cases the scoria fragments will more commonly be ragged with stringy surfaces, and more typical of the strombolian mechanism of disruption of magma.

Many eruptions will also vary in observed style from hawaiian to strombolian, and vice versa. The 1973 Heimaey eruption began with lava fountains rising 50–100 m from up to 20 vents along the length of a 1.5 km fissure (Thorarinsson *et al.* 1973, Self *et al.* 1974). Later, activity became centralised and strombolian explosions took place from three vents at the northern end of the fissure, and built a scoria cone 200 m high, from which lava continually flowed. Perhaps where detailed analysis of a deposit is possible (i.e. exposure allows many vertical sections of a deposit to be studied) the dispersal characteristics of individual fall units corresponding to different phases could be determined. However, in most cases, it is only possible to determine the finite characteristics of a deposit, which in the case of Heimaey, show it is typically strombolian (Fig. 6.11).

Finally, G. P. L. Walker (1973b) described some scoria-fall deposits with unusually high values of F as 'violent strombolian'. The scoria-fall deposit erupted from Paricutin volcano, Mexico, is of this type (Fig. 6.11). Activity during this eruption continued sporadically for nine years. Ground water possibly gained access to the magma at times, but not in sufficient amounts to produce a surtseyan fall deposit (Section 6.8).

6.4 Plinian

Plinian pyroclastic fall deposits are a common product of highly explosive eruptions of high viscosity magmas. These are generally andesitic to rhyolitic, or phonolitic and trachytic compositions, but rare basaltic scoria-fall deposits which have plinian dispersal patterns are known (S. N. Williams 1983, G. P. L. Walker *et al.* 1984). The characteristics of plinian pumice fall deposits and their eruptions are now fairly well defined, and the extensive literature about these has been reviewed by G. P. L. Walker (1981b).

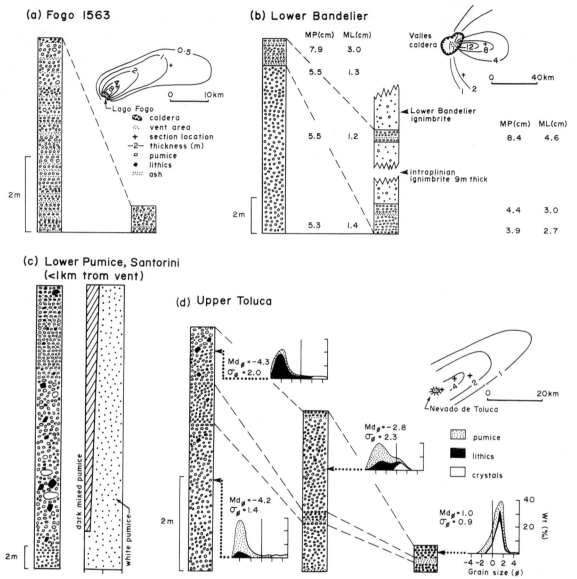

Figure 6.13 Some features of plinian pumice-fall deposits. ((a) After G. P. L. Walker & Croasdale 1971, (b) Self & Wright *unpub. data*, (c) J. V. Wright *unpub. data* and (d) Bloomfield *et al.* 1977.)

6.4.1 GENERAL CHARACTERISTICS

During plinian eruptions large volumes of pumice are ejected and extensive pumice-fall deposits are produced. These are the most impressive types of pyroclastic fall deposit found in the field (Figs 6.13 & 14). The deposits of individual eruptions may attain thicknesses near vent of 10–25 m (Fig.

6.13c); maximum thicknesses can be much smaller, but would rarely be greater. Near-vent deposits are generally homogeneous and can be very coarse, containing large pumice fragments of several tens of centimetres, and metre-sized ballistic lithic blocks. More rarely, larger pumice bombs are found, but in most cases fragmentation on impact destroys these. Away from the vent deposits become thin and fine-

grained, and change in character, and it is the documentation and analysis of these changes that can be so important in making volcanological interpretations. Plinian pumice-fall deposits are a common eruptive product of all large rhyolite volcanoes, but are also frequently found as products of a range of andesitic and alkaline stratovolcanoes.

A few plinian eruptions are known to have occurred this century, and examples that are well documented are the eruptions of Hekla, Iceland, in 1947 (Thorarinsson 1954, 1968) and Santá Maria, Guatemala, in 1902 (Rose 1972b, S. N. Williams & Self 1983). Another example is the 1932 eruption of Quizapú in the Chilean Andes (Larsson 1936), but there has not been a more recent study of the deposits of this eruption. Quite a number of plinian eruptions have occurred in earlier historic times, and the deposits of the following examples have received detailed attention: the eruptions of Askja, Iceland in 1875 (Sparks et al. 1981), Tarumai, Japan in 1667 (T. Suzuki et al. 1973), Fogo, São Miguel, the Azores in 1563 (G. P. L. Walker & Croasdale 1971; Fig. 6.13a) and, better known because of their archaeological significance, Vesuvius in AD 79 (Lirer et al. 1973) and the Minoan (1470 BC) eruption of Santorini (Bond & Sparks 1976). Historical records have complemented some of these studies, and have provided evidence of the duration of these events. As well as these, much of our data on this type of activity stems from a number of studies on older plinian deposits which abound in the Quaternary record, e.g. Booth (1973), Bloomfield et al. (1977), Booth et al, (1978), G. P. L. Walker and Croasdale (1971), G. P. L. Walker (1981c), G. P. L. Walker et al. (1981d), and see the review of G. P. L. Walker (1981b).

All of the historical plinian eruptions mentioned above seem to have taken place from central vents, and most of the older deposits studied have been mapped to 'circular vents', which are generally assumed to be located above cylindrical conduits. There is no doubt that some of these vents are aligned along linear fissures or ring fractures, but activity from different vents in many cases can be shown to be separated by long intervals, recognised, for instance, by soils between their different fall deposits. However, detailed mapping by Nairn (1981) in the Okataina rhyolitic centre in New Zealand has shown that many plinian fall deposits and associated ignimbrites were erupted in simultaneous or sequential activity from multiple vents along fissures. These eruptions were often spread along lengths of fissure more than 10 km long but, as in basaltic fissure eruptions, activity seems to have been restricted to definite points. Vent types for large explosive silicic eruptions, during which plinian activity may lead to ignimbrite formation, are discussed in Chapter 8.

When mapped out, plinian pyroclastic fall deposits are extensive sheet-like deposits. They have a large dispersal, and D is >500 km^2 (Fig. 6.2). However, fragmentation of the magma is only moderate, and F is small to medium. Sizes of ballistic lithic blocks near the vent imply that muzzle velocities of 400–600 m s^{-1} occur (L. Wilson 1976; App. I). These suggest that very high rates of magma discharge are possible which in turn lead to the 'stoking up' of very high eruption columns, and evidence suggests that column heights >20 km should be common during this type of activity (L. Wilson 1976, L. Wilson et al. 1978; Ch. 5). In the events this century, the height of the column of the 1947 Hekla eruption reached 24 km, and that of Santá Maria in 1902 reached at least 28 km (S. N. Williams and Self 1983; Table 5.1). It

◀ **Figure 6.14**   Plinian pumice-fall deposits. (a) The very impressive Lower Bandelier plinian deposit 30 km downwind from the vent. This is the section illustrated in Figure 6.13b; overlying the stratified top of the pumice-fall deposit is ignimbrite. (b) Upper Bandelier plinian deposit; note the finely stratified fall unit at the base. (c) Plinian fall deposit at the base of the Bishop Tuff, California. Darker layer is a surge deposit which is overlain by ignimbrite. (d) Plinian deposit erupted 26 000 years BP from the Okataina rhyolitic centre, New Zealand. (e) Compositionally zoned pumice fall on Tenerife; white pumice is phonolitic, dark (arrowed) is latite which is overlain by a soil. Hammerhead rests on base of the deposit (photograph by J. A. Wolff). (f) Reversely zoned basaltic to rhyolitic plinian deposit erupted 17 000 years BP from Tarawera, New Zealand. (g) Distal plinian layer deformed by soft-sediment loading; this was erupted from Hekla volcano and is stratigraphically below the two deposits shown in Plate 6.

Table 6.2 Volume estimates of some plinian deposits (highlighting some of the largest known in modern volcanic successions).

Deposit	Volume (km³)	DRE* (km³)	Composition
Shikotsu	100	24	rhyolite
Lower Bandelier	100	24	rhyolite
Upper Bandelier	70	17	rhyolite
La Primavera B (95 000 years BP)	50	14	rhyolite
La Primavera D	2	0.6	rhyolite
La Primavera E	2.6	0.7	rhyolite
La Primavera J	12	3.4	rhyolite
Upper Toluca (11 600 years BP)	9	4.0	andesite
Waimihia (3400 years BP)	29	7.1	rhyolite (mixed)
Hatepe (AD 186)	6.0	1.5	rhyolite
Vesuvius AD 79 (Pompeii pumice)	6.0	1.4	phonolite (zoned)
Askja 1875	1.0	0.2	rhyolite (mixed)
Santá Maria 1902	20	8.5	dacite (mixed)
Hekla 1947	0.4	0.1	andesite
Mt St Helens 1980	1.1	0.2	dacite

* A dense rock equivalent of 2.5 g cm^{-3} is used.
Data are taken from G. P. L. Walker (1981b, c) except the Bandelier plinian deposits (Self & Wright *unpub. data*), Santá Maria (S. N. Williams & Self 1983) and Mt St Helens 1980 (Sarna-Wojcicki *et al.* 1981). All volumes except Mt St Helens 1980 are calculated from the area plots shown in Figure 6.18 and the method of G. P. L. Walker discussed in Appendix I.

is the high eruption column in this type of activity that leads to the wide dispersal of plinian deposits. Volumes of plinian-fall deposits range from about 0.1 to >50 km³ (Table 6.2). Examples of small-volume deposits would be those from Hekla in 1947 and Mt St Helens in 1980. Much larger deposits are known further back in the record: some of the biggest are the Shikotsu pumice deposit in Japan and the Lower Bandelier plinian deposit, both about 100 km³. Volumes can be estimated from isopach maps of the deposits by various methods, and these are discussed in Appendix I.

6.4.2 INTERNAL AND LATERAL CHANGES

Many deposits at first sight appear to be fairly homogeneous, or at least their lower parts do (Figs 6.13 & 14), and this is thought to reflect continual fall-out from a downwind plume continually fed by a continuous gas blast. They are predominantly composed of juvenile material: pumice, glass shards and, when the magma is porphyritic, free crystals. However, significant departures from homogeneity can occur in detail. Reverse-grading of larger pumice fragments seems to be common, as is internal stratification. Accessory lithics, derived

from the conduit wall, can be important in certain parts of a deposit. Also, a number of plinian fall deposits are now known to be compositionally zoned, or to contain mixed pumice clasts (Figs 6.13c, 14e & f).

Reverse grading of larger pumice clasts has been described from a number of deposits (e.g. Lirer *et al.* 1973, Bond & Sparks 1976, Bloomfield *et al.* 1977). This is usually more likely to be found at distances of several to a few tens of kilometres from the vent, and outside these limits may be only slightly developed, or it may not occur. Nearer the vent (<5 km) deposits are often so coarse-grained that it is very hard to detect any grading, and more distally (>50 km) deposits may be too fine-grained. Many deposits also show an upward increase in the proportion and size of accessory lithic clasts. Reverse grading of both pumice and lithics must reflect some process occurring at the vent affecting the eruption column with time. This will be considered further in Section 6.4.3.

Internal stratification also occurs. It is usually best developed towards the vent, and is commonly and significantly found towards the top of some deposits (Figs 6.13 & 14). Further away from the vent such stratification disappears. Stratification varies from a crude internal layering to distinct mappable fall units, although dividing deposits up into fall units can be rather subjective.

There are a number of causes for the development of stratification. Eruptions are probably not truly continuous, but pulse-like. Slight fluctuations in muzzle velocity and discharge rate will cause particles of a given size and terminal fall velocity to be released from different heights. A pulsating column such as this would produce faint layering in a fall deposit (e.g. Fig. 6.14d); this type of stratification would rapidly disappear away from the vent as pyroclasts quickly mixed downwind. More-significant changes in activity are probably needed to cause distinct fall unit breaks. Activity could temporarily cease, caused, for instance, by blockage of the conduit by collapse of the vent walls. During such breaks, fine ash may settle out from the previous column as it dissipates, forming a thin, discrete fine-grained fall unit overlying the deposit of the maintained column. When activity recurs, the next fall unit may at first be rich in lithic fragments, as lithic fragments that blocked the vent are reamed out. Smaller collapses of the vent wall may just cause an increase in the amount of lithics taken into the column, and these would be recognised as a layer of lithics, perhaps within a fall unit. Obviously, there are a number of scenarios that can be considered. All of the above mechanisms involve changes at the vent affecting the behaviour of the eruption column. However, stratification can also be generated away from the vent. Rain flushing could prematurely bring down fine ash from the plume in localised areas while the eruption continues. Such fine-grained fall units will have isopachs which do not close on the vent (their distribution and thickness is related to the distribution and intensity of the rain shower), and they may contain accretionary lapilli (G. P. L. Walker 1981d).

In the above we have only discussed the generation of stratification by the 'random' processes which could occur at any time during a plinian event. The stratification and fall units found at the top of some deposits may, however, be related to more-significant changes that are developing with time as the plinian eruption continues. Sometimes pyroclastic flow and surge deposits are found interbedded between separate fall units, particularly in proximal areas (e.g. Self *et al.* 1984). The deposits of these flows can be traced laterally into fine ash-fall deposits. In many cases plinian deposits are overlain by ignimbrite, and it seems that there is a continuum from a plinian eruption column to a pyroclastic flow-forming or collapsing column. The stratification at the top of many plinian deposits may reflect instabilities in the column before wholesale collapse occurs to generate ignimbrite.

Compositionally zoned plinian pumice-fall deposits are now known to be common, and invariably these show an upward vertical increase in the proportion of a more basic juvenile component (Figs 6.13c & 14e). Rarer examples are known where this type of zonation is reversed (Fig. 6.14f). The boundary between zones can be gradational or very abrupt. This is commonly not marked by grainsize differences, showing that the discharge of magma was steady, although the composition of the magma was changing. In some cases there is an almost complete change in magma types, in others there is just a slight change in the proportion of the two types. Streaky mixed-pumice clasts are common in some deposits, and indicate mechanical mixing of the magma types before eruption (e.g. Fig. 3.22). Some deposits, although not distinctly zoned, can have a significant proportion of mixed pumice, e.g. the Askja 1875 deposit (Sparks *et al.* 1981). Aspects of the role of magma mixing in explosive volcanism have already been discussed in Chapter 3 (and see Ch. 8). It may be that injection of more basic magma into high-silica magma chambers triggered a number of plinian eruptions (Ch. 3).

Many of the overall downwind and lateral changes in the character of plinian fall deposits are now well established. These are:

(a) decrease in thickness,
(b) decrease in maximum grainsize (pumice and lithics),
(c) decrease in median grainsize (increasing M_ϕ values),
(d) increase in sorting (decreasing σ_ϕ values),
(e) changes in component population and
(f) decrease in median terminal fall velocity,

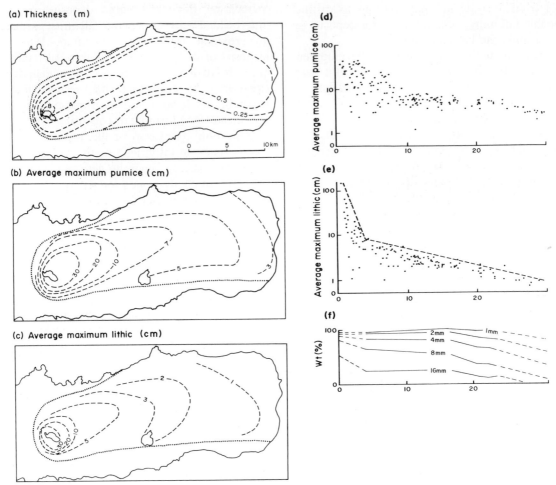

Figure 6.15   The 1563 Fogo plinian pumice-fall deposit erupted from the volcano Agua de Pau (with a caldera now occupied by Lago Fogo) on São Miguel (see also Fig. 6.13a). (a) Isopach map. (b) and (c) Maximum size isopleth maps using the average diameter of the three largest clasts for pumice and lithics, respectively. (d)–(f) Maximum pumice, lithic and total grainsize variation with distance from vent. (After G. P. L. Walker & Croasdale 1971.)

There are various forms in which these data are presented, e.g. maps or graphically (Figs 6.15–6.19). However, all illustrate the dispersal characteristics of the fall deposit.

The best way of comparing thickness, maximum grainsize and median grainsize is to make 'area plots' (G. P. L. Walker 1980, 1981b, G. P. L. Walker *et al.* 1981d; Figs 6.18 & 19). These show the areas enclosed by isopachs of thickness, or isopleths of maximum or median grainsize. These plots are useful for comparing the dispersal charac-

teristics of different plinian deposits, and can also be used to classify different types of pyroclastic fall deposits instead of using *D–F* plots. In practice, it is easier to determine the areas enclosed by maximum grainsize isopleths (Fig. 6.19) than determining *D–F* values. The value of *D* is sensitive to the choice of T_{max} (Fig. 6.1b), which may have to be extrapolated from data collected at locations some distance from the vent. G. P. L. Walker (1981b) indicated that when categorising a deposit as plinian or not, more reliability should be

(a) Thickness

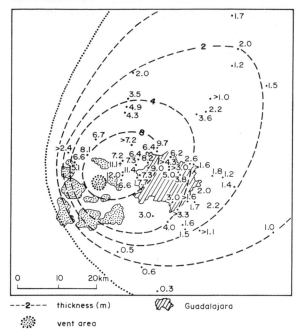

(c) Average maximum lithic

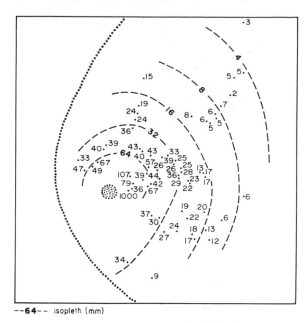

(b) Average maximum pumice

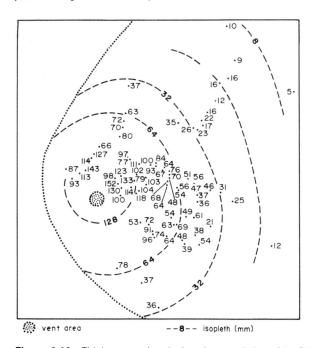

(d) Median diameter

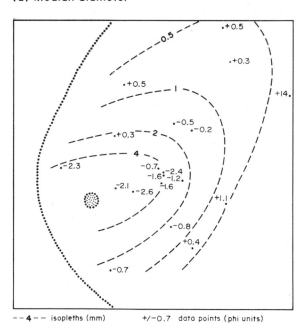

Figure 6.16 Thickness and grainsize characteristics of La Primavera B plinian pumice-fall deposit erupted from La Primavera rhyolite volcano, Mexico. (After G. P. L. Walker *et al.* 1981d.)

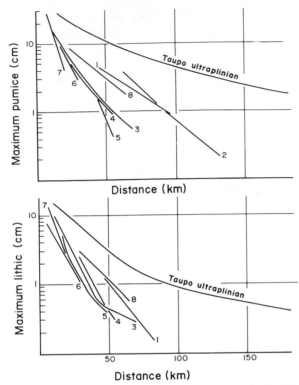

Figure 6.17 Variation in maximum pumice and lithic diameter (average of the largest three or five clasts) with distance from vent for some plinian pumice-fall deposits and the Taupo ultraplinian deposit. 1 Shikotsu; 2 Askja (1875); 3 Pompeii; 4 La Primavera B; 5 Upper Toluca; 6 Fogo (1563); 7 Fogo A; 8 Lower Bandelier. (After G. P. L. Walker 1980 and Self & Wright *unpub. data* on the Lower Bandelier plinian deposit.)

placed on the plinian field in area plots of isopleths maximum pumice (MP), maximum lithics (ML), and median (Md) grainsizes than on *D–F* plots.

6.4.3 MECHANISMS AND DYNAMICS

From observations of historic eruptions and analysis of plinian-fall deposits, coupled with theoretical analysis and modelling, a large amount is known about the mechanisms and dynamics of this type of eruption. The development of ideas on plinian eruption mechanisms can be traced in a number of papers based largely on the work of Lionel Wilson (L. Wilson 1976, 1980a, L. Wilson *et al.* 1978, 1980, Sparks 1978a, and Sparks and L.

Wilson 1976). Plinian eruptions are essentially relatively steady, high-energy events in which a continuous, turbulent flow of fragmented magma and gas is released through a conduit to the atmosphere.

We have already discussed how fragmentation of magma occurs during this type of eruption in Chapter 3 (Fig. 3.4). Gas bubbles in rising salic magma nucleate and grow until the volume occupied by bubbles has increased (by pressure decrease and gas exsolution) to a critical value of about 70–80%, when magma disrupts (Sparks 1978a). Rapid acceleration of the disrupted magma then occurs through the conduit, which is essentially a fracture propagated to the Earth's surface from the magma chamber. The maximum velocity of the mixture as it leaves the vent is a function of gas pressure at the

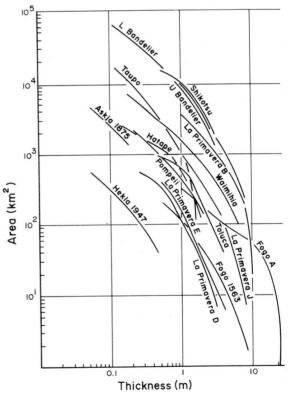

Figure 6.18 Plot of the area enclosed by each isopach against thickness for some plinian fall deposits and the Taupo ultraplinian deposit. (After G. P. L. Walker 1980, 1981b and Self & Wright *unpub. data* on the Bandelier plinian deposits.)

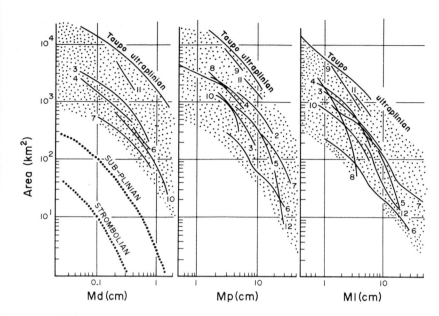

Figure 6.19 Plot of the area enclosed by isopleths of median grainsize (Md), maximum pumice diameter (Mp) and maximum lithic diameter (Ml). Stipple is field of plinian deposits. Deposits are labelled as in Figure 6.17 with the following additions: 9 Upper Bandelier; 10 La Primavera J; 11 Waimihia; 12 Minoan. (After G. P. L. Walker 1980, 1981b and Self & Wright *unpub. data* on the Bandelier plinian deposits.)

fragmentation level, which is the depth to the free surface of the magma where fragmentation is taking place (Fig. 3.4; and see L. Wilson 1980a for a detailed analysis). The theoretical models of L. Wilson (1980a) predict maximum plinian eruption velocities of 600 m s^{-1}, which would agree with maximum velocities deduced from the sizes of the largest ballistic clasts ejected in these eruptions. These exit velocities indicate that the volumetric discharge rates of magma can be as high as 10^6 m^3 s^{-1} (dense rock equivalent), which are substantially greater than in observed historic

Table 6.3 Estimated muzzle velocities and volumetric eruption rates of some plinian eruptions.

Eruption	Maximum muzzle velocity (m s^{-1})	Average volumetric eruption rate (m^3 s^{-1})
Upper Toluca	500	4.4×10^4
Minoan	330	2.8×10^4
Vesuvius AD 79	>225	1.6×10^4
Askja 1875	380	8.5×10^3
Fogo 1563	415	1.8×10^3
Santá Maria 1902	>270	1.2×10^5

Volumetric eruption rates are given in terms of dense rock equivalent.

Data from Bloomfield *et al.* (1977), Sparks *et al.* (1981), L. Wilson (1976, 1978, 1980b), G. P. L. Walker (1981b), S. N. Williams and Self (1983).

Table 6.4 Estimated durations of some plinian eruptions.

Eruption	Duration (h)	Source
Upper Toluca (11 600 years BP)	20–30	theoretical analysis
Minoan 1470 BC	20–40	theoretical analysis
Vesuvius AD 79	~24	historical records
Fogo 1563	~48	historical records
Askja 1875	6.5	historical records
Hekla 1947	1	observation
Mt St Helens 1980	9	observation

Data taken from Bloomfield *et al.* (1977), Sparks *et al.* (1981), L. Wilson (1976, 1978, 1980b).

eruptions (Table 6.3). A continuous gas blast can probably not be sustained for a long time, and from the available data typical durations vary from one hour to one day (Table 6.4). The 1563 Fogo eruption lasted up to about two days overall, but its plinian eruption phase was interrupted during this time, and the deposits contain interbedded small ignimbrite flow units and other layers. The duration of large ignimbrite-forming eruptions which sometimes follow an initial plinian phase can be much longer (Ch. 8).

As a plinian eruption proceeds, we can predict that two things will generally happen with time:

(a)   deeper, and more gas-depleted, levels of the magma chamber will be tapped and

(b) widening of the vent by wall erosion will occur.

The effect of (a) is to reduce the gas velocity of the eruption column with time. The effect of (b) is to increase the mass discharge rate with time, and this will produce a column that steadily grows in height. Both (a) and (b) cause the effective density of a plinian column to increase steadily. This can continue until at some stage the density of the column becomes greater than that of the atmosphere, when gravitational collapse will occur to generate ignimbrite-forming pyroclastic flows. Models have shown that various combinations of magmatic gas content, gas velocity and vent radius produce convecting columns and others produce collapsing columns (L. Wilson 1976, Sparks & L. Wilson 1976, Sparks et al. 1978, L. Wilson et al. 1980; Fig. 6.20), and from these we can therefore predict when eruption column collapse will occur. Columns formed from magmas with high gas contents (>5 wt% water) are likely to show convective motion, whereas those with low gas contents (<1 wt% water) will form collapsing columns. In magmas with intermediate gas contents, collapse will occur when the vent radius exceeds a value defined in Figure 6.20. However, not all plinian eruptions continue to the collapse or ignimbrite-forming stage, and there are many examples of plinian deposits without associated ignimbrites – e.g. the 1875 Askja plinian deposit (Sparks et al. 1981), the 1902 Santá Maria deposit (S. N. Williams & Self 1983) and the Upper Toluca plinian deposit in Figure 6.13d.

With this theoretical background, we can now explain two common features of plinian fall deposits that have been described: reverse grading, and stratification in the upper parts of deposits. The models of L. Wilson et al. (1980) suggest that the major cause of reverse grading in plinian-fall deposits is vent-widening by wall erosion during the eruption. As an eruption continues and the vent widens, the mass discharge rate increases and, because of the increase in mass and energy flux, increased convective velocities will raise the height of the eruption column. Particles of a given size will be taken to increasing heights in the column before

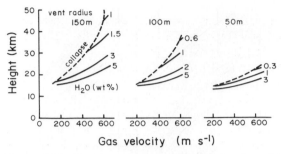

Figure 6.20 Eruption column height as a function of the muzzle gas velocity for three different vent radii. In each case curves are given for constant values of magmatic water content. Column collapse occurs for combinations of values to the left of the broken line. (After L. Wilson 1976.)

being released, and will then be transported downwind from the vent during the eruption. The increased proportion of larger clasts downwind with time will build up a reversely graded deposit. Shifts in the wind direction or speed could also have this effect, but these variables should also produce just as many examples of normally graded plinian fall deposits. This is therefore not a general mechanism to explain the common occurrence of reverse grading in many deposits. Local reverse grading could also be found in falls deposited in water, or on very steep slopes followed by secondary mass (grain) flow (see Duffield et al. 1979).

Widening of the vent, together with an increased rate of erosion during the eruption, also explains why many plinian fall deposits show a vertical increase in the proportion of accessory lithics. An estimate of the lithic content therefore indicates the amount of wall erosion and the size of conduit. For example, the Fogo A plinian pumice deposit on São Miguel contains 0.09 km^3 of lithic fragments (14 wt%), and if the magma source was at a depth of 5 km this would be equivalent to a hypothetical cored-out cylindrical conduit of diameter 78 m. However, erosion is likely to be more important near the surface, where rocks are weak and less consolidated, and flaring of the vent is therefore probably likely (L. Wilson et al. 1980).

The explanation of the stratification observed at the tops of many plinian deposits seems to be that it is caused by instabilities in a column nearing the point of collapse. Changes in wind direction and speed could cause stratified layers, but their common occurrence at this level, and the presence of interbedded pyroclastic flows and surges, suggests a more general mechanism, as with reverse grading. Any changes in gas velocity or mass discharge rate in a column verging on collapse will have pronounced effects on its behaviour. Small collapse events that generate pyroclastic flows and surges may occur, for instance, with a sudden increase in mass discharge rate. A convective column could then be re-established with a slight increase in gas velocity due to a small increase in gas content of the magma. Choking of the vent by collapse of the walls will also reduce mass discharge rate, re-establishing a convecting column, but after this lithic debris has been removed by ejection the wider vent will promote collapse of the column. A complex sequence of activity and of pyroclastic deposits could therefore be generated before massive collapse of the whole column occurs, leading to a major ignimbrite-forming eruption.

6.5 Sub-plinian

These are pumice-fall deposits which resemble plinian deposits in the field, especially near the vent, but when mapped out have a smaller dispersal and are small volumetrically. G. P. L. Walker (1973b) set arbitrary D limits for them of between 5 and 500 km^2 (Fig. 6.2). They are a common type of pyroclastic fall deposit, although only a few specific descriptions occur in the literature. This is because studies of pumice fall deposits have generally concentrated on the larger, more-dramatic examples which are usually plinian in their F and D characteristics. However, Self (1976) described a number of sub-plinian fall deposits on Terceira in the Azores (e.g. Fig. 6.21) and Booth et al. (1978) documented examples on São Miguel. Sub-plinian pyroclastic fall deposits are a product of rhyolite volcanoes and stratovolcanoes, like their larger plinian counterparts. Many form during an early explosive phase accompanying the effusion of a small rhyolite dome or coulée, as do the examples on Terceira. However, plinian deposits can also be erupted in this situation.

Sub-plinian eruptions are scaled-down plinian eruptions, and their mechanisms and dynamics can be treated as essentially the same (L. Wilson 1976, 1980b). Large lithics indicate that in some eruptions muzzle velocities are as high as in some plinian events (>400 m s^{-1}), although the lower range is 100 m s^{-1} (L. Wilson 1976). Mass discharge rate is likely to be lower for sub-plinian events, and this is the main factor controlling eruption column height and dispersal. The sub-plinian pumice-fall deposits on Terceira are well-stratified and Self (1976) suggested that there were large fluctuations in the gas exit velocity, and hence mass discharge rate. This would also inhibit the development of a fully maintained convective plume, which would therefore not attain the heights associated with plinian columns. Sub-plinian eruptions can lead to the generation of ignimbrite-forming pyroclastic flows similar to the larger plinian ones. The examples mentioned above from the Azores do not show this eruption sequence. However, it is shown in the eruption of Krakatau in 1883. A pumice-fall deposit which preceded an ignimbrite erupted at Krakatau is sub-plinian rather than plinian in its characteristics (Self & Rampino 1981).

A number of basaltic or near-basaltic scoria fall deposits are now also known to be sub-plinian in their dispersal characteristics, rather than strombolian. G. P. L. Walker (1973b) cites as an example the 1970 Hekla eruption, and another example is the scoria-fall deposit erupted with the formation of Sunset Crater (AD 1065) in the San Francisco volcanic field, Arizona (Amos et al. 1981; Fig. 6.11 & Plate 5). As well as producing a very widely dispersed scoria-fall deposit (dense rock equivalent, DRE = 0.30 km^3) the Sunset Crater eruption also built a scoria cone 300 m high (DRE = 0.15 km^3), and in this respect is more typically strombolian in its character. For such widely dispersed scoria-fall deposits, one has to envisage a fully maintained convective plume which reached greater heights than in normal strombolian eruptions. The gas thrust part of the column in the Sunset Crater eruption may have reached heights of several hundred metres, rather than the 50–200 m that is usual in normal strombolian eruptions (Amos et al. 1981). Such energetic basaltic eruption columns

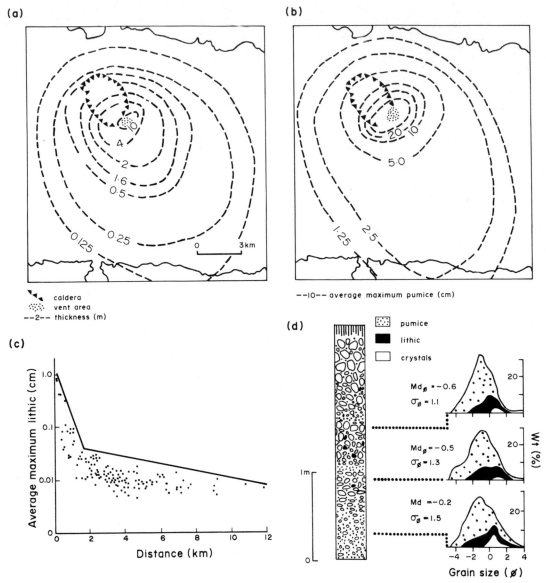

Figure 6.21 Sub-plinian pumice fall deposit from Terceira, Azores. (a) Isopach map giving thickness in metres. (b) Average maximum diameter of three largest pumice fragments in centimetres. (c) Average maximum diameter of three largest lithic fragments against distance from source. (d) Section 1.8 km SE of source on dispersal axis. Frequency curves of weight percentage against grainsize show proportions of pumice, crystals and lithics for three sieved samples, together with vertical variation in σ_ϕ and Md. (After Self 1976.)

(which in some examples are also known to have attained plinian proportions, see earlier) may have formed through a combination of a relatively high magma discharge rate and high initial magmatic gas contents.

6.6 Ultraplinian

'Ultraplinian' has been recently introduced as a separate type by G. P. L. Walker (1980) to describe the most widely dispersed plinian fall deposits.

Published data are as yet only available for one deposit, the Taupo pumice, which forms part of the eruption sequence of the AD 186 eruption of Lake Taupo, New Zealand. The products of this eruption will be discussed in more detail in Chapter 8.

The Taupo pumice is currently the most widely dispersed pyroclastic fall deposit known. *D*- and *F*-values are much higher than for normal plinian deposits (Fig. 6.2) and clasts are dispersed over a much wider area (Figs 6.17–19). The Taupo pumice only has a maximum measured thickness of 1.8 m which, by comparison with most near-vent plinian deposits, is rather thin. Another feature is that the deposit is very enriched in free crystals. This results from the high degree of fragmentation, and from the loss of a large proportion of vitric material by aeolian fractionation. Data from crystal concentration studies on the Taupo pumice show that 80%, mainly glass, fell out to sea further than 220 km from the vent.

Because of the great column height, which is estimated to be >50 km (G. P. L. Walker 1980), the deposit is also thickest 20 km downwind from the vent. Thus in these situations the isopach map must be used with caution as an indication of the vent position. From other evidence, the vent for this eruption is known to be in Lake Taupo.

6.7 Vulcanian

Vulcanian pyroclastic fall deposits from individual eruptions are thin, small volume (<1 km³), stratified ash deposits which contain large ballistic bombs and blocks near to the vent, sometimes with breadcrusted and jointed surfaces (Figs 6.22–24). In composition they are usually intermediate (basaltic-andesite, andesite, dacite). They are common products of andesite and basaltic-andesite stratovolcanoes. However, they are usually so thin and fine-grained that they are soon eroded by wind and water. When eruptions continue for a few years, bedded sequences can be built up near the vent, but these are never likely to be of great thickness, e.g. less than 2 m of ash-fall deposits accumulated just 800 m downwind of the vent on Irazú volcano, Costa Rica, from the eruptions between March 1963 and mid-1964 (Fig. 6.23).

Vulcanian activity has been observed in a large number of historic eruptions: for example; Fuego

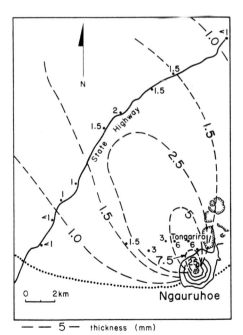

Figure 6.22 Isopach map of the vulcanian pyroclastic fall deposit erupted from Ngauruhoe on 28–29 March 1974. Note thicknesses are in millimetres. (After Self 1974.)

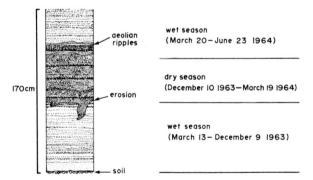

Figure 6.23 Section, dug 23 June 1964, through ash deposits accumulated from vulcanian activity at Irazú, Costa Rica, which began 13 March 1963. The location is just 800 m downwind of the vent. While many of the ash layers are the deposits of single ash falls, others represent layering produced mechanically by falling raindrops, sheet wash and aeolian reworking. For this reason ash deposited during the wet seasons appears to be better stratified. The prominent erosion surface results from a huge downpour on 10 December 1963. (After Murata *et al.* 1966.)

Figure 6.24 Ballistic blocks from the 1888–90 eruption of Vulcano. (a) Breadcrusted block. (b) Block having internal radial cooling joints.

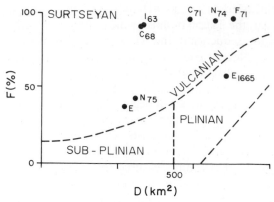

Figure 6.25 *D–F* plot showing the field of vulcanian deposits: C_{68} and C_{71}, eruptions of Cerro Negro, Guatemala, in 1968 and 1971; E, an old undated fall deposit of Mt Egmont, New Zealand; E_{1665}, fall deposit of the 1665 eruption of Mt Egmont; F_{71}, eruption of Fuego in 1971; I_{63}, eruption of Irazú in 1963; N_{74} and N_{75}, eruptions of Ngauruhoe in 1974 and 1975. (After J. V. Wright *et al.* 1980.)

of the material ejected with each explosion is not juvenile, but includes a large fraction of country rock as accessory lithics. At the onset of an eruption a plug of older, pre-existing lava may first be exploded out. However, during recently observed eruptions (for example, Fuego in 1974 and Ngauruhoe in 1975) coarser-grained scoria-fall deposits of more limited dispersal were produced during periods of more-intense, maintained explosions

volcano, Guatemala, which has had 25 vulcanian eruptions since 1944 (Rose *et al.* 1978, Martin & Rose 1981); the 1888–90 eruptions of Vulcano in the Aeolian Islands, which are the 'type example' of this activity (MacDonald 1972); the Irazú eruption (Murata *et al.* 1966); the 1976 eruption of Augustine volcano, Alaska (Kienle & Shaw 1979); and examples shown in Figure 6.25.

Activity during vulcanian eruptions proceeds as a number of discrete cannon-like explosions at intervals of commonly tens of minutes to hours. These short-lived explosions produce a series of small eruption columns (<5 to 10 km) with plumes strung out downwind of the volcano. Pyroclastic fall deposits are fine-grained with a wide dispersal (Fig. 6.25). Larger fragments in the column simply fall back around or into the vent, to be further fragmented and abraded. Commonly, a large part

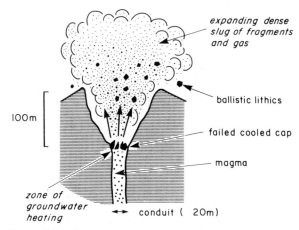

Figure 6.26 Schematic cross section through the crater of a stratovolcano at the time of a vulcanian explosion. Expanding 'cauliflower'-shaped slug rapidly decelerates during this gas thrust phase. (After Self *et al.* 1979.)

with continuous gas-streaming. Eruption columns reached heights of >10 to 20 km. This type of activity occurs intermittently with periods of short-lived explosions, and hence two types of fall deposit are formed during eruptions which have overall been termed vulcanian. The coarser scoria fall deposits seem to have similar fragmentation and dispersal indices to those deposits termed violent strombolian by G. P. L. Walker (1973b).

Small-volume pyroclastic flows are also frequently generated during vulcanian eruptions when large amounts of ejecta fall back around the vent. Very good descriptions of scoria flows associated with the 1974 Fuego and 1975 Ngauruhoe eruptions are given, respectively, by D. K. Davies *et al.* (1978a) and Nairn and Self (1978) (Fig. 6.26 and

see Ch. 5). However, not all vulcanian eruptions produce pyroclastic flows, e.g. Irazú (1963–5).

The mechanisms and dynamics of vulcanian explosions have most recently been described by Schmincke (1977), Nairn and Self (1978), Self *et al.* (1979) and L. Wilson (1980a). Self *et al.* (1979) and L. Wilson (1980a) have critically evaluated the energy equations used in previous studies for the analysis of this type of explosion (e.g. Minakami 1950, Fudali & Melson 1972, McBirney 1973). Transient explosions, typical of vulcanian eruptions, result from the sudden release of pressure in a gas due to the failure of some cap rock (Self *et al.* 1979, L. Wilson 1980a; Fig. 6.26). Because of the pressures involved, this cap rock is unlikely to be simply a layer of unconsolidated clasts. It is most

Figure 6.27 Eruption of Ngauruhoe, New Zealand, at 18.10 h on 19 February 1975. (a) Expanding slug at +8 s; large blocks are 20–30 m in diameter and breaking up with dust trails. (b) Collapse of dense interior of slug to form pyroclastic flows with air-fall plume rising above summit at +45 s. (After Nairn & Self 1978.)

likely to be the cooled, congealed cap of new magma that has risen after the previous explosion, or it could be an older plug. The pressure rise may be due to exsolution of magmatic gas, or to the heating and partial vaporisation of ground water, but not to violent mixing as in a phreatomagmatic eruption.

Gas pressure builds up until the overlying cap fails, in tension or shear. For rocks at temperatures up to 950°C, pressures of up to a few hundred bars are expected (L. Wilson 1980a). An explosion releases a vertical slug of fragments and gas, with initial velocities that may be supersonic, in which case a shock wave would also be propagated (Nairn 1976). The history of one explosion from the Ngauruhoe 1975 eruption shown in Figure 6.27 is documented in Figure 6.28. The slug of material was ejected at an estimated initial velocity of $\sim$400 m s^{-1}, and partial collapse of the column occurred at nearly 500 m above the crater rim to generate a pyroclastic flow (Fig. 6.27b).

Self *et al.* (1979) and L. Wilson (1980a) have

estimated maximum velocities of ejected debris in vulcanian explosions as a function of the pressure beneath the retaining plug at the time it fails. Their calculations suggest that previous estimates of pre-explosion gas pressures (of the order of a few kilobars) are overestimated by an order of magnitude. They also indicate that initial velocities up to 200 m s^{-1} are readily explicable by magmatic gas contents (up to a few weight per cent), and pressures up to a few hundred bars are probably consistent with material strengths. However, for initial velocities above 300 m s^{-1} the influence of external water must be postulated; even if pressures of several kilobars (which cannot be supported by rock strengths) are assumed, unreasonable magmatic water content of $>$10 wt% is implied. Heated ground water is probably a significant feature in such explosions, but not necessarily an essential feature as proposed by Schmincke (1977).

6.8 Surtseyan and phreatoplinian

These terms are used to describe pyroclastic fall deposits resulting from eruptions which have taken place in the sea or a lake, or by contact with ground water. Such eruptions are most generally called phreatomagmatic or hydrovolcanic (Chs 3 & 5). Both types of deposit have extreme fragmentation, F being nearly 100% (Fig. 6.2), and this results from the magma–water interaction. Surtseyan pyroclastic fall deposits have moderate dispersal, while phreatoplinian deposits can be extremely widely dispersed (Fig. 6.2).

There is a tendency to associate specific magma compositions with each of these two types of deposit, based on the compositions of the type examples used to define the terms originally (basaltic for surtseyan, rhyolitic for phreatoplinian; see below). However, given the right conditions, the eruption of any magma type may produce the dispersal and fragmentation characteristics defining these two types of deposits on a D–F plot. It may thus be possible to find rhyolitic surtseyan deposits, and andesitic, trachytic and even basaltic phreatoplinian deposits.

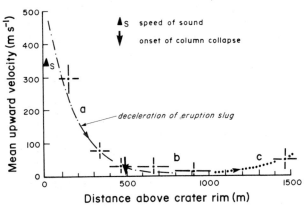

Figure 6.28 History of the 18.10 h vulcanian explosion at Ngauruhoe, New Zealand, on 19 February 1975. Data are from the analysis of still photographs; errors are shown as bars. Curved line approximates exponential deceleration of the eruption slug. The steep part of the curve (a) represents deceleration in the gas thrust phase; the flat portion (b) represents a slow, stable velocity condition while mixing with air and column collapse took place; part (c) shows a slight increase in velocity as convective recovery and rise of the plume began. This event ejected 2 × 10^8 kg of rock, although only half of this was juvenile. Half of the total volume then collapsed to form the pyroclastic flow. (After Self *et al.* 1979.)

6.8.1 SURTSEYAN ACTIVITY AND DEPOSITS

The term 'surtseyan' was used by G. P. L. Walker (1973b) to describe the type of air-fall deposit which would result from similar activity to that observed during the eruption of Surtsey in 1963. Since then, the surtseyan field has been used in a general way to group basaltic fall deposits resulting from different types of hydrovolcanic activity. Kokelaar (1983) pointed out that there may be significant differences between true surtseyan activity, where (sea) water floods into the top of an open vent, and *true* phreatomagmatic activity, involving trapped ground water. However, there is still much to be learnt about such explosive interactions, and for our purposes it is convenient to continue to use 'surtseyan' for the pyroclastic fall field on *D–F* plots irrespective of the environment of magma–water interaction or vent conditions, and 'phreatomagmatic' more loosely for all types of hydrovolcanic activity. This unfortunately still leaves unresolved problems. For instance, we use the basaltic-andesite ash fall deposited during the 1979 eruption of Soufrière, St Vincent, as an example of surtseyan activity because of its well documented phreatomagmatic characteristics and moderate dispersal. However, in detail, the high concentration of lithics is not consistent with a true surtseyan ash fall, and the deposit does not fit easily into a *D–F* pigeonhole. Perhaps classification as (phreatomagmatic) vulcanian would be better (R. S. J. Sparks, *pers. comm.*). Future studies may clarify the grainsize and dispersal characteristics produced during different types of hydrovolcanic eruption, and may lead to the definition of separate fields on the *D–F* diagram.

Phreatomagmatic activity is very common in basaltic volcanic fields, producing maars, tuff rings and tuff cones (Ch. 13). These constructional forms are largely built up from the deposits of base surges (Ch. 5), and thin ash-fall beds occur downwind (Fig. 6.29). Several examples of this type of eruption have occurred this century (Section 5.6). Eruptions of basic to intermediate magmas through small crater lakes on some stratovolcanoes have also produced phreatomagmatic air-fall deposits which

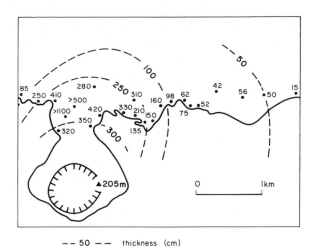

−− 50 −− thickness (cm)

Figure 6.29 Isopach map of the surtseyan ash-fall deposit from the Monte Brazil tuff ring on Terceira. (After Self 1976.)

would have to be included under our broad definition (e.g. the 1979 eruption of Soufrière, St Vincent, Shepherd & Sigurdsson 1982; Fig. 6.30).

The downwind air-fall deposits are thin, fine-grained ashes (Figs 6.29–32), and internally they may be well laminated (Fig. 6.32) because phreatomagmatic activity seems to occur as a number of short explosions. They often contain accretionary lapilli (Ch. 5), and SEM photographs of the ashes show very angular, broken surfaces due to the magma–water interaction (G. P. L. Walker & Croasdale 1972, Heiken 1974). For air-fall deposits they can be poorly sorted (Figs 6.30c & 31). Studies of the 1979 Soufrière air-fall ash layer have shown that bimodal sorting may be an important feature (Sigurdsson 1982, Brazier *et al.* 1982; Fig. 6.30c). Bimodality and poor sorting is attributed to premature fall-out of aggregated wet or damp ash in the eruption column or downwind plume. Accreted ash could occur as accretionary lapilli, or as unstructured ash clumps. During the 1979 Soufrière activity such ash clumps were observed to fall and then break on impact.

Nearer to the vent this type of air-fall deposit is found interbedded with base surge deposits. One problem in near-vent situations is distinguishing air-fall layers from planar-bedded base surge deposits, and we will suggest the criteria to discriminate between them in Chapter 7. In many cases,

(a)

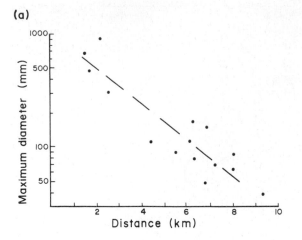

(b)

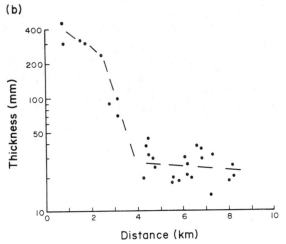

(c)

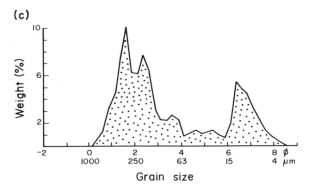

Figure 6.30 The phreatomagmatic ash fall deposit erupted in 1979 from Soufrière, St Vincent, Lesser Antilles. Variation in (a) maximum dense clast diameter and (b) thickness, both as a function of distance from vent. (c) Bimodal size distribution in a sample collected from the lower part of the deposit. (After Sigurdsson 1982.)

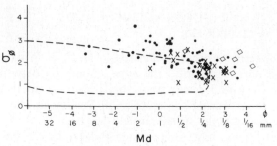

Figure 6.31 Md_ϕ/σ_ϕ for some surtseyan ash-fall deposits. Solid circles are samples collected at vent, while crosses are downwind deposits. Diamonds are ash-fall deposits from phreatomagmatic eruptions of the Quill, St Eustatius, Lesser Antilles; all of the samples were collected between 2 and 3 km from the vent. Broken line is the field of magmatic basaltic (strombolian) fall deposits (from Fig. 6.7). (After G. P. L. Walker & Croasdale 1972 and J. V. Wright *unpub. data* from St Eustatius.)

both modes of deposition may have occurred simultaneously as ash from a previous explosion, or maintained column, fell around the vent into newly generated surges. Other evidence of such ashes being wet may be entombed gas cavities (Ch. 5).

6.8.2 PHREATOPLINIAN ACTIVITY AND DEPOSITS

The term 'phreatoplinian' was introduced by Self and Sparks (1978) for the silicic analogue of surtseyan, and they described several examples, documenting in detail widespread ash layers from the Oruanui Formation (now redefined and described by Self (1983) as the Wairakei Formation), New Zealand, and from a phreatomagmatic phase of the 1875 Askja eruption (Figs 6.33–35). They also mention another example in New Zealand, the Rotongaio ash, and examples from the Minoan eruption of Santorini (Bond & Sparks 1976) and São Miguel (Booth *et al.* 1978). Self (1983) discussed the Oruanui (Wairakei) Formation in detail and G. P. L. Walker (1981a) documented the Rotongaio ash and the Hatepe ash, both of which were formed in phreatomagmatic episodes during the Taupo AD 186 eruption. We shall describe these deposits along with the ultraplinian deposit also produced by that eruption in Chapter 8. The Vesuvius AD 79 eruption also produced phreatoplinian layers, as was previously mentioned (Ch. 5).

Figure 6.32 Thinly laminated phreatomagmatic ash-fall deposit of intermediate composition erupted from the stratovolcano Mt Misery on St Kitts, Lesser Antilles (see Fig. 13.28; this ash deposit also contains accretionary lapilli (not visible)). Knife is 30 cm long.

All of these phreatoplinian deposits were produced during phreatomagmatic phases of much larger rhyolitic eruptions which involved several different styles of activity. Collectively, these include plinian, base surge and ignimbrite-forming activity (Fig. 6.33). With the exception of the examples from Santorini and Vesuvius, they all involved eruption of rhyolitic magma through caldera lakes. In the Minoan eruption, sea water gained access to the vent on Santorini. During the Vesuvius AD 79 eruption, water from a deep aquifer is thought to have gained access to the magma chamber (Ch. 5).

The extremely wide dispersal of these deposits (Fig. 6.36) indicates a high eruption column (of plinian proportions), yet all of the other characteristics of the deposits indicate that they have a phreatomagmatic origin. Deposits are very fine-grained, even close to source (Fig. 6.35) and

accretionary lapilli are common. They may be finely laminated, especially towards the vent, and near to the vent they are often associated with base surge deposits (Fig. 6.33). Deposits are poorly sorted for pyroclastic falls, especially considering their median grainsize, and their size distributions are bimodal or strongly negatively skewed, or both, indicating that they have an extended coarse tail (Fig. 6.37). In contrast, plinian deposits are positively skewed. SEM studies show that blocky shards are typical (Heiken 1972, 1974, Wohletz 1983; Ch. 3), although the phreatoplinian deposits of the Askja 1875 eruption do contain 'vesiculated' cuspate shards (Sparks *et al.* 1981), more usually associated with magmatic eruptions, as do the deposits of the Oruanui Formation (Self 1983).

Laterally, phreatoplinian deposits become imperceptibly thin and fine over wide areas. Downwind there is little sorting of the size distribution,

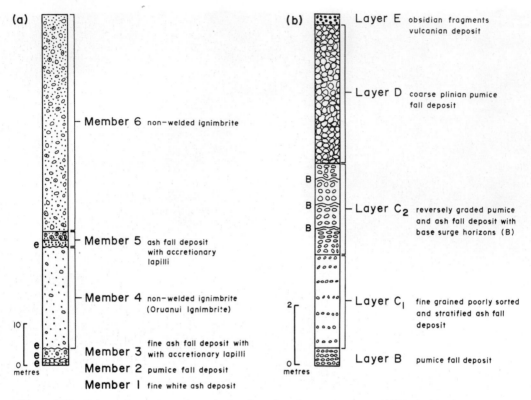

Figure 6.33 Schematic sections through the deposits of two eruptions which produced phreatoplinian deposits. (a) The Oruanui (Wairakei) Formation, New Zealand, reconstructed close to source north of Lake Taupo. Thicknesses of basal members (1–3) are exaggerated; 'e' indicates erosional breaks that occurred during the eruption. (b) The products of the Askja eruption, Iceland on 28–29 March 1875. (After Self & Sparks 1978.)

(a) **Askja 1875 phreatoplinian deposit (layer C)**

(b) **Oruanui phreatoplinian layers**

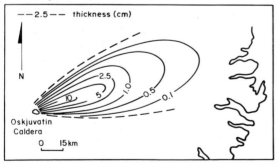

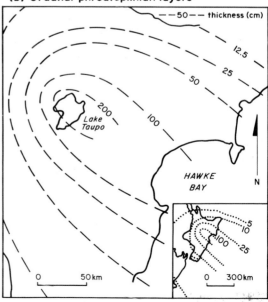

Figure 6.34 Isopach maps of the (a) Askja (1875), Iceland and (b) Oruanui phreatoplinian ash layers, New Zealand. For the Oruanui deposits this is the combined thickness of members 1, 2, 3 and 5 (Fig. 6.33). (After Self & Sparks 1978.)

Figure 6.35 Phreatoplinian members 1–3 of the Oruanui Formation, New Zealand, 25 km from source. Member 4 is the base of the co-eruptive Oruanui ignimbrite. Erosion surfaces between members can be seen. The scale is 45 cm long. (After Self & Sparks 1978.)

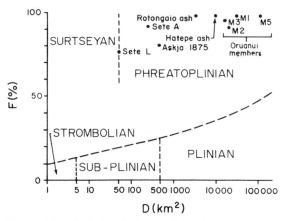

Figure 6.36 *D–F* plot of phreatoplinian deposits. (After Self & Sparks 1978 and G. P. L. Walker 1981a.)

and it is only the larger particles that are lost with increasing distance from the vent. This type of grading is *coarse-tail grading*, and contrasts with the distribution grading typical of plinian deposits where, laterally, sorting affects the whole size distribution (Fig. 6.37). Downwind sorting of fines must be inhibited in some way, and involves the bringing down of different-sized particles in aggregates or clumps to explain why these deposits are so fine-grained near source and so poorly sorted. These could fall as accretionary lapilli, but perhaps in many cases as unstructured clumps of ash. More extreme would be the wholesale water-flushing of the downwind plume, in which case ash could fall as 'mud-rain'. G. P. L. Walker (1981d) described a type of microbedding attributed to the splashing of falling water during deposition of the Hatepe ash. The source of most of the water was likely to have

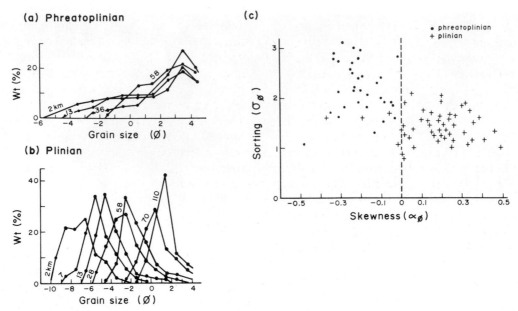

Figure 6.37 Grainsize data from the 1875 Askja deposits, Iceland, showing contrasts between the (a) phreatoplinian (layer C) and (b) plinian (layer D) deposits (Fig. 6.33). Frequency-grainsize curves are for samples collected at various distances downwind. (c) Plot of skewness against sorting. (After Sparks *et al.* 1981.)

been steam-generated at the vent by the interaction of magma and water.

Other types of pyroclastic fall deposit can resemble phreatoplinian deposits in the field. One is formed by local flushing of a downwind plinian or sub-plinian plume by atmospheric rain, and this type we have mentioned while discussing plinian deposits. Such an ash-fall deposit could contain accretionary lapilli, but an isopach map would show that the deposit only covered a very limited area. Co-ignimbrite ash-fall deposits associated with ignimbrite-forming eruptions (Chs 5 & 8) can be very widespread and, if rain-flushed, could also contain accretionary lapilli. However, these ash-fall deposits are vitric-enriched because crystals are preferentially segregated into the pyroclastic flows, while phreatoplinian deposits should contain nearer to the original magmatic crystal ratios if the magma were porphyritic. Texturally, co-ignimbrite ash-fall deposits should contain delicate shards and bubble wall fragments, suggesting fragmentation by exsolving magmatic gases. It has now also been recognised that plinian eruptions may produce substantially more fine ash than has hitherto been

suspected, and may deposit poorly sorted, bimodal ashes distally. Problems in the interpretation of distal silicic ash layers are discussed in Section 6.9.

6.8.3 MECHANISMS

We have already discussed (Chs 3 & 5) some of the physical controls of phreatomagmatic or hydro-volcanic explosions that generate surtseyan and phreatoplinian pyroclastic fall deposits. Clearly, the major contrast between these types of eruptions and magmatic eruptions is the degree of fragmentation of the magma. In magmatic eruptions disruption and fragmentation of the magma is by exsolution and expansion of its volatiles. In comparison, this produces the relatively coarse population of particles observed in strombolian and plinian deposits. However, even in the phreatomagmatic case it is likely that some vesiculation and disruption of the magma occurs by magmatic gases. SEM studies do show that shards from surtseyan and phreatoplinian deposits, although having sharp fractured boundaries, may contain small vesicles, suggesting that magmatic fragmentation could play a role (see Figs

3.18, 24 & 25; Section 3.5.1). Self and Sparks (1978) suggested a two-stage model for fragmentation in many phreatomagmatic eruptions. Magma is first partially fragmented by vesiculating magmatic gases to give a coarse population which then interacts with water. Hydrovolcanic explosions occur and a second stage of fragmentation is initiated, which is aided by the large surface area of magma presented to the water because of the initial magmatic breakage. Final grainsize characteristics of the ejecta will depend on the efficiency of mixing. From Figure 3.9 a water : magma mass ratio of 0.3 leads to the most efficient fragmentation. If almost complete hydrovolcanic fragmentation occurs, the only evidence of the role of magmatic fragmentation will be small vesicles seen within some larger shards. In less-efficient events, larger ash-sized fragments may be poorly vesicular. However, it is unlikely that abundant delicate cuspate shards and bubble wall fragments would be preserved. This two-stage model for fragmentation also explains the grainsize distribution of phreatoplinian deposits: the fine-grained unimodal population is generated by hydrovolcanic fragmentation, while the coarse-tail is of larger particles broken by magmatic processes.

Eruption columns for this type of activity are driven by rapid successions of hydrovolcanic explosions. Observations of the 1963 Surtsey eruption suggest that these occur every few seconds to tens of seconds (Thorarinsson *et al.* 1964). Each explosion at Surtsey thrust out black jets of ash and bombs, which occasionally reached heights of about 1 km. Finer particles were then taken to greater heights by convection, occasionally as high as 9 km (Thorarinsson *et al.* 1964). During the 1979 Soufrière eruption, columns from phreatomagmatic explosions rose as high as 18 km (Sparks & L. Wilson 1982).

As the steam–pyroclast mixture rises, condensation of the steam may occur in the column. This phase change requires a large change in volume, and a substantial increase in density. Although this may be partially compensated for by mixing of air from the side of the column, partial collapse of the column could occur. It is in this situation that base surges (Ch. 5) will form. Eruptions producing large amounts of steam may have lower convective plumes, because of the thermal energy lost in vaporising water (Ch. 5). However, the much finer-grained nature of the ejecta means that it will be more widely dispersed from a lower plume than it would be from an entirely magmatic plume.

There are no direct observations of phreatoplinian eruptions, but the continuous eruption of rhyolitic magma at a high discharge rate through, say, a caldera lake is likely to produce an eruptive plume some tens of kilometres high.

6.9 Distal silicic air-fall ash layers

The interpretation of distal ($> > 150$ km from the vent) air-fall ashes associated with large magnitude silicic eruptions has become problematical. Large magnitude events, in many cases ignimbrite-forming, can be very complex, and this makes understanding their distal ash layers difficult. Up to a few years ago they were generally assumed to be pre-ignimbrite air-fall deposits, a view still upheld by Izett (1981). More-recent studies have indicated that crystal-enriched ignimbrites should be accompanied by equally voluminous vitric-enriched ash falls. Many widespread ashes were subsequently interpreted as co-ignimbrite ash-fall deposits (e.g. Sparks & Huang 1980; Ch. 8). Widely dispersed phreatoplinian ash-fall deposits were then recognised, and G. P. L. Walker (1980) suggested that widespread ash layers could also be formed by littoral explosions when hot pumice flows entered the sea (see Chs 3, 9 & 10).

New studies, in which the total grainsize populations of deposits have been determined, now suggest that the volume of fine ash produced during plinian eruptions is much more than was previously supposed. Thus, the importance of widespread ash-fall deposits of plinian origin may have been underestimated.

6.9.1 WHOLE-DEPOSIT GRAINSIZE POPULATIONS

Most studies of pyroclastic fall deposits only provide data on thickness and grainsize to distances <150 km from the vent. Whole-deposit grainsize

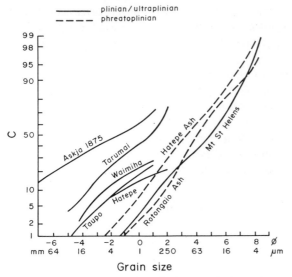

Figure 6.38 Whole-deposit grainsize populations of some pyroclastic fall deposits formed by highly explosive silicic eruptions. C is weight percentage coarser than size stated on the x-axis. (After G. P. L. Walker 1981e and Carey & Sigurdsson 1982.)

populations are important because they reflect the initial size distribution in the eruption column. The main problem has been to estimate the amount of fine-grained ash, because a proportion of this is always going to be deposited outside the limits of the minimum thickness isopach. Various methods have been employed, but usually they involve dividing the isopach map up into different segments. Grainsize data for the segments are averaged and then weighted according to the volume of each sub-unit. Crystal concentration studies are used to assess the total volume of the deposit for the area outside the minimum isopach (see App. I). The total grainsize distribution is then determined by integrating these data.

Whole-deposit grainsize curves for some deposits of highly explosive eruptions are given in Figure 6.38. What is surprising is the high proportion of fines found in some plinian deposits, and the Taupo ultraplinian deposit. These only seem to differ from the phreatomagmatic deposits in having a tail of coarser grainsizes, suggesting that fragmentation during plinian eruptions may be much greater than was previously thought. G. P. L. Walker (1981a) suggested that the markedly different appearances

of plinian and phreatoplinian deposits in the field reflect a different depositional process, perhaps water-flushing in the second case, rather than degree of fragmentation.

The data from Mt St Helens (Fig. 6.38) are difficult to assess. Near the vent the air-fall is a somewhat typical coarse-grained plinian pumice deposit. However the whole-deposit grainsize population resembles that of phreatoplinian deposits, and a very high proportion of ash was produced during the eruption. Carey and Sigurdsson (1982) suggested that interaction with external water may have been important. However, mechanisms of magma fragmentation and how they operate during explosive eruptions are little understood at present. As indicated by G. P. L. Walker (1981a), the clarification of grainsize relationships requires more work.

6.9.2 SECONDARY THICKENING AND BIMODALITY

As well as large amounts of fine-grained ash, the Mt St Helens deposit surprisingly showed secondary thickening beyond the 1 cm isopach (Fig. 5.8c). First, the air-fall deposit thinned exponentially to 1.0 cm thickness at 180 km, but then increased to 4.0 cm thickness at 300 km before thinning once again at greater distances. Carey and Sigurdsson (1982) attributed this to premature fall-out of aggregated particles in the eruption plume. Sorem (1982) described ash clusters carefully collected while settling, and shows SEM photographs of these. These ashes are poorly sorted and bimodal. Aggregation may be related to condensation of vapour in the plume, which could have been important if water was involved in the eruption, or perhaps to electrostatic charging of particles.

Secondary thickening had previously only been described in the air-fall deposit erupted from Quizapu in 1932 (Larsson 1936), but Brazier *et al.* (1983) recognised that secondary thickening and bimodality may be common features of distal ash deposits. They describe several North American examples. Some are the distal deposits of ignimbrite-forming eruptions, but others were from solely plinian eruptions.

Bimodality has been used as a criterion to distinguish co-ignimbrite ashes from precursor plinian ashes in the deposits of multi-phase ignimbrite-forming eruptions. In co-ignimbrite ashes at or near vent, bimodality has been attributed to two separate sources which contribute the different modes (Sparks & Huang 1980; Ch. 8). The results from Mt St Helens and other North American ash layers now bring into question some of these interpretations. Bimodality may also be a common feature of distal plinian deposits. Volcanological considerations still support the idea that distal ash layers have a substantial co-ignimbrite component (cf. Izett 1981), but the plinian component may be much larger than was recently thought.

It is therefore apparent that, in distal situations, clearly distinguishing phreatoplinian, plinian and co-ignimbrite ash-fall deposits is going to be very difficult. By grainsize characteristics alone this may be impossible. Proximal characteristics of the deposits and the eruption sequence will provide the best criteria.

Finally, if secondary thickening is a common feature of distal plinian deposits, it therefore has important implications for calculating their volumes. Conventional ways of estimating the volumes of plinian deposits are based on thickness–distance plots. These plots give straight-line or exponential relationships in areas near the source (<150 km from the vent). However, secondary thickening in distal deposits suggests that this method is therefore unsuitable where dispersal of ash occurs over distances greater than 150 km from the vent (cf. Froggatt 1982). The volumes quoted in Table 6.2 were obtained by a different method (see App. I).

6.10 Welded air-fall tuffs

As discussed more fully in Chapter 8 (Section 8.10.1), **welding** involves the sintering together of hot, glassy fragments, irrespective of shape and size, under the influence of a compactional lithostatic load. During welding the glassy fragments deform plastically, producing a bedding parallel

Table 6.5 Examples of welded air-fall tuffs found on modern volcanoes

Deposit	Composition	Reference
Thera welded tuff, Santorini	dacite (mixed)	Sparks & Wright (1979)
Therasia welded tuff, Santorini	dacite (mixed)	Sparks & Wright (1979)
Askja 1875, Iceland	rhyolite (mixed)	Sparks & Wright (1979)
Ruapehu, New Zealand	andesite	personal observation
Newberry caldera, Oregon, USA	andesite	personal observation
Mt Giardina, Lipari	rhyolite	personal observation
Krakatau 1883	dacite (mixed)	Self & Rampino (1981)
Mt St Helens 1980	dacite	Banks & Hoblitt (1981), Rowley *et al.* (1981)
Tenerife (several of the 'eutaxites')	phonolite	G. P. L. Walker (*pers. comm.*)
Green tuff, Pantelleria	pantellerite	J. V. Wright (1980), Wolff and Wright (1981, 1982)
Mayor Island, New Zealand	pantellerite	B. Houghton (*pers. comm.*)
Mt Suswa 'globule tuff', Kenya	pantellerite	Hay *et al.* (1979)*
Tongariro, New Zealand	basaltic	Healy (1963)
Tarawera 1886, New Zealand	basaltic	Cole (1970), G. P. L. Walker *et al.* (1984)

* The description given by Hay *et al.* (1979) of this deposit suggests it is a welded air-fall tuff. In particular, the deposit is sheet-like, mantles topography, is frequently stratified, welded tuff is found interbedded with air-fall pumice close to the vent, and the overall degree of welding shows a strong dependence on distance from source.

fabric of flattened, elongate large pumice fragments (**fiamme**) and glass shards, known as **eutaxitic** texture.

Welded tuffs are common in the geological record, and are generally called ignimbrites or ash flow tuffs, implying deposition from a pyroclastic flow. However, welded air-fall tuffs are a common volcanic product, and are known to occur on several modern volcanoes, covering nearly the full range of magma compositions (Table 6.5). They have also been recognised in ancient volcanic successions (Suthren & Furnes 1980).

The best documented examples so far are from Santorini volcano, Askja volcano, Iceland (Sparks & Wright 1979) and the island of Pantelleria in the Channel of Sicily (J. V. Wright 1980). These tuffs can be distinguished from welded rocks formed from pyroclastic flows by their geometric form, textures and field relations to non-welded counterparts. Their features indicate post-emplacement compaction and welding over a wide area, and cannot simply be ascribed to the agglutination of spatter lumps on impact. We would restrict the use of the term 'agglutination' to describe the process of deformation and sintering of air-fall pyroclasts when they impact on an accumulating bed. Flattening is thus caused by the momentum of the falling pyroclasts. This is distinct from welding, in a strict sense, which occurs under the influence of load pressure imposed by already accumulated tephra on the underlying, still hot, part of the deposit. Agglutination requires the eruption of more fluid magma and, as we have discussed, is therefore a more important process in basaltic eruptions (Section 6.3). However, extensive welded scoria-fall deposits are known (Table 6.5), and agglutination could have been an important process near the vent during these eruptions.

Welding has also been documented in '*fused*' or '*sintered*' *tuffs*, where lava has induced welding of underlying air-fall layers (Christiansen & Lipman 1966, Schmincke 1967a), but these are not formed by primary processes like the examples described here.

6.10.1 CHARACTERISTICS AND EXAMPLES

Discussion of the more general aspects of welding textures and processes will be left until we describe welded ignimbrites in Chapter 8.

In the air-fall examples that are known, welding occurs outwards to a distance of 1–7 km from the probable vent position. Ignimbrites, on the other hand, can be welded at distances of 50 km or more from the source. The basic criteria used to distinguish welded tuffs of air-fall rather than pyroclastic flow origin are:

(a) mantle bedding and deposition on steep slopes (>20°; Fig. 6.39).
(b) internal stratification and distinguishable fall units (this may be reflected by rapid fluctuations in the compaction profile) and
(c) non-welded equivalents having depositional and grainsize characteristics of airborne ejecta.

We will now briefly mention some of the examples that have been described.

The *Thera welded tuff* on Santorini forms a distinctive black, glassy, dacitic layer as much as 7 m thick (Figs 6.40–42), which in hand specimen has a well developed eutaxitic texture. It is exposed for more than 1.5 km in the caldera wall and must have originally covered at least 1 km^2. Laterally and vertically the welded tuff passes into a thick, coarse, non-welded pumice fall deposit (the Middle Pumice, Fig. 13.30) which, near the welded tuff, is thermally darkened and black in colour (Fig. 6.41). This layer mantles topography, and isopach and maximum-sized isopleth maps are typical for plinian air-fall deposits (cf. Figs 6.18 – 19). These maps indicate that the vent was near to the place where the welded tuff is thickest. On close inspection the tuff is seen to be internally stratified and to contain some conspicuous layers of coarse pink pumice. Many of these are not laterally continuous, and they are thought to have formed by the rapid local accumulation of very large pumice bombs. These produce anomalous deviations in the compaction profile, and one is shown in Figure 6.42. Grainsize analyses of the non-welded pumice deposit are typical of a fall deposit (Fig. 6.43).

The *Askja 1875 welded tuffs* were formed during

Figure 6.39 Welded air-fall tuffs mantling topography on (a) Ruapehu and (b) Tongariro volcanoes, New Zealand (both photographs by C. J. N. Wilson). Note vertical cooling joints in (b).

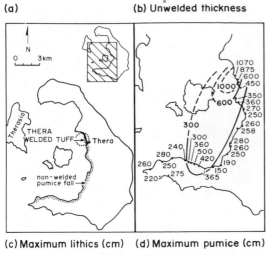

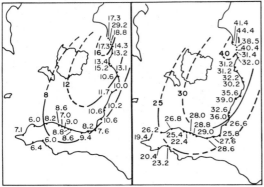

Figure 6.40 Distribution of the Thera welded tuff and its equivalent non-welded pumice-fall deposit (the Middle Pumice) in the caldera wall of Santorini, Greece. Dotted circle marks the area in which the source vent probably occurs. Isopleths show the average diameter of the five largest clasts. (After Sparks & Wright 1979.)

the same explosive rhyolitic eruption of this Icelandic volcano discussed previously in Section 6.4. The eruption produced a number of distinctive pyroclastic layers, and two of these (C and D in Fig. 6.33b) pass into welded rocks near Oskjuvatn caldera (Fig. 6.44). The most important welded tuff grades laterally into layer C_2 (Figs 6.44 & 45). This welded tuff has a maximum thickness of 4 m and covers a minimum area of 1.6 km². The welded tuff mantles topography and can be found on slopes as steep as 30°. It is stratified, and there are marked fluctuations in the compaction profile. Welded air-fall tuffs are deposited in layers, and can show

vertical variations in grainsize, sorting, components and, more importantly, accumulation rate. This can result in more-irregular compaction profiles than with ignimbrites.

The *Green Tuff* of Pantelleria is the best-exposed and most spectacular and has been shown to have been of plinian proportions (Wolff & Wright 1982), and in its original distribution it must have mantled the entire island (85 km²) with over 5 m of densely welded tuff. The tuff drapes the former topography (Figs 6.46 & 47), but is ponded in depressions due to **rheomorphism**, which is the post-depositional secondary flowage of welded tuff (Ch. 8). The overall geometry resembles that of pyroclastic surges or low aspect ratio ignimbrites (Ch. 8). However, there is no correlation between the degree of welding and the thickness, as might be expected were the thickness variation a primary flow feature.

Other welded tuffs on Pantelleria adhere to slopes steeper than 30°. A good example of a welded air-fall tuff in Figure 6.47d shows alternating layers of densely welded tuff and well sorted pumice, which are thought to reflect changes in accumulation rate during the eruption (J. V. Wright 1980).

6.10.2 CONDITIONS OF FORMATION

The critical problem posed by welded air-fall tuffs is to determine under what conditions airborne ejecta can remain sufficiently hot during flight to weld and compact after deposition. The basic controls are high discharge and accumulation rates. Field evidence suggests that accumulation rate is the critical factor. Both the Thera and Askja welded tuffs are thickest and most densely welded near the vent, where the accumulation rate was greatest (cf. welded ignimbrites which are again most densely welded where they are thickest, but where they are ponded in a depression this could be a long distance from the vent).

Accumulation rate has two main effects. First, rapid accumulation prevents radiative and convective cooling of deposited fragments. On burial, fragments are insulated, so they cool by conduction alone, which is a slow process because of the high

Figure 6.41 The Thera welded tuff, Santorini, Greece. (a) At Thera Harbour with well developed internal stratification. (b) Photomicrograph (negative) of eutaxitic texture. Note the development of perlitic cracking. Area shown is about 1 cm across. (c) Non-welded pumice fall (between arrows) which is the lateral equivalent. Note thermal darking above base.

Figure 6.42 Compaction and lithological profile through the Thera welded tuff, Santorini, Greece, where it is thickest. The minimum strain ratio assumes all ellipsoidal pumice clasts landed with their long axes parallel to bedding. (After Sparks & Wright 1979; see this paper for method of measuring strain ratio.)

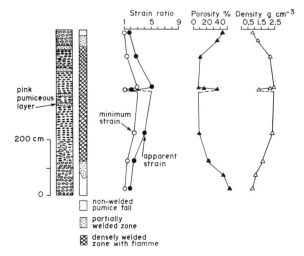

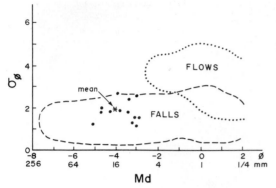

Figure 6.43 Md_ϕ/σ_ϕ plot of the non-welded part of the Thera welded tuff (for examples of grainsize histograms see Fig. 6.3). On the basis of grainsize (Ch. 12), this welded deposit and the other examples described herein would be classified as 'welded agglomerates' and 'welded lapilli tuffs'. For brevity in nomenclature, we have used 'welded tuff' as a general term to cover them all. (After Sparks & Wright 1979.)

porosity and poor conductivity of pumice. Secondly, rapid accumulation leads to lower heat losses by radiation, as a consequence of increasing concentrations of fragments per unit volume of air in the falling ejecta. At high temperature, an incandescent clast loses much of its heat by radiation. This loss is reduced if each particle is surrounded by other hot particles radiating heat in a dense cloud, as each particle absorbs radiated heat from its neighbours. The Askja 1875 eruption is sufficiently well documented to allow an estimate of the accumulation rates which produced the welded tuff (Sparks & Wright 1979). For the densely welded zone this was calculated as 20 cm min^{-1}, and at the boundary between incipiently welded and non-welded zones it was between 2 and 4 cm min^{-1}.

Both the Thera and Askja welded air-fall tuffs are also mixed pumice deposits (Sparks & Wright

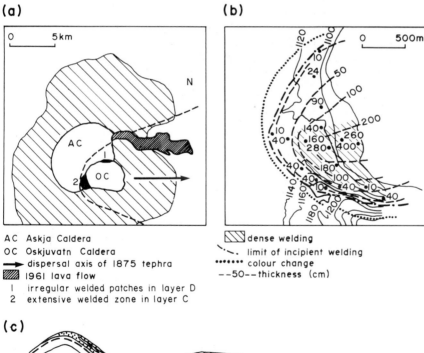

AC Askja Caldera
OC Oskjuvatn Caldera
→   dispersal axis of 1875 tephra
▨   1961 lava flow
 I   irregular welded patches in layer D
 2   extensive welded zone in layer C

▨ dense welding
·–··. limit of incipient welding
······ colour change
–-50-– thickness (cm)

Figure 6.44 (a) Location of the welded tuff horizons from the 1875 Askja eruption, Iceland. (b) Isopach map of the layer C (Fig. 6.33) welded tuff. (c) Schematic north–south profile through layer C, showing relationships between welding zones and thermal colour changes. (After Sparks & Wright 1979.)

▨ densely welded zone
▦ incipiently welded
–· colour change (white below to orange above)

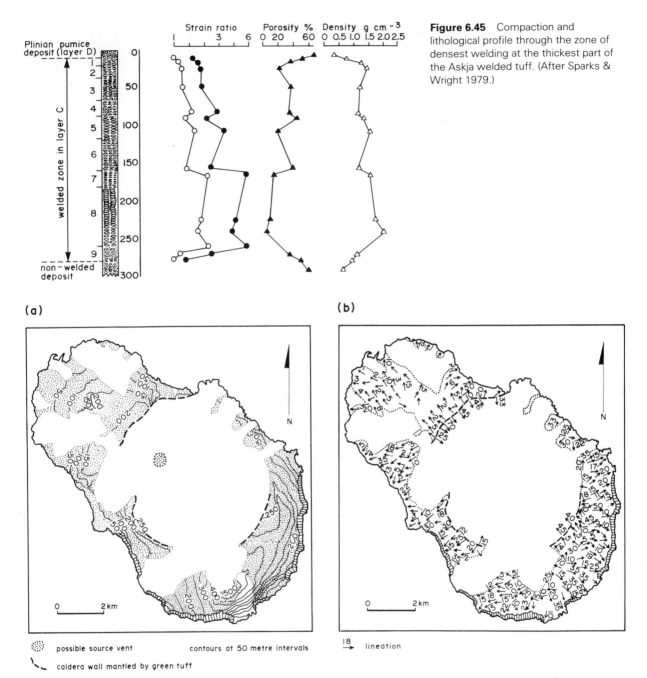

Figure 6.45 Compaction and lithological profile through the zone of densest welding at the thickest part of the Askja welded tuff. (After Sparks & Wright 1979.)

(a)

(b)

possible source vent   contours at 50 metre intervals

caldera wall mantled by green tuff

$\overset{18}{\longrightarrow}$ lineation

Figure 6.46 The Green Tuff, Pantelleria. (a) Present outcrop (stipple) and contours showing how the welded tuff mantles topography. (b) Generalised map of lineations produced by stretching of fiamme during secondary mass flowage (see Fig. 6.47b). (After J. V. Wright 1980 and Wolff & Wright 1981.)

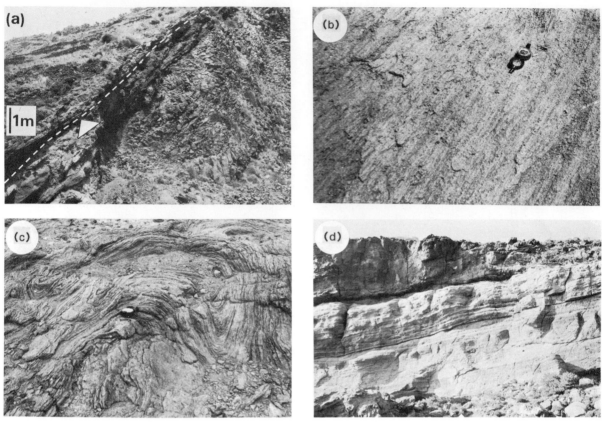

Figure 6.47 Welded tuffs on Pantelleria. (a) Two deposits mantling an older rhyolite lava dome. The uppermost (above arrow) is the Green Tuff, which has a slope of 40°. (b) Strong lineation in the Green Tuff produced during rheomorphism. (c) Refolded fold in the Green Tuff. (d) Welded air-fall tuff showing alternating layers of densely welded tuff (dark) and non-welded white pumice fall. This is overlain by a younger welded tuff (dark), which underlies the Green Tuff in (a). Section is 10 m thick.

1979). Superheating of the silicic component may have been an additional factor promoting welding. However, the importance of this is difficult to judge, because many pumice fall deposits show evidence of magma-mixing but are not welded.

Wolff and Wright (1981, 1982) suggested that compositional factors have favoured the formation of widespread and rheomorphic welded air-fall tuffs on Pantelleria. Due to the low viscosity of pantelleritic ejecta, dense welding can occur at moderate accumulation rates, and Wolff and Wright (1982) indicated a rate of the order of 1 cm min^{-1} for the Green tuff.

6.10.3 THERMAL FACIES MODEL

It is now possible to suggest a thermal facies model for pyroclastic fall deposits. This scheme is shown in Table 6.6. The lateral changes indicated also occur vertically; most of the welded examples that we have described display a range of these facies in the vertical section. What is evident from this type of analysis is the overlap between pyroclastic processes and some lava-forming processes. However, considerable uncertainty must be attached to the interpretation of clastogenic lava facies. Although this is a common facies found in the products of basaltic eruptions, it is unlikely to be found in those of more-silicic magmas. A rhyolitic agglutinate is known to us at Panum

Crater, Mono Craters, but more generally these features in high-silica rocks are only going to be found in the fluidal ejecta derived from peralkaline magmas. Pantelleritic clastogenic lavas may occur on Mayor Island, New Zealand (B. Houghton *pers. comm.*).

The rheomorphic facies again seems rare among silicic air-fall deposits. At present it has only been described from Pantelleria (J. V. Wright 1980, Wolff & Wright 1981, 1982), but other pantelleritic welded air-fall tuffs on Mayor Island and Suswa (Table 6.5) show secondary flowage. Rheomorphism has also been noted in an andesitic welded air-fall tuff on Ruapehu volcano, New Zealand (Fig. 6.39a & Table 6.5), and in some

Icelandic rhyolitic welded air-fall tuffs (H. Sigurdsson & O. Smarason *pers. comms*). These include the Askja welded air-fall tuff, which shows some incipient rheomorphic structures.

Basaltic pyroclastic fall-forming eruptions often seem to produce deposits showing a complete range of thermal facies. In contrast, pumice fall deposits are generally non-welded, with no welded facies present. However, if non-welded they may still show some thermal effects. Grey, black or brown zones of thermal darkening may occur, and care should therefore be taken when interpreting dark-coloured pumice because this may not necessarily indicate a more basic composition.

Table 6.6 Suggested thermal facies model for pyroclastic fall deposits.

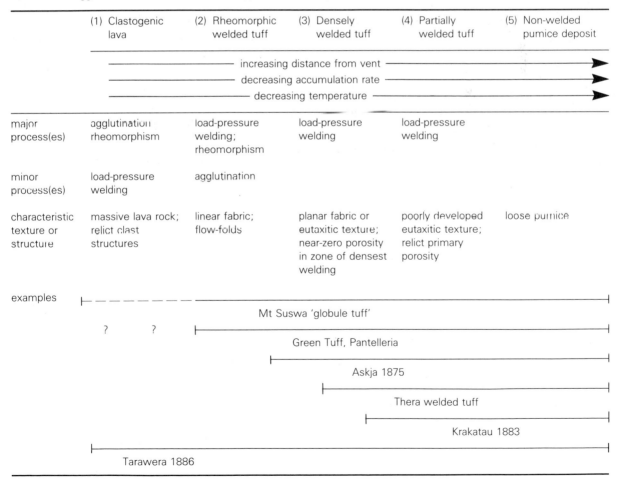

	(1) Clastogenic lava	(2) Rheomorphic welded tuff	(3) Densely welded tuff	(4) Partially welded tuff	(5) Non-welded pumice deposit
	——————— increasing distance from vent ——————→ ——————— decreasing accumulation rate ——————→ ——————— decreasing temperature ——————→				
major process(es)	agglutination rheomorphism	load-pressure welding; rheomorphism	load-pressure welding	load-pressure welding	
minor process(es)	load-pressure welding	agglutination			
characteristic texture or structure	massive lava rock; relict clast structures	linear fabric; flow-folds	planar fabric or eutaxitic texture; near-zero porosity in zone of densest welding	poorly developed eutaxitic texture; relict primary porosity	loose pumice
examples	|-—-—-—-—-—-— Mt Suswa 'globule tuff' ——————————————————————————| ? ? |——— Green Tuff, Pantelleria ——————————————————| |——— Askja 1875 ——————————————| |——— Thera welded tuff ——————| |——— Krakatau 1883 ——————| |————————— Tarawera 1886 ——|			

6.11 Further reading

G. P. L. Walker (1973b) is essential reading, and forms the basis for much of our treatment of pyroclastic fall deposits. Also highly recommended is the review by G. P. L. Walker (1981b) of plinian fall deposits. The series of papers entitled 'Explosive volcanic eruptions I–V' (G. P. L. Walker *et al.* 1971, L. Wilson 1972, 1976, L. Wilson *et al.* 1980, Sparks & L. Wilson 1982) provides a sound physical basis for understanding pyroclastic fall-producing eruptions. L. Wilson and Head (1981) is again recommended for a quantitative analysis of basaltic explosive mechanisms. Further physical treatment of pyroclastic falls is found in Allen (1982), while Fisher and Schmincke (1984) also discuss processes and the characteristics of the deposits.

Plate 7 The Taupo ignimbrite ponded in the upper Ngaruroro River valley, Kaimanawa Mountains, North Island, New Zealand. The pyroclastic flow moved from left to right (southwards), and surmounted all of the hills seen in the photograph which rise to 550 m above the level of the valley pond. (Photograph by C. J. N. Wilson.)

CHAPTER SEVEN

Transport and deposition of subaerial pyroclastic flows and surges

Initial statement

In this chapter we examine more closely the transportational and depositional processes associated with pyroclastic flows and surges (Ch. 5). Both are a type of flow, one at the dilute, low particle concentration end of the spectrum and the other at the highly concentrated end. It is shown that there is a broad spectrum of pyroclastic flow types, and that the flow mechanics can vary, as do the resultant facies. The roles of fluidisation and turbulence are evaluated. The anatomy of pyroclastic flows is also examined, and consideration is given to the characteristic depositional processes associated with the head, body and trailing cloud. We similarly look in detail at the flow and depositional mechanics associated with surges, and consider the differences between wet and dry

surges. Finally, we look at the possible relationships and differences between pyroclastic flows and surges.

7.1 Subaerial pyroclastic flows as high particle concentration flows

All workers would now agree that *pyroclastic flows are gravity controlled* and tend to move along topographic depressions. However, there has been, and still is, much debate on the details of flow mechanisms. Because of poor sorting, earlier workers thought turbulence within flows was important (R. L. Smith 1960a, Murai 1961, Fisher 1966a). The ability of some flows to surmount topographic barriers (Plate 7) led some workers to suggest that they were greatly expanded, implying dilute, turbulent flows (Yokoyama 1974, Sheridan & Ragan 1976). In contrast, from a detailed study of grainsize variations in ignimbrites, Sparks (1976) postulated that *pyroclastic flows are high concentration, poorly expanded, partially fluidised flows*, and are in many ways analogous to types of debris flow in which poor sorting is attributed to high particle concentration, not turbulence (Chs 2 & 10). Their transport and depositional mechanisms may be similar to debris flows, but in debris flows larger clasts are carried by a matrix of mud and water (cohesive flows) or of poorly sorted granular material and water (grain dominant flows), whereas in pyroclastic flows they are carried by fine ash and gas. Sparks (1976) concluded that many pyroclastic flows may be laminar in their movement, especially in the body. Several more-recent studies have concurred with this view (Sparks *et al.* 1978, Sheridan 1979, J. V. Wright & Walker 1977, 1981, C. J. N. Wilson 1980), although it is accepted that variable degrees of turbulence may be important in the head regions, and even in the bodies of violent flows.

Pyroclastic flows encompass a wide range of phenomena. At the one extreme are the small denser-clast flows frequently observed in historic eruptions (Ch. 5). These may come to rest on substantial slopes, transport very large blocks and have well developed levées (Fig. 7.1a) and flow fronts. All of these features suggest that these flows had a substantial yield strength, and that deposition was almost instantaneous by *en masse* freezing of the flows, as is generally believed to be the case in epiclastic debris flows (A. M. Johnson 1970; Ch. 10). Such flows act as simple avalanches of debris, and gas probably plays only a minor role in their movement. They are likely to flow by a combination of laminar and plug flow, typical of Bingham fluids with a high yield strength (Ch. 2). At the other extreme are the pumice flows which form ignimbrites. These generally do not transport excessively large blocks for great distances, they are generally not found on steep slopes and neither levées, nor any other surface depositional features are not commonly observed. However, pumice flows still produce poorly sorted, structureless deposits, and all of the evidence indicates they move as high

Figure 7.1 (a) Scoria-flow deposits from the 1975 eruption of Ngauruhoe with well developed levées and surface ridges. (b) Oblique view of pumice flow lobes from the 22 July and 7 August 1980, eruptions of Mt St Helens, showing their well defined surface morphology and the generally constant width and thickness of the deposits. (After L. Wilson & Head 1981.)

particle concentration flows. However, it is now recognised that there is a complete spectrum of ignimbrite types. Some of the Mt St Helens pumice-flow deposits (22 July and 7 August flows), although these produced only very small volume deposits (<0.001 km^3), have levées and surface features (Fig. 7.1b), indicating that the flows may have had higher yield strengths than are generally expected for pumice flows. Extremely *violent* pumice flows (C. J. N. Wilson & Walker 1981, G. P. L. Walker & Wilson 1983), can leave a topography mantling deposit and transport larger lithic blocks considerable distances. The best-described example is from the Taupo ignimbrite (Section 7.5; Ch. 8), which is very different in many respects from currently accepted, 'conventional' ignimbrites.

Large pumice flows can travel for distances of tens of kilometres, and their mobility has been considered spectacular because of their ability to surmount topographic barriers. For example, the Ito pyroclastic flow (Yokoyama 1974) must have surmounted a 600 m high mountain pass 60 km from source (Fig. 8.3) and the Fisher ash flow tuff (Miller & Smith 1977) crossed a 500 m barrier 25 km from source. As mentioned previously,

some workers accounted for this mobility by suggesting that pumice flows are highly expanded. However, given sufficient momentum, a high concentration pumice flow could surmount such barriers, and indeed cold rock avalanches (Ch. 10) are known to travel uphill, e.g. the Saidmarreh landslide in Iran climbed over a 600 m barrier (P. E. Kent 1966). If one allows for eruption column collapse from heights of several kilometres or more, pumice flows are found to be no less mobile than other types of mass flow (Fig. 7.2). Measurements of the heights climbed by a pyroclastic flow can be used to estimate minimum palaeoflow velocities from the simple potential energy to kinetic energy relationship

$$gh = v^2 \qquad (7.1)$$

where h is the height climbed, v is the velocity and g is the acceleration due to gravity. Absolute-minimum velocities of 60–160 m s^{-1} are inferred from several ignimbrites (e.g. Yokoyama 1974, Sparks 1976, P. W. Francis & Baker 1977, Miller & Smith 1977, Barberi *et al.* 1978). The Saidmarreh landslide must have had a velocity of at least 100 m s^{-1}, and thus such high flow velocities are not unique to pyroclastic flows.

7.2 Fluidisation

Although the mobility and momentum of pyroclastic flows can be attributed in the first instance to either the momentum acquired during collapse from a high eruption column (cf. rock avalanches, Ch. 10), or the high eruption rates and associated high exit velocities; this mobility can be enhanced by the inclusion of a lubricating fluid within the flow, especially if that fluid also provides dynamic support or uplift to the grain population, or part of it, during flow. In that way, the fluid would retard sedimentation from the flow, and so act to reduce the frictional interaction between the flow and the substrate.

Fluidisation is commonly believed to play an important role in this regard, in the transport of pyroclastic flows. As an industrial process, fluidisation was developed largely during and immedi-

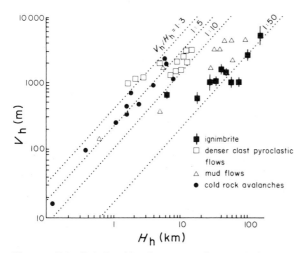

Figure 7.2 Relationship between the vertical height dropped (V_h) and the horizontal distance travelled (H_h) for types of pyroclastic flow and debris flow. V_h for ignimbrites is usually taken as the reconstructed height of the volcano before caldera collapse, and bars indicate the uncertainty. The data indicate that ignimbrites are apparently more mobile than other types of debris flow. (After Sparks 1976.)

ately after World War II. D. L. Reynolds (1954) was the first geologist to examine the chemical engineering literature and suggest that fluidisation might have geological applications, including pyroclastic flows. More recently, its role in the emplacement of pyroclastic flows has been discussed by McTaggart (1960), Sparks (1976, 1978b), Sheridan (1979) and C. J. N. Wilson (1980, 1984).

When an upward stream of gas (or liquid) is passed at increasing velocity (U) through a bed of cohesionless particulate solids, fluidisation is the condition attained at a certain critical fluid velocity (the *minimum fluidisation velocity*, U_{mf}) when the

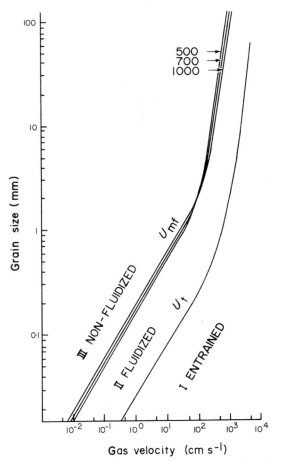

Figure 7.3 Theoretical curves of the relationship of minimum fluidisation velocity (U_{mf}) and terminal fall velocity (U_t) and grainsize. Curves are calculated for spheres of density 1.0 g cm^{-3} in CO_2 with a voidage of 0.45 at 500, 700 and 1000°C. The grainsize–U_t curve is calculated at 700°C for CO_2. (After Sparks 1976.)

drag force exerted across the bed by the fluid is equal to the buoyant weight of the bed (see C. J. N. Wilson 1980, 1984). In this state the bed no longer exists as a coherent mass, but takes on a fluid-like character. However, it is more appropriate here (see below) to use the term in a looser sense (used by chemical engineers) to cover all conditions, from very low gas velocities and packed-bed conditions ($0 < U < U_{mf}$) to high flow velocities and dilute-phase fluidisation ($U > > U_{mf}$).

Sparks (1976) showed theoretically that pyroclastic flows can only be semi-fluidised. In polydispersive systems (grain populations with a wide range of grainsizes or densities, or both), such as in pyroclastic flows, before the largest particles become fluidised the gas flow velocity exceeds the terminal fall velocity (U_t) of the smallest sizes. Such particles are entrained by the gas and carried out and lost from the system. This process is called **elutriation**, and is important industrially and in the transport of pyroclastic flows. Sparks calculated curves of U_{mf} and U_t for a wide range of particles and conditions likely to occur in pyroclastic flows (Fig. 7.3). These calculations used formulae from standard chemical engineering literature (e.g. Kunii & Levenspiel 1969). U_{mf} can be determined from the so-called modified Ergun equation:

$$\frac{1.75}{Q_s e_{mf}^3}\left(\frac{d_p U_{mf} \varrho_g}{\eta}\right)^2 + \frac{150(1 - e_{mf})}{Q_s^2 e_{mf}^3}\frac{d_p U_{mf} \varrho_g}{\eta} =$$

$$\frac{d_p^3 \varrho_g(\varrho_s - \varrho_g)g}{\eta^2} \tag{7.2}$$

where U_{mf} is the minimum velocity of fluidisation (cm s^{-1}), ϱ_g and ϱ_s are the densities of gas and solid, respectively (g cm^{-3}), d_p is mean particle diameter (cm), Q_s is the sphericity of the particle (dimensionless) (= surface area of sphere/surface area of particle where sphere and particle have same volume), e_{mf} is the voidage per unit volume at minimum fluidisation state, η is the viscosity of gas (poise = 10^{-1} Pa s) and g is the acceleration due to gravity.

U_t for a particle in a fluid is given by

$$U_t = \frac{g(\varrho_s - \varrho_g)d_p^2}{18\eta}, \qquad \text{when Re} < 0.4 \tag{7.3}$$

$$U_t = \left(\frac{4(\varrho_s - \varrho_g)^2 g^2}{225 \varrho g \eta} \right)^{1/2} d_p,$$

when $0.4 < \text{Re} < 500$ \hfill (7.4)

$$U_t = \left(3.1g \frac{(\varrho_s - \varrho g)^2}{\varrho} d_p \right)^{1/2}, \text{ when Re} > 500$$
\hfill (7.5)

where Re is the Reynolds Number (Eqn 2.6; Ch. 2).

Figure 7.3 shows that for any gas velocity it is only possible to fluidise a limited range of grainsizes that satisfy the condition $U_t > U > U_{mf}$. This led Sparks to suggest that in pyroclastic flows there must always be three phases when gas is passing through:

Phase I: particles with $U_t < U$,
Phase II: particles with $U_t \geqslant U \geqslant U_{mf}$ and
Phase III: particles with $U_{mf} > U$.

A pyroclastic flow was envisaged to comprise a matrix of particles of phases I and II, in which is dispersed particles of phase III which tend to float (pumice) or sink (lithics) depending on the density contrast with the matrix (see Section 7.3.3). Phase I particles are lost by elutriation from the matrix to form the dilute overriding ash cloud (see Fig. 5.13).

Later experimental work by Sparks (1978b), Sheridan (1979) and C. J. N. Wilson (1980, 1984) also demonstrated that pumice flows and pyroclastic

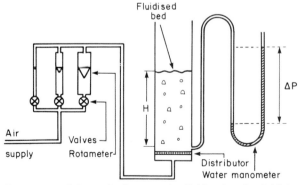

Figure 7.4 Schematic diagram of a fluidisation rig. A high pressure air supply is regulated with valves, through rotameters and the distributor, into the fluidised bed. During runs, the height of the bed (*H*) and the gas flow rates are recorded, together with the pressure drop (Δ*P*) on a water manometer. (After C. J. N. Wilson 1980.)

flows in general can only be semi-fluidised. The most comprehensive studies are those by C. J. N. Wilson (1980, 1984). Wilson reports fluidisation experiments on simple systems and on ignimbrite samples, and has supported his arguments with an extensive search into the chemical engineering literature. His results show that the fluidisation behaviour of pyroclastic flow samples differ radically from any simple system, principally because of the variable hydraulic properties of the different grain types and their resultant poor sorting.

A fluidisation rig of the type used by Wilson is illustrated in Figure 7.4. Examples of fluidisation plots of U versus $\Delta P/H$ (the pressure drop across the bed per unit thickness of bed) are given in Figure 7.5. For narrow grainsize populations of ideally smooth and spherical particles, a fluidisation plot shows two straight lines which intersect at, and define, U_{mf} (Fig. 7.5a). (The slope of the straight line for $U < U_{mf}$ can be predicted by reference to published correlations, and this information has been used in the modified Ergun equation to obtain correlations of U_{mf} versus particle and fluid characteristics.) Commonly, some degree of hysteresis is evident between the curve corresponding to increasing U and the one corresponding to decreasing U (Fig. 7.5b). This can be related to voidage changes during fluidisation, a wide grainsize distribution and/or irregular particle shape. For materials having a wide grainsize variation and irregular particle shapes, a plot is obtained with gross hysteresis (Fig. 7.5c).

Poorly sorted ($\sigma_\phi > 1.0$) materials, such as sand mixes and pyroclastic flow samples, show distinctive fluidisation behaviour (Figs 7.5c & 6). At a certain gas velocity, designated U_{ie} (the value of which cannot be predicted), some samples begin to expand, whereas at higher gas velocities, designated U_{mp}, samples begin to show segregation structures (Fig. 7.7). U_{mp} is the gas velocity at the maximum pressure-drop that can be sustained across the bed (i.e. where $\Delta P/H$ is equal to the buoyant bulk density of the bed multiplied by the acceleration due to gravity: Fig. 7.6a). U_{mp} replaces U_{mf} measured in simple systems, but cannot be predicted reliably from published U_{mf} correlations (C. J. N. Wilson 1984).

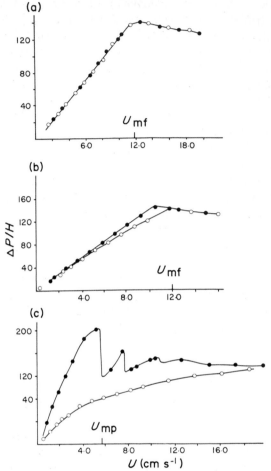

Figure 7.5 Fluidisation plots for several materials. $\Delta P/H$ is in centimetres of water per metre of bed thickness, filled circles define the runs with increasing U and open circles the runs with decreasing U. See text for the definitions of U, U_{mf}, U_{mp} and $\Delta P/H$. (a) Glass spheres, $Md_\phi = 1.40$, $\sigma_\phi = 0.22$. (b) Quartz sand, $Md_\phi = 1.68$, $\sigma_\phi = 0.32$. (c) $Md_\phi = 1.28$, $\sigma_\phi = 1.55$. (After C. J. N. Wilson 1980.)

The up-curve on fluidisation plots of pyroclastic flow samples can therefore be divided into three sections:

Section 1: $0 < U < U_{ie}$
Section 2: $U_{ie} \leqslant U \leqslant U_{mf}$ and
Section 3: $U > U_{mf}$

In section 1, no expansion of the bed occurs and there is no loss of fines by elutriation (except perhaps from the surface layer of the bed). In

section 2, although superficially similar to section 1, the bed has partially fluidised and expanded to accommodate the gas flow. Elutriation of fines is still absent. Section 3 is where an instability sets in, and part of the gas flow is concentrated into discrete channels. This leads to a sharp reduction in the $\Delta P/H$ ratio, which is accompanied by bubbling, the deposition of coarser or coarser and denser material in the channels (Figs 7.7a & b) and the elutriation of fines. At higher gas velocities, bubble-induced circulation destroys the pipes formed earlier, and irregular segregation pods settle towards the base of the bed (Fig. 7.7c). Gradually, well

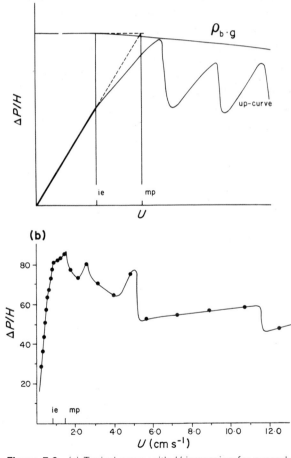

Figure 7.6 (a) Typical curve with U increasing for a poorly sorted sample such as of a pyroclastic flow to show how U_{ie} and U_{mp} are defined. (b) Curve with increasing U for an actual sample of ignimbrite fines ($Md_\phi = 1.22$, $\sigma_\phi = 1.75$). (After C. J. N. Wilson 1980, 1984.)

defined segregation layers form at the top or bottom (or both) of the bed (Figs 7.7d & e). Finer and lighter particles move to the top of the bed, while coarser and denser particles sink to the base. In extreme cases, the bed may become completely gas-sorted.

C. J. N. Wilson (1980) proposed that the fluidisation behaviour of all pyroclastic flows can be typified by a fluidisation plot similar to those in Figure 7.6. This was used as the basis for a classification of pyroclastic flow types. Types 1, 2 and 3 (Table 7.1, Fig. 7.8) were introduced to relate a pyroclastic flow to the corresponding section on a fluidisation plot. The different kinds of grading and the causative mechanisms found in pyroclastic flows are discussed further below (Section 7.3). From Wilson's results, it is evident that only in type 3 flows can the processes described by Sparks (1976) operate freely.

Rheologically, fluidised systems behave in a non-Newtonian manner. Data in the chemical engineering literature deal almost exclusively with well sorted materials ($\sigma_\phi < 1.0$) at high gas velocities ($U > U_{mf}$). Under such conditions these materials have a non-linear relationship of stress to strain rate, and a negligible yield strength. C. J. N. Wilson (1980) considered the rheology of fluidised systems at $U < U_{mf}$ as follows. At rest ($U = 0$) the bed behaves as a particulate material which, at a given depth in the bed, can support a certain differential stress before failure occurs at a stage when the yield strength is exceeded. The yield strength (S_0) increases with depth:

$$S_0 = \mu \varrho g d \qquad (7.6)$$

where μ is the tangent of the internal angle of friction, ϱ is the buoyant bulk density of the material, g is the acceleration due to gravity and d is the depth in the bed. In a partly fluidised bed ($0 < U \leqslant U_{mf}$), the passage of gas results in a pressure drop across the bed, reducing the yield strength at a given depth. At $U = U_{mf}$ the yield strength is effectively zero and, from $0 < U \leqslant U_{mf}$, the yield strength in general (S_u) varies as:

$$S_u = (1 - U/U_{mf})S_0 \qquad (7.7)$$
$$= (1 - U/U_{mf})\mu \varrho g d \qquad (7.8)$$

Once full bed support is achieved, materials with $\sigma_\phi < 1$ may be expected to have similar rheologies to published industrial examples. For materials with $\sigma_\phi > 1$, the rheology at high gas flow rates ($U > U_{mp}$) is more complex. Much of the gas flow is diverted through segregation channels in which the material is better sorted, and this means that although the bed as a whole has a higher permeability, it continues to have a yield strength.

The rheology of fluidised materials in pyroclastic flows is therefore liable to be very complex. But from the foregoing discussion we can surmise that fluidisation will effectively reduce the yield strength of a flow. This concurs with field observations of the morphological and internal features of deposits. Poorly fluidised type 1 flows show features which indicate they had high yield strengths, and a Bingham model may approximate their motion. Yield strength is an important control of the types of grading observed in flows, and highly fluidised flows have very different characteristics. Fluidisation will also induce a stable density stratification

Table 7.1 Classification of pyroclastic flow types based on fluidisation behaviour (after C. J. N. Wilson 1980).

Flow type	Fluidisation behaviour with increasing gas flow velocities	Description
1	non-expanded	Non-graded flows; lack of expansion and high yield strength inhibits gravitational coarse-tail grading. If grading of larger clasts is present, it is due to some other mechanism. Minimal loss of ash into accompanying ash-cloud deposits
2	expanded	Expansion of flow allows gravitational coarse-tail grading of pumice and lithic clasts
3	segregating	High degree of fluidisation results in extreme grading. Distinct concentration zones of pumice and lithics are found, and other gas segregation structures. Large volume of vitric ash lost into co-ignimbrite fall

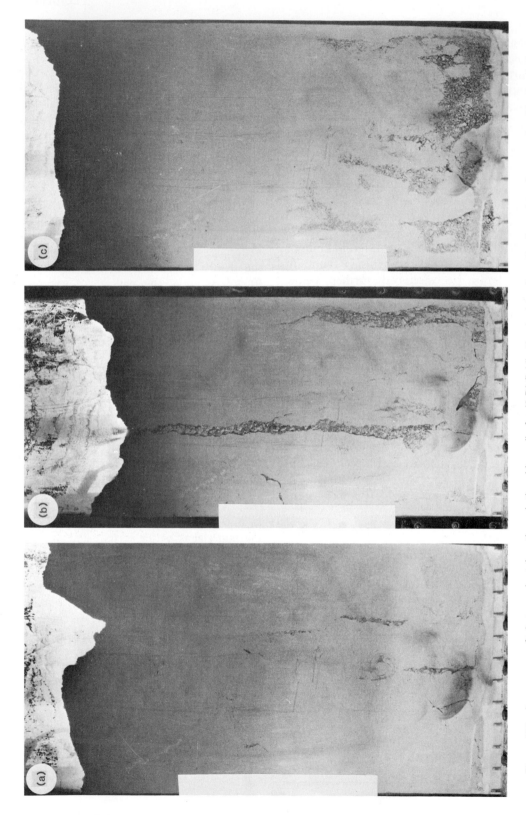

Figure 7.7 A sequence of photographs taken through the side wall of a '2-D' fluidisation bed to show the formation of segregation features in a pyroclastic flow sample ($Md_\phi = 0.75$, $\sigma_\phi = 3.2$). Scale bar is 25 cm. (a) After a few seconds of bubbling, segregation pipes start to form. (b) As bubbling continues, the pipes become larger and propagate to the bed surface. (c) As the gas velocities are increased, the pipes are broken up by bubbling-induced circulation, and irregular segregation pods sediment towards the base of the bed. (d) With time, pipes continue to propagate back towards the

surface of the bed. The material ejected by the pipes is forming a fines- and pumice-rich segregation layer at the top of the bed. (e) As higher gas velocities are reached, the bed becomes distinctly layered, being richer in lithics and crystals and poorer in fines at the bottom of the bed, and richer in pumice and fines at the top. The base of this upper segregation layer is approximately at the top of the scale bar. (f) If the gas supply is slowly reduced within the upper segregation layer, then the pipes are very thin and dominantly composed of pumice, almost all of the lithics and crystals being in the lower segregation layer. (After C. J. N. Wilson 1980.)

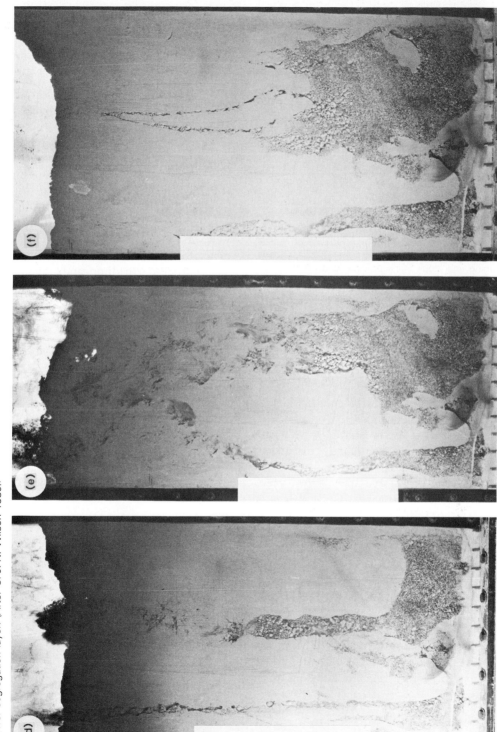

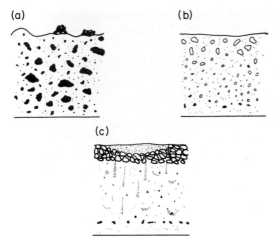

Figure 7.8 Sketch showing the characteristics of C. J. N. Wilson's three flow types. (a) Type 1: ungraded deposit, surface ridging and block trains present, basal layer poorly developed or absent. (b) Type 2: coarse-tail grading of larger pumice towards top (reverse) and larger lithics towards base (normal), well developed basal layer, concave upper surface. (c) Type 3: strongly developed coarse-tail grading with sharply bounded pumice and lithic concentration zones, segregation bodies and pipes, and a fine-grained layer of elutriated vitric ash segregated at the top of the flow. (After C. J. N. Wilson 1980.)

which will strongly suppress turbulence in the body of a moving flow (Section 7.7).

Another important feature shown by Wilson's experiments, and indicated by the above discussion, is that fluidised pyroclastic flow samples expand much less than conventional samples do, this being due to the bypassing of the gas through segregation channels. Wilson estimated that a 100 m thick pumice flow, having fairly high gas velocities, will deflate to form an ignimbrite depositional unit which is not less than 70–85 m thick. This interpretation is also supported by 'a high tide mark' found on the valley sides above some ignimbrites, e.g. in the Valley of Ten Thousand Smokes ignimbrite (Fig. 8.5) this mark rises only 50 m above the general level of the deposit.

Finally, we must briefly consider what the gas sources in pyroclastic flows are. These can be divided into:

(a) internal sources, being gases released from juvenile clasts by diffusion, breakage and attrition, and

(b) external sources, which include gas trapped during initial flow formation, air incorporated at the front of the moving flow, gases released by the combustion of vegetation and steam from heated surface water or ground water.

The relative importance and effects of these different gas sources are fully discussed by C. J. N. Wilson (1980).

The major part of the gas in pumice flows is provided by emission from juvenile fragments and by entrapment of air, both during eruptive column collapse and by engulfment at the flow-front. Sparks (1978b) modelled the diffusional loss of residual gases from juvenile ash particles during flow, and concluded that gas production rates are generally sufficient to fluidise fine and medium ash-sized particles in medium- and large-volume pumice flows. Large-volume (thick) pumice flows are likely to be substantially fluidised, and this may be important in determining their mobility. Increasing levels of fluidisation are predicted in larger flows and in flows with higher initial gas contents in the original magma. McTaggart (1960) suggested that the mobility of pyroclastic flows in general was due to the expansion of entrapped and heated air causing fluidisation of the flow. However, despite their high temperature, historic examples of small pyroclastic flows were no more mobile than cold rock avalanches (Fig. 7.2), which therefore throws some doubt on the importance of entrapped heated air.

7.3 Pyroclastic flow units and grading

As previously indicated (Ch. 5), pyroclastic flow deposits are usually composed of a number of flow units. Sparks *et al.* (1973) first proposed a layering scheme for pyroclastic flow units. The separate layers were interpreted as reflecting different depositional regimes within a pyroclastic flow (Fig. 5.13). Layer 1, the lowest layer, was thought to be the deposit of a dilute pyroclastic surge which moved in advance of the pyroclastic flow. Layer 2 was the deposit of the pyroclastic flow proper, and layer 3 was the deposit of the overriding ash cloud.

In this section we will examine layer 2. This layer forms the main portion of most flow units, and is now thought to be deposited by the 'body' of a pyroclastic flow. Some deposits now correlated with layer 1 are thought to be sedimented in or from the flow 'head', rather than from a preceding surge, and we describe these and the *form* of a pyroclastic flow in Section 7.5. Most of the following discussion is concerned with the deposits of pumice flows. The features of layer 2 to be discussed are:

thickness
basal layers

vertical grading
gas segregation structures
lateral grading
compositionally zoned pumice flow units

7.3.1 THICKNESS

Flow units may vary in thickness from <0.1 m to >100 m where ponded in depressions. As an example, to show the variation within the deposits of the same eruption, the thicknesses of flow units forming the Minoan ignimbrite on Santorini are shown in Figure 7.9a.

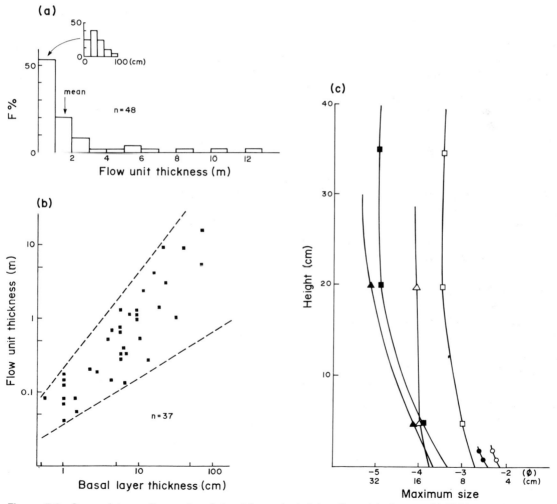

Figure 7.9 Some data on flow units of the Minoan ignimbrite, Santorini, Greece. (a) Thickness frequency histogram of flow units. (b) Relation of thickness of the basal layer (layer 2a) to the thickness of its flow unit. (c) Plot of maximum clast size with height for three basal layers (closed symbols, lithics; open symbols, pumice).

7.3.2 BASAL LAYERS

Many ignimbrite flow units have a fine-grained basal layer (layer 2a) separating them from the former ground surface (Fig. 7.10), and this is usually still present even along almost vertical valley wall contacts. The thickness of basal layers varies from a few centimetres to a metre, but there is usually only a poor correlation between their thickness and the thickness of flow units within an ignimbrite (e.g. Fig. 7.9b). The basal layer differs from the main portion of the flow unit (layer 2b), in that it lacks the largest pumice and lithic clasts. Cumulative grainsize curves of the basal layer and layer 2b therefore converge towards the finer-grained size classes (Fig. 7.11). Basal layers often

Figure 7.10 Bottom of lowermost flow unit of the Minoan ignimbrite, Santorini, Greece, with basal layer below well defined lithic concentration zone. Flow unit deposited on top of white Minoan mud flow. Scale 20 cm.

show a reverse grading of both larger pumice and lithic clasts (Figs 7.9c & 10), and these are often aligned parallel to the ground surface, sometimes producing a faint stratification.

All of these features suggest that the basal layer is an integral part of a flow unit and forms in response to boundary layer effects between the moving pyroclastic flow and the ground surface (Sparks 1976). The finer grainsize and reverse grading of clasts is generally attributed to grain dispersive pressures due to particle collisions acting away from the flow bottom (Sparks 1976), which is a zone of high shear. When particles of mixed grainsize are sheared together, the larger grains should drift away from the zone of maximum shear, i.e. away from the base and side of the flow in a valley (cf. debris flows; Chs 2 & 10). The alignment of clasts also suggests that the basal layer undergoes high shear stresses. Basal layers often cannot be distinguished in block- and ash-flow deposits, scoria-flow deposits and ignimbrite-flow units (see discussion below).

7.3.3 VERTICAL GRADING

The main portion of an ignimbrite flow unit (layer 2b) may show grading of the larger clasts (Ch. 5). The 'textbook' diagram shows reverse grading of larger pumice clasts and normal grading of larger, dense lithic clasts, forming well defined concentration zones towards, respectively, the tops and bottoms of flow units (Figs 5.14c, 7.10 & 11). Flow units with reverse grading of lithics are common, but normal grading of pumice is more rarely found, and an absence of grading is also common.

Grading processes affect only the coarse part of the grainsize distribution, and this type is called coarse-tail grading (cf. distribution grading found in *classical* turbidites, Ch. 10; Allen 1982). Cumulative frequency grainsize curves of samples collected within the same flow unit converge towards the finer-grained fractions (Fig. 7.11), indicating that this part of the distribution or 'matrix' essentially remains homogeneous throughout the flow unit. Grading is therefore a function of grainsize, and is controlled by the hindered settling velocities of particles. Only large clasts have sufficiently high

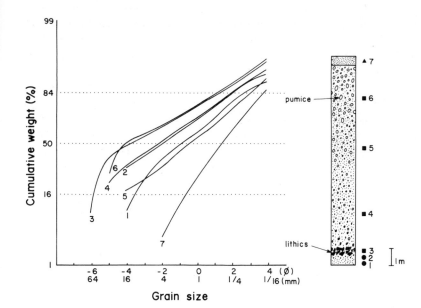

Figure 7.11 Cumulative frequency–grainsize curves for samples of the thickest flow unit of the Minoan ignimbrite. Closed circles, samples from the basal layer; closed squares, samples from layer 2b; closed triangle is a co-ignimbrite ash-fall deposit (layer 3).

settling velocities, and therefore coarse-tail grading and poor sorting result.

The example shown in Figure 7.11 is characterised by reverse grading of low density pumice and normal grading of high density lithics. This type of gravitational (or buoyancy) grading must result from the density contrast between the larger clasts and the matrix. The matrix is invariably denser than the pumice clasts (Table 7.2), and it is thought that the flotation of the larger, lower density pumice clasts causes the reverse grading (density of pumice decreases with size). This implies that the matrix *could not have been greatly expanded* in the moving

pyroclastic flow, otherwise the density contrast would have been lost, and therefore the matrix density is an upper limit for the density of the moving flow. For the example in Figure 7.11 the density of the flow must have been >0.6 to <1.05 g cm^{-3} (flow unit a, Table 7.2). Lithics, because of their high density (~2.5 g cm^{-3}) sink, and in this example the largest and heaviest clasts have formed a distinct lithic concentration zone at the base of layer 2b.

The type of grading observed within a flow unit places important controls on the properties of the moving pyroclastic flow. Normal grading of pumice may be due to more-expanded flows in which the pumice density is greater than the matrix. Reverse grading of lithics (Fig. 7.12) suggests that flows were only marginally expanded. In such flows high shear-strain rates will be imposed through their thickness and shear-induced grading of the larger clasts will result. In such flow units it is often hard to detect a separate layer 2a, because the whole flow was effectively being controlled by boundary layer effects.

Reverse grading of larger and denser clasts is similar to that commonly found in block and ash flow, and some scoria flow deposits (Figs 5.14a, & 15a & b), and again in these deposits a layer 2a is often inconspicuous or poorly developed. Nairn

Table 7.2 Comparison of the densities of pumice and matrix of four flow units of the Minoan ignimbrite.

Flow unit	Pumice density (g cm^{-3})	Matrix* density (g cm^{-3})
a	0.58	1.05
b	0.55	1.08
c	0.64	1.21
d	0.54	1.00

* Matrix is <2 mm size classes.

None of these flow units showed normal grading of pumice, which indicates that the pyroclastic flows could not have been greatly expanded. In all cases the height of the moving flows could at most have only been about twice the present flow unit thickness. Flow unit a is discussed in text and shown in Figure 7.11.

Figure 7.12 Flow unit showing reverse grading of larger lithics in the Minoan ignimbrite, Santorini, Greece. Slight erosional bench towards bottom of photograph marks finer-grained base of flow unit, but a discrete basal layer is not obvious. Scale 20 cm.

and Self (1978) stressed the importance of grain-flow (Lowe 1976; Ch. 10) and grain dispersive forces producing reverse grading in the Ngauruhoe 1975 scoria-flow deposits. However, not all reverse grading in scoria flow deposits is shear-induced, and in some deposits reverse grading of scoria clasts is controlled by gravitational grading (Fig. 5.14d). Thus, there are two mechanisms producing coarse-tail grading (within layer 2b) in pyroclastic flows in general:

(a) gravitational or buoyancy-induced grading and
(b) shear-induced grading.

These both occur in a spectrum of pumice flow types (sometimes in the same ignimbrite), and which mechanism operates is controlled by the degree of expansion of the flow.

The degree of expansion of the flow is related to the amount of fluidisation or the vertical gas flow rate upwards through the flow, and this will control grading processes. In the classification scheme of Wilson (Table 7.1, Fig. 7.8) buoyancy-induced grading is found in expanded type 2 flows, and is well developed in segregating type 3 flows. Shear-induced grading is found in unexpanded type 1 flows.

The gas flow processes involved in fluidisation produce their own grading and sorting in the finer-grained fractions. A method has been developed to analyse this type of gas grading, and the details are given in J. V. Wright and Walker (1981) and J. V. Wright (1981). The method basically assumes that high expanded flows will lose a lot of fine-grained ash (into the overriding turbulent cloud), while less-expanded flows will lose relatively smaller amounts. A qualitative assessment can be made in the field because, if the magma were porphyritic, highly expanded flows should have lost a high proportion of fine glass, and therefore the matrix should be very crystal-enriched compared with the proportion of crystals in large pumice clasts or fiamme.

The corollary is that flow units showing gravitational grading should show evidence of gas grading.

7.3.4 GAS SEGREGATION STRUCTURES

Gas segregation pipes in pyroclastic flow deposits were recognised by G. P. L. Walker (1971, 1972), who described them as 'fossil fumaroles'. He demonstrated that they were enriched in heavy components (crystals and lithics), and suggested that they formed by gas streaming through the ash matrix of a flow on deposition (Fig. 7.13). Earlier, H. Williams (1942) had described fumarole pipes in the Crater Lake succession. Since Walker's studies, they have been commonly found in ignimbrite flow units, and also in some denser-clast types of pyroclastic flow deposit (Figs 5.14, 15c, d & 16f). Pipes have also been found in ancient pyroclastic

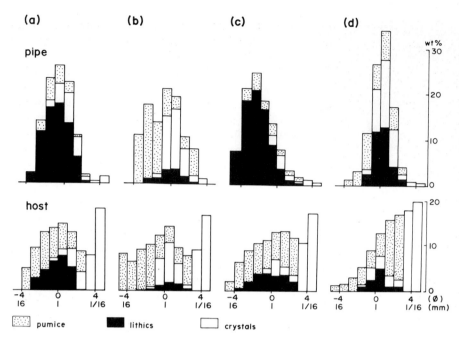

Figure 7.13 Grainsize histograms of samples from gas segregation pipes and below the host ignimbrite which contains them. (a) Vesuvius AD 79. (b) Terceira, Azores. (c) Lake Atitlan, Guatemala. (d) Campanian Tuff, Italy. (After G. P. L. Walker 1971.)

flow deposits, where they are especially important for distinguishing pyroclastic flow deposits from epiclastic debris-flow deposits (e.g. Duyverman & Roobol 1981). Water escape pipes which may appear very similar have occasionally been found in coarse-grained sedimentary debris flow deposits (Postma 1983), including the deposits of the March 1982 Mt St Helens debris-flow event (C. J. N. Wilson *pers. comm.*).

Gas segregation structures are generally up to about 50 cm in length and a few centimetres wide, but quite often they are only several centimetres long and a few grain diameters wide. They are characteristically depleted in fines and enriched in crystals and lithics. Much larger segregation pipes >2 m long occur in coarse, near-vent ignimbrite facies (Ch. 8). In cross section they may have quite irregular shapes, and are not necessarily true pipes; pod-like structures may also be found. Segregation pipes are often more common towards the upper parts of flow units; however, they may also be found at the bases (Figs 7.14, 5.14 & 15c). These variations are thought to reflect the effects of different gas sources.

As described in Section 7.2, segregation pipes and pods can be produced during fluidisation

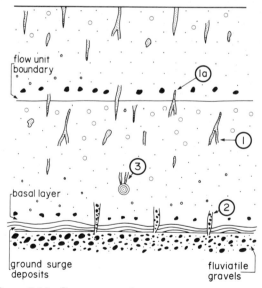

Figure 7.14 Occurrences of gas segregation structures in pyroclastic flow deposits. 1, Pipes and pods generated initially or formed entirely by intraflow gas sources during emplacement; 1a, formed by continued post-emplacement gas flow; 2, formed from heated ground water and incorporating fluviatile pebbles; 3, formed above burnt vegetation and logs.

experiments (Fig. 7.7). Such structures are very resistant to mechanical mixing, and once formed were found impossible to destroy under laboratory conditions; they have a high yield strength because U/U_{mp} is low (C. J. N. Wilson 1980). This implies segregation pipes in pyroclastic flow deposits need not always be a secondary or post-emplacement feature, but may be a primary gas-flow feature surviving from the moving flow.

It is the intraflow gas sources that are potentially the most important 'lubricating agent' and agent for the support of clasts in moving pyroclastic flows, especially pumice flows. Gas released from juvenile

Figure 7.15 Gas segregation pipes in flow units of the Minoan ignimbrite, Santorini, Greece, which are interbedded with coarse-grained flash flood deposits generated during the eruption.

Figure 7.16 Remnants of a carbonised log (by lens cap) with irregular segregation pipes rising off it. This is in the distal part of a mid-Pleistocene ignimbrite in New Zealand. (Photograph by C. J. N. Wilson.)

clasts by diffusion, and breakage and attrition, should increase systematically with height through a flow, and would therefore appear to control the formation of pipes found concentrated towards the tops of flow units. Pipes of this type are now suspected to be generated in the moving flow, although once established they may also act to concentrate post-emplacement gas flow (e.g. they may cut a later flow unit; Fig. 7.14).

Pipes found at the bottoms of flow units are perhaps generally derived from external sources, and are considered to be post-emplacement (e.g. they cut basal layers; Fig. 7.14). Good examples of this type of pipe are found in the Minoan ignimbrite, where flow units are interbedded with torrent deposits and heated ground water was the gas source (Bond & Sparks 1976; Fig. 7.15). Segregation pipes can also be found above burnt vegetation and logs (Figs 7.14 & 16).

Small segregation pipes have also been recorded from pyroclastic surge deposits. Examples are found in ash-cloud surge deposits in the Bandelier Tuffs (Fisher 1979). These cut the internal lamination of the surge deposits, and were formed by post-emplacement gas flow from the underlying parent pyroclastic flow unit.

7.3.5 LATERAL GRADING

Small, denser-clast types of pyroclastic flows may carry the largest blocks along the full length of their run-out, and show no appreciable lateral grading (e.g. P. W. Francis *et al.* 1974, D. K. Davies *et al.* 1978a).

Many ignimbrites are known to show a decrease in the maximum size of lithic clasts with distance from source (Figs 7.17 & 18). The distance at which clasts segregate out (into a lithic concentration zone at the base of layer 2b) will be dependent on:

(a) size and density of the clasts and
(b) the density or viscosity of the 'matrix' of fluidised fines.

Using the above criteria and data on the lateral grading of large lithic clasts, Sparks (1976) estimated pumice flows to have apparent viscosities in the range 10–1000 poise. Some ignimbrites show little or no lateral grading of larger lithics, and these may

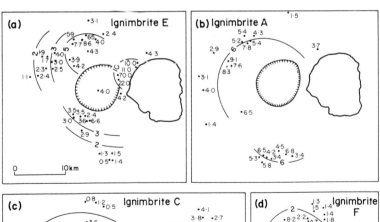

Figure 7.17 Lateral variation in the average diameter (in cm) of the three largest lithic clasts in four of the Vulsini ignimbrites. Average size decreases away from Latera caldera (left with hachures) for ignimbrites A, E and F and Bolsena caldera (right) for ignimbrite C. (After Sparks 1975)

(a)

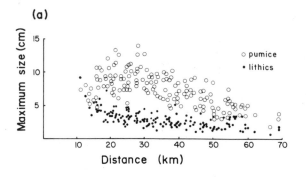

(b)

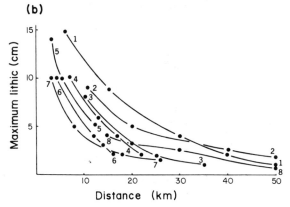

Figure 7.18 (a) Lateral variation in average maximum clast size in the Ito pyroclastic flow, Japan (after Yokoyama 1974). (b) Relationship of the average maximum lithic clast size to distance from source for several ignimbrites. 1, Towada pumice-flow deposit; 2, Ito pyroclastic flow deposit; 3, Shikotsu pyroclastic flow deposit; 4, ignimbrite C, Vulsini; 5, Lajes ignimbrite, Terceira; 6, ignimbrite F, Vulsini; 7, ignimbrite E, Vulsini; 8, Kuttyaro pyroclastic flow deposit. (After Sparks 1975.)

have had higher viscosities, e.g. the Acatlan ignimbrite, Mexico (J. V. Wright & Walker 1977, 1981).

Larger pumice clasts sometimes also show a decrease in maximum size with distance from source (e.g. Fig. 7.18). In many other cases no simple relation is found, or maximum size may even increase with distance.

7.3.6 COMPOSITIONALLY ZONED PUMICE FLOW UNITS

Many ignimbrite sheets show progressive vertical changes in mineral and chemical composition, representing the inverted sequence of magma erupted from zoned chambers. J. V. Wright and

Walker (1977, 1981) described compositional zoning within *a single flow unit*. The upper main 30 m flow unit of the Acatlan ignimbrite passes from white rhyolite pumice, through a thin mix zone (*passage zone*), into black andesitic pumice (Figs 7.19 & 20). Other ignimbrites showing a similar change within one flow unit are also known, e.g. the Hraunfossar ignimbrite in western Iceland (Sheridan 1979) and the Kaingaroa ignimbrite, New Zealand (C. J. N. Wilson *pers. comm.*). Such compositionally zoned flow units provide another way of testing the proposed emplacement mechanisms of ignimbrites. The simultaneous emplacement of light and dark pumice layers in the same flow unit with a minimum of mixing implies a laminar flow regime for the moving pumice flow. If flow were turbulent, mixing of the two pumice types would have taken place and destroyed the compositional zoning.

For the Acatlan ignimbrite, details of the grainsize distribution suggested that laminar flow was perhaps more important early in the history of the pumice flow, and later movement of the zoned flow occurred as a semi-rigid plug moving along a sheared basal layer. The question could be asked: 'why does compositional zoning not constitute evidence of layer-by-layer deposition from a turbulent, low particle concentration pyroclastic flow, as envisaged by Fisher (1966a)?' However, the massive structureless nature of the deposit, lack of bedforms and the general lack of any vertical or lateral grainsize changes (except at vent, Ch. 8) all indicate that the flow had a high particle concentration, and deposition would have been almost instantaneous by freezing of the flow, not layer by layer

Some care must also be taken in the interpretation of other colour zoned deposits, because such a change may be a post-depositional thermal effect.

7.4 Theoretical modelling of the transport of pumice flows

We have already mentioned some of the theoretical modelling by L. Wilson, Sparks and co-workers on the generation of pumice flows by collapse of plinian eruption columns (Ch. 6). Here we will

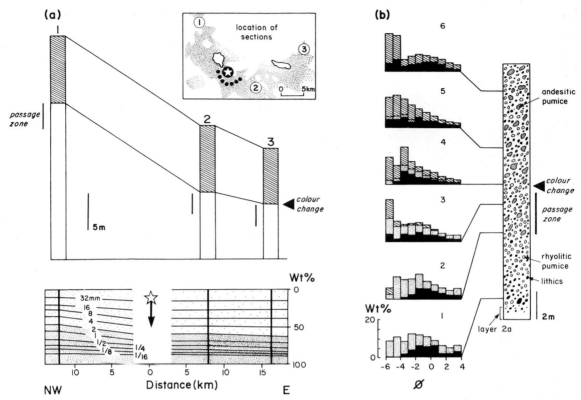

Figure 7.19 The compositionally zoned upper main flow unit of the Acatlan ignimbrite, Mexico. (a) Profiles at three locations and average grainsize distribution. Ash content highlighted by close stipple. Star marks source position; unornamented is area in which co-ignimbrite breccias occur (Ch. 8). Inset shows lateral extent of ignimbrite and positions the three locations; again, star is source vent, and large dots enclose area in which co-ignimbrite breccias are found (Fig. 8.19). (b) Details of the grainsize distribution in the profile measured at location 2. In histograms: stipple is rhyolitic pumice; diagonal rule is andesitic pumice; black is lithics. (After J. V. Wright & Walker 1981.)

briefly present some results of the modelling of the movement and emplacement of pumice flows from Sparks *et al.* (1978).

The models of Sparks *et al.* (1978) indicate that the height of eruption columns, when collapsing above the vent, is between 0.6 and 9 km for the range of values of the controlling parameters considered (gas velocity, water content and vent radius) (see Ch. 8). They then postulate that when eruption column collapse occurs, the mixture flows away from the volcano as a density current with high velocity. Initial velocities of the flows range from 60 to 300 m s^{-1} (Fig. 7.21). Flows from large eruptions may still have velocities of 100 m s^{-1} at distances of 50 km from the source. The models show that initially the mixture is likely to be a

highly turbulent, rapidly moving density current with a low particle concentration. However, a considerable proportion of the particles transported by the pumice flows are unable to be supported as a turbulent suspended load. This is because much of the grainsize distribution has terminal fall velocities well above the shearing stress velocities maintaining turbulence in a flow, even for rapidly moving flows.

The ability of turbulent flows to suspend particles is related to a parameter called the shearing stress or friction velocity, $V^\star$, defined by

$$V^\star = \left(\frac{cf}{2}\right)^U \sqrt{2} \qquad (7.9)$$

where *cf* is the drag coefficient of the ground and U

Figure 7.20 Compositionally zoned upper main flow unit of the Acatlan ignimbrite showing colour contrast between the lighter, lower rhyolitic part and the darker, upper andesitic part. Note the occasional large, dark juvenile andesitic clast in the top of the lower rhyolitic part which mark the passage zone (Fig. 7.19). Height of outcrop is 4 m.

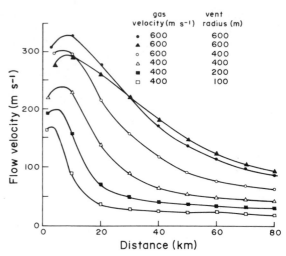

Figure 7.21 Variation of velocity with distance from a central source in a radially expanding, low density, turbulent fluid. Six sample solutions are shown, to cover a wide range of eruption column conditions. The flow velocity is calculated for the conditions of gas velocity and vent radius indicated. All of the models are for 1% water content, except for the solid circle solution, which is for 6% water content. (After Sparks *et al.* 1978.)

is the mean velocity. True turbulent suspension occurs when the terminal velocity, V_t, of a particle is less than the product of the shearing stress velocity and a constant β. Values of $\beta V^\star$ were computed at the same time as the velocity distance curves shown in Figure 7.21. Figure 7.22 shows the variation of $\beta V^\star$ with distance from the vent. $\beta V^\star$ diminishes away from the source as the intensity of turbulence diminishes. In Figure 7.23 the terminal velocities of pumice and lithic clasts are shown in fluids of several different densities, together with shearing stress velocities for flow velocities of 10–100 m s^{-1} and a drag coefficient of 0.01. These

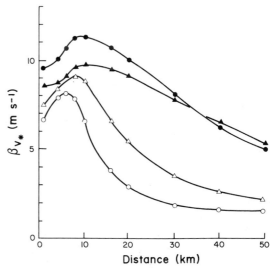

Figure 7.22 Numerical solutions to the variation of shearing stress velocity (V^*) in four model flows with distance from the source. The four examples are the same as four of those in Figure 7.21, and the legend is the same. (After Sparks *et al.* 1978.)

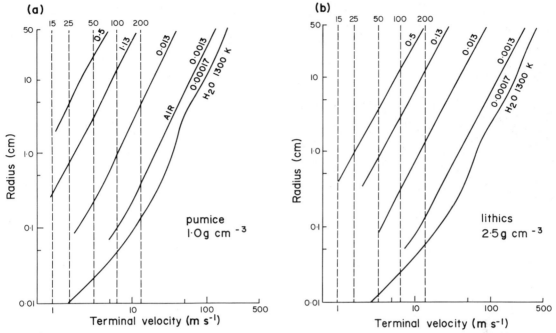

Figure 7.23 Terminal velocity–grainsize curves are given for (a) pumice clasts and (b) lithic clasts in fluids of several different densities. The curves for the densities 0.0013 and 0.00017 g cm^{-3} are for air at room temperature and steam at 1300 K, respectively. The gaseous medium in a pyroclastic flow is assumed to have a density between that of cold air ($\varrho = 0.0013$ g cm^{-3}) and that of hot steam. The curve for H$_2$O at 1300 K is an exact curve using the solutions of G. P. L. Walker *et al.* (1971), whereas the other solutions plot an approximate equation (Reynolds Number of >500). The broken vertical lines represent the shearing stress velocities of flows with velocities of 15, 25, 50, 100, and 200 m s^{-1} and a friction factor of 0.01. The intersection of any terminal velocity curve with any value of shearing stress velocity defines the size of the largest particle that can be suspended by a flow of that density and velocity. (After Sparks *et al.* 1978.)

show that even in fast flows only particles finer than about 1 mm can be transported in turbulent suspension and, in many cases, only particles finer than a few hundred micrometres can be suspended.

It was therefore deduced that pyroclastic flows develop a high-concentration basal zone within a few kilometres of the vent, as larger clasts settle to the base of the flow. In such a high concentration flow other mechanisms of particle support are dominant (see above) and the flow is capable of transporting lithic clasts of diameter several centimetres for tens of kilometres. The motion of the lower, dense flow dissociates itself from the upper turbulent cloud of fine ash and gas, which mixes with the atmosphere to form a convective plume.

7.5 Form of moving pyroclastic flows: head, body and tail deposits

The term '*form*' is used to describe the shape which a pyroclastic flow will assume when moving. C. J. N. Wilson and Walker (1982) proposed that a pyroclastic flow can be divided into a head, a body and a tail (Fig. 7.24). The head region and the body and tail region have different fluidisation states, and this controls the development of separate layers and facies within a pyroclastic flow deposit. Flows with well developed heads should produce deposits that are very different from those without heads, in which nearly all of the material is deposited from the body. It is at the head that erosion takes place, and this will be greater in flows with larger heads.

The most fluidised part of a pyroclastic flow will be the flow-head (C. J. N. Wilson 1980), into

Figure 7.24 Schematic diagram to illustrate the form of a pyroclastic flow and the division into a head, body and tail.

which large quantities of air can be ingested (Fig. 7.25). By analogy with other density currents (Allen 1971, Simpson 1972), a fast-moving pyroclastic flow is likely to develop a lobe and cleft structure at its front, with air being preferentially ingested through the clefts. Ingested air expands rapidly due to the sudden temperature increase, and this causes strong fluidisation, as well as variable degrees of turbulence. Deposits sedimented out of the flow-head should therefore be more fines-depleted and enriched in crystals and lithics than those deposited by the remaining portions of the flow (Fig. 7.25).

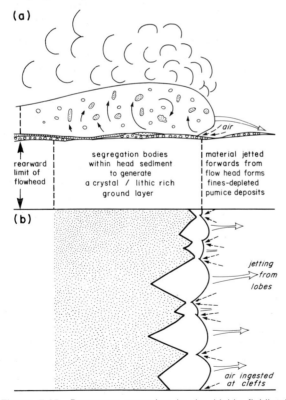

Figure 7.25 Processes occurring in the highly fluidised flow-head of a pyroclastic flow like the Taupo ignimbrite. (a) Cross section. (b) Plan view. (After C. J. N. Wilson & Walker 1982.)

Most pyroclastic flow deposits that have been studied so far appear not to show well developed head deposits, but two exceptions are the Taupo ignimbrite, New Zealand (C. J. N. Wilson & Walker 1982, G. P. L. Walker & Wilson 1983, C. J. N. Wilson 1985; Ch. 8) and the Rabaul ignimbrite, New Britain (G. P. L. Walker et al. 1981c). However, it may be that a number of crystal- or lithic-rich deposits previously regarded as ground surges originated in the heads of flows (Ch. 5). G. P. L. Walker et al. (1981a) proposed the name **ground layer** for this type of head deposit. The ground layer of the Taupo ignimbrite (Figs 7.26 & 27) extends nearly to the distal limits of the ignimbrite, although there are many short breaks in its continuity. Near the vent it is a very conspicuous bed of volcanic breccia up to 3 m thick (Ch. 8). The ground layer rests on a marked erosional surface, and erosion and deposition by the flow-head must have occurred sequentially. The top of the ground layer is separated from the base of the overlying pumiceous part of the ignimbrite by another sharp erosional contact; this is almost planar, and developed by shearing when the ground layer was overridden by the rest of the flow. In the layering scheme of Sparks et al. (1973) the ground layer is a variant of layer 1. If true ground-surge deposits are produced by surges generated from and travelling in front of a flow (Ch. 5), then a ground layer should normally also be present (see Fig. 7.29). However, neither the Taupo nor the Rabaul ignimbrites have an accompanying ground-surge deposit.

Another type of head deposit (and variant of layer 1) has also been recognised. These, in contrast to ground layers which are concentrated in heavies (crystals and lithics), are fines-depleted pumice deposits. In the Taupo ignimbrite they form a laterally discontinuous unit below the ground layer, and are separated from it by an erosive contact, described above. Fines-depleted pumice deposits are the lowest deposit of the pyroclastic flow. Thickness varies greatly, from a few centimetres to several metres, being more common and thicker in topographic depressions. This layer is thought to represent material jetted out ahead of the flow-head (Fig. 7.25). Rapid, violent expansion of air as it is

Figure 7.26 The ground layer of the Taupo ignimbrite, New Zealand, at a location about 15 km from source (Ch. 8). The ground layer is the dark lithic rich horizon (between lines) which overlies the Taupo ultraplinian pumice-fall deposit. (Photograph by C. J. N. Wilson.)

ingested into the moving flow causes portions of the pumiceous head to burst continually and to be jetted in advance of the flow proper. Large amounts of fines can be lost by this process, and in the Taupo deposit this gave rise to the distinctive **fines-depleted ignimbrite** or **FDI** (G. P. L. Walker *et al.* 1980g; Figs 7.27 & 28). At Taupo, excessive loss of fines was attributed to a very high throughput of gas, resulting partly from the ingestion of large amounts of vegetation and partly from effects of ground surface roughness on a very quickly moving pyroclastic flow, so promoting air ingestion. Data

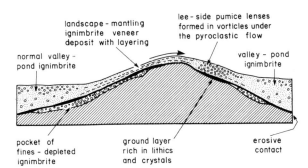

Figure 7.27 Schematic cross section showing the facies of the Taupo ignimbrite, which are thought to have been deposited by the flow head (ground layer and fines-depleted ignimbrite), body (normal valley-pond ignimbrite) and tail (ignimbrite veneer deposit) of the pyroclastic flow. (After G. P. L. Walker 1981d.)

on the height climbed by the flow suggest that the flow velocity probably exceeded 250–300 m s^{-1} near the vent (Wilson 1985). These are the highest velocities yet inferred from field observations for a pyroclastic flow, although theory suggests that they may not be unusual (Fig. 7.21). Carbonised vegetation is found at all levels in the FDI, indicating a thorough mixing by turbulence and supporting the theory of ingestion and volatilisation of large volumes of vegetation. In other fines-depleted ignimbrites, turbulence, induced locally by surface roughness, may have been the most important cause of fines loss, e.g. on St Lucia (Fig. 13.35) where pumice flows travelled down narrow, winding and heavily vegetated valleys (J. V. Wright *et al.* 1984).

The presence of jetted, as opposed to surge, deposits at the base of the Taupo ignimbrite may be related to the high velocity of the flow. Only packets of material with a high solids concentration could burst forward with sufficient velocity to be deposited before they were caught up by the flow proper. It is only where the parent flow proceeds more slowly that dilute surges have sufficient velocity to move ahead and produce a ground surge deposit (e.g. as at Santiaguito in 1973, Ch. 5). C. J. N. Wilson and Walker (1982) proposed a velocity-dependent hierarchy of conditions occur-

Figure 7.28 Fines-depleted pumice deposit formed by jetting from the flow head of the Taupo ignimbrite, New Zealand. Note the clast-supported textures.

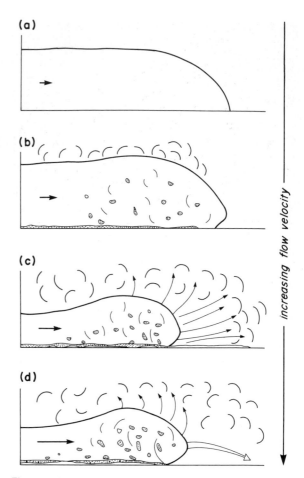

Figure 7.29 Hierarchy of conditions found at the fronts of pyroclastic flows and the formation of various layer 1 facies. (a) The flow is so slow that no significant air ingestion occurs. No layer 1 deposits are generated. (b) Minor amounts of air ingestion cause fluidisation and segregation within the head, causing the formation of a ground layer. (c) Moderate amounts of air ingestion cause dilute surges to be generated from the front of the flow, producing ground-surge deposits. Segregation within the head forms a ground layer. (d) Strong air ingestion causes the *en masse* jetting of material from the flow-front, forming jetted fines-depleted pumice deposits. Segregation within the head forms a ground layer. The estimated flow-front propagation velocities in the above groups are very approximately: (a) 0–10 m s^{-1}; (b) 10–30 m s^{-1}; (c) 30–80 m s^{-1}; (d) 80–200 m s^{-1}. At extremely high velocities (>200 m s^{-1}) the situation at the flow-front is uncertain. Evidence from the near-source outcrops of the Taupo ignimbrite implies that jetted deposits are not formed, and only a ground layer containing minor amounts of fine material is generated. What form the flow-front takes is not known. (After C. J. N. Wilson & Walker 1982.)

ring at the flow-head of a pyroclastic flow, leading to the formation of the various layer 1 facies (Fig. 7.29).

The body and tail of a pyroclastic flow give rise to layer 2 deposits. In most examples the body–tail region must have suddenly stopped, with a conventional flow unit (Section 7.3) coming to rest. Layer 2 deposits in the Taupo ignimbrite comprise two very different facies: **valley pond ignimbrite**, or **VPI** (Plate 7), and **ignimbrite veneer deposit**, or **IVD** (G. P. L. Walker *et al.* 1980b, 1981b, C. J. N. Wilson & Walker 1982, G. P. L. Walker & Wilson 1983, G. P. L. Walker 1983). VPI shows features that are typical of a normal pyroclastic flow unit. The IVD is very different, and mantles the landscape. It is stratified in localities near to the vent, and occasionally shows bedforms (Fig. 7.30), and for these reasons there has been debate whether this

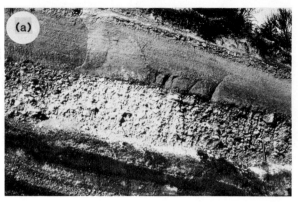

Figure 7.30 The ignimbrite veneer deposit of the Taupo ignimbrite. (a) Finely stratified IVD overlying coarse FDI; locality is 16 km from source. (b) Coarse pumice lenses in IVD on lee-side of topographic obstacle; flow direction left to right, locality is approximately 15 km from source. Note shovel for scale. (Photographs by C. J. N. Wilson.)

deposit is a type of surge or not. However, the weight of the evidence suggests that it was left behind as a trail-marker by the tail of the flow as it moved over topography. C. J. N. Wilson and Walker (1982) suggested that the tail consisted of the lower parts of the flow which, because of their proximity to the ground surface and their lower fluidisation state, were moving less rapidly than the bulk of the flow, and hence were left behind. Grainsize stratification is thought to represent the passage of waves of material in a continuous high concentration flow, each wave depositing a layer.

These layers are developed out to about 40 km from the vent, and the number decreases outwards.

Other bedforms are found on the lee-side of obstacles, where the flow jumped the ground surface. Lee-side lenses of pumice developed in turbulent vortices under the fast-moving flow, and sometimes large prograding foresets are developed (Fig. 7.30b).

IVD and VPI facies have been recognised in some other ignimbrites. They are described from the Rabaul ignimbrite (G. P. L. Walker et al. 1981c), and examples which were first called ash-

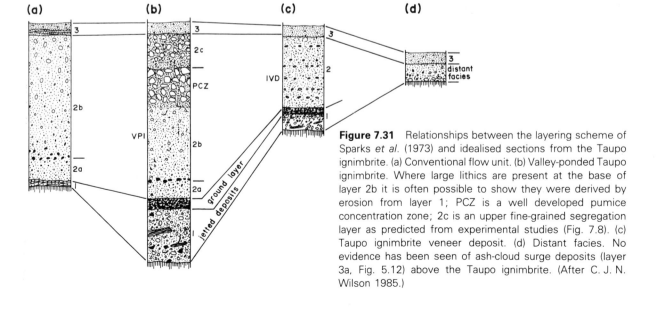

Figure 7.31 Relationships between the layering scheme of Sparks et al. (1973) and idealised sections from the Taupo ignimbrite. (a) Conventional flow unit. (b) Valley-ponded Taupo ignimbrite. Where large lithics are present at the base of layer 2b it is often possible to show they were derived by erosion from layer 1; PCZ is a well developed pumice concentration zone; 2c is an upper fine-grained segregation layer as predicted from experimental studies (Fig. 7.8). (c) Taupo ignimbrite veneer deposit. (d) Distant facies. No evidence has been seen of ash-cloud surge deposits (layer 3a, Fig. 5.12) above the Taupo ignimbrite. (After C. J. N. Wilson 1985.)

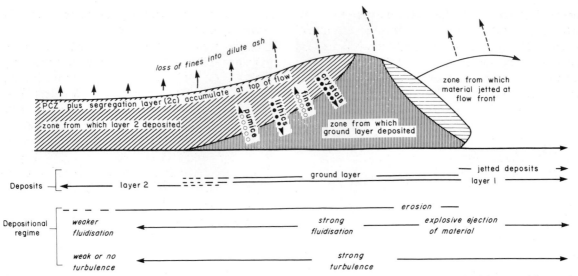

Figure 7.32 Dynamic model summarising processes which generate the different layers deposited by a rapidly moving pyroclastic flow. (After C. J. N. Wilson 1985.)

hurricanes by Roobol and Smith (1976), occur on Martinique, erupted from Mt Pelée.

Depending on the flow velocity, and the nature of the landscape over which emplacement occurs, we can envisage a complete range of pyroclastic flow forms. Slowly moving, denser-clast flows and some pumice flows only have poorly or moderately inflated heads. These produce pyroclastic flow deposits, which in section have no, or only thin head deposits. Material transported in the flow-head is dumped, forming a steep flow front, or is pushed aside by the advancing denser body of the flow which has a much higher yield strength, so forming levées (e.g. the Mt St Helens 22 July and 7 August 1980 deposits; L. Wilson & Head 1981; Fig. 7.1b). The bulk of the flow in these cases forms a conventional layer 2 deposit. Layer 1 may include a ground surge deposit (Fig. 7.31a). Other pyroclastic flows (exemplified by the Taupo ignimbrite) which are emplaced at very high velocities have highly inflated, turbulent heads, leading to the development of prominent head deposits and a highly erosive base below a ground layer. It is envisaged that as such a flow moves, material from the flow body is laterally transferred and cycled through the more strongly fluidised front of the flow. In this region bubble-induced turbulence,

due to strong fluidisation accompanying air ingestion, is important. The ground layer is then overridden by the base of the flow body, and high shear stresses result in the erosion of layer 1. These flows produce well defined body and tail deposits. The interpretation of the various facies of the Taupo ignimbrite in terms of the three-layer scheme of Sparks et al. (1973) is given in Figure 7.31. A dynamic model summarising their formation is shown in Figure 7.32. Such flows seem to stop because they run out of material rather than because they freeze. For this reason, the edge of the Taupo ignimbrite is not defined by a steep flow front. Instead a zone 3–5 km wide is found where layers 1 and 2 cannot be distinguished. This single layer is termed the distant facies (Fig. 7.31d), and is regarded as the deposit of the flow at the stage where air-ingestion fluidisation was affecting the entire flow, spreading a thin, strongly fluidised layer across the landscape (C. J. N. Wilson & Walker 1982, C. J. N. Wilson 1985). The flow presumably terminated when remaining material rose in the form of a buoyant convective plume adding to a co-ignimbrite air-fall ash.

In summary, we still have much to learn about pyroclastic flows and their deposits. There appears to be a very large spectrum of types, including dense block and ash flows at one extreme, to turbulent, low density, violent types at the other extreme. The latter, although only recently recognised, may be more common than is currently realised. Clearly, much more work still needs to be done on all aspects of pyroclastic flow processes and on the characteristics of their deposits.

7.6 Pyroclastic surges as low particle concentration flows

Pyroclastic surges are regarded as *turbulent, highly expanded, low particle concentration flows* (Ch. 5). Wohletz and Sheridan (1979) described surges as time-transient, unsteady flows of tephra that occur as a pulse or series of pulses in which the kinetic energy rapidly decays. Surges are complex *three-phase systems*, being mixtures of *solids, gases* and *water*. The proportions of these phases vary from one surge to another, and even during the flow of individual surges (Allen 1982), but volumetrically the proportion of solids (and thus their concentration) is subordinate to gas and liquid in most surges, and therefore much less than in pyroclastic flows (also see G. P. L. Walker 1983). In *hot, dry surges* the liquid phase is an insignificant component. In *wet, lower temperature surges* subequal proportions of all three phases are likely. In terms of analogy with epiclastic sediment transport processes, surges are therefore roughly akin to subaqueous turbidity currents (*but* see Section 7.11 for further critical discussion), whereas most pyroclastic flows are akin to high concentration, viscous regime debris flows (Ch. 10). Surges can result from eruptions of any magma type, and can arise from both magmatic and phreatomagmatic eruptions (Chs 5 & 6).

During the initial stages in pyroclastic surges, solid particles are widely dispersed within the fluid phase(s), and are essentially supported and driven by them through turbulence in the gaseous phase. The solids thus behave in a particulate fashion during transportation, and are free to sort themselves hydraulically. In viscous mass flows particle interaction and particle–fluid cohesion hinders sorting to a large degree. As the wetness of surges increases, particle freedom decreases due to adhesion processes (see below).

Surges have densities higher than the ambient atmospheric density, so their passage will be largely gravity-controlled. Generally, they follow topographic lows (J. G. Moore 1967, Waters & Fisher 1971) but, because of their largely turbulent nature, stemming from an initial explosive thrust or high gas content relative to solids (especially fines, see G. P. L. Walker 1983), or from both, they also have the ability to climb very significant topographic highs and to mantle them with a thin veneer of ash (e.g. Fig. 5.1; Nairn 1979).

Surge-forming eruptions are usually composed of multiple surges, frequently as pulses with negligible time intervals. Surge-forming eruptions may therefore lead to deposits which are stacks of surge-deposited beds. Successive surges may erode and rework to varying degrees the deposits of preceding surges.

7.7 Energy sources and initiation of surges

In Chapter 5 surges were subdivided into three main types:

> base surges
> ground surges
> ash-cloud surges

Each has a different origin, and may therefore contain different proportions of the three main phases (solid, gas and liquid). This may in turn lead to surges with varying physical properties, transportational and depositional modes, and facies characteristics of their deposits.

7.7.1 BASE SURGES

Base surges (J. G. Moore 1967) originate from the base of a phreatomagmatic eruption column as an outward moving, ground-hugging, turbulent cloud of fluid and ash (Ch. 5, Fig. 5.17). They develop from phreatomagmatic blasts in the vent, which eject dilute mixtures of solids, gases and steam. These expand rapidly on eruption and then spread radially, or flow along directed paths as turbulent

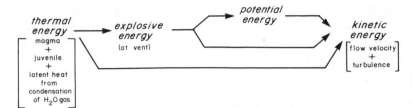

Figure 7.33 The complex energy chain involved in the initiation and flow of pyroclastic surges.

surges. Other surges originate from minor column collapse events (Ch. 5). Base surges will generally be wet, unless they are extremely hot when erupted and so contain little condensed water.

The energy for base surge initiation and flow is derived initially from the thermal energy of the rising magma (Fig. 7.33). This is translated on contact with ground water or surface water to mechanical explosive energy, as the heat from the magma is used to superheat and explosively expand the external water source (Fig. 7.33; Ch. 3). The explosive energy is translated into momentum and kinetic energy. Potential energy is also created during eruption, especially where base surges are derived from minor column collapse (Ch. 5). This potential energy, in turn, translates into kinetic energy. In addition, during at least the early stages of flow, extra thermal energy is liberated from juvenile fragments, and latent heat is released when condensation of gaseous water to liquid water droplets occurs (Fig. 7.33). The thermal energy produces turbulent convective circulation, and the large initial explosive thrust ensures high lateral transport velocities, which are responsible for turbulent flow conditions. In the Reynolds Number

criterion for turbulence (Eqn 2.5, $Re = U\varrho g/\eta$), the low viscosity of the driving fluid (gas) and the high velocity fo the surge both contribute to high Reynolds Numbers. Both the convective and velocity-induced turbulence entrain and support grains (see below).

7.7.2 GROUND SURGES

Ground surges are associated with pyroclastic flow forming processes, and are identified where they directly underlie pyroclastic flow facies. They may have several origins, as outlined in Chapter 5:

(a) directed blasts at vent,
(b) partial collapse of maintained eruption columns and
(c) projection from the head of moving pyroclastic flows.

In the first two cases the energy for initiation and the mobility of the surge originates in much the same way as for base surges. However, those that are spawned from moving pyroclastic flows may involve one further step. Their origin may be due to the ingestion of cold air in clefts at the flow front, or

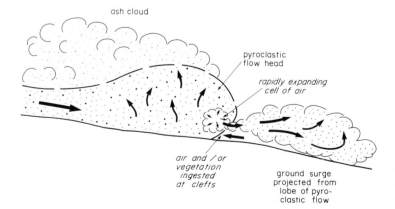

Figure 7.34 Schematic representation of the formation of a ground surge through ingestion of air or volatilisation of vegetation or both, at the head of a moving pyroclastic flow.

by erosional incorporation of excessive amounts of vegetation with its high water content (Section 7.2; Fisher 1979, C. J. N. Wilson & Walker 1982). Both of these elements would be instantly heated, and would produce rapidly expanding cells which could have enough momentum and excessive kinetic energy to be ejected forward, out of the head of the moving pyroclastic flow as a cell of highly turbulent gas and vapour with low particle concentration, which moves forward as a surge (Fig. 7.34). The surge deposit is then immediately overridden and buried under the basal facies of the following pyroclastic flow (Figs 5.13, 23a & b; Sparks & Walker 1973, C. J. N. Wilson & Walker 1982; Section 7.5).

7.7.3 ASH-CLOUD SURGES

Also associated with pyroclastic flows are ash-cloud surges (Ch. 5), forming from the trailing ash cloud which billows above and behind a pyroclastic flow, as gas and fine ash stream out of the head and body of the flow under the influence of fluidisation processes (Fisher 1979; Section 7.2). This loss of fine ash is called elutriation, and was discussed in Section 5.2.

The energy in this type of surge is derived entirely from the parent pyroclastic flow. The surge acquires its momentum and kinetic energy from the pyroclastic flow out of which it stems. Unlike the parent pyroclastic flow with high particle concentration, which is essentially flowing in laminar fashion (Section 7.2; Sparks 1976; but see above), the associated ash-cloud surge flows turbulently. This turbulence also has two sources – convective circulation and the overall high velocity of the surge, which contributes to high Reynolds Numbers, as discussed above. The turbulence supports the solids.

7.8 Transportation and grain-support processes in surges

It has been suggested above that the transportation of pyroclastic detritus in surges is dominated by turbulence in the supporting gaseous phase. This is based on direct observations of recent base surges (J. G. Moore et al. 1966, J. G. Moore 1967, Waters & Fisher 1971, Kienle et al. 1980, Self et al. 1980) and on the sedimentary structures observed in the deposits of all types of surges. Observed surges are dominated by turbulent billowing clouds of gas, steam and ash. On this basis, at least the peripheral parts of surges are turbulent. However, the unobservable inner, lower parts, or the 'body' of the surge, can also be considered to be turbulent by virtue of the tractional sedimentary structures such as waves (dunes), cross-stratification and horizontal lamination commonly found in their deposits (J. G. Moore 1967, Fisher & Waters 1970, Waters & Fisher 1971, Wohletz & Sheridan 1979, Allen 1982, Leys 1982; Figs 5.20, 22 & 23). All of these structures can only be produced by grain-by-grain tractional transport (rather than *en masse*; Ch. 10) implying low grain concentrations in the transporting medium. Under these circumstances, the *uplift force of turbulent fluid eddies* is the main grain-support process possible during the bulk of surge movement. Suspended grain-support will be maintained as long as the *drag force* of these uplifting turbulent eddies exceeds the settling velocities of grains at the appropriate Reynolds Number for the turbulence level (Allen 1982).

In wet surges, this simple analysis is complicated by the process of clumping or adhesion of particles in the very moist atmosphere of the moving surge (Allen 1982, Leys 1982). Clumping and adhesion refer to the aggregation of moist grains, especially fines, with moisture droplets, so increasing the *effective dynamic transported grainsize*. For continued suspension transport of these clumps, the upward component of the fluid drag force must therefore also exceed the settling velocity of the clump. The criterion for turbulent suspension has been discussed in Equation 7.9. Allen (1982) suggests that a simple criterion for suspension transport to be predominant is a Bagnold criterion:

$$W/V^\star \leq 1.25 \qquad (7.10)$$

where W is the particle settling velocity and $V^\star$, as defined in Equation 7.9, the shear velocity of a moving flow or current.

These conditions hold for an initial highly turbulent state and for small grainsizes. However, surge deposits are known to contain fragments several centimetres or more in diameter, which were not finally transported and deposited as ballistic blocks. They must have been at least entrained, and moved by surges as well. The actual competence of surges to transport different sizes in true suspension is not clear, in spite of the discussion related to Equation 7.9. However, it is clear that during incipient sedimentation, if not before, a large proportion of the solids are saltated (bounced), rolled and dragged over the substrate as bed load under the influence of a large lateral shear velocity (Ch. 10). More of the population will experience this mode of transport as the surge loses momentum and begins to deflate.

There is therefore very likely to be a 'grain carpet' layer at the base of most surges, out of which the bulk of sedimentation will occur, as well as a more dilute trailing ash cloud. The grain-layer population is dragged along by the shear stress operating at the base of the flow. Individual grains will propel neighbours by colliding with them and transferring momentum, which constitutes a form of inertial *dispersive pressure*. It can be speculated that a well defined, voluminous grain layer would be most common in base surges. Base surges originate as a vent blast and propel a poorly sorted grain population from which a significant coarse fraction will tend to drop out quickly. Both ground surges and ash-cloud surges consist of a population that is presorted to some degree. Therefore, in a single surge some transport will take place through turbulent suspension and some through shear-induced traction and grain collisions. As the surge loses energy and deflates, more solids, particularly the coarser fractions (including clumps) will cease being transported in a suspension mode and will become entrained in a traction mode of transport. At low grain concentrations, discrete grain traction probably occurs; at higher concentrations a grain-layer flow with high degrees of particle interaction and collision, propelled by shear-stress, will develop. The change from low grain concentration traction to high concentration, shear-induced grain layer flow, conditions should be reflected, respectively, by cross-stratification and discrete lamination, and by more massive or diffusely layered beds.

Complications to this will arise according to the wetness, or moisture content, of the surge. In very dry surges, the grain transport can be viewed as occurring in an inertial regime, i.e. free from viscous interaction with other grains or with the driving fluid, as described above. As the moisture content increases, and cohesive clumps of sediment develop, transport becomes pseudo-viscous near the base of the surge. It is still not wholly viscous, because the high shear stress operating probably continually reconstitutes wet clumps and smears them out. That is, the moisture-laden, driving gaseous fluid is distinct from the muddy clumps it is propelling, since they have not totally amalgamated into a cohesive, viscous fluid. In this sense, wet surge transport is quite different from other types of particulate tractional transport or mass-flow transport because it is a *polyphase system*. It has not been mathematically modelled. Evidence of this complex phase transportation by wet surges is found in accounts of observed base surges and their deposits. Plastering of muddy ash on the upwind side of trees and buildings was documented at Taal (J. G. Moore 1967). That these surges were still turbulent is attested by the 'sandblasting' effects in abrading and stripping of bark and foliage, and in the near-blast zone, by the total destruction of vegetation (J. G. Moore 1967; Fig 5.19).

The role of fluidisation as a grain-support process and the origins of the gaseous component in surges has been touched on by several workers (Fisher 1979, Allen 1982, Leys 1982). Fluidisation in polydispersive systems is most effective where the grain concentration is high, and an abundance of fines exists to stem the free flow of fluid through the dispersion (Allen 1982, G. P. L. Walker 1983). In pyroclastic flows it is seen as a very effective agent in maintaining the mobility of the flows (Section 7.2). In surges it is less likely to do so because of the low concentration of solids right from the outset. Any fluid which is ingested will not be concentrated along discrete channels, but will mix with the ambient fluid phases in the surge. In this sense, the ingestion of air at the head of the surge may temporarily increase the internal turbulence of the surge, particularly if the air is cold and the surge is hot, so leading to expansive heating of the ingested air. However, there will be overall loss of heat and energy from the surge. Leys (1982), following Allen (1971, 1982), suggested that because there is only a small overhang in the profile of the head or leading edge of observed surges, it is unlikely that significant ingestion of external air into the head and at the base of the head occurs. If this is so, then ingestion and mixing, and fluidisation in the head region of surges, can be discounted as significant grain-support processes. The absence of known gas and segregation pipes originating *within* surge deposits (cf. pyroclastic flow deposits, Section 7.3.4) also supports this. This is not to say that surges do not contain gas and air, which is at times reflected by post-depositional vesiculation in surge deposits (e.g. Lorenz 1974).

What, then, are the potential sources of the fluids providing grain-support in surges? As for pyroclastic flows (Section 7.2), the obvious sources are magmatic volatiles (both those initially present and those that diffuse from juvenile fragments within the surge), external water incorporated during phreatomagmatic eruptions, and volatiles derived from vegetation over which surges travel. Although the role of external air ingested into the head and fluidising the flow as a significant aid to grain support for the bulk of the solids has been discounted above, it is likely that significant mixing of air into the billowing top of the surge occurs (Allen 1982). There, the mixing will provide turbulent support for only the finest grainsizes transported by the surge.

7.9 Depositional processes in surges

Any transporting medium begins to deposit its load when the energy levels drop below the threshold level needed to maintain discrete grain support or tractional entrainment, or drop below the momentum of a mass of particles moving *en masse* as one. Surges, although initiated by an 'explosive' thrust, once in motion, behave largely as gravity controlled density currents (Allen 1982). As the turbulence which provides grain support declines in response to cooling and slowing, a changing mode from suspended transport, associated initially with an *erosion stage*, to surface traction transport will commence. The energy that generates turbulence for grain support is, as described above, of two types: convective turbulence, due to the heat in the surge, and velocity-induced turbulence in a medium of low viscosity and high lateral velocity. Turbulence due to fluidisation-induced streaming of external air is considered to be minor (Section 7.8).

Convective turbulence will be dissipated as the surge cools in transit. This will occur as latent heat is lost during condensation of steam to water and as cold air is mixed into the surge, particularly along its top. However, convective turbulence will probably be subordinate to velocity-induced turbulence. As the velocity declines, the sediment-carrying capacity of the surge will also decrease, and the surge as a density current will be dissipated. Allen (1982, p. 397) proposes four causes for the dissipation of a sediment density current with a steady head and uniform body:

(a) density reduction through sediment loss or engulfment of ambient medium,
(b) reduction in flow thickness, related to flow stretching or collapse,
(c) friction at the base and upper surface of the flow, partly involving density reduction due to mixing between flow and medium and
(d) reduction in bed slope.

Fall-out of sediment will cause a decrease in the bulk density of the surge and loss of momentum ($= mv$, mass × velocity), and hence velocity and turbulence, according to the Reynolds Number criterion. A reduction in the slope will also lower the velocity, as potential energy is being lost. Given the low density of surges, the coefficient of friction between the surge and its substrate will be low while the surge is inflated and carrying a large suspended sediment population. However, when the surge begins to deflate significant grain fall-out will produce a concentrated basal tractional grain layer. There will be a higher coefficient of friction between this grain layer and the substrate, permitting sedimentation and further slowing of the surge.

Frictional retardation of a surge can also occur by mixing of air at the head and top of the surge. Allen (1982) suggested that the dissipation rate of a

density surge is

$$\text{dissipation rate} = h_s/L \qquad (7.11)$$

where h_s, the thickness of the body of the surge, is equal to $0.01H$, H being the height of the head, and L, the streamwise length of the mixing zone, is equal to about $2-3H$. The dissipation rate is in the order of $0.01-0.001$ (Allen 1982). Mixing of air into the top probably occurs, but will have little effect on the density of the head and body. *If* surges are similar to turbidity currents in a *general* fluid dynamic sense, then the body should flow more quickly than the head, implying that once solids have reached the head they should be circulated upwards into the head proper, where mixing with the ambient medium occurs due to the growth of Kelvin–Helmholtz instability waves (billows) on the top of the head (Allen 1982, p. 400). The upper part of the head is diluted by this, and becomes less dense and therefore slower.

For relatively dry surges, grain transport will pass from suspended and surface traction modes of transport to increasing degrees of tractional transport. The *rate* at which grain transfer from the suspended to the traction mode occurs will control whether relatively low grain concentration tractional transport and sedimentation occurs (producing corresponding structures such as cross-stratification and well defined lamination), or whether high grain concentration shear-induced grain-layer flow occurs, with attendant high degrees of particle interaction, deposition by freezing of the grain mass and deposition of a massive bed. It can be speculated that with wet surges the *condensation of steam* should also be an important influence in causing sedimentation. *Adhesion* occurs as fine ash and water droplets begin to adhere to each other, producing clumps. The effective transported grainsize will be increased by this. Condensation should therefore lead to an increased rate of sedimentation compared with the pre-condensation rate of sedimentation, because the greater effective transported grainsize may exceed the prevailing competency of the surge, and because of the adhesion effects between the substrate and moist particles in the surge. The rate at which condensation takes place will therefore be important.

If surges do segregate into a lower concentrated ground layer and a more dilute suspension cloud, then some deposits should contain a separate 'body' layer and an associated fall-out layer (e.g. Fig. 7.35).

Figure 7.35   Probable distal rhyolitic base surge deposits from a Devonian tuff ring-caldera lake succession, Snowy River Volcanics, eastern Victoria. The massive layers represent surge 'body' deposits separated by accretionary lapilli-rich 'co-surge' ash-fall layers (finger).

7.10 Facies characteristics of surge deposits

The facies characteristics of surges can be readily discussed in terms of some of the essential facies parameters introduced in Chapter 1, including:

geometry
grainsize
sorting
shape and vesicularity
composition
depositional structures

7.10.1 GEOMETRY

The geometry of surge deposits will depend on the type of surge, the topographic control and post-depositional erosion. *Base surges*, being a vent-related facies, will build up an annulus around the vent which is wedge-shaped in cross section, thinning radially away from the vent. Base surges form one of the principal facies of tuff rings and maar volcanoes (Ch. 13). The deposit from a single surge will be a thin sheet, with minor variations in thickness being controlled by pre-depositional topography and the surface bed-form (e.g. dune forms). However, base surge deposits are almost invariably composed of multiple layers representing multiple surge events (e.g. Figs 5.21 & 22). Such a pile may be tens of metres thick around the vent (e.g. Crowe & Fisher 1973, Schmincke *et al.* 1973, Sheridan & Updike 1975, Fisher 1977). Individual layers may be up to a metre thick near the vent, although usually closer to 50 cm or less, and distally away from the vent, only millimetres to several centimetres thick. Base surges probably do not flow further than 10 km from the vent, and usually only several kilometres. As such, a succession of base surge deposits in the rock record is indicative of proximity to the vent. They may contain intercalated air-fall layers, as well as thin near-vent pyroclastic flow deposits (e.g. Fisher *et al.* 1983). Successive base surge deposits may be separated by thin co-surge ash-fall deposits up to several centimetres thick, derived from a trailing ash cloud (cf. co-ignimbrite ash-fall deposits).

Ground surges, usually being expulsions from the head of a pyroclastic flow will usually deposit thinner facies intervals than base surges do (e.g. Fig. 5.23). They may be no more than several metres thick, and usually a metre or less, and should be directly overlain by the basal facies of the parent pyroclastic flow. Being ejected from a moving pyroclastic flow, the geometry and extent of ground-surge deposits will largely be controlled by the topography into which the pyroclastic flow has moved.

Ash-cloud surge deposits should occur as a thin (up to several metres thick) sheet of fine ash, mantling the deposit of the host pyroclastic flow behind which it trailed (e.g. Fisher 1979; Fig. 5.21). However, the preservation potential of ash-cloud deposits is low because of the effects of post-depositional erosion. Unless quickly buried beneath further eruptive products, they are likely to be stripped off shortly after emplacement.

7.10.2 GRAINSIZE

The grainsize of surge deposits reflects both the degree of fragmentation at the time of eruption *and* the competency of surges to carry particular grainsizes (Ch. 1). The coarseness and variance in the grainsize is thought to be a reflection of the levels of turbulence (G. P. L. Walker 1983). As Crowe and Fisher (1973) pointed out, because surge-forming eruptions are highly pulsatory and of variable explosive strength, there may be considerable changes in the flow-power and in the grainsize characteristics of successive surges, and even within single surges. Near to the vent ballistic fragments and air-fall materials may also be incorporated into surge deposits. However, both *base-surge* and *ground-surge* deposits, because of their high energy state and because their sources at the vent and the heads of pyroclastic flows may contain considerable coarse debris, are capable of carrying, and do carry, significant amounts of large lapilli-size clasts. However, *ash-cloud surges*, being the products of continued elutriation processes in the head and body of pyroclastic flows, will be low energy systems and their deposits will be fine-grained, rarely containing lapilli, and then probably only highly vesicular, low density ones.

7.10.3 SORTING

The sorting characteristics of the different surge deposits will also be variable, and will be controlled by the degree of fragmentation, the density variations in the debris being carried, the turbulence levels, adhesion processes and the time available for sorting. As discussed in Chapter 1, the sorting is an hydraulic process, and in populations containing grains with a variety of size, density and shape, variation in grainsize will not necessarily reflect poor hydraulic sorting. Nevertheless, given the origins of the three surge types, it is likely that *ash-cloud surge* deposits will be better sorted than *base-surge* or *ground-surge* deposits. In addition, because of the relatively low particle concentration in surges, the sorting will be significantly better than in most pyroclastic flows, but less so than in fall deposits (Fig. 7.36). However, field observation shows that major grainsize variations occur between the layers in a single surge deposit, especially base-surge deposits, suggesting a highly pulsatory mode of flow and grain transport within a single surge. Secondly, sorting is generally still much poorer than in most normal sedimentary systems transporting equivalent grainsizes, even in turbidites, the nearest sedimentary analogue.

7.10.4 SHAPE AND VESICULARITY

The shape and vesicularity of grain populations in the deposits of the three surge types will be more a

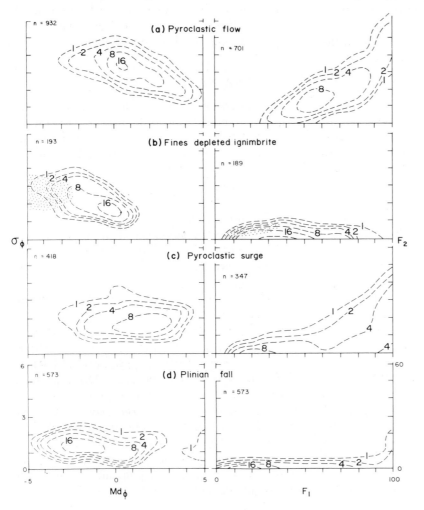

Figure 7.36 Plots of Md_ϕ/σ_ϕ and of weight percentage finer than $\frac{1}{16}$ mm (F_2), versus weight percentage finer than 1 mm (F_1) for pyroclastic flows and three kinds of fines-depleted pyroclastic deposits. In (b) various kinds of fines-depleted facies associated with ignimbrite are included: ground layers, fines-depleted ignimbrite and gas segregation pipes. The dotted area contains co-ignimbrite breccias not included in the contoured plot. Note that fields in (b), (c) and (d) overlap, showing that grainsize parameters alone are not good criteria of origin. (After G. P. L. Walker 1983.)

reflection of the mode of fragmentation of the magma during eruption than of flow processes. Hence, base-surge deposits, being of phreatomagmatic origin, will be dominated by poorly vesicular, blocky fragments where the erupting magma is poorly vesiculated (Ch. 3), as will the deposits of ground and ash-cloud surges whose host pyroclastic flows have been associated with phreatomagmatic eruptions (e.g. Self 1983). However, ground and ash-cloud surge deposits derived from eruptions driven by magmatic explosions (Ch. 3) will contain abundant vesicular fragments, although concentration processes during the initiation of ground surges may concentrate denser lithics and crystals in the surges (Section 5.7).

7.10.5 COMPOSITION

The composition of the erupting magma and products has an indirect relationship to surge types. Basaltic pyroclastic eruptions essentially do not produce pyroclastic flows. As a result, essentially all basaltic surge deposits found in the rock record can be inferred to be near-vent base surge deposits. The converse is not true, however. Intermediate and silicic eruptions can give rise to all three surge types: base, ground and ash cloud. Accidental clasts may be a significant element in a surge deposit due to explosive incorporation at the vent, and particularly in phreatomagmatic base surge deposits. Accessory clasts may be picked up in transit.

7.10.6 DEPOSITIONAL STRUCTURES

Depositional structures are diverse in surge deposits, and have been recognised as a response to varying flow and physical conditions ever since surges were recognised as a pyroclastic transportational and depositional agent (J. G. Moore et al. 1966, J. G. Moore 1967, Fisher & Waters 1970, Waters & Fisher 1971, Heiken 1971, Crowe & Fisher 1973, Schmincke et al. 1973, Mattson & Alvarez 1973, Sheridan & Updike 1975, Fisher 1977, Wohletz & Sheridan 1979, Allen 1982, Leys 1982, Fisher et al. 1983, Edney 1984, Edney & Cas in prep.). Many of these authors have also recognised lateral changes in the nature of bedforms or associated internal depositional structures, or both, with distance from the vent, as a response to changing flow conditions as dissipation of surge energy occurs.

As with normal sedimentary structures, structures in surge deposits can be classified as:

pre-depositional,
syn-depositional and,
post-depositional.

Pre-depositional structures

Pre-depositional structures include *erosion gullies* and *U-shaped channels* (Ch. 5) carved out of the depositional surface and infilled by surge deposits. The origin of these depressions may be normal epiclastic erosional processes or erosion by pyroclastic surges (e.g. Fisher 1977, Richards 1959, J. G. Moore 1967; Ch. 5.7, Fig. 5.21c & d). Others include *planar slide surfaces*, usually on the inner crater wall where segments of the unconsolidated, frequently bedded pile of pyroclastics forming the upper crater wall collapse and slide back into the crater (Fig. 7.37). Younger pyroclastic surge and fall deposits may accumulate on this slide surface.

Syn-depositional structures

Syn-depositional structures include wave-like (or dune-like) structures called *dune forms* or dunes, and their internal *cross-stratification, massive beds* without structure and *planar beds*. In a discussion of the lateral facies changes of surges from vent to

Figure 7.37 Slide surface sloping into the crater of the Lake Purrumbete maar tuff ring, Western Districts, Victoria, Australia. Younger surge deposits have been deposited on this slide surface. Flow direction is from right to left.

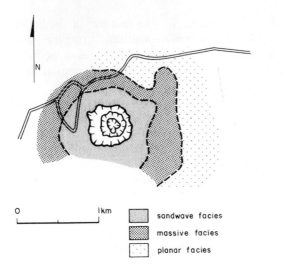

Figure 7.38 Distribution of major surge facies relative to distance from vent, Ubehebe Crater, California, USA. (After Wohletz & Sheridan 1979.)

equivalently structured ash facies, presumably similar to those of Wohletz and Sheridan (1979). The massive to diffusely layered lapilli facies at least, is considered by Edney and Cas (*in prep.*) to represent a proximal near-vent surge or surge grain-layer underflow, perhaps deposited by the head of the surge as suggested by Leys (1982), or both. $Md_\phi/\sigma\phi$ plots (Fig. 7.41) show that the massive lapilli facies are better sorted than pyroclastic flows (cf. Fig. 7.36) and that all facies are coarser than the previously defined facies of Wohletz and Sheridan (cf. Fig. 7.39). We suggest that, given the complexities of surge mechanics (dry, hot, wet, cold, condensation, intensity of fragmentation, variation in degree of explosive thrust), it may yet be premature to propose simple proximal to distal and upsequence facies changes and models similar to the Bouma facies model for turbidites. Lapilli and ash-surge facies have quite different hydraulic equivalence, and the relationships are bound to be complex.

The terms dune form, dune and antidune have

distal settings, Wohletz and Sheridan (1979), using many of the concepts presented by Sheridan and Updike (1975), document the occurrence of these syn-depositional structures relative to distance from the vent, based on studies of many volcanic centres in the western USA. Facies intervals dominated by wave- or dune-like bedforms and cross-stratification occur closest to the vent, facies intervals with dominantly massive, structureless beds occur at medial distances and facies intervals dominated by planar beds are most distal (e.g. Fig. 7.38).

Wohletz and Sheridan (1979) suggest that the ordering of these facies with distance from vent can be related to the changing flow conditions within a surge with time and distance, and the implication is that a facies model, much like the Bouma sequence for turbidites, can be considered. The documented grainsize characteristics for these facies (Wohletz 1983, Sheridan & Wohletz 1983) indicate that all are dominated by ash-sized material (Fig. 7.39). However, Edney (1984) and Edney and Cas (*in prep.*) have recognised that there is also an extensive assemblage of lapilli-dominated base-surge facies in the hydrovolcanic basaltic centres of western Victoria, Australia, including **massive to diffusely layered lapilli facies** (Fig. 7.40), **cross-bedded lapilli facies** (Fig. 5.21a) and **planar-bedded lapilli facies** (Fig. 7.40). These are associated with

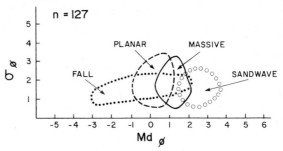

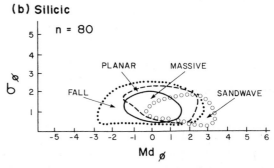

Figure 7.39 Grainsize characteristics of the principal surge facies of Wohletz and Sheridan (1979) compared with associated fall deposits. (After Wohletz 1983.)

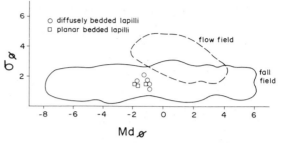

Figure 7.41 Grainsize characteristics of massive and diffusely layered lapilli surge facies, Tower Hill Volcanic Centre, western Victoria. (After Edney 1984.)

Figure 7.40 Three depositional facies of base surges: dune form (top), planar bedded (middle) and massive to diffusely layered lapilli (bottom). The upward succession from massive to planar bedded to dune-form facies reflects progressively lower concentration surges passing this point. The corresponding decrease in grainsize corresponds with increasing degrees of fragmentation and explosiveness or decreased surge competency, or both. Note the thin lenses of very low angle cross-stratified ash within the planar bedded facies representing incipient dune form formation. The dune-form facies interval (top) contains cross-beds ranging from less than angle of repose, to steeper than angle of repose, reflecting the wet cohesive nature of the phreatomagmatic ash. Flow is obliquely into the photograph from right to left. Holocene Tower Hill Volcanic Centre, Western Districts, Victoria, Australia.

all been applied to the wave-like structures that characterise **dune-form facies** (both ash and lapilli deposits, Figs 5.21, 22 & 7.40). Allen (1982) points out that 'dune' and 'antidune' have been used almost invariably synonymously with *aqueous* epiclastic wave forms. However, the use of these terms

for surge bed-forms may be incorrect, because the physical processes involved in the formation of pyroclastic dune forms and aqueous dunes and antidunes are quite different. Dune-like forms may be symmetrical or asymmetrical. Asymmetry may be either upflow or downflow. Dune-form bedding set thickness decreases approximately logarithmically away from the vent (Wohletz & Sheridan 1979) and average wavelength also decreases away from the vent (J. G..Moore 1967, Waters & Fisher 1971, Wohletz & Sheridan 1979). Wavelength and wave height are logarithmically proportional to each other (Allen 1982).

Cross-stratification angles are highly variable (Figs 5.21, 22 & 7.40). Although they approach the angle of repose in some instances, in most instances they are considerably less than the angle of repose, suggesting that the origin of the cross-stratification is not analogous to low flow regime, aqueous, dune formation, in which cross-beds form by *passive*, inertial gravitational avalanching of grain layers down the downstream face of the dune. Such a process also operates with the formation of aeolian dunes, and some ripples and cross-stratification. A similar process may be responsible for high angle cross-stratification in surge dune forms, probably associated with dry surges.

Low angle cross-stratification is due to high velocity, current-driven grain layers being sheared over an irregular, waved surface which itself has a relatively low relief due to high bed shear stress operating. Low angles of cross-stratification in surge deposits have frequently been equated with aqueous antidunes, irrespective of whether the dip direction has been towards the source (normal with aqueous antidunes) or away from the source. The implication of this is that upper flow regime conditions, analogous to those in aqueous unidirectional flow systems, were operating. However, as Allen (1982) points out, this direct analogy is not valid. In the aqueous system we are dealing with cohesionless sediment entrained within an aqueous medium with a density of 1 g cm^{-3}. In the pyroclastic surge situation we are dealing with particulate matter driven by a fluid which has a density very much less than that of water.

In the case of *dry surges*, which are simple, *two-phase gas–solid systems*, the processes may be analogous to normal aeolian processes. With dry surge dune forms, normal asymptotic, angle of repose cross-strata of avalanche origin may occur but, even so, the competency of surges in transporting coarse debris is well beyond the limits of normal surface winds, and low angle cross-stratification is common. In the case of *wet, three-phase surge systems* there seems to be no direct analogy with normal sedimentary processes and systems, so implications of analogous flow regime conditions and processes should strictly be avoided. With wet surges, added complications arise because of the existence of a changeable three-phase system, the adhesion processes taking place in transit and the cohesive nature of the substrate over which the surge travels. The last two of these are likely to be responsible for the often very fluidal nature of the layering in surge dune forms. The depositional process will involve adhesion interaction between entrained detritus and substrate, and deposition will take place under the influence of a 'smearing' bed shear force.

Further support for the view that flow conditions in surges and normal sedimentary systems are not analogous comes from the very significant grainsize variations between the layers of the same surge cross-bed set, suggesting a much more pulsatory non-equilibrium flow system. This grainsize variation can be from ash to lapilli sizes. Nevertheless, there is conceptual, and probably some fluid dynamic, equivalence between aqueous antidunes and surge dune forms, given that the low angle cross-stratification requires high shear stresses at the bed surface – much higher than that associated with low flow regime, high angle of repose cross-stratification.

Nevertheless surge dune forms display sufficient regularity in form for them to be used in determining surge transport directions. Observations on modern volcanoes have shown that dune crests are continuous and are orientated perpendicular to the surge flow direction (J. G. Moore 1967, Waters & Fisher 1979, Allen 1982; Fig. 5.19d). Internal cross-stratification can also be used to infer flow directions (Figs 5.21–23 & 7.40). Allen (1982) related the documented variations in bedforms and cross-stratification relationships in base surge deposits to sedimentation rate, surge temperature and moisture content (Fig. 7.42).

In this scheme he has recognised three categories of surge dune bedform: *progressive*, *stationary* and *regressive*. In progressive bed-forms (types A$_1$, A$_2$ and B, Fig. 7.42) the crests of the dune forms migrate in the direction of surge flow. Where the sedimentation rate is high, the internal arrangement of cross-strata resembles normal sedimentary climbing ripples and cross-stratification (type A$_1$). Progressive dune forms are considered by Allen to be the result of surges that are relatively dry or hot, or both. Stationary dune forms (type F) are distinguished by crests which migrate neither upstream or downstream. Regressive dune forms are distinguished by an upstream migration of the crest of the dune form, as represented in the crestal point of successive sigmoidally-shaped cross-strata in a single set. Allen suggests that regressive dune forms are the products of wet or cool surges. Although the crests are retreating upstream, the through-flow of solids within individual layers is downstream. In regressive bed-forms, the depositional process involves cohesive interaction between the transported debris and the substrate. Coarser fragments occur on the upstream side of the crest (cf. progressive dune forms). 'Climbing' effects, in this case upstream, result from both high sedimentation rates and adhesion–plastering due to wetness. Allen suggests that stationary dune forms result from surges with temperatures at the condensation or boiling point of water (Fig. 7.42). In support of his scheme, Allen (1982) points out that

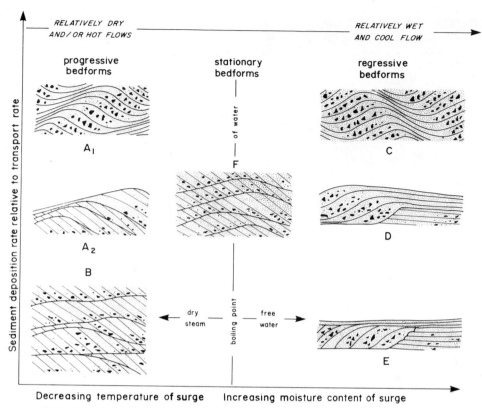

Figure 7.42 Classification of base surge bedform and internal cross-stratification variations related to depositional rate (relative to transport rate; vertical axis) and surge temperature and moisture content (horizontal axis). (After Allen 1982.)

only deposits with stationary or regressive dune forms are known to contain accretionary or armoured lapilli, or both, as well as vesicles which are due to post-depositional expansion of trapped hot volatiles in wet, impermeable ashes.

Small amplitude, irregular adhesion ripples, similar to those formed in wind-swept snowfields, are another, less common, form of dune forms associated with surge deposits (Allen 1982, Leys 1982).

The cross-stratification in surge dune form facies can be very variable in form. In high angle of repose sets it can be upward concave, asymptotic and wedging downstream. In other situations it is much more fluidal, showing significant thickness variations and, unlike normal sedimentary cross-stratification, individual layers may be continuous from one dune form into the next, showing pinch and swell geometry, and thinning over the crests of

waveforms. The dips in cross-beds may vary. Where it is greater than the angle of repose (>30–35°), adhesion processes or soft-state plastic deformation, or both, perhaps under the shearing influence of the host surge, are almost certainly responsible. Angle of repose cross-stratification may also be due to these influences or due to lee-side avalanching from a dry surge or from part of one. More commonly, however, surge cross-stratification is characteristically of a low angle, indicating the influence of significant bed shear or wet state smearing influences, or both, rather than grain avalanching. Although the cross-beds in individual sets may be conformable with each other, very low angle, and irregular, truncations may be common, and these reflect the highly pulsatory nature of surges, even during deposition.

Massive ash facies (cf. massive lapilli facies, above) are considered by Wohletz and Sheridan

Figure 7.43 Very low angle cross-stratification interval (middle) within planar bedded facies of uncertain but probable surge origin. Eocene–Oligocene Waiareka–Deborah Volcanics, Oamaru, New Zealand.

(1979) to be transitional between dune form and planar bedded facies. They describe beds of massive facies as being frequently lens-shaped and occurring on the lee-side of dune forms. They are ungraded, and generally unstructured internally except for occasional planar to wave-like diffuse layering or pebble trains. Wohletz and Sheridan (1979) consider massive beds to be deposited by a deflating, highly concentrated stage of surge flow, perhaps involving some fluidisation. The massive character is indicative of high particle concentrations and very rapid rates of deposition, probably involving

en masse freezing of the grain population. The diffuse layering probably reflects internal shearing at the time of deposition in the highly concentrated grain aggregate.

Planar bedded facies consists of layers millimetres to several centimetres thick which are planar, slightly wavy or even locally lensoidal. Just as in the cross-stratified sets of dune forms, major grainsize variations are common from layer to layer and between different sets. Some are ashes, others lapilli deposits (Edney 1984, Edney & Cas *in prep.*). Layers are generally laterally extensive, and reverse grading is common. The grainsize variation between layers again suggests a highly pulsatory depositional process during the formation of a set of planar beds. Reverse grading indicates that a layer of grains is shearing laterally over the substrate, larger grains migrating upwards out of the zone of maximum shear at the interface between the surge and the substrate. Grain fall-out from suspension is not indicated by this reverse grading. The reverse grading, being of a shear origin, also implies that the surge is in a deflated, highly concentrated state at the time that this, the most distal facies recognised, forms. Planar-bedded facies may be difficult to distinguish from bedded fall deposits without evidence of lateral transport (scours, cross-stratification, low angle truncations).

Figure 7.44 Cross-stratified surge deposits viewed perpendicular to flow direction showing wavy layering, and low angle cross-stratification and truncation. Mt Leura, western Victoria, Australia.

Planar beds may be gradational with cross-stratified intervals, and in some instances may have a *very* low angle inclination (Fig. 7.43) appearing to be planar but being incipient cross-beds (W. Edney *pers. comm.*). However, in this regard the perspective is important. Exposures that are perpendicular to flow direction will expose wavy, parallel to subparallel layering of cross-bed sets (Fig. 7.44).

Post-depositional structures

Post-depositional structures commonly associated with surge deposits include *bomb sags* (Fig. 5.21g & h), formed when ballistic blocks and bombs lob into wet, unconsolidated surge deposits producing *soft-sediment plastic deformation*. *Flame structures* at the base of surges have also been described (Crowe & Fisher 1973) and *soft-sediment oversteepening* of dune form cross-strata under the shearing influence of the succeeding surge is common. Such oversteepening may bring strata that are normally at lower than angle of repose inclinations (<30°), to angle of repose (30–35°) or even steeper. The layering is usually still intact, but is stretched and squeezed due to plastic deformation while in a wet, cohesive state. Similarly, normal soft-sediment slumping of wet, cohesive ash is common (Fig. 5.21f).

7.11 Surges compared with turbidity currents

In Section 7.6 it was suggested that, as a crude generalisation, surges are akin to turbidity currents in the normal sedimentary sphere, whereas pyroclastic flows are akin to viscous debris flows. This general analogy is based on the fact that surges and turbidity currents are turbulent, gravity-controlled mass flows with low particle concentration, whereas pyroclastic flows and debris flows are generally both viscous, laminar-plug flow systems. There the analogy ends, however.

Whereas turbidity currents begin to flow because of their potential energy, surges are initiated by an explosive thrust, with the exception of ash-cloud surges. Thereafter there is a complex energy chain, as discussed in Section 7.7. Whereas in turbidity currents the particulate population is essentially cohesionless, this is clearly not the case with surges (Sections 7.8–10). Turbidity currents are simple two-phase solid–liquid systems in which the density contrast between the solids and the supporting liquid is relatively small compared with that in the complex three-phase (solid–liquid–gas) system of most surges. The relative proportions of the three phases may change in surges, as may the effective dynamic granulometry, due to adhesion processes leading to clumping.

Because of all these differences, hydrodynamic equivalence between turbidity currents and surges cannot be considered valid (Allen 1982). However, once initiated, both move as gravity-controlled, turbulent, cloudy masses, which seem to produce a systematic succession of facies during deceleration and deposition. The facies are not the same, and in both cases not all facies elements need be produced by all flows. Much work still needs to be done on the facies and facies relationships produced by surges.

7.12 Pyroclastic surges and pyroclastic flows – relationships

The discussion of pyroclastic flows and surges so far has emphasised their unique characteristics, the former being described as poorly inflated, non-turbulent, concentrated mass flows which maintain their kinetic energy over long distances and long periods (relatively), whereas the latter highly inflated, turbulent, low particle concentration, short-term phenomenon which rapidly dissipates. However, it has also been shown (Section 7.5) that there are some extremely violent pyroclastic flows which are at least partly turbulent, and are capable of producing wavy stratification, bedforms and low angle truncations and cross-stratification (Fig. 7.30) in their veneer deposits. The Taupo ignimbrite (G. P. L. Walker *et al.* 1981b, C. J. N. Wilson 1985), the Rabaul ignimbrite (G. P. L. Walker *et al.* 1981c) and the ignimbrite of the 1815 eruption of Tambora (Self *et al.* 1984) are key examples. This brings into question the degree to which pyroclastic flows and surges are distinct entities

(Section 5.6). Each is a type of flow, but is there a spectrum from dense pyroclastic flows (e.g. nuée ardentes, block and ash flows) at one end to dilute surges at the other? G. P. L. Walker (1983) and G. P. L. Walker and McBroome (1983) touched on this problem. In both these publications Walker poses the possibility that two of the better-documented so-called surge events of modern time – those of Mt Pelée in 1902 and the 18 May 1980, cataclysmic eruption of Mt St Helens, were not surges, as has been asserted in the literature to date (Fisher *et al.* 1980, A. L. Smith & Roobol 1982: Mt Pelée; Hoblitt *et al.* 1981, J. G. Moore & Sisson 1981: Mt St Helens; Section 5.6), but highly violent pyroclastic flows. Waitt (1981) observed that the characteristics of the 18 May Mt St Helens deposit were similar to neither purely surge nor pyroclastic flow deposits, but shared the characteristics of each. Hoblitt *et al.* (1981) and J. G. Moore and Sisson (1981) also noted anomalies compared with the documented characteristics of surge deposits.

The diverse descriptions and subdivisions of the so-called surge deposits by different authors make it difficult to extract a consensus stratigraphy but, following Walker and McBroome, it seems that three principal layers can be identified.

Layer 1 is a relatively well sorted gravelly or sandy layer with rapid lateral changes in thickness and grainsize and some lateral discontinuity. Grainsize and thickness decrease outward from the source, and significant fines depletion is evident. Layer 2 contains two facies, a massive one which 'has the aspect of a pyroclastic flow deposit particularly where it occurs in valley ponds' (G. P. L. Walker & McBroome 1983, p. 571), and one showing stratification and dune forms. The massive facies is homogeneous, occurs in valley pond settings, and contains gas escape pipes where it is thick. Layer 3 is very fine, contains accretionary lapilli, is rarely more than several centimetres thick and is agreed by all authors to be an ash fall-out of dilute trailing ash clouds derived from the 'blast-

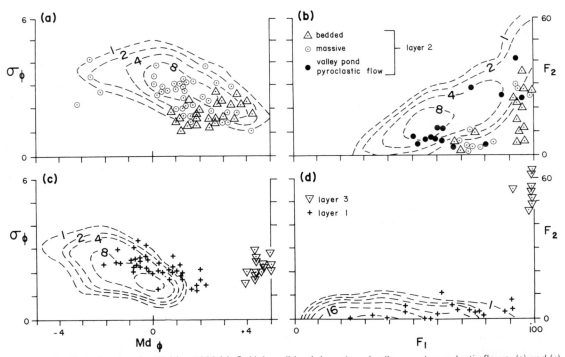

Figure 7.45 Grainsize data for 18 May 1980 Mt St Helens 'blast' deposit and valley-pond pyroclastic flows. (a) and (c) Md_ϕ/σ_ϕ plots, (b) and (d) plots of weight percentage finer than $\frac{1}{16}$ mm (F_2) versus weight percentage finer than 1 mm (F_1). In (a) and (b) contoured fields for pyroclastic flows are shown, and in (c) and (d) contoured fields for samples from fines-depleted facies of ignimbrite (G. P. L. Walker 1983) are shown. (After G. P. L. Walker & McBroome 1983.)

surge' body. The relationship between layers 1 and 2 is generally abrupt, but locally their interface is gradational (Moore & Sisson 1981, Waitt 1981). The total thickness of this assemblage of layers rarely exceeds 1 m, except in hollows.

Based on σ_ϕ versus $Md_\phi/\sigma\phi$ plots and plots of F_2 (the weight percentage finer than 1/16 mm) versus F_1 (the weight percentage finer than 1 mm) plots, G. P. L. Walker and McBroome (1983) show a close relationship between the layer 1 and 2 samples and pyroclastic flow fields (Fig. 7.45). However, samples show a significant depletion in fines, indicating better sorting than most pyroclastic flow deposits.

Similarly, G. P. L. Walker and McBroome (1983) point out that the surge deposits of Fisher *et al.* (1980) and Fisher and Heiken (1982) associated with the 1902 Mt Pelée eruption consist of three layers: a basal gravelly to sandy layer, a stratified to massive middle layer and a fine, thin, accretionary-lapilli bearing, ash layer. Grainsize characteristics are similar to those of Mt St Helens.

G. P. L. Walker and McBroome (1983) and G. P. L. Walker (1983) interpret both the Mt St Helens and Mt Pelée deposits to be the products of very violent, low aspect ratio pyroclastic flows. In the Mt St Helens case this is based on the grainsize affinities with pyroclastic flows, and the doubt that dilute surges could travel 30 km and maintain turbulent suspension of clasts over that distance. Layer 1 is interpreted as the fines-depleted ground layer deposited from the fluidised head. The dune forms and their internal wavy layering seem ill-defined relative to most true surge deposits, suggesting weak, low levels of turbulence (G. P. L. Walker 1983). The main layer, layer 2, seems very similar to valley pond ignimbrite, and also occurs on gently sloping valley walls, as a thin veneer deposit (Ch. 8; G. P. L. Walker & McBroome 1983).

It is possible then that two modern catastrophic events which had been interpreted as surges (Section 5.6), may have been very violent, even partly turbulent pyroclastic flows, similar to the historic ones erupted from the Taupo, Rabaul and Tambora volcanic centres. It is therefore implicit that a spectrum of pyroclastic flow types exists, from very dense, high particle concentration ones to relatively low concentration, but violent, ones. There is an implication in this that there is also a spectrum of grain-support processes from perhaps inertial grain flow avalanching through fluidised, laminar semi-plug flow (Sparks 1976, C. J. N. Wilson 1980), through to fluidised, turbulent flow (C. J. N. Wilson 1980). Given that surges are turbulent, but that during deflation they may undergo shear-induced laminar flow, there is not much of a quantum gap between dilute types of pyroclastic flow and surges, implying a nearly complete spectrum of processes and deposits. However, on that point G. P. L. Walker (1983) maintains that discreteness is required because, to maintain integrity and to be capable of transporting large dispersed clasts, pyroclastic flows are dependent on fluidisation, which can only be maintained while *fines are retained* in order to act as resistance to, and to channel, fluidising ingested air. In surges, fines loss is inherent because of the high level of turbulence and the consequent removal of large volumes of fines by elutriation, which essentially discounts the possibility of fluidisation. On the other hand, however, fines loss or depletion can also occur in certain situations related to pyroclastic flow processes (and not just surge transport; G. P. L. Walker 1983; Fig. 7.46) without affecting the overall coherence of the pyroclastic flow. It is clear that much is yet to be learned about the processes operating in pyroclastic flows and surges and how these two flow processes are related. This is more than vindicated by the diverging views on the G. P. L. Walker and McBroome (1983) interpretation, as shown by the responses and reply to that paper (Hoblitt & Miller 1984, Waitt 1984, G. P. L. Walker & Morgan 1984)!

7.13 Further reading

The papers by Sparks (1976), Sheridan (1979), Fisher (1979), C. J. N. Wilson (1980, 1984) and C. J. N. Wilson and Walker (1982) are essential reading on aspects of flow and depositional processes in pyroclastic flows, even though they have been summarised here. On surge processes and facies

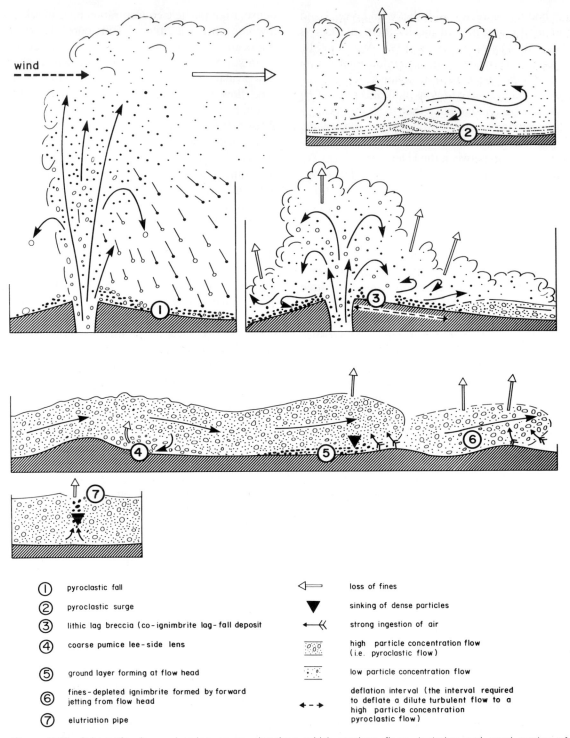

Figure 7.46 Schematic views showing seven situations which produce fines depletion and good sorting of pyroclastic deposits. (After G. P. L. Walker 1983.)

characteristics, the review by Wohletz and Sheridan (1979) is extremely useful, as are the earliest papers on surge deposits by J. G. Moore (1967), Fisher and Waters (1970), Crowe and Fisher (1973) and Sheridan and Updike (1975). On the problem of the distinction between violent pyroclastic flows and turbulent surges, the papers by G. P. L. Walker (1983) and G. P. L. Walker and McBroome (1983) are recommended.

Plate 8 Canyon wall cutting through the Upper Bandelier Tuff showing a common sequence of deposits produced during ignimbrite-forming eruptions: plinian fall deposit (base is at same level as top of the geologist's head), overlain by a pyroclastic surge deposit and massive ignimbrite. This sequence overlies at least two smaller, sub-plinian(?) pumice-fall deposits resulting from earlier eruptions of the Valles caldera complex.

CHAPTER EIGHT

Ignimbrites and ignimbrite-forming eruptions

Initial statement

Ignimbrites are the most voluminous of volcanic products. Some are the largest single eruptive units known, covering thousands of square kilometres and having volumes of more than 1000 km³. Although man has never witnessed an eruption giving rise to such large volume units, they must be the most cataclysmic of all geological phenomena. Even small-volume, historic ignimbrite-forming eruptions are awesome: e.g. the 1470 BC eruption of Santorini, which has been linked with the rapid decline of the Bronze Age Minoan civilisation centre on Crete; the AD 79 eruption of Vesuvius, in which the towns of Pompeii and Herculaneum were destroyed; the eruption of Krakatau in 1883 which set in motion tsunamis killing more than 30 000 people on neighbouring islands; Tambora, which in 1815 caused the deaths of more than 90 000 people either directly, or as a result of tsunamis and an ensuing famine; Mt St Helens in 1980; and El Chichón in Mexico in 1982, after which 2000 people were missing.

This chapter examines in detail the geology of ignimbrites: definition, occurrence, volume, types of vent from which erupted, eruption sequence, a depositional facies model for ignimbrite-forming eruptions and welding. We also highlight the AD 186 Taupo eruption, because recent studies of its deposits have greatly affected current thinking on ignimbrite-forming events. Transportation and depositional mechanisms are discussed in the previous chapter.

8.1 Enigma of ignimbrites

Ignimbrites pose many problems but, as recently stated by G. P. L. Walker, unravelling their origin has been 'one of the outstanding success stories of modern volcanology'. They vary widely in lithology, and few products of volcanism have been misinterpreted for so long. They vary from incoherent ash deposits texturally similar to mud flows, to hard, densely welded tuffs which may be difficult to distinguish from lava flows. Indeed, welded ignimbrites were generally regarded as lava flows until the mid-1930s, and even later. They vary from rock composed almost entirely of sub-millimetre particles to that in which the largest clasts may be over a metre in diameter. Many ignimbrites are completely non-welded, but the importance of these was probably not recognised until the late-1960s and early-1970s.

It will be apparent to the reader who has researched the volcanological literature that the term 'ignimbrite' has been used in many different ways. Marshall (1935) first introduced the term into the geological literature. Confusion, partly attributable to the imprecision of Marshall's definitions, has arisen and still exists because 'ignimbrite' is sometimes used in a lithological sense to mean welded tuff, and sometimes in a genetic sense to mean the rock or deposit formed from pyroclastic flows. Very often the term is used in both senses, or it is used in ways that are never clearly defined. However, it is illogical to use it as a lithological term solely for welded rocks, since ignimbrites generally also have non-welded zones. The occurrence of welding has occasionally been cited as sufficient evidence of a pyroclastic flow origin. However, even many large ignimbrites are entirely non-welded, e.g. the Los Chocoyos ash-flow tuff in Guatemala (Table 8.1). Also, welded *air-fall* tuffs are believed to be a common volcanic rock (Ch. 6). Welding cannot therefore be considered to be a fundamental characteristic of ignimbrites.

Following Sparks *et al.* (1973) 'ignimbrite' is defined here as: the rock or deposit formed from pumiceous pyroclastic flows irrespective of the degree of welding or volume. 'Pumice-flow deposit' is an equivalent term (Ch. 5).

Table 8.1 Bulk volume estimates of some ignimbrites*.

Deposit	Volume (km^3)
USA	
Timber Mountain Tuff	
Rainier Mesa Member	1200
Ammonia Tanks Member	900
Paintbrush Tuff	
Topopah Spring Member	250
Tiva Canyon Member	1000
Nelson Mountain Tuff	500
Carpenter Ridge Tuff	500
Fish Canyon Tuff	3000
Sapinero Mesa Tuff	1000
Upper Bandelier Tuff	200
Bishop Tuff	500
Peach Springs Tuff	90
John Day pyroclastic flow deposit	75
Crater Lake pumice-flow deposit	25
Valley of Ten Thousand Smokes ignimbrite	12
Central and South America	
Rio Caliente ignimbrite	30
Acatlan ignimbrite	5
Los Chocoyas ash-flow tuff	200
Apoyo pyroclastic flow deposit	5
Puricipar ignimbrite	100
El Yeso ignimbrite	40
Cerro Galan ignimbrite	1000
Lesser Antilles	
Roseau ignimbrite†	30
Average Belfond pumice flow deposit on	
St Lucia	0.2
Iceland	
Skessa welded tuff	4
Roydarfjordur tuff	2.5
Matarhnjkur welded tuff	1
Mediterranean	
average of six ignimbrites erupted from	
Vulsini volcano	6
Vesuvius AD 79	4
Campanian ignimbrite	100
Minoan ignimbrite†	30
Japan	
Ito pyroclastic flow deposit	110
Aso III pyroclastic flow deposit	175
Kutcharo welded tuff	90
Akan welded tuff	60
Hakone pyroclastic flow deposit	15
Hijiori pyroclastic flow deposit	1.4
Kuttara welded tuff	10

Deposit	Volume (km^3)
Mashu pyroclastic flow deposit	5
Numajiri welded tuff	1.2
Tazawara welded tuff	150
Tokachi welded tuff	60
Tokachidake 1926 pumice-flow deposit	0.02
Agatsuma pyroclastic flow deposit	0.1
Asama pumice-flow deposit I	2
Asama pumice-flow deposit II	1
Komagatake 1929 pyroclastic flow deposit	0.15
Nantai pumice-flow deposit	0.8
Shikotsu pyroclastic flow deposit	80
Towada pyroclastic flow deposit I	25
Towada pyroclastic flow deposit II	25
Indonesia	
Toba Tuff	2000
Bali ignimbrite	20
Tambora 1815	25
Krakatau 1883†	12
New Zealand	
Ongatiti ignimbrite	190
Whakamaru ignimbrite	150
Matahina ignimbrite	100
Taupo ignimbrite	30

* There are many problems with such a survey of the literature. First, data from which estimates have been made vary from detailed maps and volcanological studies to simply gross estimates. Secondly, a significant or major part of some ignimbrites is welded tuff, but the proportion of welded to non-welded material is generally not given; therefore, no allowance is made for compaction although this is an important factor when making size comparisons. Thirdly, published studies have tended to describe the larger, more spectacular ignimbrites. Note also that the volume lost into a co-ignimbrite ash fall deposit is not included in these estimates.

† Major part deposited in sea as subaqueous 'pyroclastic flow' deposits and ash turbidites (Ch. 9).

Based on the supposed dominance of ash-sized particles in pyroclastic flows, R. L. Smith (1960a, b) and Ross and Smith (1961) introduced the term 'ash-flow tuff', which is extensively used throughout the American literature. However, as the modal grainsize in many pumice-flow deposits is in the lapilli or bomb size range, such deposits would therefore not strictly come under the definition of ash-flow tuff. For this reason the 'English school' has tended to retain the term 'ignimbrite'. It is unlikely that either term will be dropped from the literature.

Also, 'nuée ardente' has sometimes been used for 'ignimbrite', as well as for other types of pyroclastic flows (Chs 5 & 12). If used, 'nuée ardente' should be restricted to those small-volume block and ash flows produced by the collapse of an actively growing lava flow or dome, as originally described by La Croix from the 1902 eruption of Mt Pelée (La Croix 1904). Nowadays we tend to avoid the term, not only because it has become ambiguous, but also because there are other more-specific terms, and most of the 'glowing cloud' forms ash-cloud surge and ash fall deposits.

8.2 Occurrence, composition and size

Ignimbrites are common volcanic products, being found in all volcano-tectonic settings: oceanic islands (e.g. Iceland, Azores and Canary Islands), island arcs (Lesser Antilles), microcontinental arcs (New Zealand), continental margin arcs (Andes) and continental interiors (Western USA). Rhyolite, dacite and andesite are the most common compositions. Many of the most voluminous ignimbrites are rhyolites, some of which are compositionally zoned, indicating in some cases the tapping of large zoned magma chambers (R. L. Smith 1979). Ignimbrites can have alkaline compositions, e.g. in the Azores, Tenerife, Philippines and Italy (Ch. 13). Pantelleritic ignimbrites have been recorded from a number of areas, e.g. the East African Rift, Gran Canaria and the Western USA. Ignimbrites have been recognised in geological formations of all ages.

In size, ignimbrites range over at least five orders of magnitude of volume. The largest are restricted in their occurrence to continental margins and interiors, and large islands. They typically result from eruptions of silicic calc-alkaline magmas, and tend to form extensive sheets or shields (Ch. 13). The smallest are found in all settings and are a common product of stratovolcanoes, although not restricted to this type of centre (Ch. 13). These tend to form valley fill deposits, and their distribution may be localised and their stratigraphy very complex. Figures 8.1–8.7 illustrate the distributions of some selected ignimbrites. Note the differences in

(a) Nelson Mountain Tuff (26.5 Ma)

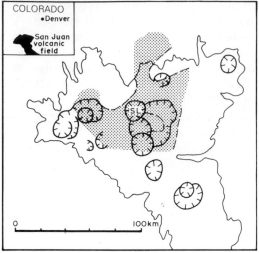

SL San Luis Caldera

(b) Carpenter Ridge Tuff (27.2 Ma)

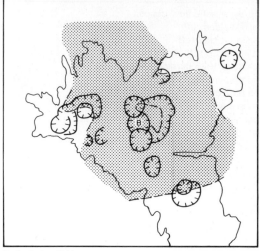

B Bachelor Caldera

(c) Fish Canyon Tuff (27.8 Ma)

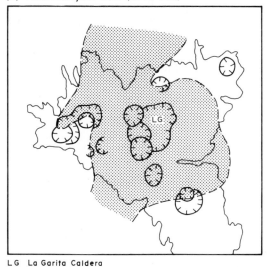

LG La Garita Caldera

(d) Sapinero Mesa Tuff (28.1 Ma)

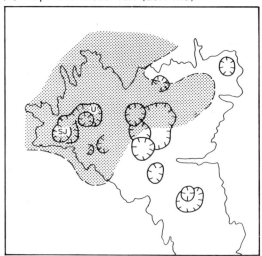

U and SJ   Uncompahgre–San Juan caldera complex

Figure 8.1 Examples of large ignimbrite sheets and associated calderas in the Tertiary San Juan volcanic field, Colorado. For the most part they are densely welded tuffs. Note that these maps only show schematic original distribution, not present outcrop pattern, which is very complex because of erosion. (After Steven & Lipman 1976.)

scales between maps; there is a 24-fold difference between maps showing examples from the San Juan volcanic field in Colorado and the Mt St Helens 1980 deposit.

Table 8.1 lists the estimated volumes of a number of ignimbrites. Most of the largest ignim-

brites described are those associated with large calderas in the Western USA, and in many cases are, for the most part, densely welded tuff. The world's biggest is the Fish Canyon Tuff in the San Juan volcanic field (Fig. 8.1c) which is thought to have a minimum volume of 3000 km^3. An enormous

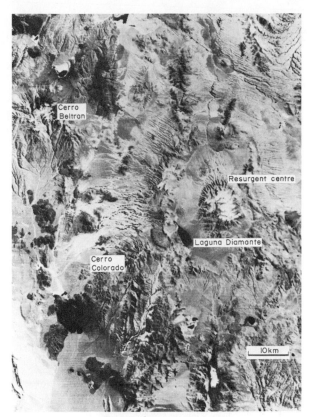

Figure 8.2 A LANDSAT image showing the giant Cerro Galan ignimbrite and caldera in northwestern Argentina. Extensive ignimbrite shield is seen as pale-toned, regularly dissected smooth expanses, best displayed on the northern flanks. The snow-capped resurgent dome (6100 m above sea level) is conspicuous in the middle of the caldera. Laguna Diamante is a relic of a much larger lake which may have filled the caldera before resurgence. Cerro Colorado and Cerro Beltran are pre-caldera andesitic stratovolcanoes. Dark areas west of Cerro Colorado are basaltic andesite scoria cones and lava flows broadly contemporaneous with the ignimbrite. (After P. W. Francis *et al.* 1983.)

part of these large deposits occurs as thick **intra-caldera ignimbrite**, while smaller volumes are found as **outflow sheets**. Later updoming of the intracaldera pile may form a resurgent dome (Fig. 8.2, Ch. 13).

In Table 8.2, the maximum distance travelled from the source vent is given for a number of ignimbrites. The data demonstrate the ability of pumice flows to travel large distances, often over gently sloping ground and sometimes over topographic barriers (also see Figs 7.2, 17, 18 & 21).

(a)

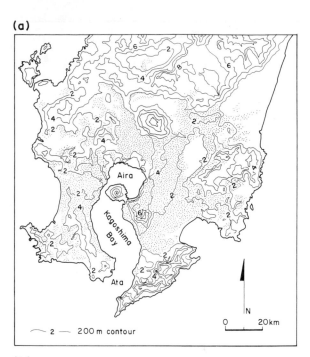

(b)

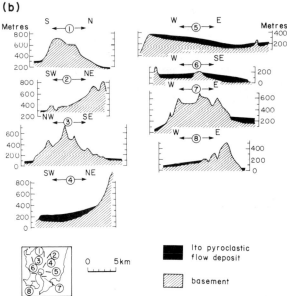

Figure 8.3 Ito pyroclastic flow deposit from Aira caldera, Japan, erupted about 22 000 years BP. Its eruption is thought to have produced the Aira caldera. (a) and (b) show the complex distribution pattern of the ignimbrite, and the topography that the (one) pumice flow was capable of climbing; cross section 4 shows a good example of depositional ramping (Section 8.8). The ignimbrite is largely non-welded. (After Yokoyama 1974.)

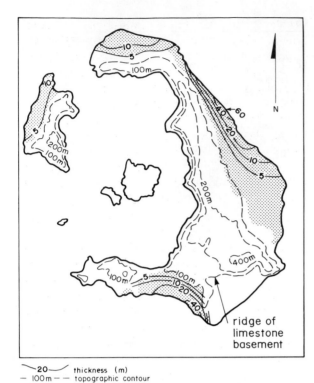

```
~20~      thickness (m)
— 100m — —  topographic contour
```

Figure 8.4 Isopach map of the Minoan ignimbrite, Santorini.

Table 8.2 Maximum distances travelled from source by some ignimbrites*.

Ignimbrite	Distance (km)
Timber Mountain Tuff	
Rainier Mesa Member	76
Paintbrush Tuff	
Topopah Springs Member	60
Nelson Mountain Tuff	50
Fish Canyon Tuff	100
Sapinero Mesa Tuff	110
Upper Bandelier Tuff	30
John Day pyroclastic flow deposit	75
Crater Lake pumice-flow deposit	58
Valley of Ten Thousand Smokes ignimbrite	22
Aniakchak ash-flow tuff	50
Fisher ash-flow tuff	30
Mt St Helens 1980	8
Rio Caliente ignimbrite	20
Acatlan ignimbrite	20
Puricipar ignimbrite	35
Cerro Galan ignimbrite	70
Ignimbrite A Vulsini volcano	17
Ignimbrite C Vulsini volcano	27
Ito pyroclastic flow deposit	80

Table 8.2 *continued*

Ignimbrite	Distance (km)
Aso III pyroclastic flow deposit	70
Hakone pyroclastic flow deposit	18
Ata ash-flow tuff	60
Shikotsu pyroclastic flow deposit	40
Bali ignimbrite	40
Whakamura ignimbrite	48
Morrinsville ignimbrite	225
Taupo ignimbrite	80

* Erosion probably makes many of these measurements underestimates.

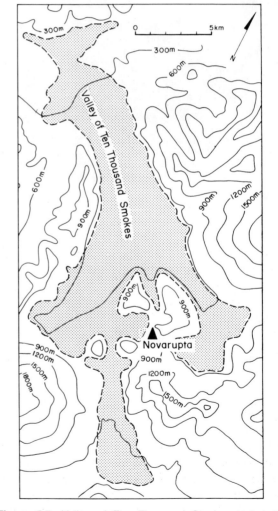

Figure 8.5 Valley of Ten Thousand Smokes ignimbrite (after C. N. Fenner 1920 and Curtis 1968).

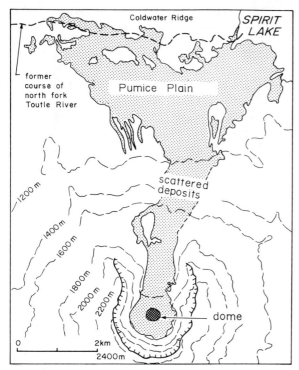

Figure 8.6 Distribution of Mt St Helens 1980 pumice-flow deposits. (After Rowley *et al.* 1981.)

8.3 Eruption sequence and column collapse

Most studies of the stratigraphic relations within ignimbrites and interpretive analysis of their eruptions have been made on small- to medium-volume (<1 a few hundred cubic kilometres) deposits. These often indicate a common sequence of activity (Sparks *et al.* 1973; Plate 8):

(a) plinian phase producing a pumice-fall deposit (Ch. 6),

(b) pyroclastic flow-phase producing ignimbrite and pyroclastic surges and

(c) effusive-phase producing lava (Ch. 4).

This sequence is thought to represent the tapping of deeper and less gas-rich levels of the magma chamber. A consequence is that a plinian eruption will be driven towards an ignimbrite-forming one as its column overloads and collapses to generate pyroclastic flows (Fig. 8.8). Theoretical modelling

has shown that the change from plinian to ignimbrite-forming activity varies with vent radius, gas velocity and H_2O content (L. Wilson 1976, Sparks & L. Wilson 1976, L. Wilson *et al.* 1978, 1980; Ch. 6). Collapse occurs at a critical level of magma water content, given a constant vent radius and muzzle velocity. Widening or flaring of the vent during the eruption will also drive the column towards collapse conditions. The fountain height of the collapsing column as a function of gas content and vent radius is shown in Figure 8.9. Scenarios of three model ignimbrite-forming eruptions (Fig. 8.10) show the variations in column height and gas velocity that take place as the conduit is eroded and gas content decreases during the course of an eruption. In the first case (Fig. 8.10a), vent radius changes while exsolved water content remains constant at 3.45 wt%. The eruption column height grows steadily until sudden collapse occurs and a low ignimbrite-forming fountain is established. Eruption velocity in the vent slowly increases throughout the eruption. In the second case, vent radius increases and exsolved gas content decreases. Variation in height of the eruption column is similar to that found in the first case, though fountain height decreases after the collapse event. Eruption velocity decreases throughout the eruption, due to the falling gas content. In the third case, vent radius remains fixed and gas content decreases during the eruption. There is a much smaller variation in plinian column height than in the first two cases. Again, fountain height gradually decreases after the collapse event.

In Figure 8.11a, combinations of the controlling parameters are plotted and fields of convecting plinian and collapsing columns are shown. From a knowledge of the stratigraphy of a deposit, and by estimating values of the controlling parameters, it is now possible to track the progress and changes in energetics of an eruption to a first-order approximation. Three examples are shown in Figures 8.11b, c and d. Eruptions which terminate before column collapse conditions are reached will only produce plinian fall deposits (Fig. 8.11b; Ch. 6). S. N. Williams and Self (1983) suggested that the Santá Maria 1902 eruption resulted from release of volatile-rich magma at the top of a volatile stratified

(a)

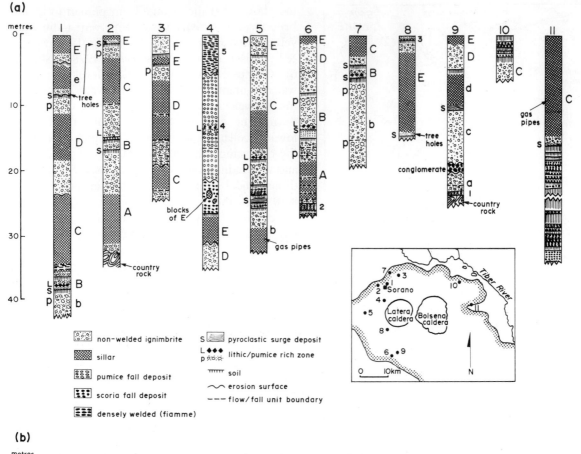

non-welded ignimbrite

sillar

pumice fall deposit

scoria fall deposit

densely welded (fiamme)

S pyroclastic surge deposit

L lithic/pumice rich zone
p

soil

erosion surface

--- flow/fall unit boundary

(b)

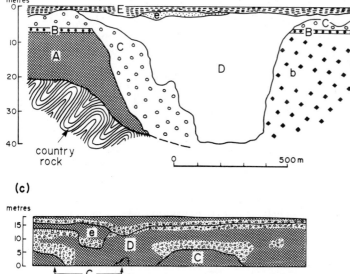

Figure 8.7 (a) Stratigraphy of the ignimbrites in the Vulsini area, Italy. Ignimbrites are named by letters on right-hand side of sections, A being oldest. Sillar refers to coherent ignimbrite and its formation is discussed in Section 8.10. Most of the Vulsini sillars result from vapour-phase crystallisation. (b) and (c) show some of the stratigraphic complexities with ignimbrites filling in valleys cut into older ignimbrites. Repeated infilling and excavation of the same valley is commonly observed with ignimbrites. Because of compaction effects fluvial erosion often quickly restores the pre-existing drainage pattern. (After Sparks 1976.)

(c)

non-welded   sillar

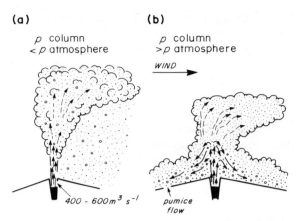

Figure 8.8 Models of eruption columns formed during ignimbrite volcanism: (a) plinian and (b) ignimbrite-forming.

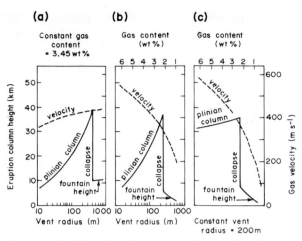

Figure 8.10 Scenarios of three model ignimbrite-forming eruptions (see text). (After L. Wilson *et al.* 1980.)

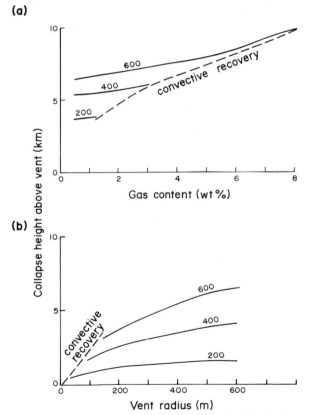

Figure 8.9 Fountain height of collapsing ignimbrite-forming eruption columns. (a) Plotted as a function of gas content for three vent radii for a constant gas velocity of 600 m s⁻¹. (b) Plotted as a function of vent radius for three gas velocities. The collapse height is only slightly affected by water content, but the computations shown here refer to a value of 1%. (After Sparks *et al.* 1978.)

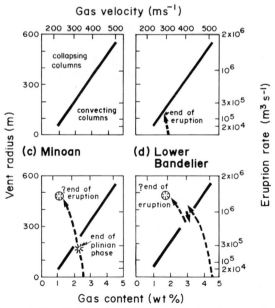

Figure 8.11 (a) General model showing plot of vent radius, gas content, eruption rate and gas velocity, relating these parameters to convecting plinian and collapsing ignimbrite-forming eruption columns (after L. Wilson *et al.* 1980). (b)–(d) Changing conditions during three eruptions (after S. N. Williams & Self 1983, L. Wilson 1980a and Self & Wright *unpub. data*).

magma chamber. As the eruption proceeded, less volatile-rich magma was encountered and structural instabilities caused collapse of the conduit walls, abruptly shutting off the magma supply. The remaining magma was unable to generate sufficient volatile pressure to clear the conduit, and the plinian phase ended; it was 20 years later when volatile-poor magma reached the surface as a lava. The Minoan plinian column was maintained until collapse conditions developed (Fig. 8.11c), although a phase of phreatomagmatic activity between the plinian and pyroclastic flow-forming phases may have complicated the transition between the two. As the Lower Bandelier plinian column neared the point of collapse (Fig. 8.11d), slight fluctuations in gas velocity and mass discharge rate caused insta-

bilities in the column and small-volume intraplinian flow units to form (Fig. 6.13b). Catastrophic column collapse ensued. Instabilities in eruption columns nearing collapse conditions would account for the stratification found at the tops of many plinian deposits (Ch. 6).

Some deposits show much more complex eruption sequences than indicated above, and ignimbrite flow units are interstratified with plinian fall units, e.g. the Rio Caliente ignimbrite Mexico (Fig. 8.12) and the Roseau Tuff, Dominica (Ch. 9). The Rio Caliente can also be described as an **intraplinian** ignimbrite because it is sandwiched between two co-eruptive plinian fall deposits. Mechanisms to produce such complex sequences and changes in style of activity include a large increase in gas content, increasing proportions of low molecular weight volatiles, and sudden closure of the vent by faulting and its relocation. Choking of the vent by wall collapse may also produce alternations from ignimbrite-forming to plinian activity, especially if column conditions are close to the boundary between convection and collapse (Ch. 6). Very complicated eruption sequences can also be produced if water is involved and collapse of phreato-plinian eruption columns takes place (Sheridan *et al.* 1981, Self 1983; Fig. 6.33a).

It must be stressed that not all ignimbrite-forming eruptions have an earlier plinian phase, e.g. the Cerro Galan ignimbrite (P. W. Francis *et al.* 1983; Fig. 8.2). This suggests that this very large ignimbrite formed by a rather different eruption mechanism than has been proposed for the well described small- to medium-volume deposits. The mammoth Fish Canyon Tuff, on the other hand, is underlain by a pre-ignimbrite pumice fall deposit and a thick sequence of pyroclastic surge deposits. The implications in terms of type of vent for the very large ignimbrite eruptions is discussed in the next section.

Observations at Mt St Helens have shown that even during small eruptions a high eruption column is not an essential prerequisite for generating a pumice flow (Ch. 5; Fig. 5.12). In such cases the venting eruptive mixture may be so dense that a well defined, maintained eruption column does not form but, instead, a low maintained pyroclastic

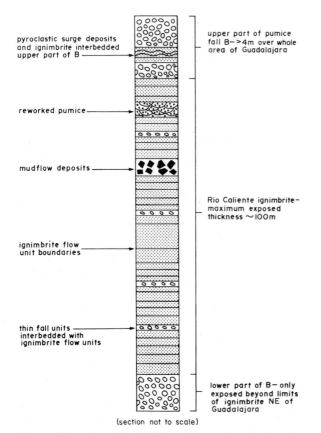

pyroclastic surge deposits and ignimbrite interbedded upper part of B

upper part of pumice fall B—>4m over whole area of Guadalajara

reworked pumice

mudflow deposits

Rio Caliente ignimbrite— maximum exposed thickness ~100m

ignimbrite flow unit boundaries

thin fall units interbedded with ignimbrite flow units

lower part of B— only exposed beyond limits of ignimbrite NE of Guadalajara

(section not to scale)

Figure 8.12 Generalised stratigraphic section showing the complex eruption sequence of the Rio Caliente ignimbrite, Mexico, and co-eruptive plinian falls (designated La Primavera B, Ch. 6). (After J. V. Wright 1981.)

fountain, with the appearance of 'a pot boiling over', may form. Another possibility is that rapid lateral expansion of the erupting mixture (exit pressures >1 bar; L. Wilson *et al.* 1980) causes it to move away immediately as a flow.

8.4 Source vents

Several types of source vents for ignimbrites have been proposed. The most voluminous ignimbrites have generally been associated with either linear fissure eruptions (*Valley of Ten Thousand Smokes type* of H. Williams & McBirney 1979) or continuous ring fissure eruptions (*Valles type*). All observed ignimbrite-forming eruptions have issued from what seem to be essentially point sources, or central vents. Studies over the past ten years on a number of smaller- and medium-volume ignimbrites have suggested that these ignimbrites are commonly erupted from central vents.

Ignimbrites are often associated with calderas, a fact which has been recognised for a long time, and was very well documented in the classic paper of H. Williams (1941). Many cases are known where single ignimbrite forming eruptions seem to have resulted in the creation of large calderas (e.g. the Aira caldera in Japan) (Fig. 8.3); Crater Lake, Oregon (Figs 8.16, 13.25 & 26); and Santorini caldera (Figs 8.4, 13.30 & 31)). For example, in the San Juan volcanic field the ignimbrite to caldera ratio is low, suggesting that many eruptions are monogenetic outpourings with each forming a caldera. Figure 8.13 shows that there is a crude relationship between caldera dimensions and the size of the associated ignimbrite. The largest caldera shown is La Garita, which is the source of the Fish Canyon Tuff (Fig. 8.1c). The size of calderas is often assumed to approximate the dimensions of magma chambers at depth. In the case of the San Juan volcanic field each ignimbrite is believed to chronicle the emplacement of successive segments or stocks of an underlying batholith (Steven & Lipman 1976).

Caldera collapse takes place when the lithostatic pressure on the roof of the magma chamber exceeds the chamber pressure by the compressive strength

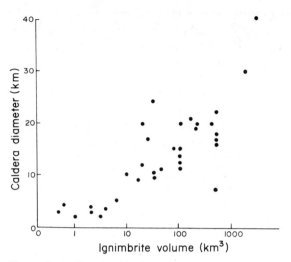

Figure 8.13 Relationship between size of ignimbrite and associated caldera.

of the overlying rock. Whether or not caldera collapse occurs during or after an eruption will be an important control on the type of vent.

8.4.1 LINEAR FISSURE VENTS

The fissure hypothesis originated from the observation of lines of fumaroles in the Valley of Ten Thousand Smokes ignimbrite after the 1912 eruption at Katmai in Alaska (Fig. 8.5). This led C. N. Fenner (1920) to conclude that the ignimbrite erupted from linear fissures in the floor of the valley. However, more recently, from the closure of contours on isopach maps of pyroclastic fall deposits produced in the same eruption, Curtis (1968) located the central vent of Novarupta as the source. The fumarole lines are now thought to be reflections of basement faults along which ground waters moved.

Since Fenner's observations at Katmai, the importance of fissure eruptions for the formation of ignimbrites has been assumed by a number of workers, and has been regarded as the way to account for the more voluminous deposits. Van Bemmelen (1961) considered that ignimbrites were erupted from major fissures in a similar manner to flood basalts. Korringa (1973) documented a linear vent system for the source of the Soldier Meadow Tuff in Nevada, and this is often quoted as an

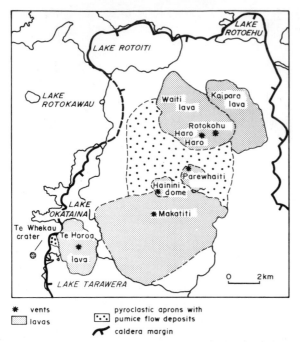

Figure 8.14 Linear fissure vent system of the Mamaku eruption, Okataina rhyolitic centre, New Zealand. (After Nairn 1981.)

example. However, Greene and Plouff (1981) have suggested that this is related to part of the margin of a larger caldera. In many other described examples field evidence for linear fissure vents is meagre or controversial.

Convincing examples are described by Nairn (1981) from the Okataina rhyolitic centre, New Zealand (Ch. 6). All of the post caldera-forming eruptions (<20 000 years BP) are thought to have occurred along fissures from multiple vents in simultaneous or sequential activity. These produced plinian deposits which are interbedded near source with pumice flow and surge deposits and associated lavas, forming thick multiple-bedded deposits. These are exemplified by the Mamaku eruption (7500 years BP), during which at least five magmatic vents and one phreatic vent were active, these being spread along 14 km of an underlying fissure (Fig. 8.14). No major time breaks between the various deposits are apparent, and ^{14}C ages available for two pyroclastic deposits erupted from separate vents are not significantly different. As in basaltic

eruptions which begin along fissures, localisation to a number of points seems to have occurred (Chs 4 & 6).

8.4.2 RING FISSURE VENTS

In another popular model ignimbrites are thought to be erupted along a continuous ring fissure, around which caldera collapse may have been occurring during the eruption (R. L. Smith & Bailey 1968). The ring fissure model has been popular in the USA because it explains the radial distribution of thin outflow sheets associated with very much thicker intracaldera ignimbrites, which are thought to have accumulated by ponding of pumice flows erupted around the ring fissure bounding a continually subsiding caldera floor block (Fig. 8.15). However, there have been very few studies that have quantitatively mapped the different facies of larger volume ignimbrites to locate the source vent, as has been done for many smaller-volume ignimbrites.

At Cerro Galan, eruption from around the ring fissure was perhaps the important mechanism throughout. The absence of a pre-ignimbrite plinian deposit could suggest that the radius of the vent widened almost instantaneously. This would make a convecting column unstable, because of the very high eruption rate (Fig. 8.11). P. W. Francis *et al.* (1983) indicated that very large eruption rates would be produced by sinking of the caldera floor block into the magma chamber along outward dipping ring fractures. Inward-dipping ring fractures, as generally suggested in previous models of ring fissure eruptions (Fig. 8.15), would inhibit very high eruption rates. The present thickness of

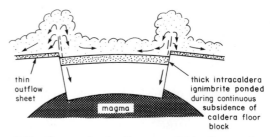

Figure 8.15 The popular ring fissure model for the eruption of ignimbrites.

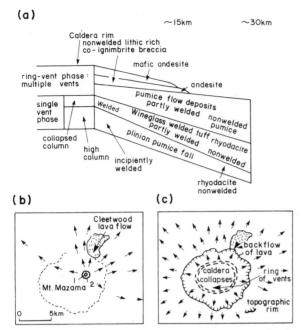

Figure 8.16 (a) Stratigraphic relations and vent type for the products of the 6845 years BP Crater Lake ignimbrite-forming eruption, Oregon. Note also the last erupted pumice flows were mafic andesites in contrast to the earlier rhyodacites, and spectacular compositional zonation is shown. (b) Single vent phase of the eruption; 1 indicates vent for plinian eruption and 2 the enlarged vent during eruption of the pumice flows that deposited the Wineglass welded tuff. Arrows show direction of movement of pumice flows, which were confined to topographic low areas. Mount Mazama is the name given to the pre-collapse stratovolcano. (c) Ring vent phase during which pumice flows surmounted most topographic barriers. The ring fissure is shown inboard of the present topographic rim which would indicate scarp-retreat by landsliding and slumping of the caldera wall during or after caldera-collapse. Field evidence demonstrates the Cleetwood lava flow must have been erupted just before the ignimbrite-forming eruption, and after caldera-collapse backflow down the caldera wall occurred. (After Bacon 1983.)

intracaldera ignimbrite exposed by the resurgent dome is 1200 m (Fig. 8.2), and there is no doubt that caldera collapse must have been concurrent with eruption.

Detailed field studies of the smaller Crater Lake pumice flow deposit have confirmed that eruption from around the ring fracture must have occurred while caldera collapse was taking place (Bacon 1983). The eruption sequence can be divided into two stages: single vent and ring fissure vent phases. The initial plinian phase was erupted from a single vent. Collapse of this column generated a series of pumice flows deposited to the north and east of Mt Mazama (Fig. 8.16). Approximately 30 km^3 of magma was expelled before collapse of the roof of the magma chamber occurred, and the inception of caldera formation ended the single vent phase. Later pumice flows are believed to have erupted from multiple vents around the ring fracture as caldera collapse continued. Composition of lithic clasts within proximal co-ignimbrite breccias (Section 8.5) varies around the rim and certain types can be correlated to probable vent locations. During the ring vent phase another 20 km^3 of magma may have been erupted. Druitt and Sparks (1985) and Druitt (1985) have also proposed such a two-stage model for caldera-related ignimbrite-forming eruptions, based on studies on Santorini.

One noteworthy study of an ancient ring fissure vent system is that by Almond (1971) of the Late-Precambrian–Cambrian Sabaloka cauldron, Sudan. Almond has described welded ignimbrite dykes which he believes fed their eruption around a ring fracture.

8.4.3 VENT SYSTEM FOR THE FISH CANYON TUFF

The eruption sequence of the Fish Canyon Tuff may suggest an alternative type of vent system for the largest ignimbrites (Self & Wright 1983). Where studied, the outflow sheet is underlain by a thin pre-ignimbrite plinian layer, and a relatively thick sequence of pyroclastic beds bounded at the top by a surface having an unusually large wave-like form (Fig. 8.17). These megawaves indicate that there may have been an enormous, sudden release of energy accompanying the eruption. The explanation suggested is that the thin crust over a high-level magma chamber collapsed inwards, effectively opening the magma chamber to the surface and triggering an enormous release of energy, which was translated into lateral blasts or surges. Failure of the roof seems to have occurred soon after the beginning of the eruption, perhaps because the upper part of the magma chamber had

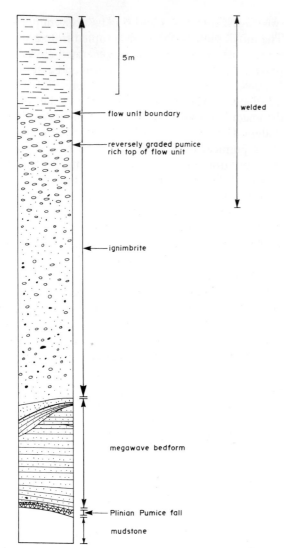

Figure 8.17 Measured stratigraphic section through the outflow sheet of the Fish Canyon Tuff at a location 35 km south-east of the centre of La Garita caldera (Fig. 8.1). (After Self & Wright 1983.)

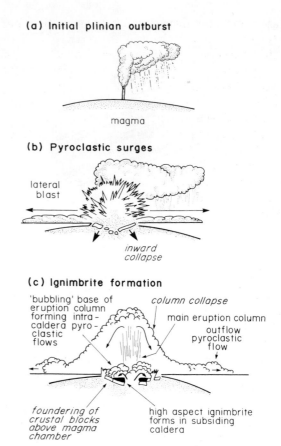

Figure 8.18 Possible eruption sequence for the Fish Canyon Tuff and other very large ignimbrites. (After Self & Wright, *unpubl. data.*)

been quickly drained by a high intensity, initial plinian phase. Figure 8.18 shows schematically the sequence of events. After the surge blasts, the eruption continued to generate ignimbrite-forming pumice flows. As a result of the continued evacuation of magma, failure of the crust over the whole of the present La Garita caldera area may have occurred. Many vents could have opened around foundering crustal blocks over a large area, rather than simply around a ring fracture. It is possible to envisage these being aligned on a complexly organised tensional fracture system and, in a sense, these would be of the linear fissure type. As the eruption continued, more-localised vents may have developed.

With gradual subsidence of crustal blocks into the magma chamber a thick (>1.4 km) intracaldera ignimbrite accumulated. The outflow ignimbrite sheet (20–500 m thick) is composed of a number of flow units, which are similar to smaller volume ignimbrite flow units in their characteristics. The fact that these were able to travel up to 100 km from source suggests they acquired much of their kinetic energy by collapse of a vertical eruption column, probably at least 10 km high. The intracaldera flows (very densely welded and devitrified) could be more complex in their origin, being composed not only of flows formed by column collapse, but also perhaps generated by frothing over of material at vent without column collapse. Such flows would not be greatly expanded or travel very far, and would build up a thick, high aspect ratio ignimbrite (Section 8.7) within the caldera. Losses of vitric ash should therefore also be minimal. The intracaldera Fish Canyon Tuff, like many other intracaldera ignimbrites, is very crystal-rich, but examination of their juvenile clasts often seems to show that this is a property of the magma, not extreme fractionation by flow processes (Ch. 11).

8.4.4 CENTRAL VENTS

A number of Quaternary ignimbrites have been related to central vents by several methods, including mapping their distribution, mapping the distribution of co-eruptive pumice fall deposits and identifying proximal facies (Section 8.5). These include ignimbrites with associated caldera formation. For example, the Rio Caliente ignimbrite (Fig. 8.31) not only contains a chaotic near-vent facies with co-ignimbrite breccias which can be used to locate a central vent, but also a thicker intracaldera ignimbrite which is partly due to caldera collapse during the eruption. The Minoan eruption sequence has been interpreted in terms of a central vent which progressively widened during the eruption (Fig. 8.11). If this is correct, then caldera collapse, forming Santorini's present caldera, had to happen near the end of or after the Minoan eruption. Similarly, at Krakatau, caldera collapse seems to have taken place very late in the eruption (Self & Rampino 1981). The ultraplinian- and ignimbrite-forming Taupo AD 186 eruption has been related to a central vent. Sudden collapse of the vent area is thought to have greatly widened the vent, and to have triggered the eruption of the violent Taupo ignimbrite, perhaps in a manner somewhat similar to that suggested for the Fish Canyon Tuff, but on a smaller scale and much later in the eruption. Some central vent eruptions have apparently been unrelated to any pre-existing volcano. Just as Parícutin was 'born in a cornfield' in 1943, so the Acatlan ignimbrite was erupted

from a 'hole in the ground' vent (located by the near-vent occurrence of a co-ignimbrite breccia and lava dome) on flat ground (J. V. Wright & Walker 1977; see below).

From the foregoing discussion it will be apparent that the vent type and eruption mechanisms for the largest ignimbrites remain unresolved. Models of linear fissure and ring fissure eruptions must be tested by the analysis of eruption sequences and facies, and some of the answers are likely to be found by further examination of ancient volcanic terrains where the roots of large centres and calderas are exposed. Vent type may change as an eruption proceeds (e.g. Druitt 1985). Even if eruption is initiated along a fissure, or if one or more fissures develop due to later caldera-collapse, localisation to multiple vents or a central vent may be a natural progression. During the initial stages of a highly explosive eruption the widest part of the fissure will offer least frictional resistance to the flow of gas and magma. Most-rapid erosion will take place at this point, further accentuating the flow until the eruption is confined to this point. Note that in the theoretical column-collapse models (Figs 8.9, 10, 11 & 6.20) a circular vent is assumed, but the results can be applied to a fissure eruption by replacing the vent radius with the half-width of the fissure.

8.5 Co-ignimbrite breccias

Proximal ignimbrite breccias generated in the eruption column and as part of an ignimbrite are now known to be common. J. V. Wright and Walker (1977) first described a co-ignimbrite lag-fall deposit from the Acatlan ignimbrite (Fig. 8.19). This coarse, lithic-rich deposit was identified as part of the ignimbrite because it showed the same compositional zoning as the ignimbrite (Figs 7.19 & 20). This was the key to the interpretation, because it showed that the breccia accumulated synchronously with the formation of the ignimbrite, together possibly representing only one flow unit. J. V. Wright and Walker (1977, 1981) envisaged that the deposit formed at or near the site of continuous column collapse, and consisted mainly of pyroclasts

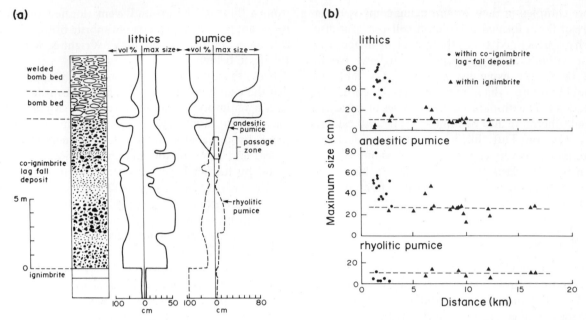

Figure 8.19 (a) Measured stratigraphic section through the Acatlan lag-fall deposit, Mexico, approximately 1.5 km from the most likely source vent; the conduit now covered by a rhyolite lava dome. (b) Average maximum diameter of the five largest clasts plotted against distance from source in the main, compositionally zoned ignimbrite flow unit and the lagfall deposit. (After J. V. Wright & Walker 1977.)

that were too large and heavy for the column to support. The term 'lag-fall' deposit was proposed, because the accumulation of lithics was a residue left behind by the pumice flows – that is, a lag deposit. The lag deposit was of air-fall type as shown by the stratification. The absence of fine-grained fall units (ash grade <2 mm) and of discrete bedding planes was thought to be evidence for rapid accumulation from a continuous, vigorous eruption column with only minor variations in eruption intensity.

Similar breccias have since been recognised as a near-vent facies of several ignimbrites (Figs 8.20 & 21). However, it is now realised that there are different types, and a better general term to use for these rocks is simply co-**ignimbrite breccia**. Many of these examples did not simply accumulate by the fall of lithics out of the eruption column, as originally envisaged for the Acatlan ignimbrite, but were emplaced by flow processes. Such types of deposit include clast-supported lithic breccias containing a small amount of ignimbrite matrix. In some cases they may grade into matrix-supported

breccias and coarse pumiceous ignimbrite containing large lithics (Figs 8.21 & 22). These deposits can contain large segregation pipes and structures enriched in lithics and depleted in fines, which indicate a high degree of fluidisation in the proximal pumice flows (Ch. 7). Complex interrelations between breccias and ignimbrite can be found (Fig. 8.21).

Recent work shows that there are two types of proximal ignimbrite breccias emplaced by flow. In the first type, segregation of lithics takes place through the body of the flow, and this is found above a basal layer. This type is the near-vent equivalent of the lithic concentration zone commonly observed at the bottom of layer 2b in ignimbrite flow units (Ch. 7). Towards the vent this type of breccia should grade back into a co-ignimbrite lag-fall of the Acatlan-type, in which lithics settled out of the eruption column. Druitt and Sparks (1982) have logically proposed that both these cases of body segregation breccias should be termed co-**ignimbrite lag breccias**. In the second type, the segregation of lithics appears to have

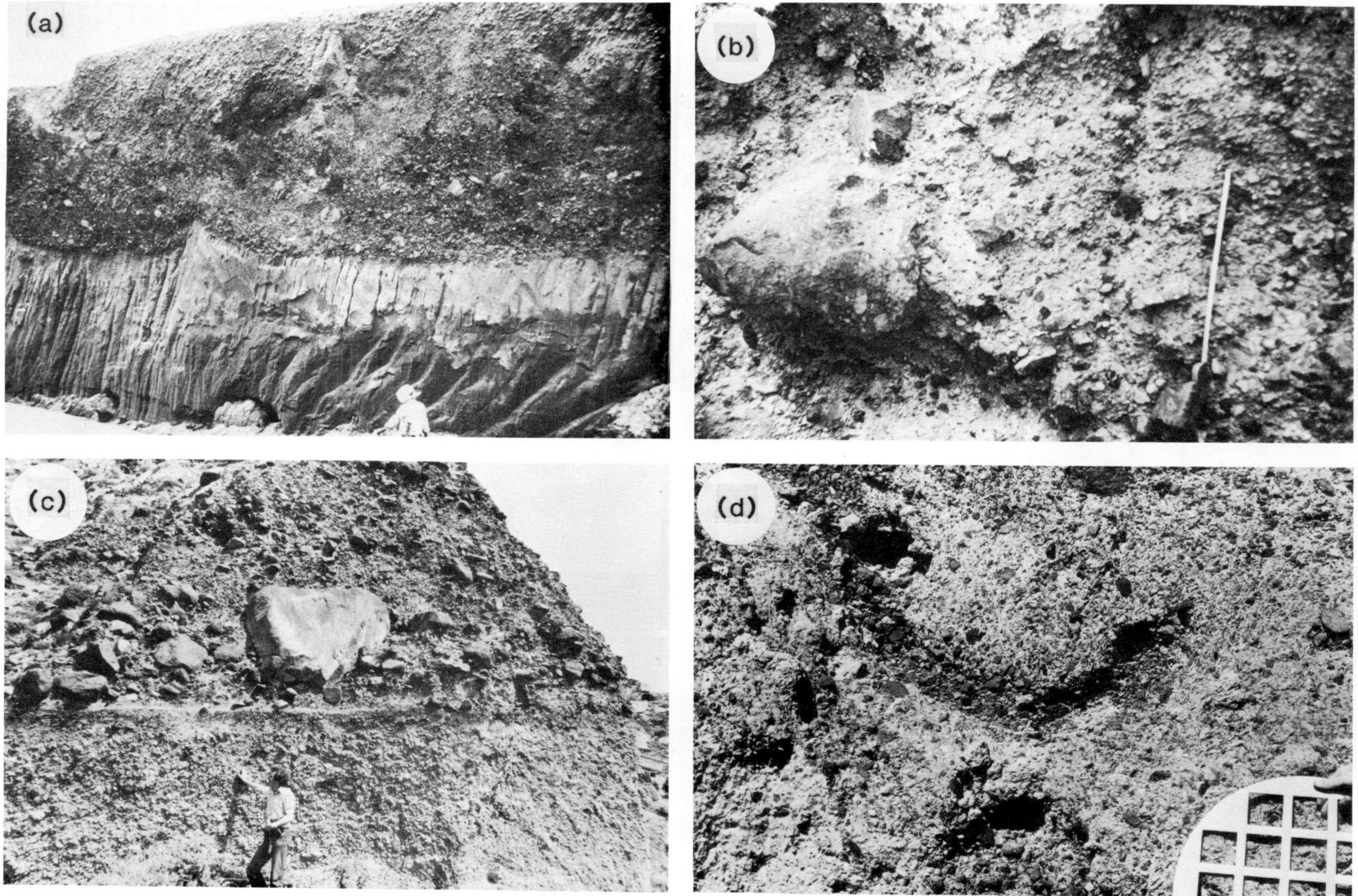

Figure 8.20 Examples of co-ignimbrite breccias. (a) The Kamewarizaka breccia a proximal facies of the Ito pyroclastic flow deposit found close to the edge of Aira caldera, Japan (Fig. 8.3). Largest lithics are 20–30 cm in diameter. Lighter coloured lower half is the older Tsumaya pyroclastic flow deposit (photograph by S. Yokoyama). (b) Lag-breccia in the Rio Caliente ignimbrite, Mexico. Note segregation pipes. Largest lithic is a block of welded Rio Caliente ignimbrite which has been 're-erupted', indicating considerable vent widening, or migration, during the eruption. (c) Lag-breccia of the 18 500 years BP eruption of Santorini, Greece (Fig. 13.31). The breccias overlie a basal plinian pumice fall deposit (against which the geologist is resting his hand) and show normal grading (after Druitt & Sparks 1982). (d) Lenticular segregation pod in lag breccias of the 18 500 years BP Santorini eruption (photograph by T. H. Druitt). (e) Wall of the Valles caldera showing a 5 m thick co-

(continued)

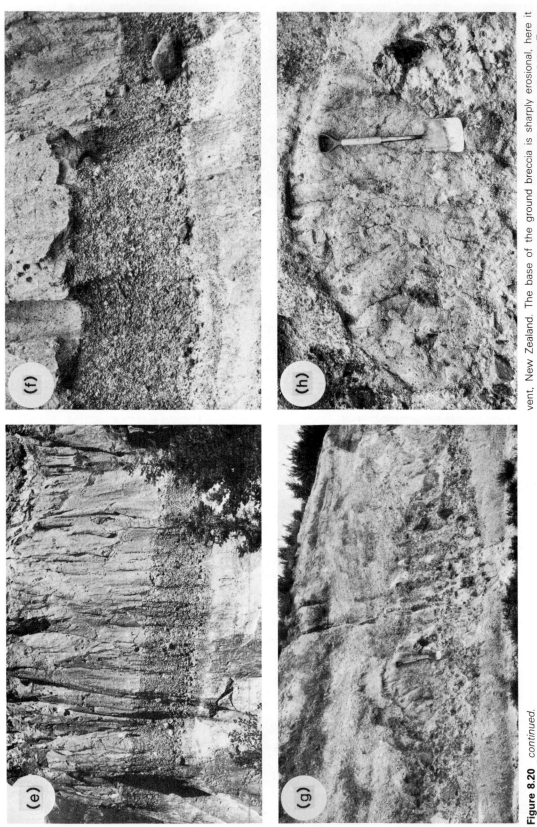

Figure 8.20 *continued.* ignimbrite breccia bed in the Lower Bandelier Tuff, New Mexico. (f) Close-up of Lower Bandelier co-ignimbrite breccia showing largest clasts up to nearly 2 m. (g) Ground breccia of the Taupo ignimbrite at a location approximately 8 km from the vent, New Zealand. The base of the ground breccia is sharply erosional, here it overlies a local early ignimbrite flow unit (Section 8.12). (h) Detail of the Taupo ground breccia with large, well-rounded mudstone block. Spade is in same position seen in (g).

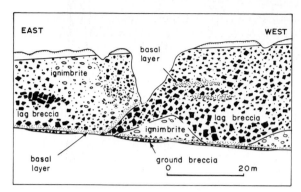

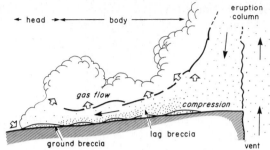

Figure 8.21 Field sketch at an exposure in the Santorini caldera wall showing co-ignimbrite breccias of the 18 500 years BP eruption, Santorini, Greece. This illustrates the complex relationships that can be found between such proximal breccias and ignimbrite. (After Druitt & Sparks 1982.)

(a)

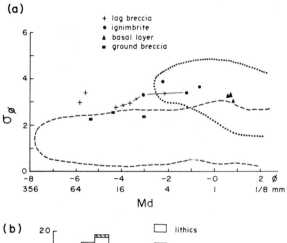

(b)

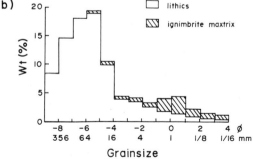

Figure 8.22 (a) Md_ϕ/σ_ϕ plot for proximal deposits of the Santorini 18 500 years BP eruption, Greece. Tie-lines with arrows connect samples taken from a gradational section measured on the eastern side of location sketched in Figure 8.21. (b) Grainsize histograms of a typical lag-breccia. The deposit is bimodal with a mode in the >0.5 mm fraction due to a high proportion of crystals. (After Druitt & Sparks 1982.)

Figure 8.23 Model for the proximal segregation of lithic clasts in pumice flows generated by column-collapse. Ingestion of air at the flow-head causes strong fluidisation, and the sedimentation of the ground breccia. Compression of the particle-gas mixture at the base of the collapsing column generates high pore pressures within a high particle concentration flow body. Decompression of the gas phase as the flow moves away laterally causes strong fluidisation with the body, and the segregation of lithics to generate the lag breccia. Open arrows schematically depict the passage of gas through the flow system. (After Druitt & Sparks 1982.)

taken place in the more strongly fluidised head of the moving pumice flow. This layer is then overridden by the rest or body of the flow (with layers 2a and 2b). This type has been termed a **ground breccia**, and the best example that has been described is associated with the Taupo ignimbrite (Figs 8.20g & h). This is the proximal equivalent of the ground layer (G. P. L. Walker *et al.* 1981a; Ch. 7). Both types of breccia can be found within the same flow unit. For their formation, Druitt and Sparks (1982) have suggested the model illustrated in Figure 8.23. Druitt (1985) has suggested that the sudden appearance of lag breccias between normal ignimbrites above and below, on Santorini, records the change from a single point source eruption point to caldera collapse. This leads to an increased discharge rate, erosion of the collapsing roof materials and eruption of the eroded blocks together with ignimbrite from multiple eruption points along the caldera collapse fracture system, as multiple, localised deposits of lag breccia. The succeeding normal ignimbrite records a return to a single, stable eruption point.

Other types of breccia may be generated during eruption of ignimbrites and are found closely associated with them near vent. Lipman (1976) has described caldera-collapse breccias within intra-caldera ignimbrites in the San Juan volcanic field as meggabreccias including clasts metres to hundreds of metres in size (e.g. Thompson 1985). These are thought to have formed by landslides from the walls of calderas as collapse took place (see caption to Fig. 8.16c and Ch. 13). Flash-flood breccias are a prominent feature of the Minoan ignimbrite (Bond & Sparks 1976; Fig. 7.15). G. P. L. Walker (1985) has comprehensively reviewed the origin of coarse lithic breccias near ignimbrite source vents.

8.6 Co-ignimbrite ash falls

In recent years it has become apparent that large-volume ignimbrites should have associated with them large and very widespread air-fall ashes which are synchronous with emplacement of the ignimbrite (Ch. 5). Ignimbrites commonly show a marked concentration of free crystals in the matrix compared with the juvenile magmatic content. This may be attributed to the large amount of vitric dust that has been lost during eruption and pyroclastic flow, and is then deposited in an associated co-ignimbrite ash-fall deposit (Sparks & Walker 1977; Chs 5 & 11). From crystal concentration studies (App. I), ignimbrites show average vitric ash losses of at least 35% from the total erupted magma. Thus, for an ignimbrite having a volume of 200 km^3, there may be more than another 100 km^3 accompanying this as widely dispersed air-fall ash.

The significance of co-ignimbrite ashes is even greater when the volumes of plinian deposits are considered. Very few plinian deposits are known to be greater than 25 km^3 in volume (Table 6.2). Although dispersal is not necessarily related to volume, this at least shows that extensive, very widely dispersed volcanic ashes are more likely to be associated with the formation of the ignimbrite, rather than with a preceding plinian phase or individual plinian eruptions. However, the volcanological interpretation of distal silicic air-fall ash layers is not easy. Other types of eruption will produce widely dispersed ash layers, and the plinian component can be substantial (Ch. 6).

During the eruption of ignimbrites, two sources from which fine ash could be carried high as convective plumes are (a) above the collapsing column and (b) above the moving pyroclastic flow.

(a)　　　　　　　　　　　　　　　　　　　　　　　　**(b)**

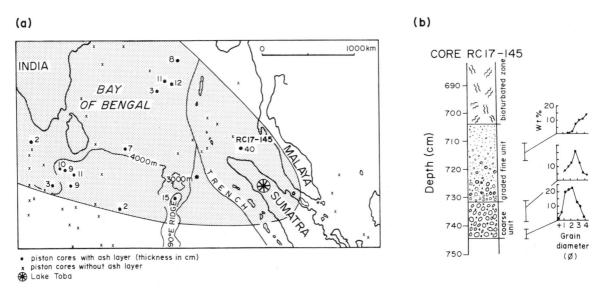

Figure 8.24 (a) Map showing the distribution and thickness of the Toba ash-layer in deep-sea cores. This is a product of an eruption from Lake Toba, Sumatra, 75 000 years BP, which produced a 2000 km^3 ignimbrite (Table 8.1), and is the largest magnitude eruption documented in the Quaternary. (b) Section showing the ash layer in the most proximal core. The lower, coarser unit is interpreted as a distal plinian ash, and the upper graded fine unit the co-ignimbrite ash. (After Ninkovich et al. 1978.)

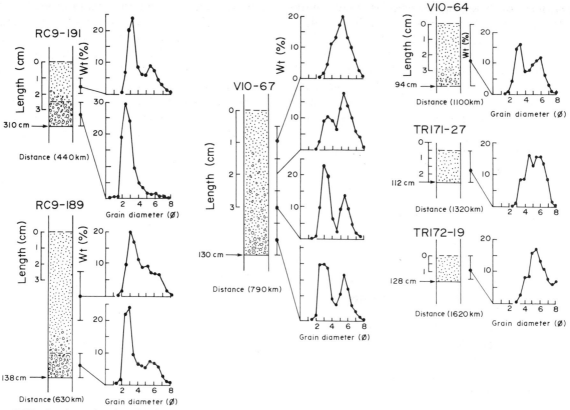

Figure 8.25   Sections showing the Campanian (Y–5) ash layer in six deep-sea cores from the eastern Mediterranean Sea at increasing distance from the Bay of Naples. This is correlated with the Campanian ignimbrite, distributed over a large area in the Napolitan region (Table 8.1, Ch. 13). (After Sparks & Huang 1980.)

When pyroclastic flows are generated by column collapse a convective column rises above the vent, producing a high eruption column, and lower clouds of fine ash rise above the moving pumice flows (Fig. 8.8). It is the preferential loss of vitric material into convecting plumes which leads to a complementary increase in the proportion of crystals in the ignimbrite. Sparks *et al.* (*in press*) have also suggested that some flows may become completely buoyant at their distal flow margins, and then rise into the atmosphere producing a major mushroom-like plume of ash. This may be the source of very large volumes of co-ignimbrite air-fall deposits. This work is based on analysis of space and ground photographs of the 18 May 1980 Mt St Helens blast event, which suggests the growth and rise of a major ash cloud at a point well away from the vent.

Many of the volcanic ashes found in deep-sea cores are considered to be co-ignimbrite ash layers (e.g. Figs 8.24 & 25). The largest known is that associated with the Toba Tuff, erupted from Lake Toba, Sumatra. Nearer to the source, cores also pick up the distal plinian ash below the co-ignimbrite ash; in all the described examples this is a thinner, coarser layer which peters out with distance. In more-proximal cores through the Campanian ash layer of the eastern Mediterranean (Fig. 8.25), this ash is clearly divisible into a coarse lower unit separated by a sharp boundary from a fine graded upper unit. In core RC9-191 the lower unit only shows a coarse mode, whereas the normally graded top possesses a coarse and a finer mode. The normally graded unit in core V10-67 has a bimodal distribution throughout most of its thickness. Bimodality disappears downwind, because the coarser mode decreases in diameter with distance from source, but the finer mode shows no lateral variation. During the plinian phase of an eruption the coarse-grained lower unit, and at least part of the coarse mode found in bimodal ashes, is formed. The fine-grained upper unit is the co-ignimbrite ash. The higher convecting plume above the collapsing column may also contribute to the coarse mode in this layer. However, bimodality in distal silicic ash layers is not unique to co-ignimbrite ashes and can be produced by other mechanisms (Ch. 6). Observations on the size grading of crystals in deep ash layers are also important, and can be used to estimate eruption durations (Ch. 9).

In the geological record, many of the stratigraphically important bentonite layers and 'tonsteins' may be co-ignimbrite ashes, perhaps derived from late stage, buoyant plumes originating from the distal margins of flows as suggested by J. G. Moore *et al.* (*in press*).

8.7 Depositional facies model

For our purpose here, a *facies* can be considered as an eruptive unit, or part thereof, having distinct spatial and geometrical relations and internal characteristics (e.g. grainsize and depositional structures; Ch. 1). A *facies model* is a generalised summary of the organisation and associations of the facies in space and time.

C. J. N. Wilson and Walker (1982) described the different facies that could be found in an ignimbrite and related them to the depositional regimes or 'anatomy', of the moving pyroclastic flow. However, much of this discussion was based on the study of one ignimbrite – the excellently preserved Taupo ignimbrite. This has exceptionally well developed facies contrasts, related to the extreme '*violence*' or velocity with which it was emplaced (Ch. 7). Indeed, the Taupo ignimbrite can be considered to be at one end of a spectrum of ignimbrite types. Ideally, when constructing facies models of depositional systems it is important to study many examples. It is only after the local effects of a number of examples are '*distilled away*' that a generalised model results (Ch. 14). However,

unlike many sedimentary systems, there are still relatively few ignimbrites for which different facies and lateral relationships are known. One example where lateral relationships have been evaluated are in the Laacher See tephra in Germany, where Freundt and Schmincke (1985) have been able to identify proximal, medial and distal facies.

The model presented here is largely developed from a study of the Bandelier Tuffs (J. V. Wright *et al.* 1981). Seminal works on many of the characteristics of ignimbrite volcanism originated from studies of these deposits (R. L. Smith 1960a, b, Ross & Smith 1961, R. L. Smith & Bailey 1966, 1968). The Bandelier Tuffs are therefore already models for this type of volcanism, and we consider them to be a good norm for comparison, which is an important property of any facies model (R. G. Walker 1984; Ch. 14). To illustrate the extremes in ignimbrites, we can also examine and compare the Rio Caliente and Taupo ignimbrites, which seem to be at the opposite ends of the spectrum of ignimbrite types. Readers are also referred to the facies model of Freundt and Schmincke (1985) for ignimbrites of the Laacher See volcano in Germany.

8.7.1 BANDELIER TUFFS AND MODEL

The Lower and Upper Bandelier tuffs of New Mexico are the products of two voluminous, rhyolitic ignimbrite-forming eruptions, dated 1.4 and 1.1 Ma BP, respectively. Associated with these eruptions was the formation of a large caldera complex, including the Valles caldera and the Toledo caldera (Figs 8.26, 13.42 & Plate 13). Toledo caldera was believed to be the source of the earlier ignimbrite, and Valles to be that of the later ignimbrite. Stratigraphically the Lower and Upper Bandelier tuffs are separated by erosional surfaces, epiclastic sediments, soils and a sequence of up to six pumice-fall (sub-plinian or plinian) deposits (Plate 8) erupted from rhyolite domes in Toledo caldera.

Both tuffs have similar eruption histories. Extensive plinian fall layers occur at their bases (Plate 8, Figs 6.13b, 14a & b). The Lower Bandelier plinian deposit with its volume of 100 km³ and the Upper plinian deposit with its volume of 70 km³

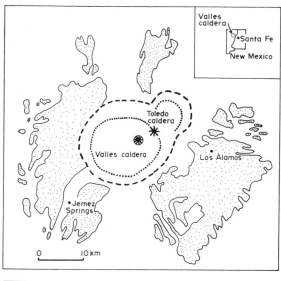

Figure 8.26 Location map of the Bandelier Tuffs, New Mexico. Vent positions are from Self & Wright (*unpubl. data*).

are two of the most voluminous yet recognised (Ch. 6). Each has a lower, thick, homogeneous fall unit, produced by the continuous gas-blast of a high intensity, maintained, stable eruption column. The homogeneous units are overlain by finely bedded fall units, which are locally interstratified with pyroclastic surge and flow deposits. These are indicative of instabilities in the column, documenting fluctuations in gas exit velocity, mass discharge rate, column height and minor collapse events. Dispersal patterns of fall units within the plinian deposits indicate the location of a central vent for each eruption (Fig. 8.26). The sequence of deposits suggests a continuum of events leading to a major collapse (Fig. 8.11). Also, the initial pumice flows at least seem likely to have been erupted from the same central vents. Eruptions along a ring-fissure vent system may have developed later.

The two ignimbrites both have volumes of about 200 km³, and consist of a number of flow units. Flow unit thickness varies from <1 m to >20 m, although in some of the deep palaeocanyons where

the ignimbrites are ponded and densely welded (see Fig. 8.42, below), individual flow units could have had uncompacted thicknesses of more than 100 m. For most of the two sheets, any one section only shows a small number (two to four) of massive flow units (Plate 8). Basal layers, normal grading of lithic clasts and reverse-grading of pumices are often evident.

Lithic breccias are interbedded, and grade into ignimbrite flow units in the caldera wall. These are co-ignimbrite lag breccias. The breccias form thick segregation layers (Figs 8.20e & f), segregation pods and large pipes. The appearance of lag breccias associated with the ignimbrites of the caldera wall may herald the onset of caldera collapse and the opening of ring fracture vents.

The ignimbrites are underlain by ground surge deposits (Plate 8, Figs 5.23a & b) and we have already discussed the possibilities for the generation of this facies (Chs 5 & 7). Also, directly below the base of the lowermost flow unit of the Lower Bandelier ignimbrite, large pumice clasts occur in dune bed-forms (height 0.5–1.5 m, wavelength 5–10 m; Fig. 8.27) and in discontinuous lenses (*pumice swarms*). These pumice swarms have also been found in the thin flow units at the base of the Upper Bandelier ignimbrite. More rarely, such pumice concentrations have been found at the base

Figure 8.27 Pumice dune below base of lowermost flow unit of the Lower Bandelier ignimbrite. Height of crest is 50 cm.

Figure 8.28 Thin, very pumice-rich, reversely graded flow units and ash-cloud surge forming a distal facies association of the Upper Bandelier ignimbrite. Deposits are indurated by vapour-phase crystallisation (Section 8.10).

of flow units higher up in the ignimbrites, but they are certainly best developed in the lowermost one. They represent jets of turbulent, highly fluidised material which burst forward from the flow-head (Ch. 7). This process was best developed in the initial pumice flow because its flow-head was ingesting cold air which would greatly expand, enhancing fluidisation. Later flows ingested considerably hotter, and probably ash-charged, air which would expand less.

Ground- and ash-cloud surge deposits also occur interbedded with ignimbrite flow units (Figs 5.23c & d). Ash-cloud surge deposits of the Upper Bandelier ignimbrite have been described by Fisher (1979). These are best developed at the edge of the ignimbrite sheet associated with a number of relatively thin, very pumice-rich, reversely graded flow units (Fig. 8.28). These features suggest that, distally, pumice flows were splitting up into separate lobes, ash-cloud surges were well developed and pumice was being dumped at the flow-fronts. These processes are also all recognised at Mt St Helens, but on a different scale.

Fine-grained ash from the eruption of the Lower Bandelier Tuff is known to have been dispersed 500 km (Fig. 8.29). This distal ash deposit is divisible into two layers. There is a lower, coarser ($Md\phi = 2.9$), crystal-rich (36 wt% free crystals)

layer and this is regarded as distal Lower Bandelier plinian fall. An upper, finer-grained ($Md\phi = 4.1$), vitric (14 wt% free crystals) layer is regarded as a co-ignimbrite ash fall. Crystal concentration studies of the Lower Bandelier ignimbrite show that approximately 100 km^3 of ash should have been deposited as a co-ignimbrite ash-fall, and for the Upper Bandelier ignimbrite one of comparable volume is indicated.

In Figure 8.30 the different depositional facies are placed together schematically as a model for the products of ignimbrite-forming eruptions.

8.7.2 RIO CALIENTE AND TAUPO IGNIMBRITES

Both of these ignimbrites have vesiculated volumes of about 30 km^3, which makes them an order of magnitude smaller than each of the Bandelier ignimbrites. The Rio Caliente ignimbrite, from Mexico, was formed from a low intensity (low magma discharge rate) eruption, forming a multiple-flow unit deposit (Fig. 5.16a) with a high aspect ratio (1 : 300) (J. V. Wright 1981). At the other end of the spectrum, the Taupo ignimbrite, from New Zealand, was erupted with an extremely high magma discharge rate and travelled with an extremely high velocity. It formed a low-aspect ratio (1 : 70 000) ignimbrite, consisting of a single-flow unit, and travelled radially outwards in all directions, almost regardless of topography, climbing

Figure 8.29 Distal air-fall ash layer from the Lower Bandelier eruption at Floydada, Texas.

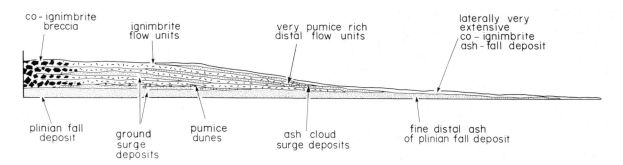

Figure 8.30 Depositional facies model for the products of ignimbrite-forming eruptions based on the Bandelier Tuffs. (After J. V. Wright *et al.* 1981.)

hills up to 1500 m above the vent in Lake Taupo (Ch. 7).

Maps of the two ignimbrites are shown in Figures 8.31 and 32. The distribution of facies in the Taupo ignimbrite is strikingly different from those in the Rio Caliente ignimbrite and our Bandelier Tuff model. Several important differences can be highlighted:

(a) An outstanding feature of the Taupo ignimbrite is that it occurs in two contrasting forms, each having quite different relationships to the pre-existing land surface. One is a landscape mantling ignimbrite veneer deposit (IVD), while the other fills in depressions like 'conventional' ignimbrites do, and is called valley-ponded ignimbrite (VPI; Figs 8.32a & 7.27). The IVD is stratified and occasionally shows bedforms. All of the evidence suggests this

was left behind as a 'trail-marker' by the rapidly moving Taupo flow (Ch. 7). Such types of layers have not been found in the less mobile Rio Caliente ignimbrite, which is composed of normal flow units. However, the Rio Caliente ignimbrite seems to have been erupted onto gentle topography, and if hills were present maybe local ignimbrite veneer deposits could have formed; such a local veneer deposit has been found at one locality in the Upper Bandelier ignimbrite. What is important is they could never have been an important facies, as in the Taupo ignimbrite.

(b) Co-ignimbrite breccias in the Taupo ignimbrite extend much further from the vent. They are of the ground breccia type, generated in the extremely fluidised head of the Taupo flow. Laterally they pass into the Taupo ground layer and the >10 cm average

(a)

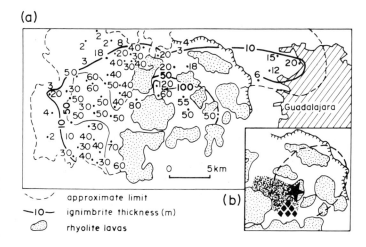

approximate limit
—10— ignimbrite thickness (m)
rhyolite lavas

(b)

Figure 8.31 (a) Isopach map of the Rio Caliente ignimbrite, Sierra La Primavera volcano. The base of the ignimbrite is rarely seen and so recorded values are minimum values. Within the caldera area (Ch. 13) the ignimbrite is now known from drilling to be as much as 360 m thick, suggesting that caldera-collapse was also concomitant with the eruption. (b) Location of vent position (star); solid diamonds represent exposed proximal ignimbrite facies with co-ignimbrite breccias and fines-depleted ignimbrite; broken line is >1 m lithic isopleth for the co-eruptive plinian fall (Fig. 8.12). Also shown is the distribution (close stipple) of some localised mixed-pumice flow units which travelled only a short distance from the vent. (After J. V. Wright 1981.)

(a)

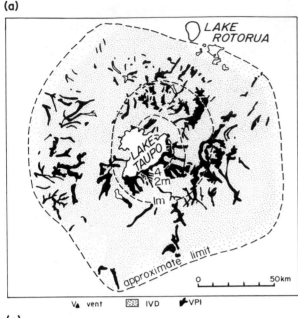

V▲ vent   ▨ IVD   ⚲VPI

(b)

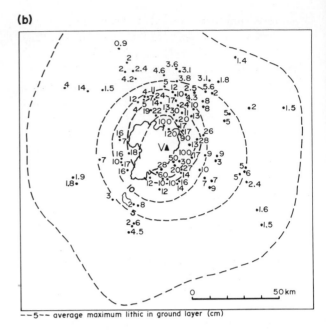

––5–– average maximum lithic in ground layer (cm)

(c)

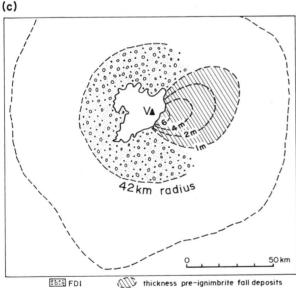

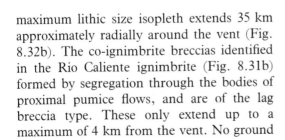

▨▨ FDI   ⬭⬭⬭ thickness pre-ignimbrite fall deposits

Figure 8.32 (a) Map of the Taupo ignimbrite distinguishing ignimbrite veneer deposit from valley-ponded ignimbrite. Isopachs show the average thickness of IVD. (b) Isopleth map of the average maximum diameter of the three largest lithic clasts in the Taupo ground layer. (c) Distribution of fines-depleted ignimbrite and pre-ignimbrite fall deposits. (After G. P. L. Walker *et al.* 1980a, 1981a,b.)

maximum lithic size isopleth extends 35 km approximately radially around the vent (Fig. 8.32b). The co-ignimbrite breccias identified in the Rio Caliente ignimbrite (Fig. 8.31b) formed by segregation through the bodies of proximal pumice flows, and are of the lag breccia type. These only extend up to a maximum of 4 km from the vent. No ground breccias have been found in the Rio Caliente ignimbrite.

(c) Another quite remarkable facies of the Taupo ignimbrite is fines depleted ignimbrite (FDI). This is very distinctive in the field, and is characterised by large rounded framework pumice clasts, with few fines and abundant charcoal fragments (Fig. 7.28). The FDI is restricted to an area where vegetation was not covered by earlier co-eruptive fall deposits (Fig. 8.32c; Section 8.12). The FDI is thought to be material jetted out *en masse* from the flow head (similar to the pumice dunes), but in this case mixed with large amounts of vegetation which also generated large amounts of gas, carrying off even more fines. This fines depleted facies extends 42 km from the vent. Ignimbrite depleted in fines is found in the Rio Caliente ignimbrite, but this only occurs in proximal locations closely associated with co-ignimbrite breccias (Fig. 8.31b). This is again a facies consistent with a high degree of agitation–fluidisation near the vent in proximal pumice flows, as they are segregating from the collapsing column.

(d) Following on from the above, total loss of vitric material is very much greater in the Taupo ignimbrite than the Rio Caliente ignimbrite. Although both ignimbrites have similar volumes, the quantity of ash occurring as a co-ignimbrite ash-fall facies is quite different for each. The Taupo co-ignimbrite ash has an estimated volume of 20 km³ and the Rio Caliente co-ignimbrite ash only 7 km³. This was clearly controlled by the degree of expansion and mobility of the moving pumice flows (Section 7.3).

8.7.3 IGNIMBRITE FACIES AND ERUPTION RATE

The most important control on the overall geometry and internal characteristics, and therefore facies, of the Rio Caliente and Taupo ignimbrites is believed to be magma discharge rate. J. V. Wright (1981) drew the analogy between multiple and single flow unit ignimbrites, and compound and simple lava flows (Fig. 8.33). It was G. P. L. Walker (1970, 1973a) who first suggested that magma discharge rate was perhaps the single most important factor governing lava flow morphology and the distances that lavas travelled (Ch. 4), and this appears to be the same for ignimbrites. For most ignimbrites it is not possible to determine discharge rate. However, J. V. Wright (1981) and C. J. N. Wilson and

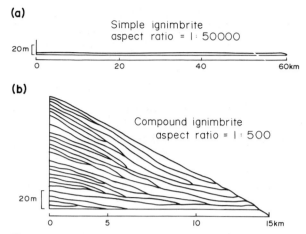

Figure 8.33 Simple and compound ignimbrites. Relationships between thickness, number of flow units and distance travelled are shown schematically for two ignimbrites with similar volumes. (After J. V. Wright 1981.)

Walker (1981) managed to make an estimate for the Rio Caliente and Taupo ignimbrites, respectively, and these are used as the basis for Figure 8.34, together with other data available in the literature.

From the scenario suggested in Figures 8.33 and 34 we can make an assessment of where the Bandelier ignimbrites, and therefore our facies model derived from them, are positioned within this spectrum of ignimbrite types, and also an order of magnitude estimate of their magma discharge rates. Consideration of aspect ratio (1 : 500), number of flow units and proportional volume of

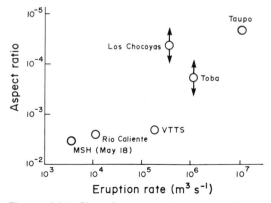

Figure 8.34 Plot of aspect ratio against the average volumetric eruption rate of magma for a number of ignimbrites. Arrows for the Los Chocoyas and Toba ignimbrites indicate uncertainty in measuring aspect ratio.

fine ash lost suggests that both of the Bandelier ignimbrites are towards the lower end of the discharge rate spectrum, and that they have discharge rates of about 10^5 m^3 s^{-1}.

8.8 Palaeocurrent indicators

Flow direction has been determined for a number of ignimbrites by measuring the preferred orientation of elongate crystals and lithic fragments (K. Suzuki & Ui 1982), tree logs (Froggatt *et al.* 1981), welding fabrics (Elston & Smith 1970, Rhodes & Smith 1972, Kamata & Mimura 1983) and magnetic fabrics (Ellwood 1982). K. Suzuki and Ui (1982) have shown that flow orientation was strongly controlled by the pre-existing topography for the Ata pyroclastic flow deposit. Flow lineations oriented radially away from source were only obtained for samples collected from the surface of the ignimbrite sheet. Samples collected from the bottoms of valley-fill deposits had lineations parallel to the axis of the valleys. Samples collected from valley walls tended to be parallel to the slope of the valley. Froggatt *et al.* (1981) also noted that in the Taupo ignimbrite the orientation of tree logs in areas far from the source was parallel to valley axes, even though these could be perpendicular to the general outflow direction.

Wolff and Wright (1981) showed that directional fabrics in welded tuffs are strongly controlled by local palaeoslope. The fact that ignimbrites generally slope away from their source volcano has suggested to some workers that these directional fabrics are likely to reflect flow direction. However, above, we have shown that this line of reasoning is not valid. This short discussion therefore suggests that caution is needed when palaeocurrents of ignimbrites are measured, especially in ancient examples, where they are used to locate source vents. Contouring average maximum lithic clast sizes is the alternative method.

K. Suzuki and Ui (1982) also noted large depositional ramps on the original surface of the Ata pyroclastic flow deposit which provide an independent flow direction indicator. These are shown by the asymmetric distribution of elevation and thickness of deposits in valleys and basins, with deposits ramped up higher on vent-facing slopes. Depositional ramps are also recognised in cross sections through the Ito pyroclastic flow deposit (Fig. 8.3).

8.9 Secondary deposits

Some ignimbrites are obviously interbedded with and overlain by secondary deposits derived from the ignimbrite (e.g. Figs 5.16b & 7.15). These include variously pumiceous mud flows, torrent deposits, lacustrine sediments, reworked wind deposits (Ch. 10) and locally generated phreatic ashes.

Pumiceous mud-flow deposits can closely resemble ignimbrite, and be difficult to distinguish. J. V. Wright (1978) showed that mud-flow deposits interbedded with the Minoan eruption sequence on Santorini could be distinguished from ignimbrite flow units by the absence of a thermal remanent magnetism. Segregation pipes indicating that they had a continuous gas phase, may also be used to identify ignimbrite deposits. However, if mud flows have been generated by water flooding into still-hot ignimbrite, then similar pipes could form but, in these, clasts may be mud-coated and entrapped vesicles may be present in the muddy matrix. Note also that Sheridan *et al.* (1981) have indicated that some pumiceous mud flows could be *primary* products formed at vent during hydrovolcanic phases of ignimbrite-forming eruptions. Torrent deposits that are rich in large lithic blocks may be difficult to distinguish from co-ignimbrite breccias, and pumiceous examples may be hard to differentiate from surge deposits. Angle of repose cross-stratification (Ch. 10), if present, indicates an alluvial origin. Lacustrine deposits will form where river valleys have been blocked by debris. Deposits could include ash turbidites and fines-free assemblages of coarse, very well rounded pumices representing pumice rafts which formed a floating layer on the surface of the lake (Ch. 10). Wind reworking of the top of an ignimbrite could produce stratified and low angle cross-stratified layers which would closely resemble ash-cloud

Figure 8.35 Phreatic explosion craters in the 18 May pumice flow deposits on the western part of the Pumice Plain at Mt St Helens (Fig. 8.6). Craters are approximately 5–25 m in diameter. (After Rowley *et al.* 1981.)

surge deposits. However, wavelength and size would not be related to pumice-flow transport direction. On a larger scale, wind erosion can strip large volumes of ignimbrite, adding this to a loess blanket as has happened in New Zealand. Without knowing the regional stratigraphy, such a deposit is probably difficult to distinguish from an air-fall ash deposit at some outcrops (Fig. 10.32b; Section 10.3).

Where large quantities of water gain access to a still-hot ignimbrite, steam explosions can be triggered. These form rootless explosion craters like those observed at Mt St Helens (Fig. 8.35), where local phreatic surge and ash-fall layers were also generated. Similar stratified deposits formed by

steam explosions occur above the Taupo ignimbrite (C. J. N. Wilson & Walker 1985) and Crater Lake pumice flow deposit (Bacon 1983).

8.10 Welding and post-depositional processes

Ignimbrites are emplaced at high temperatures and welding, recrystallisation and alteration may occur during the period of cooling. Three main processes can be recognised:

welding
vapour-phase crystallisation
devitrification

8.10.1 WELDING

Welding is the sintering together of hot pumice fragments and glass shards under a compactional load (R. L. Smith 1960a, b, Ross & Smith 1961). The most important controls are:

(a) glass viscosity (dependent on temperature and composition) and
(b) lithostatic load (dependent on the thickness of the deposit).

Lithic content in a deposit will also affect the development of welding (Eichelberger & Koch 1979). Experimental studies indicate that welding begins between about 600 and 750°C for rhyolitic compositions, depending on load pressure and H_2O content of the glass (I. Friedman *et al.* 1963, Bierwirth 1982; Fig. 8.36).

Characteristically, when welding approaches completion, three zones of dense welding, partial welding and non-welding are produced (Fig. 8.37; R. L. Smith 1960b). In the welded zones flattened, often glassy juvenile clasts called **fiamme**, and glass shards define a planar foliation or eutaxitic texture (Figs 8.38 & 3.23b). In most cases fiamme seem to be flattened pumice clasts, but sometimes they may have originally been unvesiculated juvenile clasts (Gibson & Tazieff 1967). Welding is often associated with distinctive colour changes, which are due to different oxidation states of iron. In densely

(a)

(b)

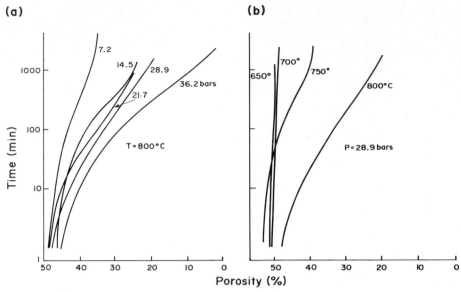

Figure 8.36 Experimentally derived compaction curves for anhydrous rhyolitic ash (<2 mm) from the Upper Bandelier ignimbrite. (a) At a constant temperature of 800°C and load pressures of 7.2, 14.5, 21.7, 28.9 and 36.2 bars. (b) At a constant load pressure of 28.9 bars and temperatures of 650, 700, 750 and 800°C. (After Bierwirth 1982.)

welded zones columnar cooling joints are often well developed (Fig. 8.39). Lithophysae may also be present (Fig. 8.37; Ch. 5). A densely welded tuff, which in hand specimen has a glassy appearance, is sometimes called a **vitrophyre**.

Many workers have recorded systematic changes with height in bulk density and porosity in welded ignimbrites (Fig. 8.40). Typically, bulk density is at a maximum, whereas porosity is at minimum in the lower central half of the deposit, and this corresponds to the zone of dense welding. Ragan and Sheridan (1972) showed in the Bishop Tuff that these features are related to the degree of flattening

of the pumice clasts, which can be considered a measure of the compactional strain (Fig. 8.41). Strain ratio (see also Figs 6.42 & 45) can be measured in welded tuffs by applying techniques developed by structural geologists to measure tectonic strain (Ramsay 1967, Dunett 1969, Elliot 1970, Lisle 1977, B. Roberts & Siddans 1970, Ragan & Sheridan 1972, Sheridan & Ragan 1976, Sparks & Wright 1979, Wolff & Wright 1981).

R. L. Smith (1960b) classified ignimbrites that showed such simple welding variations as **simple cooling units**. Riehle (1973) found that these variations could be predicted from theoretical

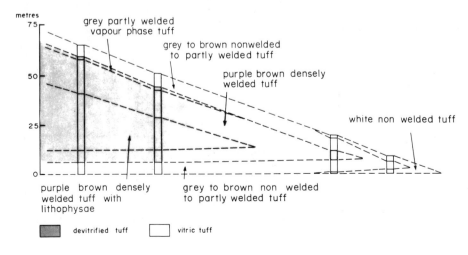

Figure 8.37 Schematic cross section based on four measured sections of the Yucca Mountain Tuff showing welding and devitrification zones in an ignimbrite. (After Lipman & Christiansen 1964.)

Figure 8.38 Welded ignimbrites with eutaxitic textures defined by fiamme. (a) Ignimbrite erupted from Platoro caldera (part of Treasure Mountain Tuff), San Juan volcanic field, Colorado. (b) Very densely welded ignimbrite near La Piedad, Central Mexico.

Figure 8.39 Columnar jointing in welded ignimbrites. (a) Amealco ignimbrite (Quaternary), near San Juan del Rio, Central Mexico. (b) Bishop Tuff, California. (c) Rubicon Rhyolite (Devonian), Central Victoria, Australia.

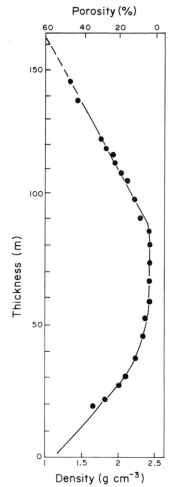

Figure 8.40 Density–porosity profile for the Bishop Tuff. (After Ragan & Sheridan 1972.)

foliation of fiamme they show a well defined lineation (Fig. 8.43). This indicates secondary mass flowage of the tuff during welding, and is termed **rheomorphism**. Flow folds are sometimes also well developed. Schmincke and Swanson (1967) and other workers (e.g. Chapin & Lowell 1979) have suggested that stretching and welding of fiamme may be a primary flow feature during the final stages of pyroclastic flow. Wolff and Wright (1981) argued that it must be a secondary process involving first compactional deformation and then subsequent flow on a slope. Many (but not all) examples of rheomorphic welded tuffs are peralkaline rhyolites having unusually low glass viscosities compared with calc-alkaline rhyolites.

Ignimbrites which show well developed zones of dense welding, and in which the proportion of welding is high, can be termed **high-grade** ignimbrites, in contrast to **low-grade** ignimbrites, which are totally non-welded or of which only a small proportion is welded. The temperature of emplacement of ignimbrites is therefore very variable. R. L. Smith (1960a) considered that processes within the eruptive column must be responsible, and Sparks *et al.* (1978) explained this by the eruption column collapse model. Eruptions with low gas content and low gas velocity will lead to low collapse heights and little heat loss during collapse,

analysis of the cooling of a sheet emplaced with uniform temperature. However, some ignimbrites have several zones of dense welding and partial welding. Such ignimbrites are classified by Smith as **compound cooling units** (Fig. 8.42). These suggest that the upper parts of earlier flow units of the ignimbrite must have been partially or wholly cooled before the emplacement of later flow units. Some care must be taken when describing compound cooling units, as they may even consist of a number of ignimbrites separated by long intervals between eruptions.

Most welded rocks only show compactional flattening of fiamme and glass shards (Fig. 8.41). However, in some welded tuffs fiamme are seen to be stretched, and in sections parallel to the plane of

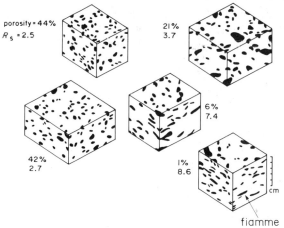

Figure 8.41 Cut blocks of Bishop Tuff, showing increase in flattening of pumice clasts with decreasing porosity. Values of strain ratio (R_s) show the increasing compactional strain. Note block (c) is cut slightly oblique to the foliation. (After Ragan & Sheridan 1972.)

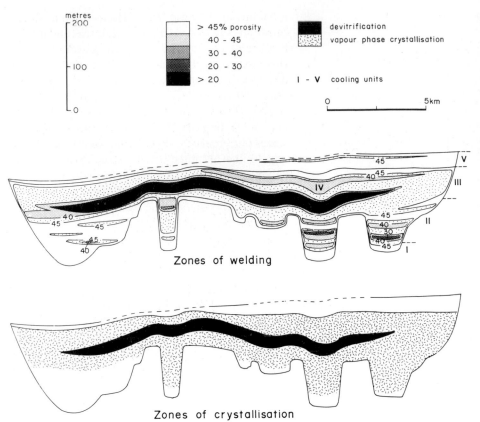

metres
200

100

0

> 45% porosity
40 - 45
30 - 40
20 - 30
> 20

devitrification
vapour phase crystallisation

I - V   cooling units

0   5km

45
40 45
45
40
45
30
45
40
45
45

V
III
IV
II
I

Zones of welding

Zones of crystallisation

Figure 8.42 Cross section through part of the Upper Bandelier Tuff, showing welding and crystallisation zones. The ignimbrite is a compound cooling unit, and shows an upward increase in the degree of welding in cooling units I–III. Recognisable flow units are much thinner in units IV and V, and nearer the source they pass into densely welded tuff, continuing the trend towards higher temperature of emplacement of successive pumice flows that is more clearly shown by units I–III. Note, the topography that the ignimbrite fills in is cut into older ignimbrites and basement, including Precambrian (cross section is approximately normal to movement direction of the pumice flows). (After R. L. Smith & Bailey 1966.)

which will favour formation of a densely welded ignimbrite. This condition will also lead to less-expanded flows which should lose only minor amounts of vitric ash, as seems to be the case for many densely welded intracaldera ignimbrites (Section 8.4; Ch. 11). Flows formed from high collapse heights will be emplaced at relatively lower temperatures, and may form non-welded deposits. Some large ignimbrites show an upward increase in welding due to the emplacement of flow units of successively higher temperature. The Upper Bandelier ignimbrite shows such a sequence, with flow unit temperatures changing from the order of 550–800°C (R. L. Smith & Bailey 1966; Fig. 8.44). Widening of the vent and a decrease in gas content during eruption and column collapse could both explain increasing emplacement temperature.

The time taken to produce welding has been estimated theoretically by Riehle (1973) and Kono and Osima (1971). However, their methods are based on the assumption that viscous strain occurs, but this is incorrect for porous volcanic ash because it does not allow for the effects of pore space (porosity) and packing geometry. Bierwirth (1982) considered the effects of porosity as shown by models of sintering and hot pressing processes used in the ceramic and metallurgical industries. The strain rate for porous materials is given by

$$\dot{\varepsilon} = Kf(\varrho)P^n e^{-Q/RT} \qquad (8.1)$$

where, T is temperature, P is load pressure, Q is activation energy, n is a stress factor, $f(\varrho)$ is a density function, R is the universal gas constant and K is the strain rate constant. Using the results of his Bandelier Tuff experiments (Fig. 8.36), values of Q, n and K were determined and a nearest-fit equation assigned to $f(\varrho)$. After converting strain to density, Equation 8.1 was integrated giving an equation defining compaction rates of anhydrous ash from the Bandelier Tuff:

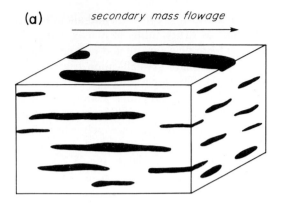

secondary mass flowage

(a)

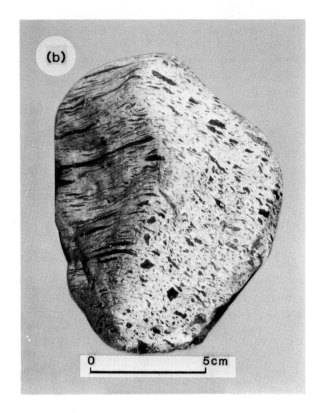

(b)

Figure 8.43 Lineation produced by stretching fiamme during rheomorphism.

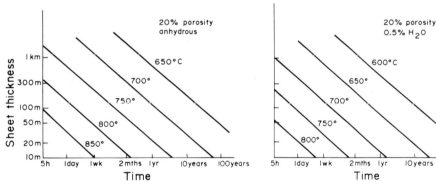

Figure 8.44 Time required to produce dense welding of rhyolitic ash (a) anhydrous and (b) with 0.5 wt% water for various ignimbrite thicknesses and emplacement temperatures. Calculated from Equation 8.2 based on experimental studies of the Upper Bandelier ignimbrite. The model is least reliable for thin flows and long times, due to the effects of cooling. Sheet thickness was calculated for porous ash (1 bar = 7 m). (After Bierwirth 1982.)

$$19.7\varrho^{11} + 0.04p^{-1} - 0.092 = (6.3 \times 10^{15} P^{1.7} e^{-5.6} \times 10^{4/T})t \quad (8.2)$$

where ϱ is fractional density, t is time in seconds, P is in bars and T in K.

Values are reliable for fractional densities less than 0.9 (<2.4 g cm^{-3}, $>10\%$ porosity). Equation 8.2 produces values which correlate well with experimental results, and enables predictions to be made beyond the time range of experiments. For an ignimbrite 100 m thick and emplaced at 850°C, dense welding ($<20\%$ porosity) would occur within one week (Fig. 8.44a). For even a 1 km thick intracaldera ignimbrite at 650°C, dense welding would only be achieved after one year, assuming that crystallisation of the groundmass did not prevent compaction. Welding temperatures will be lowered by the effect of water content in the glass, and the times shown in Figure 8.44a can be regarded as maximum values. Bierwirth indicated that 0.5 wt% water will lower welding temperatures by approximately 60°C compared with anhydrous conditions (Fig. 8.44b).

8.10.2 VAPOUR-PHASE CRYSTALLISATION

Vapour-phase crystallisation results from the percolation of hot gases through ignimbrites during cooling. The most important gas sources are probably diffusion from juvenile vitric particles and heated ground water.

The main products of vapour phase crystallis-

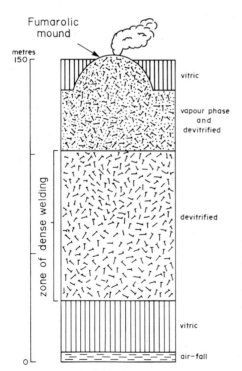

Figure 8.45 Zones of vapour-phase crystallisation and devitrification in the Bishop Tuff. Fumarole mounds project from top of vapour-phase zone through non-welded ash. (After Sheridan 1970.)

ation are tridymite, cristobalite and alkali feldspar, which occur as drusy infills of matrix and pumice cavities, so forming a cement and reducing the pore space. Vapour-phase crystallisation can produce a coherent rock. The term sillar (a Peruvian word first applied by C. N. Fenner 1948) is often used for such rocks, but the term has also been applied to incipiently welded tuffs in which ash grains are barely sintered at point contacts and show no other deformation of juvenile clasts (Fig. 8.7). Vapour-phase crystallisation can occur in separate zones in compactionally welded ignimbrites, and is commonly found towards the top of a sheet (Figs 8.37, 42 & 45). Sheridan (1970) described fumarolic mounds and ridges of sillar above the Bishop Tuff (Figs 8.45 & 46a). Similar palaeofumarolic features and peculiar 'steam pipes' occur in the Rio Caliente ignimbrite, which are thought to have formed where it was deposited in a shallow lake or on marshy ground (Figs 8.46b, c & 47). Fumarolic pipes have also been found in the Bandelier ignimbrites (Fig. 8.46d).

8.10.3 DEVITRIFICATION

Devitrification involves the sub-solidus crystallisation of metastable glass (Ross & Smith 1961, Lofgren 1970). The main products are cristobalite and alkali feldspar. Devitrification tends to be more prevalent in densely welded tuffs, and particularly in thick intracaldera ignimbrites because of the protracted cooling (Figs 8.37, 42 & 45). Nevertheless, more-porous ignimbrites may also be devitrified. (Note that devitrification is also common in coherent glassy lavas and shallow intrusives.) Devitrification is discussed further in Chapter 14 (Section 14.3.2).

8.11 Chemical analyses?

Post-depositional chemical alteration is a common feature even of *modern* ignimbrites. There are several processes which can produce alteration. One is leaching by ground water, and studies show that metastable glass is easily leached and Na, K and Si are often removed (e.g. Noble 1967, Scott

Figure 8.46 (a) Fumarole mounds on the surface of the Bishop Tuff. (b) and (c) Small 'steam pipes' in the Rio Caliente ignimbrite. Vapour-phase minerals are predominantly clinoptilolite and heulandite which suggests low temperature fumarolic activity (compared with tridymite, cristobalite and alkali feldspar, which are the usual vapour-phase minerals found, e.g. in the Bishop Tuff). This is as might be expected if the ignimbrite were locally deposited in a shallow lake or on marshy ground and vaporised ground water was important. (d) Fumarole pipes in the Upper Bandelier Tuff.

(a)

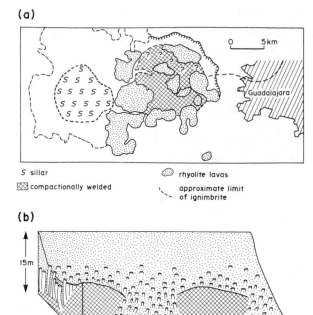

S sillar
⊠ compactionally welded
◉ rhyolite lavas
`\,` approximate limit of ignimbrite

(b)

Figure 8.47 (a) Map showing distribution of sillar and palaeofumarolic area in the Rio Caliente ignimbrite, New Mexico. (b) Field sketch showing relations in the area shown in Figure 8.47a. Pipes are radially distributed about massive zones of sillar, which seemed to have been the sites of most intense vapour-phase activity.

1971). Vapour phase and fumarolic activity and hot springs related to the regional geothermal system (Fig. 13.48) is the second important source of alteration (see the previous section).

Both Ui (1971) and G. P. L. Walker (1972) warn against the acceptance of whole rock analyses of ignimbrites as guides to the composition of the parent magma. Such analyses are subject to considerable error, because of the presence of xenoliths in the lithic component (Ui 1971). Perhaps more important, though, is the concentration of crystals that is found in ignimbrites due to the selective loss of fine vitric ash during eruption and emplacement. The composition of the ignimbrite will be enriched in those elements occurring in higher proportions in the crystals, and will be depleted in those occurring in higher proportions in the glass.

The above questions the value of chemical analyses of ignimbrite (especially whole rock analyses of ancient welded ignimbrites), and indicates the care needed in sampling.

8.12 The great Taupo AD 186 eruption

The studies by G. P. L. Walker, C. J. N. Wilson and co-workers on the products of this ignimbrite-forming eruption have greatly stimulated volcanology, yielding new insights into explosive volcanism and extending the range and scale of known volcanic phenomena. A brief eruption narrative is, therefore, a fitting finale to this chapter. The eruption produced a great variety of pyroclastic products, including the most powerful plinian and most violent ignimbrite-forming events yet documented (Chs 6 & 7; Section 8.7). It is the youth and excellent preservation of the deposits that enabled the eruption sequence to be examined in such detail.

The products of the Taupo eruption (Figs 8.48 & 49) are dispersed over a large part of the North Island of New Zealand. They consist of: an initial plinian pumice-fall deposit, the Hatepe pumice; two phreatoplinian ashes, the Hatepe ash and the Rotongaio ash, which are separated from each other by an erosional break; an ultraplinian pumice-fall deposit, the Taupo pumice; and, finally, the

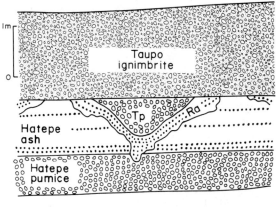

Tp Taupo pumice
Ra Rotongaio ash

Figure 8.48 Stratigraphy of the products of the Taupo AD 186 eruption, New Zealand. (After G. P. L. Walker *et al.* 1981a.)

Figure 8.49 The Taupo eruption sequence as explained in Figure 8.48, showing prominent infilled erosion gullies between the Hatepe and Rotongaio ashes, and strongly discordant erosion surface between the Taupo ignimbrite and Taupo pumice. This location is 20 km ESE of the vent. (Photographs by C. J. N. Wilson.)

climactic Taupo ignimbrite. The total erupted volume was about 100 km^3 (dense rock equivalent), of which 24 km^3 is found in the Taupo ultraplinian pumice-fall deposit, 30 km^3 in the Taupo ignimbrite and another 20 km^3 in a co-ignimbrite ash fall. In this section we briefly consider some of the exceptional features of this sequence.

From various lines of evidence, the vent position is inferred to be at or near the Horomatangi Reefs in Lake Taupo. Horomatangi Reefs may be lava domes or the remnants of a pyroclastic cone. The

sequence of phases can be interpreted in terms of tumescence of the vent area or loss of lake water, so exposing a dry-land vent, and subsequent detumescence submerging the vent during the eruption (Fig. 8.50).

We will now consider the principal phases of the Taupo eruption:

early air-fall phases
Taupo ultraplinian fall deposit
Taupo ignimbrite

8.12.1 EARLY AIR-FALL PHASES

The Hatepe pumice is a coarse, relatively homogeneous and well sorted layer which shows mantle bedding. Its dispersal (Fig. 8.51a) and grainsize characteristics are typically plinian.

The two phreatoplinian layers (Figs 8.51b & c) are fine-grained, stratified, relatively poorly sorted ash deposits (G. P. L. Walker 1981a). Both show very little change in grainsize and sorting with distance from source. They both show many signs of being deposited as 'wet' ash. For example, the Hatepe ash locally shows a type of microbedding attributed to the splashing of falling water; both deposits contain vesicles formed when air was trapped by ash falling as sticky mud; the Rotongaio ash, where thick, often slid as small mud flows into the gullies cut into the Hatepe ash; and soft-sediment deformation structures are found (Fig. 8.52). However, both lack accretionary lapilli, but very wet accretionary lapilli similar to mud-rain could have easily splashed on impact. The poor sorting of the deposits is attributed to water-scavenging of the ash cloud. Although they only show limited fractionation with distance, both layers show a regular exponential thinning from source (Figs 8.51b & c), indicating that the scavenging water must have been derived from the source, Lake Taupo. The water was therefore an integral part of the eruption column and ash cloud, and G. P. L. Walker (1981a) envisaged a fully water-charged column erupting out of Lake Taupo, with both ashes actually falling as mud.

The deep-gullied erosion surface separating the two ash layers is significant, and there is a similar,

(a) Hatepe pumice

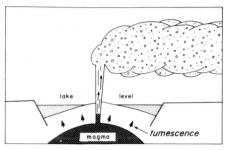

(b) Hatepe ash ('putty ash')

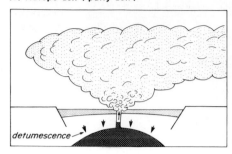

(c) 'Volcanogene waterspout'

(d) Rotongaio ash

(e) Taupo pumice

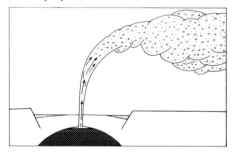

(f) Taupo ignimbrite

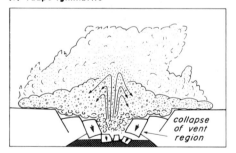

Figure 8.50 Pictorial representation of the different phases of the Taupo eruption.

but less prominent, erosion surface within the Hatepe ash. Features of the gullies indicate very rapid erosion in two brief episodes during the eruption. The depth of gullying, and therefore the amount of erosion, decreases systematically away from Lake Taupo (Fig. 8.53). This strongly suggests that the gully erosion was somehow related to the eruption column, and Walker believes that this was when a large part of the lake was erupted as a giant *'volcanogene waterspout'* (Fig. 8.50c).

8.12.2 TAUPO ULTRAPLINIAN FALL DEPOSIT

The Taupo pumice is extensively described by G. P. L. Walker (1980), who used this deposit to define the term *'ultraplinian'*, reflecting its extremely wide dispersal (Fig. 8.54; Ch. 6). It is a coarse pumice fall deposit (Fig. 8.49), which is internally stratified, especially near vent, shows mantle bedding and otherwise resembles a normal plinian deposit on a local scale. Although the maximum lithic and pumice isopleths demonstrate that the

(a) Hatepe pumice

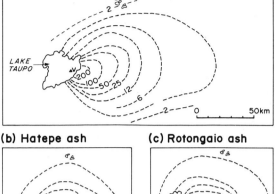

(b) Hatepe ash   (c) Rotongaio ash

Figure 8.51 Isopach maps illustrating the distribution of the early air-fall phases. (After G. P. L. Walker 1981a & c.)

vent lay within the present Lake Taupo, the isopachs of the deposit close about a point about 20 km east of the lake, a fact which led earlier workers to place a vent on-shore. However, this secondary thickening away from the vent can be explained by the combination of a very high eruption column, partly attributable to a very high eruption rate (estimated at 10^6 m^3 s^{-1} of magma), and a strong westerly wind shifting the eruption column sideways (Fig. 8.50e). Also, at many of the

near-source outcrops the Taupo pumice has obviously been eroded by the Taupo ignimbrite which followed, and therefore thickness measurements are anomalously thin.

The maximum measured thickness is only 1.8 m, which reflects the great dispersal of the deposit although, again, even where it is thickest there is a marked erosional discordance between it and the ignimbrite. Despite its *unimpressive* thickness (cf. other plinian deposits, Ch. 6), calculations by G. P. L. Walker (1980) from the concentration of free crystals in the deposit show that the total volume erupted was 24 km^3 (6 km^3 dense rock equivalent (DRE)), which is as big as some of the

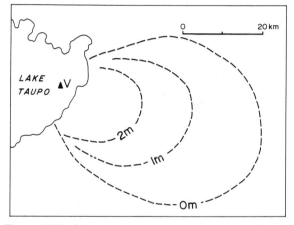

Figure 8.53 Map showing approximate extent of gully erosion in the interval between the eruption of the Hatepe and Rotongaio ashes with isopleths of gully depth. (After G. P. L. Walker 1981a.)

Figure 8.52 Soft-sediment deformation structures between the Hatepe (pale) and Rotongaio (dark) ashes. The contact separating the two is partly erosive (Fig. 8.49). (Photograph by C. J. N. Wilson.)

(a) Thickness (cm)

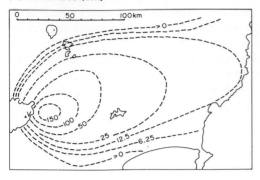

(b) Maximum pumice (cm)

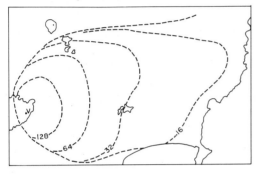

(c) Maximum lithic (cm)

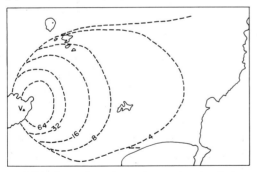

Figure 8.54 Isopach and maximum-size isopleth maps (average diameter of the largest clasts) for the Taupo pumice. (After G. P. L. Walker 1980.)

larger plinian deposits known (Table 6.2). Of this volume only about 20% fell on land, and 80% (mainly finer vitric ash) fell out to sea more than 220 km from source. Independent methods indicate that the height of the eruption column must have been in excess of 50 km, and that this phase lasted 6–17 h.

8.12.3 TAUPO IGNIMBRITE

We have already discussed at length many of the features of this spectacular ignimbrite (Ch. 7; Section 8.7). The abrupt switch from fall- to flow-forming activity is thought to have been caused by a drastic increase in discharge rate (Fig. 8.34), and cannot simply have been due to the collapse of the very high ultraplinian column. C. J. N. Wilson and Walker (1985) suggest that the high eruption rate during the ultraplinian phase drained the upper part of the magma chamber, leaving its roof unsupported. This eventually led to major collapse of the vent region (Fig. 8.50f). This greatly widened the vent, and the ignimbrite-producing flow was then formed by instantaneous column collapse, the discharge rate being suddenly much greater than that which would allow for the maintenance of a stable convective column (Fig. 8.11). The lithic content of the pre-ignimbrite air-fall deposits (1 km^3) implies that vent widening by erosion was important and would eventually have led to column collapse if the eruption had followed a normal course of events. Indeed, occasional partial collapses of the ultraplinian column did occur, generating a number of early ignimbrite flow units found interbedded with the Taupo pumice near the source. A sudden increase in the volume (2 km^3) and sizes of lithics in the Taupo ignimbrite support the idea that there was a drastic change in eruption conditions.

Field data suggest that the parent flow of the Taupo ignimbrite erupted over only 400 s in batches of material which gradually coalesced so that from about 40 km outwards the flow was a single wave of material (C. J. N. Wilson & Walker 1981). During most of its passage, the flow consisted of a head (strongly fluidised by ingested air) which generated layer 1 deposits, and a body plus tail which generated layer 2 deposits (Ch. 7). The flow moved at very high velocities over a locally mountainous terrain at speeds probably exceeding 250–300 m s^{-1} near the vent, and which remained high (locally >100 m s^{-1}) to the outer limits of the flow. The flow-head was highly erosive and locally, almost unbelievably, scalped and overturned the floor over which it rode (Fig. 8.55). A variety of

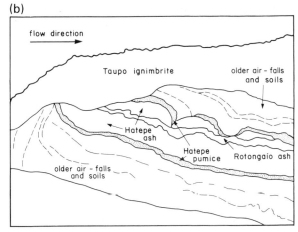

Figure 8.55 The Taupo ignimbrite at this location scalped and overturned air-fall layers of the preceding phases in the eruption and older air-fall deposits and soils. This outcrop is approximately 15 km east of the vent; the Taupo pumice is absent because it was presumably stripped off before the erosive overturning event occurred.

processes acting in response to fluidisation or flow kinetics operated within the flow to produce the great variety of facies.

8.12.4 OVERVIEW

The total thickness of air-fall deposits amounts to over 5 m near the vent, nearly all of the material having been blown to the east of the vent by strong south-west to westerly winds (Figs 8.51 & 54). Air-fall deposits more than 10 cm thick were deposited over an area of 30 000 km², while an area of about 20 000 km² was devastated by the Taupo ignimbrite

flow. In addition, mud flows and floods reached far beyond the limits affected by primary eruption products. The total duration of the eruption could have been as short as a few days or as long as months, depending on the length of the time gap between the Hatepe and Rotongaio phreatoplinian phases. If the Taupo eruption were repeated, there can be no doubt how catastrophic its effects would be.

8.13 Further reading

R. L. Smith's classic papers published in the early-1960s (R. L. Smith 1960a, b, Ross & Smith 1961) are still essential reading, having been, for the past two decades, the foundation for the advances in our understanding of ignimbrites. The Geological Society of America has honoured this pioneering work with publication of the Special Paper entitled 'Ash-flow tuffs', edited by Chapin and Elston (1979). This contains some of the more up-to-date information and reviews, including one by Smith himself (R. L. Smith 1979) which looks in some detail at physicochemical aspects of ignimbrite magma chambers. G. P. L. Walker (1983) reviewed ignimbrite types and problems, and in this very useful article 'takes stock of what is known, and also what needs to be known before ignimbrites are well understood'. In addition Walker (1985) has reviewed the origins of ignimbrite-associated breccias. Although much of the Taupo AD 186 story is scattered in a large number of papers, C. J. N. Wilson and Walker (1985) have pieced together an eruption narrative, and C. J. N. Wilson (1985) presents his complete story for the Taupo ignimbrite in an outstanding publication. One other, slightly older, paper worthy of special note is Yokayama (1974), on the Ito pyroclastic flow deposit (Fig. 8.3). An important volume on caldera formation and caldera geology is the special issue of *Journal of Geophysical Research* (1984, volume 89 B10), edited by Lipman, Self and Heiken. The papers by Lipman (1984) and G. P. L. Walker (1984a) in this volume provide interesting contrasts, Lipman high-lighting the complexities of the 'standard' caldera collapse model, including the changing nature and position of the vent system, whereas Walker puts the 'standard' caldera model into perspective by identifying significant deviations from the standard model in many volcanoes. Fisher and Schmincke (1984) have also given a comprehensive summary of the characteristics of pyroclastic flow deposits.

Plate 9 The Minoan ignimbrite sloping towards the southern shoreline of Santorini. Entrance of pumice flows into the sea during the 1490 BC eruption could have deposited substantial accumulations of subaqueous pyroclastic flow deposits and ash turbidites in submarine basins surrounding the island.

CHAPTER NINE

Subaqueous pyroclastic flows and deep-sea ash layers

Initial statement

The modes of formation of subaqueous volcaniclastic rocks can be very diverse (Ch. 3), and
recognition of the origins is often not simple. This
chapter specifically deals with those volcaniclastics
considered to be primary pyroclastic products of
highly explosive eruptions. However, one is never
far removed from the problems of differentiating
primary from redistributed, or epiclastic, volcanigenic sediments. We examine the concept of
subaqueous pyroclastic flows and ignimbrites, discussing the types of deposits that have been
described, the terminology used and the controversial subjects of subaqueous welding and subaqueous eruption. We suggest that, in general, such
deposits are the lateral equivalents of subaerial
eruptions from island volcanoes rather than wholly
subaqueous deposits of submarine eruptions, and
that welding in ignimbrites is a feature found in
those deposited on land or in very shallow water.
We also consider many examples of widely dispersed deep-sea ash layers, which result from the
distal fall-out of large-magnitude continental silicic
eruptions. Marine processes which affect their
distribution and accumulation are discussed, as
well as a method for determining eruption duration
from graded bedding. Finally, we briefly speculate
on whether subaqueous base surges can form.

9.1 Introduction

In this chapter we address one of the most difficult
and speculative areas in volcanology: subaqueous
pyroclastic flows and subaqueous welding. These

aspects are controversial because, unlike subaerial phenomena, subaqueous pyroclastic flows have not been observed, and because the deposits that have been interpreted as subaqueous pyroclastic flow deposits, especially welded ones, have not to our mind been convincingly proven to be such. Major unresolved problems exist with regard to subaqueous pyroclastic flows. Very little is known about the physical interaction between hot, gas-supported particulate pyroclastic flows and a cold ambient body of water and about the boundary conditions between them. Can hot, subaerial, gas-supported, pyroclastic flows retain their integrity and heat when they flow into water, and continue to flow underwater, uninterrupted, as a hot, gas-supported pyroclastic flow? Will such a flow be capable of producing a welded deposit? Or, will the combination of a relatively low bulk density, high flow velocity, the essentially frictionless nature of the flow, the surface tension of the water and the low angle of incidence of a flow on to the water surface cause deflection of the flow so that it skips or flows over the water (cf. skip bombs)? Can, in fact, hot pyroclastic flows be generated from shallow, subaqueous pyroclastic eruptions?

All of these problems are of fundamental importance to those of us who have to make palaeoenvironmental interpretations of volcanic successions. Until recently, welded ignimbrites were generally accepted to be good indicators of subaerial, to at best shallow marine, environments. However, there are now several papers in the literature that claim to have found subaqueous welded ignimbrites, even in relatively deep-water settings.

Given the uncertainties and the questions posed above, we wish to some degree to play the devil's advocates. This is because we feel strongly that *all* of the essential pieces of evidence necessary to prove the case for submarine pyroclastic flows and welding have not been presented in any study. We hope that our discussion may provoke further consideration of the strength of the evidence available, and of some of the problems involved in generating subaqueous pyroclastic flows.

The essential evidence that we feel needs to be established in each case is briefly summarised as follows. First, the facies characteristics of the host sedimentary succession in immediate contact with the suspected pyroclastic flow deposits should be fully documented. It is insufficient simply to state the preferred depositional setting without proving it, given the controversial nature of the subject. Secondly, the evidence for a hot state of emplacement (welding, gas segregation structures, columnar jointing) has to be very convincingly documented *and* illustrated in each case. The only unequivocal evidence for welding is plastically deformed, flattened, annealed cuspate glass shards (burst vesicle walls) or preserved *plastically* collapsed, flattened vesicles in larger glassy fragments (pumice), or both, *at the microscopic scale*. Without this, welding is not proven. In old, tectonised, altered successions this may be difficult. In such situations deformation and alteration can produce elongation and flattening of clasts, especially original glassy clasts that have been altered to phyllosilicates. This can produce a *pseudo-eutaxitic texture* that can be mistaken for welding and the existence of pumice.

It should also be noted (also see Ch. 10) that the presence of shards in a deposit is not proof of a pyroclastic flow depositional origin. Shards indicate a pyroclastic mode of fragmentation originally (Ch. 3), but thereafter the pyroclastic debris, without any further reworking or sorting, could be redeposited and transported long distances from the source by subaqueous (and, for that matter, subaerial) epiclastic mass-flow processes (Ch. 10). The deposit will be a very juvenile aggregate of pyroclastically fragmented debris, but such a history does not involve transport as a subaqueous pyroclastic flow.

9.2 Types of subaqueous pyroclastic flow

A large proportion of volcanic rock transported as pyroclastic flows from island volcanoes must be deposited into the ocean (Plate 9). In a broad sense, the subaqueous equivalents belong to, and can be described by, the terminology of subaqueous clastic mass-flow facies associations (e.g. Mutti & Ricci Lucchi 1978, R. G. Walker 1984; Ch. 10). Two main types of deposit formed by the movement of pyroclastic flows underwater have been described in the literature:

subaqueous pyroclastic flow deposits
ash turbidites

9.2.1 SUBAQUEOUS PYROCLASTIC FLOW DEPOSITS

Subaqueous pyroclastic flow deposits should, by virtue of their name, be the subaqueous equivalents of subaerial pyroclastic flow deposits. They may be the lateral equivalents of subaerially erupted pyroclastic flows, or they may be subaqueously erupted flows of the type envisaged by Fiske (1963) and Fiske and Matsuda (1964) (Section 9.5). In the literature the term is generally applied to massive depositional units which show features akin to high concentration types of mass flow (Figs 9.1–4). Some individual depositional units are very thick, and examples described by Fiske and Matsuda (1964) from the Tokiwa Formation in Japan are up to 50 m thick (Fig. 9.1d). Recent examples found in the Grenada Basin, which can be shown to be the submarine equivalents of pumice flow deposits forming the Roseau Tuff on Dominica, are much thinner, but again show sedimentary features suggesting deposition by high concentration mass flows (Carey & Sigurdsson 1980; Figs 9.3 & 4).

Carey and Sigurdsson (1980) suggest that the term 'subaqueous pyroclastic flow deposit' should only be used if there is evidence of a high emplacement temperature. They prefer to use the term *subaqueous pyroclastic debris* flow deposits' to describe the submarine equivalents of the Roseau ignimbrite. This is supposed to indicate that deposition was from high-concentration debris flows of pyroclastic material, without implying that they were hot. We feel that the term 'pyroclastic debris flow', although well intended, is potentially confusing and contradictory. If *'pyroclastic'* is to be used in the term at all, there should be evidence that the deposits have both a demonstrable pyroclastic mode of fragmentation *and mode of transport.* If this is not the case, and if there is a reasonable possibility that normal epiclastic mass flow of juvenile volcanic debris was involved, then terms such as 'volcaniclastic debris-flow deposit' or 'volcaniclastic granular mass-flow deposit' should be used (Chs 10 & 12).

In the case of subaerially erupted pyroclastic flows, the distance that a flow can be maintained underwater as a gas–particulate dispersion is unknown. If ingestion of water through the flow-front and mixing into the body of the flow (Allen 1971, Simpson 1972) were important, a water–particulate debris flow or granular mass-flow (Ch. 10) could quickly be generated and, eventually, with extensive dilution, a turbidity current. The lower massive unit of the Dali Ash (Figs 9.1e & 2) contains foraminifera dispersed through its thickness, indicating that this example was certainly deposited from a water–particulate system. Also, bed-forms are developed at the base of the lower massive unit, and there is a particularly distinctive thin (4 cm), planar-laminated, crystal-rich layer at the very base. It has been suggested that such tractional features formed in the more dilute, more turbulent head of the flow (J. V. Wright & Mutti 1981), indicating that this must have been actively ingesting water. The inflated flow head would have allowed an early stage of traction plus fall-out from a more dilute dispersion, followed by *en masse* deposition of the overlying and more concentrated body of the flow (see also Sparks & C. J. N. Wilson 1983). The crystal-rich layer may correspond to the ground layer found in some subaerial pyroclastic flow deposits (Ch. 7). Traction structures are found at the base of some clastic mass-flow deposits (Lowe 1982), but the possibility that these could be flow-head deposits has not been previously suggested.

Deciding whether these massive types of volcaniclastic sedimentary mass-flow deposit are primary eruptive products can be *very* difficult (J. V. Wright & Mutti 1981). This is especially so in ancient submarine volcaniclastic sequences where pyroclastic layers cannot be correlated with land-based studies, as has been done, for example, in studies in the Lesser Antilles island arc (Carey & Sigurdsson 1980; Fig. 9.3). Therefore, in most cases such deposits should be described simply as volcaniclastic debris flow deposits. To stress this point, although it is tempting to interpret the Dali Ash as a primary eruptive product, this is by no means proved, and it is possible that the whole deposit is remobilised pyroclastic debris, including shards, slumped, for example, off the sides of a

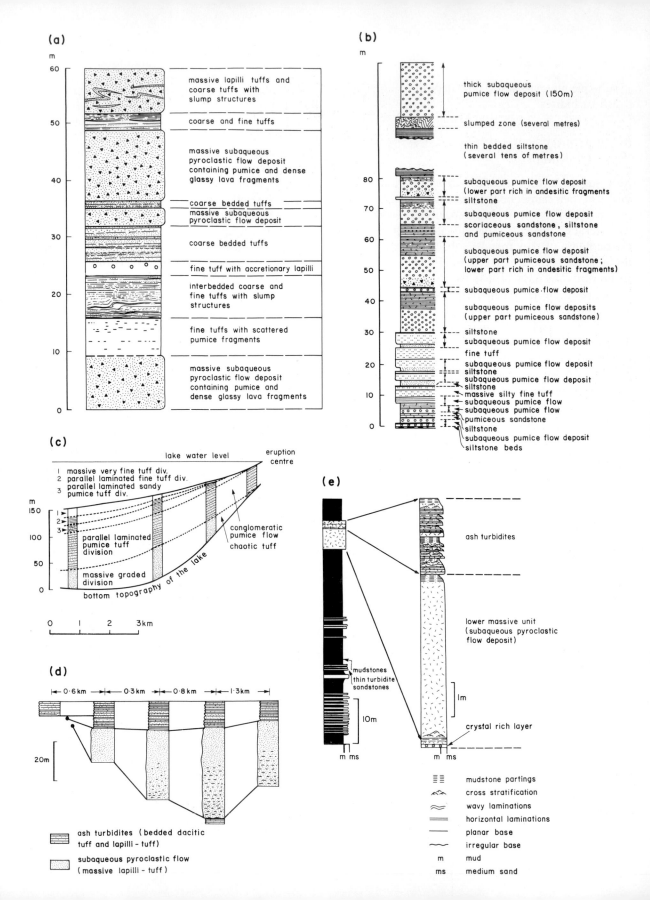

Figure 9.2   The Dali Ash deposit showing a lower massive unit (5 m thick) overlain by bedded ash turbidites (2.5 m thick). The lower massive unit was deposited by a high concentration turbidity current, and the ash turbidites by dilute turbidity currents. Foraminifera are dispersed throughout the deposit and indicate that all of the sedimentary mass flows were cold water–particulate systems. If the Dali Ash is the lateral equivalent of a subaerial ignimbrite the lower massive unit may be interpreted as a large pumice flow that continued subaqueously, and the ash-turbidites as smaller, later flows generated during the eruption, or redeposited slumps of the lower massive unit off a submarine slope. (After J. V. Wright & Mutti 1981.)

shallow-marine rhyolitic tuff ring complex. Even the Roseau volcaniclastic debris flow units in the Grenada Basin could have been generated by slumping of primary or redistributed Roseau deposits from higher up the submarine slope or shelf.

In a similar example, Cas (1979, 1983a) demonstrated that the massive sedimentation units of the Devonian Merrions Tuff of southeastern Australia consist of very juvenile volcanic detritus (Ch. 11). However, in the absence of evidence for primary pyroclastic transport processes, he concluded that the material was finally transported and deposited by huge epiclastic granular mass flows, perhaps originally derived from subaerial pyroclastic activity (Cas 1983a).

◀ **Figure 9.1**   Some examples of deposits which have been termed 'subaqueous pyroclastic flow' deposits in the literature. All of the examples are Tertiary in age. (a) From the Ohanapecosh Formation, Washington, described as examples of subaqueously erupted pyroclastic flows deposited in a marginal freshwater–marine basin (after Fiske 1963). (b) and (c) are examples thought to have been erupted subaerially and then deposited in a caldera lake at Onikobe in Japan. On entering water they flowed as a turbidity current and a vertical sequence comparable to that of clastic turbidites is found, although, sometimes considerably thicker; the example shown in (c) is the uppermost subaqueous pumice flow deposit indicated in (b) (after Yamada 1973). (d) Sections of the Wadaira Tuff (unit D) in the Tokiwa Formation, Japan, described as having been erupted subaqueously and deposited in a marginal marine basin (after Fiske & Matsuda 1964). (e) The Dali Ash, Rhodes. Although quoted as a deep-water welded tuff, it is now known to be non-welded, and thought (but not proved) to be the submarine lateral equivalent of a subaerially erupted ignimbrite (see Fig. 9.2) (after J. V. Wright & Mutti 1981).

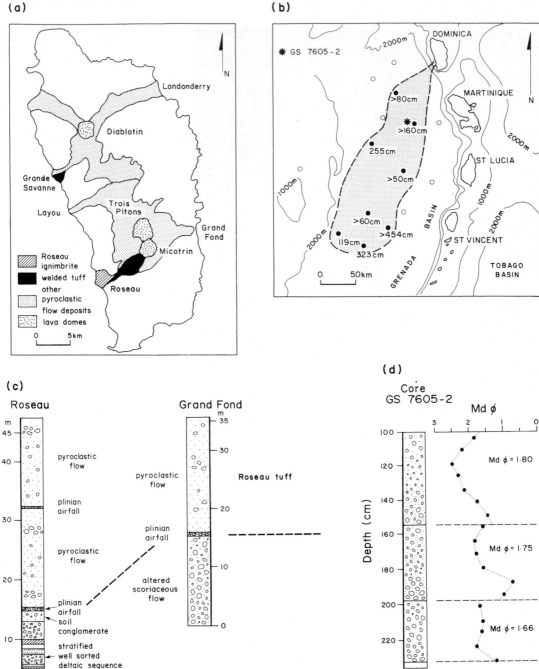

Figure 9.3 The Roseau ignimbrite on the island of Dominica and its submarine equivalents in the Grenada Basin. (a) Subaerial distribution of the major young pyroclastic flow deposits and associated lava domes on Dominica. The Roseau ignimbrite includes the more southerly of the two welded outcrops; the northern outcrop is another stratigraphic unit. (b) Distribution of the Roseau subaqueous pyroclastic flow deposits. Solid circles indicate piston cores in which subaqueous deposits of pyroclastic debris have been found, with their thicknesses, while open circles correspond to cores which do not contain them. (c) Subaerial stratigraphy at two locations. (d) Submarine stratigraphy and grainsize profiles of three subaqueous deposits from the piston core located in (b). (After Carey & Sigurdsson 1980.)

Figure 9.4 Submarine volcaniclastic deposits in piston cores from the Lesser Antilles arc. Scale bars are in centimetres. (a) Roseau ash-fall layer in a core from the Atlantic Ocean, 700 km east of its source on Dominica. (b) Roseau 'subaqueous pyroclastic flow' deposit in a core from Grenada Basin, 270 km west of Dominica. Note large pumice clasts. (c) Subaqueous deposit with large clay rip-up clast in Grenada Basin. (d) Epiclastic sand layer interbedded with hemipelagic sediment in a core from the Grenada Basin. This thin deposit was possibly emplaced by a small turbidity current which transported sand-sized volcaniclastic detritus derived from the narrow shelf between islands of the arc. (After Sigurdsson *et al.* 1980.)

9.2.2 ASH TURBIDITES

Ash turbidites are the deposits of turbidity currents in which volcanic ash predominates. 'Tuffaceous turbidite', 'pyroturbidite' and 'igniturbidite' are other terms that have been used in the literature. These deposits show sedimentary structures akin to classic turbidites, and can be described using the Bouma sequence (Figs 9.1 & 2; Ch. 10). In some cases, ash turbidites may represent the lateral equivalents of dense subaqueous mass flows of juvenile debris after mixing with water and extensive dilution has produced a low particle concentration turbidity current. This may occur by mixing water directly into the body of the flow by ingestion at the base of the flow-front, or by ablation of the upper surface of the flow-front by shearing and erosion, so generating an upper turbulent zone of mixing which would continue as a turbidity current after the debris flow has come to rest (Hampton 1972).

A number of so-called subaqueous pyroclastic flow deposits pass vertically into bedded sequences of ash turbidites (Fig. 9.1). Some of these show **double grading** (Fig. 9.5). Fiske and Matsuda (1964) interpreted this as evidence for a waning submarine eruption column (Section 9.5). The example detailed by Yamada (1973; Fig. 9.1c) also shows this type of grading, but he interpreted this in terms of a normal turbidity current mechanism. In the Dali Ash, the upper sequence of ash turbidites may represent several later, smaller pumice flows which mixed with more appreciable amounts of water and were deposited by lower concentration flows than the lower massive unit. Another alternative is that they are slumps off the submarine slope of the lower massive unit, which itself could have been redeposited, as indicated in the foregoing discussion.

However, not all ash turbidites are the lateral equivalents of pyroclastic flows. They are more

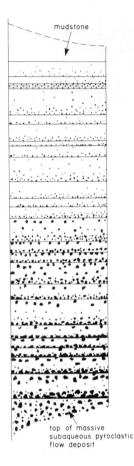

mudstone

Figure 9.5 Schematic representation of double grading in ash turbidites in the upper bedded part of Waidaira Tuff D (Fig. 9.1d). Each bed is graded, and the beds at the base of the sequence contain coarser and denser juvenile fragments than the beds towards the top. In the field this sequence actually contains about 200 beds, not 25 as illustrated here. (After Fiske & Matsuda 1964.)

top of massive subaqueous pyroclastic flow deposit

likely to be composed of redeposited material from turbidity currents generated by slumping of older volcanic debris, although even this slumping could be a co-eruptive event. This happened during the 1902 eruption of Soufrière, St Vincent, when pyroclastic flows built out a delta into the sea and then slumped into the Grenada Trough to generate an ash turbidite (Carey & Sigurdsson 1978). A better non-genetic term to use for these rocks is therefore 'volcaniclastic turbidite'.

9.3 Hot subaqueous pyroclastic flows and subaqueous welding of ignimbrites

There has been much debate recently on the transport of pyroclastic flows into and under water, and particularly on whether such flows can retain enough heat to weld. The debate was stimulated by a reinterpretation of some of the Caradocian (Ordovician) volcanic rocks of North Wales (Howells *et al.* 1973, 1985, E. H. Francis & Howells 1973), and has been heightened by the comprehensive, thought-provoking field guide for the Welsh volcanic–sedimentary successions by Kokelaar *et al.* (1984), and the more detailed descriptions of the Ramsey Island succession by Kokelaar *et al.* (1985).

Ignimbrites of the Capel Curig Volcanic Formation can be traced from an uninterrupted sequence of subaerial, densely welded tuffs into three separate, welded, submarine ignimbrites which pass upwards and laterally into current-bedded reworked tuffs and tuffaceous sandstones (E. H. Francis & Howells 1973; Figs 9.6–8). Welding is considered to have occurred underwater in the submarine equivalents, and Francis and Howells suggested that as a flow entered water an insulating carapace of steam formed around it. However, at the Capel Curig location, where the ignimbrites are believed to have been welded subaqueously, palaeontological evidence from the associated epiclastic sediments only indicates near-shore marine conditions, while sedimentological evidence such as bimodal cosets of large-scale cross-stratification, suggests a tidal to subtidal shallow-marine environment. P. Sutcliffe (*pers. comm.*) even reports mudcracks in epiclastic sediments between the Garth Tuff and Racks Tuff at this location, suggesting periodic exposure and desiccation. However, a more accurate estimate of water depth is a problem. One method is to use heights of sets of large-scale cross-stratification, assuming that these represent the minimum heights of dune bed-forms, and the results of such an exercise suggest palaeo-water depths of <10 m (Table 9.1). This approach can be criticised for several reasons, but the estimate agrees with water depths found in comparable present sedimentary environments. For example, in a non-barred, high energy, near-shore environment on the southern Oregon coast, Clifton *et al.* (1971) found that dunes, formed in <5 m of water, had heights between 30 and 100 cm, which is about the range in heights of sets of cross-strata found at Capel Curig (Table 9.1). At Capel Curig, we therefore feel there might be an alternative explanation, namely, that these thick ignimbrites

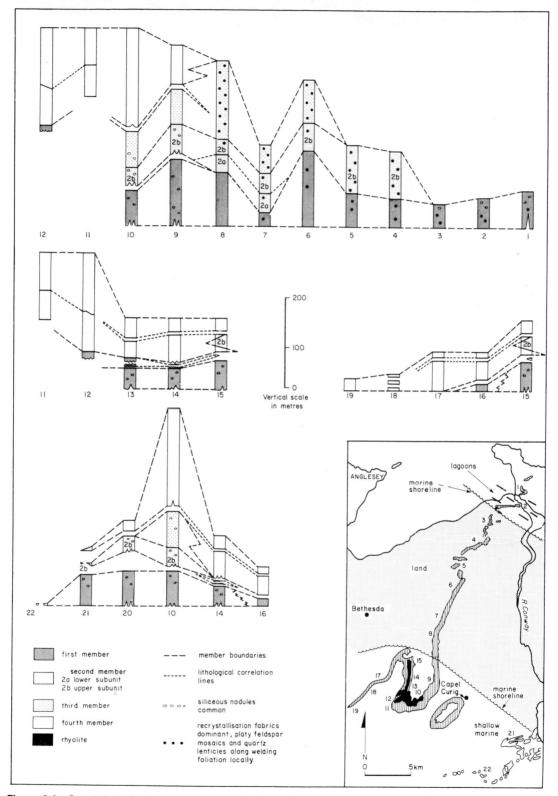

Figure 9.6 Correlation of tuff members within the Capel Curig Volcanic Formation, Wales, and the depositional setting. Welded ignimbrites form Members 1, 2b and 3. Note the close association of siliceous nodules and the marine environment. (After Howells *et al.* 1979.)

Table 9.1 Water palaeodepths of shallow-marine sediments intercalated with Caradocian welded ignimbrites in Snowdonia, North Wales.

Sediments	No. of sets of cross-strata measured	Range in heights of cross-strata (cm)	Range in water depth (m)	Mean water depth (m)
below Garth Tuff (Fig. 9.7)	12	10–70	3.2–11.1	5.4
between Garth Tuff and Racks Tuff (Fig. 9.7)	10	10–50	2.3– 8.4	3.1
below Pitts Head Tuff (Fig. 9.9)	21	15–90	4.0–14.0	5.4
above Pitts Head Tuff (Fig. 9.9)	12	12–60	2.9–10.3	3.1

Water palaeodepths were estimated from the maximum heights of sets of large-scale cross-stratification which are assumed to represent heights of dune bedforms using the relationship (determined by Allen, 1968) $H = 0.086d^{1.19}$, where H is the height of the dune bed-form and d is the water depth. 95% confidence limits were placed on Allen's (1968) data, and the range in water depth was determined from the depth minima and maxima.

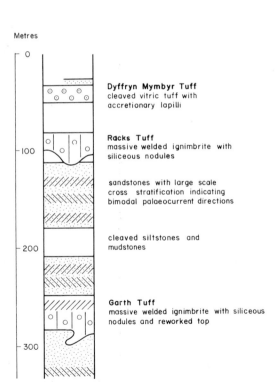

Figure 9.7 Generalised stratigraphic section of the Capel Curig Volcanic Formation at Capel Curig, Wales. The Garth Tuff and Racks Tuff are Members 1 and 2b, respectively, in Figure 9.6, and the Dyffryn Mymbyr Tuff is Member 4. (After E. H. Francis & Howells 1973.)

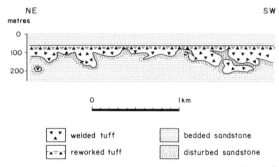

Figure 9.8 Schematic cross section through the Garth Tuff at Capel Curig, Wales. (After E. H. Francis & Howells 1973.)

blanketed the marine shelf but were not submerged. Welding took place on top of shallow-water sediments, but not necessarily underwater. The tuffs do show some notable differences from subaerial welded ignimbrites which are compatible with our interpretation. The highly irregular bases (Fig. 9.8) may have formed as a result of loading of the underlying water saturated sands, superheating of the interstitial water to steam and the liquefaction and fluidisation of the sands. Siliceous nodules in the ignimbrites are composed of drusy quartz (Figs 9.6 & 7) and are thought to have infilled original vesicles which formed as a result of the upward migration of large amounts of steam.

Howells *et al*. (1985) have also described the occurrence of spaced, isolated pods of 'subaqueous welded ash-flow tuff' as the lateral equivalents of

(a)

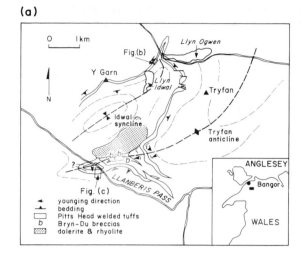

(c)

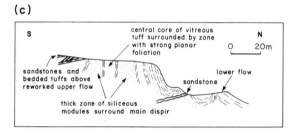

(b)

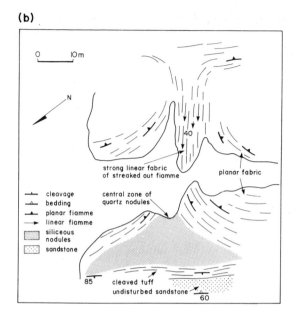

Figure 9.9 Diapiric structures interpreted as explosive rootless vents in welded ignimbrites at two locations in the Pitts Head Tuff Formation, Wales. Note again the association of siliceous nodules. (After J. V. Wright & Coward 1977.)

the continuous ignimbrites described above. These pods are up to several tens of metres in diameter, and are spaced hundreds of metres apart. Although the evidence is not presented, they are apparently completely welded through to their bases, and have been interpreted as being due to the disruption of a subaqueous ash flow by interaction with water during flow. The enclosing sediments are not adequately described but are considered by Howells *et al.* to be tidal to subtidal, shallow-marine in origin. This study is problematic, in that there are several unexplained, undocumented aspects of major importance, especially given that the paper is trying to establish the existence of subaqueous pyroclastic flows and subaqueous welding. First, the facies characteristics of the enclosing sediments are not described and documented, which seems necessary to establish the depositional context. Secondly, the assumed welding textures are not described and illustrated. Finally, is it possible that the end of the continuous sheet, and the appearance of the pods, represents the point where the host

pyroclastic flow became wholly buoyant because of its low bulk density, and began to flow *over*, rather than under, the sea. Pods of pyroclastic debris may have periodically dropped out of the base of the moving pyroclastic flow, perhaps associated with local disruptive phreatic explosions, and sagged into the shelf sediment sequence to form discrete pods.

Elsewhere in North Wales, J. V. Wright and Coward (1977) concluded that ignimbrites of the Pitt's Head Tuff Formation (again Caradocian in age, but stratigraphically higher than the Capel Curig Volcanic Formation; Fig. 9.9a) were emplaced in a shallow, gently shelving sea where water depths were insufficient to prevent welding. Data on water depths estimated from the heights of large-scale cross-stratification are given in Table 9.1, and again suggest a palaeowater depth of <10 m. J. V. Wright and Coward (1977) described diapiric structures (Fig. 9.9) within these welded ignimbrites, and concluded that they were rootless vents formed from steam trapped below the welding ignimbrites, which periodically burst through as small, secondary phreatic explosions. These might have produced small pyroclastic craters on the surface of the ignimbrite, like those made by secondary explosions at Mt St Helens (Fig. 8.35). Also associated with the rootless vents are zones of siliceous nodules, and sometimes these nodules occur in bubble trains extending parallel to the margins of the diapiric structures, indicating the upward migration of steam. Rootless vents and the zones of nodules are thought to be good indicators of ignimbrite deposited in a shallow-water environment (J. V. Wright & Coward 1977).

Other examples of welded ignimbrites emplaced in shallow-water environments can be found in the literature. Stanton (1960) and Dewey (1963) described welded ignimbrites in the Mweelrea Group (Caradocian) of Murrisk, Ireland, which they thought had been deposited in a shallow water deltaic environment. Again, here, such shallow water depths perhaps would not prevent the ignimbrites from welding. In a discussion of densely welded ignimbrites in the Lower Palaeozoic rocks of Newfoundland, Lock (1972) describes a 'sludge-flow subfacies' for rocks formed by the rapid quenching, collapse and liquefaction of a pyroclastic flow passing from a subaerial to a submarine environment.

The Dali Ash on the island of Rhodes, Greece (Figs 9.1e & 2), has been cited as an example of a hot subaqueous pyroclastic flow deposit because of the references made to welding textures by Mutti (1965). Unlike the welded ignimbrites described from North Wales, the Dali Ash occurs in deep-water sediments, and therefore subaqueous pyroclastic flows must have remained hot for appreciable distances underwater. However, it is now known that these textures were misidentified, and there is no textural evidence of welding or palaeomagnetic evidence of a high emplacement temperature (J. V. Wright & Mutti 1981). The deposit contains abundant angular and cuspate shards (Fig. 3.23a), but these are not plastically flattened or even incipiently welded. The Dali Ash may be the lateral, subaqueous equivalent of several flow units of a subaerially erupted ignimbrite but, as discussed in Section 9.2, this is not certain.

Supposedly deep-water welded ignimbrites have also been described from the Fishguard Volcanic Group (Lower Ordovician) in South Wales (Lowman & Bloxam 1981). Several welded ignimbrites are associated with graptolite-bearing shales and basaltic pillow lavas. Vesicles in pillow lavas indicate water depths ranging from tens of metres to 2 km, and Lowman and Bloxam (1981) have suggested a depth of 1–2 km. However, these measurements, although made within the mapped Fishguard Volcanic Group, are located 8 km away from the welded tuffs. Both areas show differing volcanic rock associations, and the stratigraphy could be more complex than has been considered to date. Lowman and Bloxam suggest that a steam jacket insulated the moving subaqueous pyroclastic flows, and the hydrostatic pressure of sea water maintained cohesion and retained heat and volatiles within the flows.

Lower Ordovician deep-water welded ignimbrites are also reported from Ramsey Island, just off the South Wales coast (Kokelaar et al. 1984, 1985). Welded shard textures are not evident in thin section because of recrystallisation, but pumice fragments show flattening parallel to the dip of the ignimbrite sheets. The main unit of interest, the Cader Rhwydog Tuff also has a normally graded top, which contains strongly flattened pumices in its lower part. This has been interpreted as a subaqueous welded fall tuff, and Kokelaar et al. (1984) interestingly proposed that the eruption, emplacement and welding sequence of the whole unit occurred underwater. The unit is 186 m thick, consisting of a lower massive interval of 161 m, and an upper 25 m of unbedded, normally graded lapilli tuff to fine tuff above. The base overlies an erosional surface and contains rounded to angular clasts of rhyolite, as well as intraclasts of mudstone. In one place, a load cast-like feature 10 m deep protrudes into the underlying mudstone from the base, and nearby, rhyolite pebble horizons occur in the tuff. The depositional context is demonstrated to be submarine, below wave base, although no absolute water depth can be determined. However, the evidence for the welding is more equivocal. It is essentially based on the presence of 'strongly flattened, ragged fragments of porphyritic rhyolite', which are interpreted to be flattened, welded pumice. However, it has not been demonstrated that these fragments were pumice (are highly vesicular textures preserved?) or that the flattening was caused by plastic, hot state, compactional welding, producing highly strained, but intact, vesicle walls. If they were pumices, could they have been flattened in a more brittle fashion during normal burial compaction, and subsequent shape moulding during alteration, metamorphism and deformation? Although the presence of 'crude columnar joints' is cited as further evidence of a hot state of emplacement, crude 'columnar-like jointing' can develop in many rock types, and lack of illustration again makes this aspect of the proof equivocal.

Is it possible that the Cader Rhwydog Tuff is a huge, water-supported mass flow of pyroclastic and/or quench-fragmented debris transported by epiclastic mass-flow processes, as has been suggested for the Devonian Merrions Tuff (Cas 1979, 1983a; Ch. 11)? Many of the described features of the Cader Rhwydog Tuff are consistent with this. For example, the massive, large-scale graded aspect, the erosional base with load cast-like features, the included rounded clasts and intraclasts, the former sometimes in distinct horizons. The flattening of rhyolite clasts could be due to burial compaction and subsequent shape moulding of the altered, ductile clasts during alteration, metamorphism and deformation, as also appears to have happened with the Merrions Tuff.

Yamada (1984) has also indicated that deep-water welded ignimbrites occur in the Neogene deposits of Japan. These occur within the Green Tuff, a sequence also important for its association with Kuroko-type massive sulphide deposits (Chs 13 & 14). The welded tuffs are thought to be proximal facies of subaqueous pumice flows; typically the Green Tuff consists of distal equivalents intercalated with fossiliferous marine siltstones and sandstones. This study also lacks adequate documentation and photographic illustration of the depositional context, i.e. of the facies characteristics of the host sedimentary succession, and of the microscopic evidence for welding.

Another reported deep-sea welded tuff, dredged from the Grenada Basin (Sparks & Carey 1978), is now known to have been misidentified (Sparks et al. 1980a). The only other claimed example of a subaqueous welded tuff known to us comes from the Philippines. Fernandez (1969) noted a welded ignimbrite associated with a sequence of pillow lavas and tuffaceous sediments (Cretaceous to Palaeogene). However, whether the stratigraphic evidence is sufficiently substantial to conclude that subaqueous welding occurred is debatable, particularly given that pillow lavas can form in very shallow water (Ch. 4).

Hot, subaqueous pyroclastic flow deposits have been described from the Donzurubo Formation (Miocene) of Japan (Kato et al. 1971, Yamazaki et al. 1973). Although none of the pyroclastic flow deposits shows any textural evidence of welding, it was concluded from the stabilities of the natural remanent magnetism and Curie points that they contained a thermal remanent magnetism and were deposited at about 500°C. Such high temperatures preclude mingling of the flow with water. Stratigraphic evidence suggests a water depth of 50 m but documentation of associated facies is lacking. Some workers have doubted this evidence and, indeed, have even suggested that the Donzurubo subaqueous pyroclastic flow deposits could be subaerial (E. Yamada pers. comm.). Hot pyroclastic flows (they are pink in colour) of mid-Miocene age on Santa Cruz Island in the Californian borderland are thought to have been deposited in a near-shore subaqueous environment (Fisher & Charleton 1976).

Recent studies of the Krakatau 1883 eruption (Self & Rampino 1981) suggest that a hot, submarine ignimbrite could have been formed. Shallowing of the sea floor up to 15 km north of Krakatau is thought to have been caused by subaqueous pyroclastic flows which largely infilled parts of the Sunda Strait that had been 20–60 m deep (Fig. 9.10). The subaqueous deposits have not been cored, so it is not known whether the flows were actually dense gas–particulate flows, dense water–particulate flows or turbidity currents, or if there were any vertical and lateral facies changes. After the eruption, temporary islands and shallow banks of pumice had formed, and it seems the flows (of

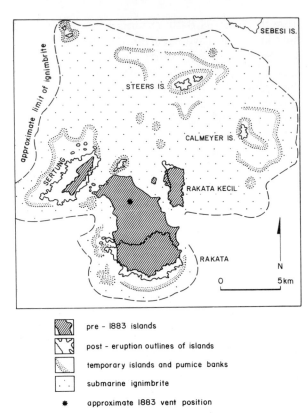

pre - 1883 islands

post - eruption outlines of islands

temporary islands and pumice banks

submarine ignimbrite

✱   approximate 1883 vent position

Figure 9.10 Submarine distribution of the 1883 ignimbrite at Krakatau, Indonesia. (After Self & Rampino 1981.)

whatever type) gradually ponded until water depths became very shallow, in which case the last erupted pumice flows would probably have maintained their essential character. Secondary phreatic explosions produced small craters on Calmeyer Island (Fig. 9.10). These formed by explosive vaporisation of sea water, either below one of the last erupted pumice flows or in contact with hot deposits at depth. If the latter occurred, the possibility exists that welding might also have been occurring in the lower part of the deposit. The loosely consolidated deposits forming the islands were eroded, and these quickly vanished below the sea.

Recently, Sparks *et al.* (1980a, b) have presented geological and theoretical models, both for the entry of pyroclastic flows into water and for subaqueous welding. They suggest that welding of an ignimbrite is more favourable underwater than subaerially. Essentially, this is because there can be

a substantial reduction in glass viscosity after emplacement at moderate to deep depths, due to the high solubility of steam in rhyolitic glass at pressures of tens of bars. However, there are several points of contention with their studies. The most important is the geological evidence for the entry of pyroclastic flows into the sea and their behaviour subaqueously, this evidence being based on the observations of a dense-clast block and ash flow erupted on Dominica. Such types of flows are generally much less expanded and have a higher yield strength than most pumice flows which form ignimbrite. There would seem to be no problem for block and ash flows (with a flow density perhaps as high as 2.0 g cm^{-3}) passing into water and maintaining their identity. However, this type of pyroclastic flow does not form welded tuff (Ch. 5). With pumice flows, which have a much lower bulk density, it is uncertain what will happen when they flow into a standing body of water. Will they maintain their integrity and continue to flow uninterrupted underwater as a hot gas-particulate system? Or, given the low density contrasts between the pumice flow and the water, and the enormous thermal potential between the pyroclastic flow and the cold water, will the interaction be more dynamic? We suspect that in most pumiceous pyroclastic flows where their density is nearer to that of water, even if they get underwater, ingestion of water and inflation of the flow-front will be important and they will quickly lose their identity as gas-particulate systems. The major problem for pumice flows will be to make that smooth transition underwater. In shallow water the interaction of water with hot ash at the flow boundary is potentially explosive, and could lead to the destruction of many flows. At Krakatau, it has now been suggested that many of the separate explosions known to have occurred during the eruption were caused by hot pyroclastic flows entering the sea. Such violent explosions would lead to the formation of a widespread ash fall (G. P. L. Walker 1979), and could trigger the formation of water supported mass flows (Fig. 9.11), as suggested for the Merrions Tuff (Ch. 11; Cas 1979, 1983a).

The low bulk density of pumice flows suggests that many flows may be buoyant enough to flow

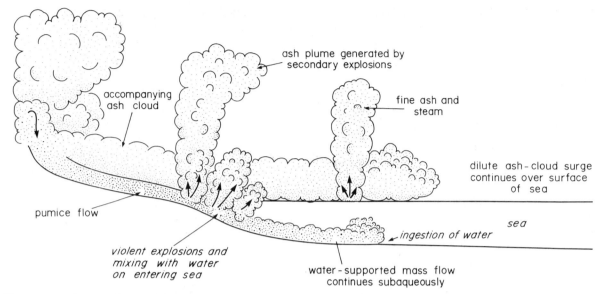

Figure 9.11 One possible model for what occurs when a subaerially erupted pumice flow enters the sea. The density of the pumice flow depicted is greater than 1.0 g cm^{-3}. It has been almost destroyed as a gas–particulate flow by violent explosions on entering the sea. The accompanying ash-cloud surge with a density much less than 1.0 g cm^{-3} has ridden across the surface of the sea. As turbulence becomes less effective, the larger clasts in this must drop out into the sea, and eventually the surge is depleted in solid material. Presumably something similar happens when a pumice flow with a density less than that of water enters the sea, although in detail, particle-support mechanisms may be very different (Ch. 7).

over water, as was suggested earlier to account for isolated pods of ignimbrite within the Capel Curig Formation of Wales. This certainly appears to be the case for ignimbrite deposited on islands some distance away from the 1883 Krakatau eruption in Indonesia (Self & Rampino 1981), as well as the distribution of the Plateau ignimbrite on several islands of the Dodecanese group in the Aegean Sea (Fig. 13.44). The surface tension of the water together with the high velocity and frictionless nature of the flow may act as a sufficient physical barrier to prevent intimate mixing of the flow and water, although phreatic explosions need not be excluded.

On these grounds, we therefore speculate that it is unlikely that pumice flows will in general form subaqueous welded ignimbrites because of the innate difficulties there seem to be in maintaining integrity both at the land–water interface, as well as underwater. Are there conditions where we can envisage a welded submarine ignimbrite forming? One would have to invoke the eruption of a very

thick, very poorly expanded, unusually dense pumice flow which could maintain its identity through the land–water interface by very rapid passage into deep water. Another situation, discussed further in Section 9.5, is where a submarine caldera forms in relatively shallow waters, accompanied by high magma discharge rates, rapid caldera subsidence and rapid caldera infilling (e.g. Busby-Spera 1984). Welded ignimbrites can also be envisaged forming in shallow-water environments where pyroclastic flows might essentially blanket part of the shelf, so making water depths insignificant, as perhaps happened in North Wales. However, it is important to state that nobody has yet traced a Recent welded subaerial ignimbrite into a welded submarine equivalent. The Roseau ignimbrite on Dominica is welded in places, but its submarine equivalent shows no evidence of a high emplacement temperature (Carey & Sigurdsson 1980; Figs 9.3 & 4).

In summary, it seems imperative to establish on the *soundest* sedimentological grounds the palaeoenvironmental significance of the sedimentary facies intercalated with suspected subaqueous pyroclastic flow deposits. In the past insufficient attention seems to have been given to the sedimentary facies. Equally important, the clearest evidence for welding must be presented, especially at the microscopic scale.

9.4 Submarine eruption of pyroclastic flows?

A model for the submarine eruption of subaqueous pyroclastic flows was developed by Fiske (1963) and Fiske and Matsuda (1964), based on studies of the Ohanapecosh Formation (Eocene to Oligocene?), USA, and the Tokiwa Formation (Miocene), Japan, respectively. According to this model, at the start of an eruption vesiculating silicic magma is erupted into cold sea water, and a submarine eruption column begins to rise above the vent (Fig. 9.12a). During the climax of the eruption (Fig. 9.12b) large volumes of ejecta form a fountain above the vent, and the submarine eruption column carries much fragmented pyroclastic debris high into suspension. The eruption column could also burst through the sea surface into the air. Intense sorting splits the debris into various fractions: buoyant pumice floats; denser semi- and non-vesicular juvenile fragments and large crystals settle around the vent and slough laterally in a subaqueous flow of pyroclastic debris; but most ash remains in suspension. At the end of the eruption the amount of material falling out of the submarine eruption column gradually decreases, and is insufficient to maintain a steady debris flow (Fig. 9.12c). The flow is therefore replaced by repeated turbidity currents. Because the volume of material raining down around the vent progressively decreases and becomes finer grained and less dense, the later turbidity currents become more infrequent and carry finer and less dense ash. The sequence of turbidites that is deposited shows double grading (Fig. 9.5). Much fine ash probably remains in suspension and is dispersed by ocean currents.

Wadaira Tuff D (Fig. 9.1d) and other examples in the Tokiwa Formation show a fully developed

(a) Beginning of eruption

sea level

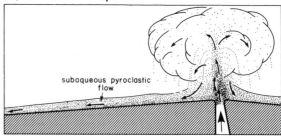

(b) Climax of eruption

subaqueous pyroclastic flow

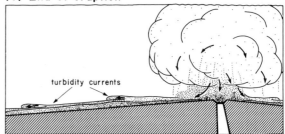

(c) End of eruption

turbidity currents

Figure 9.12 Model for the submarine eruption of 'subaqueous pyroclastic flows'. (After Fiske & Matsuda 1964; see text.)

submarine eruption sequence. In the Ohanapecosh examples (Fig. 9.1a) doubly graded units are not found, and Fiske (1963) suggested that this was because the Ohanapecosh vents were in shallower water, and the erupted debris was not thoroughly sorted as it settled back to the flanks of the submerged volcanoes. Doubly-graded depositional sequences would therefore only appear to be characteristic of the deeper submarine eruptions. Foraminifera indicate water depths of up to 500 m for the Tokiwa Formation.

Similar depositional sequences have since been

described from other areas, and workers often interpret them in terms of the model of Fiske and Matsuda, e.g. the examples of Howells *et al.* (1973) from the Caradocian of North Wales. The physical basis of the model was questioned by McBirney (1971) on the grounds that from the available data (Fig. 3.7b) it is apparent that salic magmas can only vesiculate *significantly* at water depths of tens of metres or less. Thus, in general salic pumice eruptions can only occur in very shallow waters, and at such water depths a phreatoplinian eruption (Ch. 6) will almost certainly occur. Collapse of the eruption column could lead to the formation of pyroclastic flows which might then continue sub-aqueously. At present not enough is known about salic eruptions involving water, and their deposits, to decide whether subaqueous eruption of ignimbrite is likely to occur. Submarine eruption of vesiculated salic magma certainly does occur, as is evident from pumice rafts, such as those observed in 1979 between the Tonga and Fiji islands in the South Pacific.

Intuitively, it would be expected that an eruption of the type proposed by Fiske and Matsuda would produce dense to pumiceous, quench-fragmented blocky hyaloclastite fragments of silicic glass, and hyaloclastite debris flow deposits. The examples described from the Ohanapecosh and Tokiwa Formations do contain a large proportion of glassy, non-vesiculated juvenile fragments.

Busby-Spera (1984) has suggested a more likely circumstance by which subaqueous pyroclastic flow deposits may be formed. In describing the character of the Triassic–Early Jurassic Mineral King roof pendant in the southern Sierra Nevada, Busby-Spera describes voluminous ash-flow tuffs which she interprets to have accumulated in a submarine caldera. They contain interstratified marine sedimentary rocks. Contemporaneous caldera collapse has led to the ponding of ash-flow tuffs greater than 0.5 km. If eruption was in shallow enough water (Ch. 3), the magma discharge rate was high enough, and caldera subsidence was rapid enough, then it seems feasible for volumes of ignimbrite to be preserved subaqueously, although it is surprising that more explosive interaction with water did not take place.

9.5 A model for the passage of pyroclastic flows into subaqueous environments

A schematic model for the transportation of volcaniclastic material as pyroclastic flows from island volcanoes into ocean basins is presented in Figure 9.13. We suggest that, in general, if truly subaqueous pyroclastic flows and ignimbrites occur, then they are distal equivalents of subaerial deposits, rather than the wholly subaqueous products of submarine eruptions. We also suggest that welding in pyroclastic flow deposits is essentially a feature of subaerially erupted ignimbrites deposited on land, or in very shallow water, where water depths would be insufficient to prevent welding. Exactly what happens to a pyroclastic flow on entering the sea is, though, perhaps as controversial as ever. What occurs in a particular case will be controlled by a number of factors. These include characteristics of the flow, such as thickness, discharge rate and, most importantly, the degree of expansion (which will determine its density). In addition, environmental factors will be important, and these include the slope, which will determine the water depth, the width of the shallow water shelf and, hence, the degree of interaction with water at the critical point at which entry into the sea is made.

Figure 9.13a shows the eruption of a large, single pumiceous pyroclastic flow forming an ignimbrite. On land the deposit had sufficient heat and compactional load to weld. On reaching the sea, the thick flow spread across a shallow shelf. The flow had sufficient momentum and density to continue subaqueously into deeper water, but rapidly, and perhaps explosively, mixed with water and chilled. The cooled water–particulate subaqueous debris flow continued along the sea floor and deposited a non-welded massive subaqueous deposit of originally pyroclastically fragmented debris. Gradually, mixing with more and more water occurred, until at some point the mechanism became that of a turbidity current and deposited an ash turbidite. On the shallow shelf, steam generated by the hot, thick pyroclastic flow was trapped at the base of the flow, as consolidation and welding occurred. Secondary steam explosions periodically burst

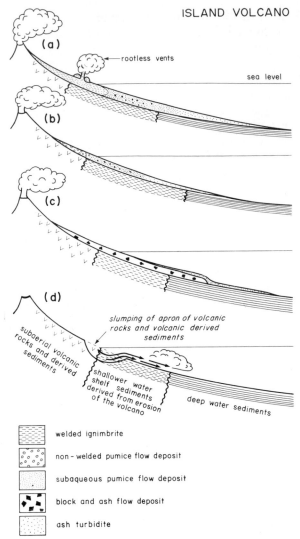

ISLAND VOLCANO

(a) rootless vents
sea level

(b)

(c)

(d)
slumping of apron of volcanic rocks and volcanic derived sediments

subaerial volcanic rocks and derived sediments

shallower water shelf sediments derived from erosion of the volcano

deep water sediments

welded ignimbrite

non-welded pumice flow deposit

subaqueous pumice flow deposit

block and ash flow deposit

ash turbidite

Figure 9.13 Schematic model for the passage of pyroclastic flows into the sea, and the transportation of volcaniclastic material as mass flows from island volcanoes into ocean basins (see text).

through the welding flow to produce rootless vents with associated pyroclastic craters.

The eruption of a small pumiceous pyroclastic flow is represented in Figure 9.13b. On land it formed a non-welded deposit. On reaching the sea, the pyroclastic flow did manage to continue subaqueously initially, but mixing with water quickly changed the flow mechanism to that of a turbidity current which deposited an ash turbidite.

In contrast, small, dense block and ash flows may move into water and continue subaqueously without losing their essential character (Fig. 9.13c). Ablation of the flow-front may generate an upper turbulent zone of mixing, which may continue as a turbidity current after the block and ash flow has come to rest, to deposit an ash turbidite.

However, the simplest, and undoubtedly the most important, way by which volcanic material is transported into ocean basins as flows is by epiclastic processes and slumping (Fig. 9.13d; Chs 10, 11, 13 & 14). Coupled with seismicity, the steep slopes of many island volcanoes are obviously unstable and material consequently readily slumps into the ocean basins, generating high concentration debris flows and low concentration turbidity currents. As well as small-scale slumping, large-scale sector collapse of volcanoes could occur. In the Lesser Antilles island arc the so-called Qualibou caldera on St Lucia may be such a gravitational slide structure, and similar structures have been recognised on Dominica and St Vincent (Roobol *et al.* 1983; Fig. 13.35). Flows generated by such slumping should be heterolithic and generally much more heterogeneous than those which are the lateral equivalents of primary pyroclastic flows. The deposits are volcaniclastic debris-flow deposits and volcaniclastic turbidites.

9.6 Deep-sea ash layers

Megascopically visible ash layers preserved in sediment cores have often been found several hundreds of kilometres away from their potential sources (e.g. Figs 8.24, 9.14 & 15). Many of these layers are the distal air-fall equivalents of large silicic eruptions that have taken place on both island and continental volcanoes (Chs 6 & 8). Until relatively recently, widespread volcanic ash layers in deep-sea sediment cores were only used for stratigraphic purposes. Early studies were designed for application to marine investigations requiring a stratigraphic framework (e.g. Worzel 1959, Ninkovich & Heezen 1965, Ninkovich *et al.* 1966, Hays & Ninkovich 1970). Now studies go far beyond being a stratigraphic correlation exercise,

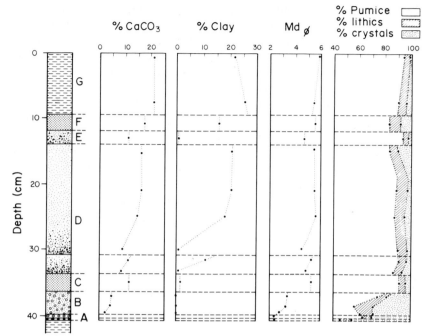

Figure 9.14 Section showing lithological variations in the Minoan ash-layer in core TR 172–9, taken approximately 170 km south-east of Santorini. From bottom to top: units A and B are interpreted as the plinian fall; C, as a fine ash fall; D, as two repeated ash turbidites; E and F, as laminated ash layers winnowed by bottom currents; G, as pelagic muds admixed with ash. (After Watkins *et al.* 1978.)

and have made an important contribution to understanding volcanic processes and volcaniclastic sedimentation in marine basins (e.g. Huang *et al.* 1975, 1979, Watkins & Huang 1977, Watkins *et al.* 1978, Carey & Sigurdsson 1978, 1980, Ninkovich *et al.* 1978, Ledbetter & Sparks 1979, Sigurdsson *et al.* 1980, Sparks & Huang 1980). In this section we concentrate strictly on ash layers that result from fall-out of airborne ash. However, as discussed in the foregoing sections, substantial accumulations of subaqueous pyroclastic flow deposits and ash turbidites that are ponded in basins around island volcanoes may be in part their proximal equivalents.

Both volcanic and marine processes will control the accumulation of any deep-sea air-fall ash layer. Their interpretation and internal stratigraphy can be as complex as the subaerial products of the eruptions (e.g. Fig. 9.14). More distally, the problem of distinguishing the different types of silicic air-fall ash also becomes more complex (Section 6.11).

Several marine processes may affect the thickness of air-fall ash deposits:

(a) ocean currents,
(b) secondary slumping,

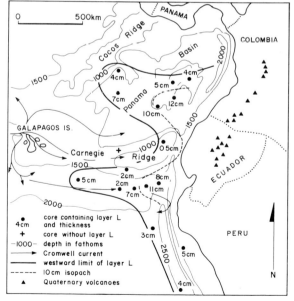

Figure 9.15 Distribution patterns of the Worzel L ash layer (230 000 years BP), probably erupted from a source in northern South America. (After Ninkovich & Shackleton 1975.)

(c)   bioturbation and
(d)   compaction.

Although wind transport is the dominant dispersal mechanism, the strength and direction of *ocean currents* will have an important effect on the distribution of the ash in the marine environment. Ninkovich and Shackleton (1975) attributed the W-shaped pattern of the Worzel L ash layer to the effect of the strong equatorial undercurrent, or Cromwell current, moving eastwards across the Pacific at a depth between 100 and 300 m, and a speed of 100–150 cm s^{-1} (Fig. 9.15). As well as surface and near-surface currents, bottom currents can cause reworking and redistribution, and in terminology applied to clastic sediments, such layers would be called **contourites** (Stow & Lovell 1979).

Ash which settles-out on top of submarine topographic highs will be susceptible to *secondary slumping*, generating turbidity currents. The consequence of such processes is an increase in ash-layer thickness in basins, and a thinning or removal of ash layers from highs and steep slopes. In the basins a succession of ash turbidites may be deposited by repeated slumping during or after an eruption. Figure 9.14 shows the lithology of several

units within the Minoan ash layer, cored in a small basin south-east of Santorini in the Aegean Sea. The two units labelled D are both interpreted as turbidites as they have erosional basal boundaries, show normal grading and contain abundant carbonate, clay and some rounded terrigenous quartz grains mixed with the ash (Watkins *et al.* 1978). Crystal-poor, fine-grained air-fall ash derived from the phreatomagmatic phase in the eruption, or co-ignimbrite ash, probably accumulated on submarine highs and then slumped into the small deep-water basin where this core is located. Close inspection of Figure 9.15 also suggests that slumping could at least be partly responsible for the distribution pattern of the Worzel L ash layer.

Bioturbation of ash layers mixes the ash with other sediments and can reduce visible ash layer thicknesses. In some cores through the Minoan ash layer, Watkins *et al.* (1978) found that over 50% of the original ash had been dispersed upward by this process.

Post-depositional *compaction* can substantially reduce the thickness of the ash layer. A reduction factor of 50% is usually adopted when recalculating isopach contours to represent the thicknesses of freshly fallen ash (e.g. Watkins *et al.* 1978).

One of the most exciting new applications of the

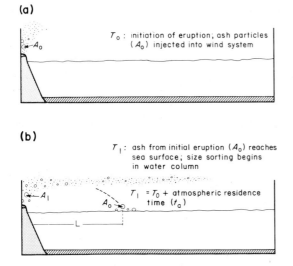

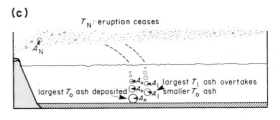

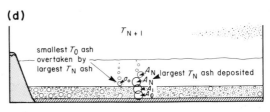

(e)

T_{N+1}: largest T_N ash (A_N) & smallest T_0 ash that is overtaken (σ_0) are both deposited

$T_{N+1} = T_N + t_a +$ settling time of A_N ash (TA_N) &

$T_{N+1} = T_0 + t_a +$ settling time of σ_0 ash (Ta_0)

$Ta_0 + TA_N = T_N - T_0 = \underline{\text{duration}}$

Figure 9.16 Model for the duration of an explosive volcanic eruption deduced from graded bedding in deep-sea ash layers. (After Ledbetter & Sparks 1979.)

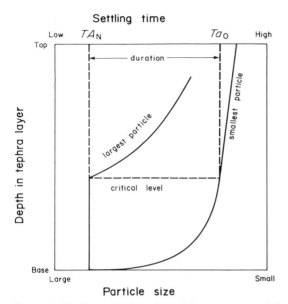

Settling time

Figure 9.17 Theoretical grading of the coarsest and finest particles in a deep-sea ash layer. Explanation is in the text. The duration is estimated from the difference in settling times of the coarsest and finest particles at the level where a marked inflection occurs in the grading of coarse particles. (After Ledbetter & Sparks 1979.)

work on deep-sea ash layers is in estimating the duration of large-magnitude explosive eruptions. A model has been developed by Ledbetter and Sparks (1979) that estimates the duration of these eruptions from the vertical size-grading of feldspar phenocrysts near the base of deep-sea ash layers. The size-grading is a function of the release time of the particles from the atmosphere, their settling velocity, the water depth at the site of the ash layer, and the duration of the eruption (Fig. 9.16). This model predicts a zone at the base of the deep-sea ash layer where the size of the coarsest particle remains constant with height (Fig. 9.17). Above this zone the size of the coarsest particle decreases with height. The prominent break in the coarsest particle size-to-height curve represents the deposition of the last largest particle ejected at the end of the eruption (A_N). The finest particle (a_0) deposited in conjunction with the last largest particle must have been erupted at the beginning of the eruption. The difference in settling times between the largest and smallest particles at the critical level is equal to the duration of the eruption.

There are several assumptions that have to be made with the duration model. The most important is that size of the largest particles arriving at any downwind site remain the same throughout the eruption. For this to occur, the range of particle sizes ejected at the vent must not vary substantially, the eruption column height must not change substantially, and the wind direction and dispersal direction must not vary. If these conditions are not met, the vertical size distribution of particles would not show the pattern predicted in the model, and more-complex patterns of grading would result. It is also necessary that the ash layer be recognised as a primary air-fall deposit. The normal size grading of an ash turbidite may approximate the size grading predicted by the model for the base of a primary ash-fall layer. It is therefore important to search for other distinguishing sedimentary features.

Results from the study of the Worzel D ash layer are shown in Figure 9.18. This layer is the distal counterpart of the Los Chocoyos Ash, which is the product of a major rhyolitic ignimbrite- and caldera-forming eruption associated with Lake Atitlan in Guatemala, as well as being the largest-volume Quaternary eruption in the Central American region (Drexler *et al.* 1980; Ch. 8). The estimated duration of this eruption is 20–27 days. The model has also been tested on the Toba deep-sea ash layer (Fig. 8.24), and indicates that the duration of the eruption was about 10 days. Knowing the approximate volume of the total products in these two eruptions, it is possible to estimate the volumetric eruption rate of magma (Fig. 8.34). This provides a means of comparing large-magnitude explosive eruptions with *relatively* small-scale ones.

Deep-sea ash layers are potentially the best record of age, location, duration and magnitude of explosive volcanism. Methods for high resolution dating of sedimentary cores have been refined so that the dating of a discrete event can be readily carried out (e.g. Ninkovich & Shackleton 1975, Ledbetter 1985). However, the full potential of the study of ash layers in the ocean basins for further resolving eruption histories and characteristics remains to be exploited.

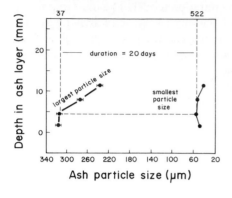

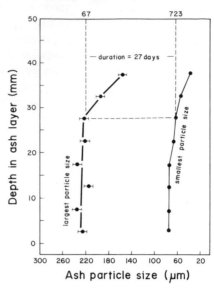

Figure 9.18 Duration model applied to the Worzel D ash layer in two cores from the eastern equatorial Pacific. (a) Core TR 163–7 in water 3435 m deep. (b) Core TR 163–10 in water 3500 m deep. (After Ledbetter & Sparks 1979.)

9.7 Subaqueous base surges?

Conceptually, the likelihood of subaqueous base surges forming does not seem feasible because of the resistance to flow by the water body. Surges are low density media, with probably somewhat lower densities than that of water at 1 g cm^{-3}. The turbulent, dispersed mixtures of solids and water which make up turbidity currents are considered to have densities of 1.1–1.4 g cm^{-3} (Allen 1982), so allowing them to flow as gravity driven undercurrents in a body of water. The average density of a surge is probably below 1 g cm^{-3} because the fluid medium of the surge is essentially gaseous and supports condensed water droplets, and solid glass, lithic fragments and congealing magma clots. Because of its vesicularity, this solid component is likely to have a density lower than the normal nonvesicular grains transported by most turbidity currents. Overall, therefore, surge densities should be less than 1 g cm^{-3}, perhaps in the range greater than 0.6 g cm^{-3}. Very wet, condensed, deflating surges may have densities greater than 1 g cm^{-3}. Nevertheless, it seems that if the surge density is less than that of water, then the resistance to flow by a body of water should be too great for surges to

travel through it *as surges*. Initial surge-like thrusts may translate into turbidity currents.

However, Kokelaar (1983) has suggested that in Surtseyan eruptions with their near-surface, yet submerged, vents, the cone of water within the submerged crater contains a partly fluidised slurry of pyroclastic debris and water, through which steam, generated from the magma in the vent, is turbulently streaming. The water in this cone must also be close to its boiling point. Periodically, during the pulsatory magma rise and explosive eruptions which characterise Surtseyan eruptions, it is possible that the cone is evacuated of a continuous body of water and contains a dense atmosphere of steam, water droplets and particulate matter, through which surges could be propagated. This speculation is being further investigated.

9.8 Further reading

J. V. Wright and Mutti (1981) is important reading because it identifies the major kinds of problems encountered when interpreting so-called 'subaqueous pyroclastic flow' deposits. Cas (1979) is a useful example of the value of applying a sedimentological approach to such deposits. The excellent paper by Sigurdsson *et al.* (1980) is a very good guide to the spectrum of volcaniclastic depositional processes and sediment types that can occur in submarine basins around intermediate to silicic island volcanoes. Papers by Carey and Sigurdsson, and by Fisher (both in Kokelaar & Howells 1984) are also helpful reviews. This special publication also includes other relevant papers by Yamada and by Kokelaar *et al.* to which we have referred. To appreciate fully the processes and products of volcanigenic sedimentation that occur in, and feed subaqueous basins, Chapters 10 and 11 are now essential reading.

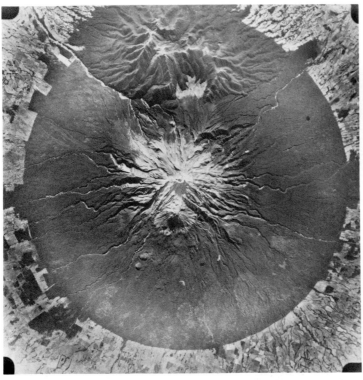

Plate 10 (a) Aerial view of the incised stratovolcano, Mt Egmont, New Zealand. The diameter of near- circular vegetated area is about 20 km. Egmont is a multi-vent centre, the principal visible ones are at the centre of the cone, the parasitic vent, Fanthams Peak, just south of the central vent, and those associated with the small domes. An older, more degraded volcano lies to the north. The rugged, ampitheatre-like area between the central vent and Fanthams Peak is thought to have resulted from sector collapse producing a major debris flow unit, the Opua Formation (Neall 1979). (b) Erosional canyon on the flanks of the stratovolcano Mt Egmont, New Zealand, cut into fluviatile sheet flood, lahar, air-fall deposits and lavas.

CHAPTER TEN

Epiclastic processes in volcanic terrains

Initial statement

Volcanic eruptions are short-lived and occupy only a small time interval in the total history of a volcano, especially on large volcanoes. Their principal effect is aggradational, building up the surface relief against the base level to which erosional processes work. Repose periods between eruptions occupy the bulk of the time interval covering the history of a volcano. Given the effects of aggradation above erosional base level, the development of high slope angles overall or locally, and the general abundance of loose erodible debris, it is not surprising that the long-lived repose periods are dominated by normal epiclastic surface processes: erosion, transportation and redeposition. In this chapter we review the types of epiclastic processes that are likely to occur in both subaerial and subaqueous volcanic terrains. In particular we examine transportational and depositional processes and the basic principles underlying these processes, and we consider the characteristics of resultant products.

10.1 Introduction

Volcanic terrains have rarely been looked at from the point of view of epiclastic sedimentology. They have most commonly been considered to be the realm of igneous processes. Yet given the abundance of loose debris and the overall steep slopes, the processes of erosion, transportation and redeposition of fragmental debris constitute a major phase of activity in the majority of volcanic terrains.

In this chapter we examine the types of epiclastic

processes that operate in different volcanic terrains. Epiclastic processes are here taken to include weathering, erosion, and transportation and deposition of fragmental sediments at the surface of the lithosphere in both subaerial and subaqueous environments. The epiclastic processes which affect volcaniclastic debris are no different from those that affect non-volcanic terrigeneous detritus. However, given the greater diversity of densities due largely to variable degrees of vesiculation in volcanic debris, and the effects of hydraulic sorting, volcaniclastic deposits of epiclastic origin *may* be more poorly sorted, and therefore slightly different texturally and in terms of their sedimentary structures. On the whole, however, the facies should clearly reflect their epiclastic origins.

By reviewing epiclastic processes and the characteristics of the resultant deposits, we will highlight the features that make epiclastic deposits in the rock record distinguishable from other facies in volcanic terrains (also see App. II). Our review of sedimentological principles and processes cannot be comprehensive. We will summarise basic concepts and, by way of examples from volcanic settings, show how epiclastic processes are as relevant in volcanic settings as in non-volcanic ones.

Relatively little will be said about the processes of weathering of volcanic material. The reader is referred to relevant sources (Paton 1978, Press & Siever 1978, Blatt *et al.* 1980). Because of the metastable nature of volcanic glass and the highly weatherable labile mineralogy of many volcanic successions, chemical weathering and physical disintegration into particulate and soluble material is likely to occur at high rates given the right climatic conditions *and time*. However, the time factor in most volcanic terrains is insignificant because the high slopes and highly erodible nature of volcanic terrains ensures rapid transit of detritus through the sediment cycle to the site of deposition, so short-circuiting chemical weathering. In low relief terrains, the effects of chemical weathering may be more pronounced, but may also be postponed until volcanism has ceased, because of the emplacement of new volcanic deposits associated with successive eruptive episodes.

Physical processes may be very significant in liberating particulate matter for subsequent transportation and deposition. Important physical processes include gravitational collapse or mass-wasting, and the abrasion effects of waves (marine and lacustrine shorelines), running water (rivers, rain water and melt water on subaerial slopes), moving mass-flows (subaerial and subaqueous), moving ice (in high altitude and/or high latitude volcanic terrains), the effects of thawing and expansion, and of a permafrost (also in relatively high altitude and/or high latitude volcanic terrains).

10.2 Importance of erosion and sediment transport in volcanic terrains

The fact that volcanism generally produces positive topographic relief may suggest that erosion has little or no effect in volcanic terrains during the active life of a volcano. The lifespan of active volcanoes is highly variable, ranging perhaps from a few weeks or months for basaltic cinder cones to a million or more years for stratovolcanoes or rhyolitic volcanic centres (Ch. 13). For cinder cones, the eruptive periods span nearly the entire, if short, active life of the volcano. However, for stratovolcanoes and rhyolitic centres there are significant repose periods between eruptions (Ch. 13). During these periods of eruptive quiescence, normal surface processes operate at very high rates. In fact, in such centres the total of all the repose periods far exceeds the total time over which active eruptions take place. In such terrains, therefore, it is not that erosion is an insignificant process during the active life of the volcanic centre, but rather that the enormous volumes of rock that are erupted and the rates at which they are erupted during eruptive episodes far exceeds the huge volumes removed and the very high rates of removal during quiescent intervals. As in normal sedimentary successions, the time interval represented by the boundaries between depositional units is greater than that represented by the depositional units (e.g. Sadler 1981). However, once the active life of the volcanic centre has ceased, erosion continues to operate and in a geologically short period is capable of eliminating the majority of the surface relief of volcanic piles.

E. H. Francis (1983) summarised known erosion rates within the context of volcanic terrains. He cites figures of 0.1–1.0 m per thousand years as determined by erosion downcutting rates in areas of high relief (after Young 1969), and figures of 1–2 km Ma^{-1} in the Andes of South America (after Drake 1976). He also points out the effects of high ratios of loose debris to lavas, the loss of vegetation cover in accelerating the erosion rate, and suggests that over a geologically significant period (1–2 Ma) the proximal, highest near-vent area of a stratovolcano would be eroded away, leaving only the core complex and part of the voluminous, epiclastic-dominated ring-plain sequence. These erosion rates accord well with those calculated by Mills (1976) for the stratovolcano Mt Rainier of the Cascades arc. Rates based on dissection of landforms are 1.1 m per thousand years, and those based on river sediment load are 3–4+ m per thousand years, leading to a span of 1–3 Ma for complete erosion of the 3 km stratovolcano. These erosion rates are higher than those inferred for scoria cones based on the studies of Wood (1980b) and Kieffer (1971) (Section 13.4), and those calculated by G. P. L. Walker (1984a) for basaltic highlands in Iceland. Walker calculated downcutting rates of 58 m Ma^{-1} from his Icelandic study, in a terrain with an average elevation of 400 m. These figures are only a general guide to rates, and variations will occur for a variety of reasons.

Vessell and Davies (1981) documented the close temporal relationships between volcanic and epiclastic processes in the Guatemalan chain, based on their study of Fuego volcano, which has erupted many times in historical and recent times. They suggest that the history of the Guatemalan volcanic chain is marked by cycles of activity consisting of four phases:

Phase 1, which *lasts between 80 and 125 years*, is the inter-eruption phase, and is marked by low rates of sediment deposition, by erosional incision of meandering rivers and by reworking of deltas.

Phase 2 is the eruptive phase. It is marked by eruption of lavas, air falls and pyroclastic flows and *lasts for less than one year*.

Phase 3, which is called the fan-building phase, is dominated by debris flows and coarse fluvial sedimentation, which occurs in response to intense rainfall after Phase 2. It lasts up to two years. Alluvial fans characteristically develop around the margins or base of stratovolcanoes during this phase.

Phase 4, the braiding phase, results from the influx of large volumes of volcanic sediment into stream systems, transforming them from incised meandering systems to flood-prone braided systems. Rapid progradation of deltas also occurs during this stage. This phase lasts from 20 to 30 years.

It is clear from this summary that the duration of epiclastic processes far exceeds that of the actual eruptive phases, and that epiclastic processes are rejuvenated in response to eruptive episodes.

Kuenzi *et al.* (1979) provided a valuable account of the relationship between volcanism, erosion and sedimentation resulting from the 1902 eruption of the andesite stratovolcano Santá Maria, also in southern Guatemala. Santá Maria, along with its near neighbour Santiaguito, is drained by the headwaters of the Samalá River, which fall from an altitude of over 2500 m around Santá Maria down to the Pacific shoreline over 60 km away. Following the major 1902 eruption, sediment supply to the Samalá River increased dramatically. The eruption resulted in a mantle of pyroclastic debris 20–30 m thick on the slopes within the watershed of the Samalá River. Given the typical torrential rainfall levels (>381 cm year^{-1}) and the associated total destruction of vegetation by the eruption, it is not surprising that the sediment load of the Samalá increased so much. Kuenzi *et al.* calculate that between 1902 and 1922 the bed of the Samalá rose 10–15 m. In so doing, it dammed up several major pre-1902 tributaries, producing elongate valley ponded lakes (Fig. 10.1) and re-activating older (?Pleistocene) alluvial fan sedimentation. In addition, a significant delta of volcanic sediment was built at the mouth of the Samalá, extending the shoreline nearly 7 km seaward, and involving 4 km^3 of sediment (Fig. 10.1). Subsequently, reduced sediment supply has led to retreat and

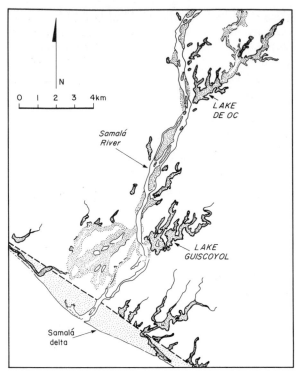

Figure 10.1 Detailed representation of the depositional systems of the Samalá River on the Guatemalan coastal plain following the 1902 eruption of Santá Maria. (After Kuenzi *et al.* 1979.) Note the valley-ponded lakes in the tributaries of the Samalá River and the major deltaic lobe at the mouth of the Samalá.

stabilisation of the shoreline. In addition, the active braided channel system of the Samalá has experienced major shifts on the coastal plain (Fig. 10.1). The Guatemalan volcanic chain can thus be seen to be a very dynamic system, not only from the point of view of its volcanic activity, but equally from the point of view of the contemporaneous epiclastic processes. Epiclastic volcanic sediments therefore represent a significant facies group, and are intimately associated with primary volcanic facies.

Richards (1965) described the types of erosional processes, their rates and effects in marine volcanic terrains, in discussing the erosional history of the volcano Bárcena in the Pacific Ocean in 1952. The principal agents of erosion were gravity, wind, shoreline wave erosion and surface runoff following several periods of heavy rain. Bárcena is a cone of

trachytic tephra and lesser lavas. Over two months 2.7×10^8 m^3 of tephra were erupted. The tephra consisted largely of ash, and lesser proportions of scoria lapilli, blocks and bombs. Gravitational grain-flow avalanching of tephra occurred during the eruption period, producing a radiating system of downslope furrows on the cone surface. Wind erosion was prominent until the first rains resulted in a surface crust, but recurred during subsequent dry periods. Rains produced significant erosional gullies, which in time rounded the knife-like crater rim, markedly modified the initial furrow system and commenced the infilling of the crater and the burial of the crater dome. Wave erosion of both tephra and a lava delta caused major changes to the position of the shoreline and removed large volumes of debris, which were then presumably redeposited subaqueously offshore. In the four months after eruption began, erosion of tephra sea-cliffs occurred at a rate of 1.7 m day^{-1}, and about 0.71×10^6 m^3 of tephra were removed by wave erosion over a 40 day period. A lava flow was eroded at a rate of 0.12 m day^{-1} over a 158 day period in 1953. Similarly, Thorarinsson (1967) records the major impact of storm wave erosion on Surtsey and its satellite centres from 1963–66.

Subaqueous epiclastic processes and rates from modern volcanoes are not well studied or documented, but ancient examples from the rock record are abundant. J. G. Jones (1967b) and Mitchell (1970) have shown that the subaqueous parts of marine stratovolcanoes are dominated by redeposited mass-flow volcaniclastics (debris flows, turbidites, etc.). In these cases, much of this debris was probably fed from the subaerial and shallow marine parts of the volcanoes, and was eroded by subaerial and shallow marine processes such as those in the cases described by Vessell and Davies (1981), Kuenzi *et al.* (1979) and Richards (1965) and discussed above. At the other extreme, wholly subaqueous volcanoes are likely to be less dominated by epiclastic processes and facies than are subaerially exposed volcanoes, because several of the main agents of erosion (wind, running water and wave action) are absent. In such subaqueous settings, gravitational collapse is the main epiclastic process operating, but may nevertheless be significant. However, if such volcanoes occur in restricted, small basins, then extra-formational epiclastic sediment transported from shallower basin margins will also be important.

In summary, epiclastic processes are extremely significant in volcanic terrains, both in terms of their duration and in terms of the volumes of rock matter that may be transported. Their activity seems to be accelerated in direct response to eruptive activity. In ancient successions, therefore, erosional discordances and variable proportions of epiclastic volcanic sediments should be expected, depending on the balance between erosion and sedimentation during the history of the volcanic terrain.

10.3 Epiclastic sediment transport

Sediment transport in both volcanic and non-volcanic terrains takes place in two basic forms: in *particulate* fashion, whereby each particle behaves individually, or by *mass-movement*, whereby a packet of sediment is moved as one, essentially instantaneously. In particulate sediment movement, particles have the freedom to move as individuals in response to the forces acting on them. In mass-movement, because a large population of fragments is moving *en masse*, particle freedom is largely inhibited, and collisions and dynamic interaction between fragments is common. As well as subdividing sediment movement on the basis of it being particulate or by mass-movement, further subdivision can be made on the basis of the involvement of an interstitial medium as follows (Table 10.1):

> no necessary interstitial medium
> ice as an essential interstitial medium
> water as an essential interstitial medium
> air as an essential interstitial medium

The function of an interstitial medium is to drive sediment movement, provide grain-support or act as a lubricant.

Table 10.1 A classification of sediment transport processes.

Epiclastic	Nearest primary volcanic analogue
1 Sediment transport not dependent on an interstitial medium	
particulate	
(a) particle free fall	air-fall (Chs 5 & 6)
(b) particle creep	
mass-movement	
(c) rock fall	
(d) slides	
(e) avalanches	nuées ardentes, block and ash flows (Chs 5 & 7)
(f) grain flow	
2 Sediment transport in which ice is an essential interstitial medium	
particulate	
(a) ice rafting	
(b) glaciers	
mass-movement	
(c) glaciers	
(d) permafrost creep	
3 Sediment transport in which water is an essential interstitial medium	
particulate	
(a) traction (bed load, saltation)	
(b) suspension	
(c) in solution	
mass-movement	
(d) fluvial torrent flow, sheet flood	
(e) subaqueous granular mass flow (e.g. turbidity currents)	
(f) mud flows, debris flows	lahars (herein)
(g) slumps	
(h) soil creep	
4 Sediment transport in which air is an essential interstitial medium	
particulate	
(a) traction } windblown	surges (Chs 5 & 7)
(b) suspension }	eruption columns, plumes (Chs 5 & 6)
mass-movement	
(c) air-lubricated rock avalanches	pumice, scoria, ash, block and ash flows (Chs 5 & 7)

10.3.1 SEDIMENT TRANSPORT NOT DEPENDENT ON AN INTERSTITIAL MEDIUM

In this situation there is no interstitial medium to drive sediment movement, provide grain-support or act as a lubricant for sediment movement. The only force causing sediment movement is gravity, and the gravitational potential of the sediment. This can occur either in *particulate* fashion, and includes:

> *particle free fall,*
> *particle creep,*

or under *mass-movement* conditions, and includes:

> *rockfall,*
> *slides,*
> *avalanches* and
> *grain flows.*

Particle free fall

Here a fragment is dislodged from an elevated position above a local base-level and free falls either through air or water, or both, and/or tumbles downslope. Although the fragment falls through air or water, neither medium contributes to the motion of the particle. Examples include dislodgement of individual fragments or blocks from cliffs or steep slopes, with deposition either subaerially or subaqueously, and free rafting of debris in vegetation

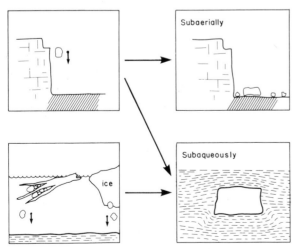

Figure 10.2 Various situations of particle free fall and resultant characteristics of deposits.

root systems in a lake or ocean, with fall-out from the root system through the water column (Fig. 10.2). Deposition onto a hard substrate will produce isolated angular fragments, and repeated events will produce an unstructured pile of fragments. Where deposition occurs onto a soft, unconsolidated substrate, soft-sediment deformation and impact structures may occur. Subsequent deposition of normal ambient sediment should produce drape structures (Fig. 10.2). Impact structures are also produced by explosively ejected ballistics (Figs 5.20g & h), and by fall-out after ice rafting (Section 10.3.2). In the former case there is usually a directional asymmetry resulting from the directed trajectory of the ballistics. In the latter case other evidence for the contemporaneous influence of ice and glacial activity should be apparent.

Particle creep

Particle creep involves the slow, almost imperceptible, downslope movement of individual particles, again solely under the influence of gravity. The particle rolls or slides bit by bit. Strictly, particle creep is unaided by any medium but, in practice, on exposed slopes rainfall, splash erosion, surface water sheet flow and the effects of wind will almost certainly be periodic causes of movement as well as gravity. There may also be an alternation between creep, tumbling and free fall and, where a large number of particles move together, mass-movement sliding, avalanching or grain flow will obviously occur. Particle creep can occur on any slope, whether it be subaerial or subaqueous, where a particle has gravitational potential. Particle creep will be widespread in volcanic terrains where unconsolidated debris lies on slopes. Its influence in ancient deposits may be almost impossible to recognise, since it produces no unique structure.

Rockfall

This form of mass-movement is a direct extension of particle free fall discussed above, but involves multiple particles or a large volume of rock matter. Again, a volume of rock is dislodged, and then free falls. After impact at a break in slope it may avalanche or tumble very rapidly downslope under its own momentum and the influence of gravity,

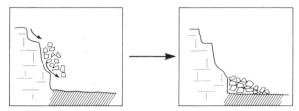

Figure 10.3 Mass-movement rockfall and avalanche producing an unstructured, base of slope talus deposit.

without the assistance of any medium. The process can take place subaerially or subaqueously, and is well documented from non-volcanic terrains (A. M. Johnson 1970, Voight 1978), but is also common in volcanic terrains. The volume of rock matter may be deposited as an unstructured pile of rubble at the base of slope (Fig. 10.3), or may translate into another form of mass-movement such as slide, air-lubricated rock-avalanche, or viscous debris flow (e.g. the Olokele rock avalanche on the island of Kauai, Hawaii, when it flowed into a river

valley, Fig. 10.4; B. L. Jones *et al.* 1984). The Olekele avalanche originated from a 60° slope, 800 m high at the head of a canyon entrenched into the plateau-like summit of a shield volcano. The caldera margin collapse breccias of Lipman (1976) (Ch. 8) are also examples of this mode of epiclastic erosion and transport, and lead to progressive caldera scarp retreat and build-up of a caldera margin talus deposit.

Slides

In slides (or landslides or rockslides) the dislodged mass of rock maintains contact with and shears over the substrate. It moves as an essentially coherent mass under the influence of gravity. If it acquires enough momentum it may even climb and slide up slope, as appears to have been the case with the Vaiont slide in Italy. In most cases slides will be initiated on relatively steep slopes. Their scale may be enormously variable. They may be very small, local, insignificant events, or they may be major catastrophic events involving millions of tonnes of rock. Examples of major subaerial slides have been adequately documented in the volume edited by Voight (1978). Subaqueous slides (e.g. D. G. Moore 1978) should be just as common given the extra lubrication provided by an aqueous environment, even though water is not an essential influence in the movement of slides. The source areas for slides should be marked by major scars.

In volcanic terrains slides should be very common, given steep slopes and the interbedding of lithologies of very diverse mechanical properties. Such slides may be very local in significance, or may involve large-scale sector collapse of the volcano (e.g. Ui 1983; Fig. 10.5 & Plate 10). One of the best known slides of recent times in volcanic terrains was the spectacular one initiating the devastating 18 May 1980 eruption of Mt St Helens. The growth of a high-level magma reservoir within the core of the stratovolcano caused gradual oversteepening of the northern flank of the volcano. On 18 May, apparently triggered by an earthquake, the oversteepened slopes collapsed, causing a huge landslide which unroofed the magma reservoir, and initiated a simultaneous phase of pyroclastic activity (Fig. 10.6a; Section 5.4; Christiansen & Peterson

Figure 10.4 The scar of the Olekele rockfall–avalanche–debris flow event, Olekele Gorge, Kauai, Hawaii. (After B. L. Jones *et al.* 1984.)

Figure 10.5 Major erosion scar on the flanks of the island volcano Stromboli resulting from a major slide which continued into the surrounding submarine environment. (After Korgen 1972.)

1981, Voight *et al.* 1981). Several discrete initial slide blocks were identified from time-spaced photographs, but under the influence of the simultaneous explosive blast, and interstitial hot water and steam, the slide masses moved away in dyamic fashion, leaving a huge horseshoe shaped, concave, amphitheatre-like erosion scar (Fig. 10.6b). Many similar volcanic analogues are known (Voight *et al.* 1981, Ui 1983; Section 5.4). The characteristics of the final deposit will be described in the next section, on avalanches, since the slide evolved into a mobile avalanche.

True slides will therefore be relatively coherent masses retaining some element of original stratigraphy. Some degree of internal deformation may be apparent (faults, folds – some soft sediment ones), and the mass will be separated from the substrate by a sole thrust. The source area will be marked by a major scar, which is frequently concave in the direction of movement (Fig. 10.7). Stranded detached slabs of the slide mass may mark the trail of the slide back to the scar. Slumps (Section 10.3.3) represent near-source, incipient slides.

Several possible major submarine slide masses have been proposed by J. G. Moore (1964) on the flanks of the Hawaiian ridge, based on detailed bathymetric surveys. The proposed source region is marked by a topographic scar, and the slide bodies are marked by irregular topography. One mass may be as large as 150 km long and 50 km wide.

Avalanches

In an avalanche the dislodged rock mass flows, rather than slides, and the frictional interaction with the substrate is very low (Hsü 1975, 1978, Ui 1983, Siebert 1984). The term 'flow' implies a degree of relative motion and freedom between clasts, indicating that pervasive internal deformation occurs within the flow. Avalanches are therefore not coherent, and there is considerable interaction between clasts in the flow. Hsü (1975, 1978), interpreting the work of Albert Heim (e.g. 1932) on sturzströms, or catastrophic rock debris

Figure 10.6 (a) The initial landslide of the 18 May 1980 Mt St Helens eruption, also showing the initiation of the simultaneous pyroclastic blast event. The slide mass is moving away to the right at the foot of the volcano, leaving a major concave erosion scar (after G. Rosenquist in Voight 1981). (b) The concave, amphitheatre-like erosion scar resulting after the major slide and pyroclastic explosion event of 18 May 1980, Mt St Helens (after Lipman & Mullineaux 1981).

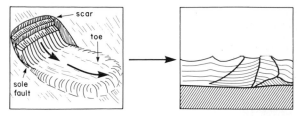

Figure 10.7 Initiation and major elements of a coherent slide, and the resultant internal deformed deposit.

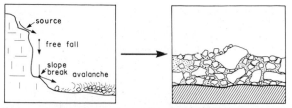

Figure 10.9 Formation of rock avalanches from rock-falls, and characteristics of avalanche deposits.

streams, equated the flow mechanics of avalanches with inertial Bagnoldian grain flow. In this, grain support, grain momentum, and flow momentum and mobility are all maintained by grain-to-grain collisions leading to a progressive transfer of momentum through the flow. That is, if an interstitial medium is present it plays no necessary, active, dynamic role in the motion of the avalanche, nor does it buffer collisions between clasts, these being essentially cohesionless. Fluids may nevertheless be present. For example, many documented avalanches (including the 18 May Mt St Helens one) have included volumes of snow and ice, which melt during flow as a result of impacts. Air may also be trapped within the flow, and fluidisation by air (Ch. 7) may also occur, although it is not essential to the mechanics of avalanches as seen from interpreted lunar avalanches (Howard 1973, Hsü 1975). In the past there has been some speculation

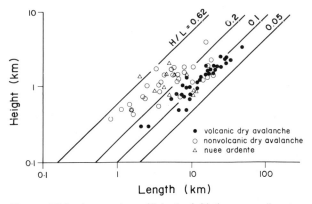

Figure 10.8 Apparent coefficient of friction according to the Coulomb law for sliding friction for volcanic rock avalanches, non-volcanic rock avalanches and ignimbrites. The apparent coefficient is defined as the vertical height moved to the projected horizontal distance travelled, *H/L*. Also see Figure 7.2. (After Ui 1983.)

that rock avalanches owed their mobility to a trapped cushion of compressed air on which the avalanches rode (e.g. Shreve 1968). In this model the air is trapped at a break in slope or hydraulic jump, where the avalanche changes from a largely fall trajectory to a lateral flow trajectory.

Mechanically, avalanches can be distinguished from more-viscous modes of transport, such as debris flows, by their greater speeds, and by the lower frictional interaction with the substrate. Speeds of 200–300 km h^{-1} or more are not uncommon and are based on well constrained calculations (Voight 1978). Huge blocks, metres in diameter, hurled a kilometre or more on ballistic trajectories by avalanches, also attest to amazing speeds (e.g. Plafker & Ericksen 1978). The apparent coefficient of friction can be as low as 0.06 (Ui 1983), compared with, for example, a block of rock sliding according to Coulomb's law of sliding friction, with an apparent coefficient of friction of 0.62 (Figs 10.8 & 7.2; Hsü 1975, 1978). Avalanches in non-volcanic terrains most frequently originate from rockfalls (Fig. 10.9). In volcanic terrains they may also do so (e.g. Fahnestock's 1978 account of rock avalanches on Mt Rainier), but they may also evolve from slides associated with collapse of a major sector of a volcano accompanied by simultaneous explosive release, as described in the Mt St Helens case above (Fig. 10.6a; Voight *et al.* 1981, Ui 1983) and in many other situations on stratovolcanoes (Ui 1983, Siebert 1984, Francis *et al.* 1985). Hence, a dynamic slide of volcanic debris can translate into a significant rock debris avalanche.

Deposition takes place by frictional freezing of the avalanching rock mass as it loses momentum, implying a relatively high bulk viscosity for the avalanche flow, and a high degree of particle

Figure 10.10 Rock avalanche deposit on the Emmons Glacier high on the Cascades stratovolcano Mt Rainier. The megablock with the person on top is 40 m wide and 18 m high, and is estimated to be more than 45 000 tonnes in weight. Note the predominance of blocks. The matrix is of sand-to-gravel size. (After Fahnestock 1978.)

interaction. Avalanche deposits form well defined lobes with steep margins. Marginal ridges including levées, internal pressure ridges and surface pile-up mounds have been described and suggest a significant yield strength (Ch. 2). Because of these significant topographic variations on the surface of the flow, thicknesses are very variable. Internally avalanches are massive and extremely poorly sorted, with fragment sizes ranging from huge megablocks thousands of cubic metres in size through to gravel, sand and mud sizes (Figs 10.9 & 10). Rare individual blocks up to 500 m in diameter have been claimed (Ui 1983). Individual blocks are frequently broken in transit, but pieces can be fitted back together again in jigsaw puzzle like fit. Frequently, original stratigraphy or lateral relationships or order of lithology as documented in the source can be pieced together, suggesting that relatively little internal mixing of clasts has occurred. The avalanche can thus be viewed as a mobile, fluidal, close packing of blocks and fragments which jostle, bump, push, collide and fragment each other in transit. The matrix between blocks can vary greatly in character from deposit to deposit. It often includes material such as soil, alluvial gravel and vegetation eroded in transit.

However, much of it is abraded in transit from clast collisions.

Few detailed granulometric analyses are available for avalanche deposits and there are obvious problems in representing the many large blocks. Analyses of the matrix for the Mt St Helens avalanche deposit average 4, 11, 42 and 43 wt%, respectively, for clay, silt, sand and gravel sizes. The deposits are therefore very granular, perhaps more so than might be expected of viscous debris flow deposits. Herein lies the problem with avalanche deposits. Are they distinguishable in the rock record from debris flow deposits? This problem is touched on by Voight *et al.* (1981) in comparing the granulometry of the Mt St Helens avalanche deposit with other known volcanic avalanche deposits (the Mt Rainier ones referred to above) and coarse deposits from both Mt St Helens and Mt Rainier which had previously been interpreted as lahars (i.e. debris flows of volcanic composition). The overall morphology of both is similar, which suggests a generally similar rheology (Ch. 2). Possible criteria include the presence of megablocks (many metres to tens of metres or more in diameter) and the lack of a *pervasive* muddy matrix in avalanche deposits. Debris flows frequently develop a basal laminar, shear zone in which crude bedding develops, and grading of clasts occurs. Open framework – that is, matrix-support of large clasts – might also be expected in debris-flow deposits.

Grain flow

Grain flow involves the spontaneous, passive downslope rolling and sliding of a population of cohesionless grains that together constitute a moving grain layer. The grains move solely under the influence of gravity, and not under the influence of an external shear stress, induced, for example, by flowing water or wind. However, grain-to-grain collisions which produce a transfer of momentum are an important mechanism in maintaining the mobility of the entire grain-flow layer. These collisions produce a reaction force, called dispersive pressure. Given that the grains are cohesionless and that no viscous interstitial fluid is involved, the dispersive pressure or collisions operate in an inertial regime. Grain flows can also occur subaqueously in more viscous water, and grain collisions also occur in very viscous fluids such as mud (e.g. in debris flows). In those situations the collision events are buffered by pressure gradients set up in the viscous fluids during near approaches of grains just before collision. These pressure gradients also contribute to the reaction force affecting the colliding grains, so that they are subjected to viscous dispersive pressure developed in the fluid between the colliding grains, as well as the inertial dispersive pressure of the actual collision between grains.

Given these constraints, it is not surprising that true grain flow only occurs on steep slopes, those steeper than the maximum natural angle of stability or repose for a pile of grains. If this angle is exceeded, then downslope movement will eventually be initiated, the angle at which this occurs being called the angle of initial yield (Allen 1982). After flow, the angle of slope will be adjusted to a lower, more stable one equal to or less than the angle of repose, which Allen (1982) calls the residual angle after shearing. The natural angle of repose for an aggregate of grains depends on the grainsize, sorting, packing, grain shape and grain surface texture. For most sands and gravels it is between 30° and 35°, but may be steeper for coarse, very angular aggregates.

Grain flow is different from rock avalanches discussed in the previous section, in that in grain flow there is considerable particle freedom and

(a) Ripples and dunes

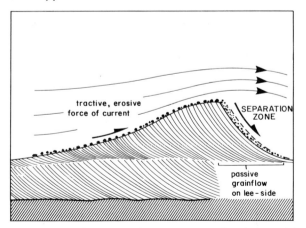

(b) Scree slopes

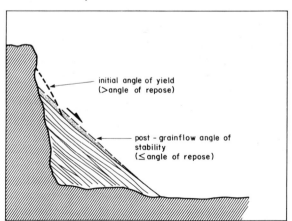

Figure 10.11 Two common occurrences of grain flow. (a) Lee-side cascading of grain-flow layers in ripples and dunes so forming internal cross-stratification and resulting in downcurrent migration of the ripple or dune. A carpet of grains is eroded off the up-current side of the ripple and moved to the crest by the tractive force of the current. Oversteepening occurs at the crest and a grain-flow layer cascades down the lee-side face, into the space called the separation zone over which flow streamlines pass. (b) Grain flow on scree slopes initiated by oversteepening to the angle of initial yield. Grain-flow restores the slope to a more stable attitude equal to or less than the angle of repose.

grain mobility within the grain-flow layer, and there is considerable frictional contact or interaction with the substrate. In grain flow the mobility of the grain-flow layer is largely due to the momentum of the individual grains. In rock avalanches the mobility is largely due to the momentum of the whole rock avalanche mass.

Grain flow can occur in many natural situations and at different scales. It is responsible for the formation of angle of repose cross-stratification in ripples and dunes by the passive cascading of sand grain-flow layers down the steep, protected lee-side of the ripple or dune structure (Hunter 1985, Fig. 10.11a). It also occurs in the formation of relatively fine-grained sandy to gravelly scree slopes (Fig. 10.11b).

In volcanic terrains grain flow will be significant in both types of grain flow cited here. On the small scale it will be manifest in cross-stratification produced by epiclastic tractional sediment transport of volcanic sediments (Section 10.3.3), and on a larger scale, on steep slopes strewn with pyroclastic or epiclastic debris, or both, may give rise to scree slopes. The steep slopes of cinder or scoria cones and stratovolcanoes (Ch. 13) are not just coincidentally sloping at about angle of repose. Steeper slopes are prone to gravitational collapse, producing either slides or grain flows. Grain flow is particularly prominent on cinder or scoria cones (Ch. 6), and often produces a radiating downslope system of furrows (e.g. Richards 1965), as well as maintaining the slopes at or about angle of repose during pyroclastic eruptions (Ch. 13). Scree slopes are common features adjacent to steep scarps made of semi-consolidated debris. They often form inside craters, rifts (e.g. Fig. 10.12a) or against fault scarps, and again the slope angles are nearly angle of repose. Reverse-graded pumice grain-flow layers frequently accompany pumice falls on steep slopes (Duffield et al. 1979).

Grain-flow deposits will be marked by steeply inclined stratification (30° to 35°), internally diffuse, relatively thin stratification caused by the high frictional interaction between the grain-flow layer and the substrate, and by the development of internal shear layering. Individual layers will be thin (centimetres to a few tens of centimetres) and may develop a reverse grading produced by intense

Figure 10.12 (a) Scree slopes in the rift vent formed by the 1886 eruption of Mt Tarawera, New Zealand. The older rhyolitic dome complex of Tarawera was mantled by basaltic tephra (dark) of the 1886 eruption and both sources are contributing to the scree slope debris. The trails of individual grain flows are represented by the streaks on the scree slope surfaces. (b) Pyroclastically erupted and deposited (air-fall) debris of Pukeonake cone, New Zealand. Following air-fall deposition the debris was moved downslope by grain flow, producing a series of slightly irregular, mostly reverse-graded grain-supported layers. Reverse-grading is especially well developed in the upper coarsest layer, but also in lower layers.

shearing at the contact with the substrate (Figs 10.12b & 6.6b; Lowe 1982).

10.3.2 SEDIMENT TRANSPORT INVOLVING ICE AS AN ESSENTIAL INTERSTITIAL MEDIUM

The principle here is that sediment particles are trapped within or on top of significant bodies of ice which act as transporting agents. The ice thus provides grain support in a solid state. Where the ice floats, it and its sediment load are buoyed up, usually by slightly denser sea water, and movement occurs in response to currents. Where the ice is glacial or permafrost, the ice–sediment mass has gravitational potential and the lubrication provided by the base of the ice mass ensures relatively rapid downslope movement under the influence of gravity. Ice can move sediment in either particulate fashion by:

ice rafting
glaciers,

or in mass-movement form by:

glaciers and
permafrost creep.

Ice rafting

The principles involved in deposition of sediment after being ice-rafted are the same as those discussed above for particle free fall (Section 10.3.1). An iceberg carrying sediment, and perhaps originating from a glacier flowing into sea water or from shelf-ice, begins to melt and, bit by bit, releases its sediment load. Dropped boulders with impact structures and draped sediment layers should be expected (Fig. 10.2). However, not all ice-rafted dropstones necessarily produce impact structures, especially if the dropstone still has attached to it a large volume of ice, so reducing the total bulk density of the dropstone and ice, and so very markedly reducing the fall velocity. Volcanic terrains in high latitudes with associated marine environments should therefore be the source for, and site of, deposition of dropstones of volcanic composition (Fig. 10.13), as well as those of other compositions (e.g. Lisitzin 1962).

Glaciers (particulate movement)

Glaciers are relatively mobile huge masses of ice which creep, flow or slide down slope due to their gravitational potential at speeds that vary from the imperceptible to greater than 100 m day^{-1}. The latter occurs where frictional coupling between the

Figure 10.13 Inferred dropstone of latite composition. Permian Kiama Sandstone Member, Broughton Formation ('Gerringong Volcanics'), Sydney Basin, New South Wales, Australia. Note ill-defined draping over top right of boulder. Enclosing volcanic sandstones are considered to be turbidites deposited in a storm-dominated shelf setting. As part of Gondwanaland, the Sydney Basin was located in a periglacial setting during sedimentation of the Kiama Sandstone.

Figure 10.14 (a) The stratovolcano Mt Rainier of the Cascades arc of western North America. Note the Emmons Glacier, the rock avalanche burying the toe of the glacier, Little Tahoma Peak, the source of the avalanche 600 m above the glacier, the moraine deposits of the glacier, and the White River, the braided outwash river of the glacier (after Fahnestock 1978). (b) Thorsnarsk Glacier, Iceland and its associated fluvial outwash system. (c) Snow cap and summit glacier, Mt Kilimanjaro in the equatorial zone of the East Africa Rift Zone, Africa (photograph by Frances Williams).

glacier and the substrate has been reduced by basal melting (Raymond 1978). Glaciers have enormous erosive capacity, as evidenced by the U-shaped valleys they gouge out, and the huge volumes of sediment they deposit in ground, lateral and terminal moraines. Glaciers transport their sediment as base load, interior load or surface load. Base load is eroded from the substrate at the base and edges of the glacier. Interior load may originally have commenced transport as base load, but because internal flow paths are often forward and upward, it ends up within the interior, well above the base. Surface load results from material falling onto the surface of the glacier, for example by rockfall or rock avalanche as discussed above (Fig. 10.10), or from wind transport or volcanic air fall.

Glaciers develop at relatively high altitudes, where snowfall is common and accumulates for a substantial part of the year. This is not restricted to high latitudes, and glaciers are even known at high altitudes in the equatorial zone. Since volcanism occurs all over the globe independent of climate, it is not surprising that volcanoes and glaciers are frequent associates. For example, glaciers drain off the volcanic chains of Alaska, the Cascades (Fig. 10.14a), the Andes and Antarctica, to name but a few. Glaciers are even major epiclastic agents on the Atlantic mid-oceanic spreading ridge system in Iceland (Fig. 10.14b). Mt Kilimanjaro, in the equatorial belt of the East Africa rift system, also has glaciers draining off its summit (Fig. 10.14c).

The sediment load of glaciers is dumped at the base, margins and terminus of glaciers in irregular piles or ridge-like accumulations called ground, lateral and terminal moraines, respectively. Moraines consist of unsorted debris varying from particles as fine as clay, resulting from the intense grinding abilities of the glacial ice mass, to huge blocks that have been dragged and rafted along. The terms 'till', 'tillite' and 'diamictite' are used for these moraine deposits, the first being for unconsolidated moraines and the other two for lithified moraines. 'Diamictite' is a more general, non-genetic textural term for poorly sorted aggregates such as till/tillite and debris flow deposits. The term should be used in this sense. The provenance of till depends on the source rocks in the drainage area of

the glacial system. If these are volcanic, then volcanic till will result (e.g. Crandell & Miller 1964, 1974, Fahnestock 1963). Sediment particles drop out one by one as melting of the ice occurs (Fig. 10.15). If the terminus of the glacier is on land, then the moraine material may be reworked to varying degrees by the high energy outwash river systems (Section 10.3.3; Figs 10.14a & b) fed by the melting of ice at the terminus and the base. These systems are frequently of the braided type. Given that most glacial systems have both short-term and long-term histories of advance and retreat, the upsequence facies variations through a pile of glacial sediment may be complex (Eyles et al. 1983). Facies differences also occur according to whether the glaciers are temperate, sub-polar or polar, which affects their thermal regime and the degree of basal melting. Interested readers are referred to Eyles et al. (1983) and Eyles and Miall (1984).

The stratovolcano Mt Rainier and the Cascades arc have already been referred to with regard to rock avalanches in volcanic terrains (Section 10.3.1). Mt Rainier also has a developed glacial system including the Emmons Glacier (Fig. 10.13a), lateral and terminal moraine deposits, and a fluvial outwash system (Fig. 10.14a; Fahnestock 1963, 1978, Crandell & Miller 1964, 1974). All are sourced within the volcanic pile of Mt Rainier. Mt St Helens also had a complex of summit glaciers before 1980, much of which was destroyed by the 18 May eruption (Brugman & Meier 1981).

Glaciers (mass-movement)

Glacial transport also essentially involves mass-movement as well as particulate transport, because

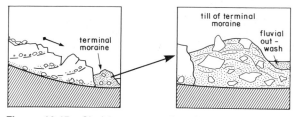

Figure 10.15 Glacial transport of sediment as bed load, interior load and surface load, and textural character of glacial till of the terminal moraine, which is being reworked by the fluvial outwash system.

in addition to carrying individual eroded fragments, glacial transport involves the carrying of aggregates of rock, perhaps picked up as base load from fluvial valley fill sedimentary piles over which the glacier is moving, or as surface load, that is deposited on the glacier surface by a rock avalanche as discussed above (Fahnestock 1978). The principles of transportation, and of deposition of aggregates of rock by glacial transport and the characteristics of the resultant deposits, are the same as in the previous case).

Permafrost creep

Permafrost creep involves the very slow downslope creep of a frozen surface ground layer. That layer may be soil or loose surface aggregate in which interstitial ground water has become frozen, and due to position on slope, the whole frozen layer of rock matter and ice acquires an increased gravitational potential which drives its slow passage down slope. On melting, an unstructured pile of material is deposited, perhaps preserving some original stratigraphy, and perhaps containing less finely ground or milled clay size material than in true glacial till.

10.3.3 SEDIMENT TRANSPORT INVOLVING WATER AS AN ESSENTIAL INTERSTITIAL MEDIUM

Again, the sediment transport can occur in both particulate and mass-movement forms. If the sediment movement is particulate or involves the mass-movement of a cohesionless population of grains (i.e. grains which have no significant attraction or repulsion to each other) in low viscosity, non-cohesive, relatively pure water, then the transport is said to take place in a non-viscous or inertial regime. If mass-movement involves cohesive particles (e.g. electrostatically charged clay platelets) and/or a cohesive aqueous fluid (e.g. a viscous clay–water, mud mixture), then the transport is said to take place in a viscous regime. In the non-viscous or inertial regime the settling of the cohesionless particles is not significantly retarded by the interstitial fluid (water), and is essentially controlled by the momentum of the particles, their settling velocities in water and their degree of interaction (collision) with other particles.

In the different modes of sediment transport to be discussed here, water performs all of the different functions of an interstitial medium discussed previously. That is, it can drive sediment movement, it can provide grain-support or it can act as a lubricant (further discussed below in the relevant sections). However, its role as a sorting agent is also extremely important in controlling the textural characteristics of the resultant deposits. As discussed in Chapter 1, with volcanic sediments lack of uniformity of grainsize is not necessarily a reflection of an hydraulically poorly sorted sediment. Fluids, especially water, sort clastic aggregates according to their hydraulic properties, the most important of which will be shape and density. Given the variable density of volcanic detritus due to variable vesicularity, hydraulically well sorted aggregates may be very poorly sorted according to grainsize. The other effect of water is to lower the effective or dynamic transportable mass of sediment particles where these are immersed in water. Given that mineral and rock fragments have densities of around $2.5+$ g cm^{-3}, immersion of such grains in water (density of 1 g cm^{-3}) lowers the effective transportable mass of such fragments by 40% relative to their mass in air, making erosion an easier prospect.

In particulate form, sediment transport takes place in the following ways:

traction,
suspension and flotation,
in solution,

and in mass-movement by:

fluvial torrent flow, sheet flow,
subaqueous granular mass flow (e.g. turbidity currents),
mud flows, debris flows, lahars,
slumps and
soil creep.

Traction

Tractional sediment transport produces a very important group of sedimentary structures called, appropriately, *tractional sedimentary structures* (e.g.

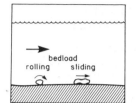
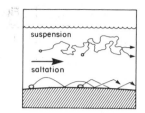

Figure 10.16 The two sediment transport modes that together constitute tractional sediment transport: bed load (rolling and sliding) and saltation (jumping). Smaller grains may also be lifted into continuous or intermittent suspension.

ripples, dunes, cross-stratification, horizontal or plane-bed lamination, antidunes). Tractional structures can be produced where any fluid (water, air) flows over a bed of cohesionless grains and initiates movement of the grains as *bed load* or in *saltation* (Fig. 10.16). Tractional reworking will also remove most clays, so producing granular clastic aggregates with little or no mud matrix.

Bed load refers to the rolling or sliding of particles along the bed (Fig. 10.16). Bed load movement occurs because the flowing fluid exerts a lateral shear stress at the bed surface which is great enough to overcome the weight of the particle and the frictional resistance at the bed surface. At this stage the fluid has an erosive capacity, its velocity is greater than the minimum threshold velocity for erosion of the grainsize in question (e.g. according to the Hjulström curve), and the flow conditions in the fluid are turbulent, so that Reynolds Numbers are high (Ch. 2). Erosion does not take place under low velocity laminar flow conditions.

Saltation refers to a downstream bouncing or jumping mode of transport, involving short-term suspension of particles (Fig. 10.16) that may be induced either by the impact of other particles falling back to the bed, or by high enough velocities (and therefore bed shear stress) to lift the particles into suspension. This is effected by the deflection of streamlines over the crest of particles, concentrating the streamlines and producing a pressure gradient in the fluid above the grain. A low pressure zone develops on the lee-side of the grain. The resultant Bernoulli effect produces an uplift force in a downstream direction, which may be sufficient to lift the particle from the bed.

The net effect of continuous bed load and saltation (i.e. tractional) transport of a large number of cohesionless grains, simultaneously under the influence of a unidirectional water flow, is to produce a range of tractional bed forms or sedimentary structures (Fig. 10.17). For any particular grainsize population under steadily accelerating or decelerating flow conditions, a regular succession of structures should occur. However, the succession is different for different grainsizes (Fig. 10.17). Ripples, sandwaves and dunes are all wave-like bedforms that form in sequence as velocity is progressively increased. They are all asymmetrical structures produced by unidirectional current flow, the steep face being on the downstream, or downcurrent, side (Fig. 10.18). Internally they are marked by cross-stratification (cross-laminae if the cross-strata <1 cm thick, cross-beds where cross-strata >1 cm thick; Figs 10.19a–c), which dips in the direction of flow. This cross-stratification is typically at or just below angle of repose (25–35°) at its upper end, and forms by the passive grain flow of packets of granular sediment down the downcurrent face under the influence of gravity (Fig. 10.12a). Little or no lateral current-induced bed surface shear stress is involved in this passive lee-

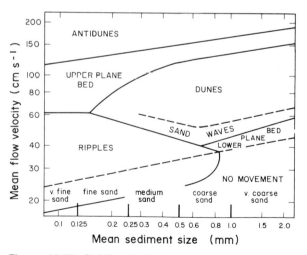

Figure 10.17 Stability fields for the common tractional structures as a function of mean water flow velocity and mean grainsize. Data points represent many experimental flume runs using different grainsizes. Note that for different grainsizes different sequences of bed-forms develop as flow velocities increase or decrease. (After Harms *et al.* 1982.)

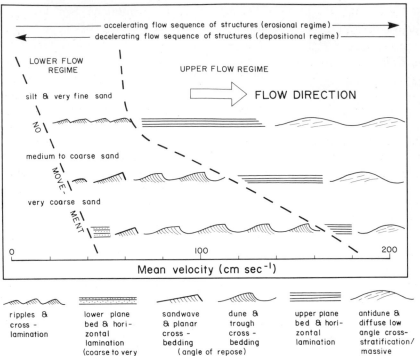

Figure 10.18 Schematic representation of succession of tractional bed-forms and corresponding internal sedimentary structures formed under conditions of unidirectional water flow, with flow either accelerating (left to right) or decelerating (right to left). During acceleration the succession of bed-forms is eroded, under deceleration the succession of bed-forms and flow stages may be preserved upsequence because net aggradation (deposition) will occur at the bed. The velocity–grainsize relationships hold only for dense mineral and lithic grains. No data are available for low density, vesicular detritus.

side grain flow but, where it is involved, the effect is to progressively lower the angle of cross-stratification below the angle of repose with increasing bed shear stress. Sandwaves (or large two-dimensional ripples) have straight to slightly sinuous continuous crests and long wavelength crests, and produce tabular planar cross-strata. Ripples and dunes have diverse morphologies (Allen 1982, Blatt *et al.* 1980) and most commonly produce upwards-concave sets of trough cross-strata. Ripples are less than 4 cm in amplitude. Lower plane bed lamination forms in place of ripples in coarse sediments (Figs 10.17 & 18) under low flow velocity conditions. Ripples do not form in coarse sediment.

If the flow velocity increases further, dune bed-forms are smeared out by the current under very high bed shear stress conditions, and are replaced by a flat or horizontal planar bed-form (Figs 10.17 & 18) which, internally, is represented by horizontal, planar, parallel laminated sands (Fig. 10.19d). This bed-form represents the transition from relatively low energy conditions known as the *lower flow regime*, to higher energy and velocity flow

conditions known as the *upper flow regime*, which is represented by a bed-form called antidunes (Fig. 10.18). Lower and upper flow regimes are distinguished by a parameter called the Froude Number (Fr):

$$Fr = U/\sqrt{gh} \qquad (10.1)$$

where U is the velocity, g is the acceleration due to gravity and h is the flow depth. Where $Fr < 1$, lower flow regime bed-forms and structures occur, and this corresponds to a fluid flow state of subcritical turbulence. Where $Fr \sim 1$, antidunes form, and this corresponds to supercritical turbulent flow, which means it is a very dynamic, high energy flow system. Antidunes are marked internally by a poorly defined, low angle ($<20°$) cross-stratification that most commonly dips upstream, but may wrap over the crest of the structure, or rarely dip downcurrent. Antidunes have no clear-cut asymmetry. The low angle cross-stratification (Fig. 10.19e) results from the smearing of sediment over the bed surface under conditions of very high lateral bed shear stress, as discussed above. However, antidunes are rarely preserved. That flow stage is more commonly represented by massive beds of cohesionless sediment, if represented at all. Massive, internally structureless beds result from the rapid deposition (mass-dumping) of large volumes of sediment out of suspension. True tractional sediment transport has probably been surpassed by this stage.

Figures 10.19a–e are a selection of photographs of the common, hydrodynamically significant tractional structures discussed above, taken from a variety of volcaniclastic sediment successions.

The complete succession of structures reflecting the change in flow regime conditions during one flow cycle is only likely to be preserved from the decelerating part of the cycle, since only then will sedimentation occur. As a result, antidunes are frequently destroyed by reworking, during lower, declining flow regime conditions. Upward fining cycles and sequences of structures from fluvial channel-point bar depositional settings (especially meandering river successions) are most frequently depicted as showing the complete succession of flow regime level structures (e.g. R. G. Walker 1984). As discussed further below, subaqueous granular mass flows can also produce such successions of sedimentary structure intervals, reflecting declining flow regime levels during the decelerating or waning stages of flow.

Although many other features of tractional sediment transport and sedimentary structures could be discussed, space precludes this. However, some further comment should be made about cross-stratification and its diverse origins (see Allen 1982 for a comprehensive discussion). Cross-stratification can be subdivided into two essential types: high angle, which is at or near the angle of repose (25–35°), and low angle (usually <20°). Only high angle cross-stratification can be confidently interpreted as the product of lower flow regime tractional transportation and deposition. It can occur in sets of cross-laminae only centimetres thick to sets of cross-beds many metres thick. Some surge cross-stratification can be high angle, but there are other criteria for recognising this origin, as discussed below and in Chapter 7. Low angle cross-stratification can have many origins, including:

(a) Tractional, representing the assymptotic toeset portion of a once larger-scaled set of cross-beds, the upper part of which approached the angle of repose. Such a set would be associated with a migrating dune or sandwave, the top of which was truncated by contemporaneous flow erosion. The downstream converging nature of the assymptotic layers should be definitive and should also distinguish such layers from upper flow regime horizontal lamination for which they could be mistaken in large-scale, broad sets.

(b) Upper flow regime antidune cross-stratification, as discussed above.

(c) Nested scour and fill trough shaped sets, which in sections perpendicular to flow direction (Fig. 10.19f) give apparent low-angle truncations. These are end-on views of trough cross-bedded sets formed by downcurrent migration of large dunes by normal tractional processes. Compare this with the flow-parallel section in Figure 10.19b.

(d) Two-way cross-stratification associated with symmetrical ripples formed by oscillating, near-shore wave reworking (Fig. 10.19g).

(e) Low profile domal sets of cross-strata, called hummocky cross-stratification (Fig. 10.19h), and associated broad, open, low profile trough-like sets called swaley cross-stratification. These are produced when the bases of storm waves surge over the bed in storm-affected shelf settings. Each layer represents a short-lived surge event rather than passive angle of repose cross-stratification. Shelf turbidites may be present (R. G. Walker 1984).

(f) Beach face low angle cross-stratification marked by very low angle truncations (<10° to 15°) between sets which are made up of broad planar to open trough sets. The sets are produced by onshore wave surges onto the beach face, in the swash–backwash zone.

(g) Pyroclastic surge cross-stratification (Chs 5 & 7). This is usually very fluidal in form, with set layers wrapping over dune crests and thickening and thinning. Sorting is poor, and major grainsize variations occur between set layers. Accretionary lapilli may be present.

Figure 10.19 (a) Ripple cross-lamination in pumiceous sands, 5 m terrace Lake Taupo, New Zealand. Current flow from left to right. Note the lenses of large pumice trapped in the much finer-grained sands suggesting that because of the very vesicular, low density nature, and its flotational abilities, such large pumice is hydraulically equivalent to very much smaller, denser grains. (b) Angle of repose trough cross-bedding in coarse pumiceous fluviatile sediment. Locality as for (a) current flow from right to left. (c) Planar angle of respose cross-bedding, in fluvio-lacustrine volcanic sandstone, Lower Devonian Snowy River Volcanics, W–Tree, Victoria. Current flow from left to right. (d) Planar, horizontal lamination and massive bedding (= antidune stage) in fluviatile volcanic sandstones of the Cretaceous Strzelecki Group, Gippsland Basin, Victoria, Australia.

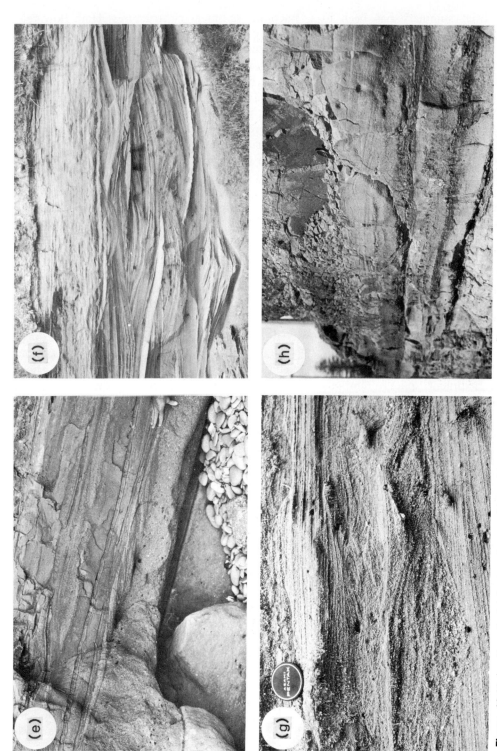

Figure 10.19 (continued) (e) Low-angle, possible antidune cross-stratification in the submarine Eocene–Oligocene Waiareka–Deborah volcanics, Oamaru, New Zealand. Current flow from left to right. (f) Scour and fill structures (=? swales) Pleistocene (?) Makariki Tuff, Rewa Hill, New Zealand. The view is end-on to sets of trough cross-bedded pumiceous sands. (g) Two-directional cross-stratification and symmetrical ripples formed by near-shore wave oscillation, ancestral Lake Taupo shoreline, Puketapu, New Zealand. (h) Mound of hummocky cross-stratified marine volcanic sandstone. Permian 'Gerringong Volcanics'. Broughton Formation, Kiama, New South Wales, Australia.

The aim of this brief summary has been to highlight those primary, current-generated tractional sedimentary structures that have hydrodynamic significance, so that some assessment of the energy regimes operating during sedimentation can be made. However, interpretation of the palaeoenvironment where tractional structures are found has to be made with care, because aqueous tractional structures can form in any setting where water drives sediment. This occurs in a subaerial landscape where surface waters have gravitational potential (e.g. rivers), or in standing bodies of water (lakes, oceans) where water movement is induced by winds, tides, Coriolis force, or temperature or density differences. Tractional sedimentary structures can therefore form in a whole range of environments and water depths (e.g. alluvial fans, rivers, deltas, shorelines, shelf, deeper-marine) affected by bed-surface currents. They can also result from sub-aqueous (i.e. relatively deep water), non-viscous granular mass flow such as turbidity currents. Although these are initiated by gravitationally controlled downslope movement of sediment, this moving sediment quickly entrains ambient water. When the turbidity current slows and sediment is deposited, the entrained water has momentum. As the water decelerates it can therefore drive and rework the depositing sediment into tractional structures, which often form the upper parts of turbidite deposits (see below for further discussion). Tractional structures are therefore rarely indicative of any specific depositional environment, which thus has to be worked out from the overall facies characteristics of the succession.

Suspension and flotation

Suspension involves the long-term support of grains above the bed by the fluid (Fig. 10.16). Where the grains are silt and coarser, this takes place during turbulent flow conditions. Suspension of particulate granular sediment can occur during normal tractional sediment transport. Whereas coarser grains are moved in tractional form as discussed above, finer grains are lifted into suspended transport. During current deceleration, fall-out from suspension occurs and the sediment returns to the traction mode, resulting in standard tractional sedimentary structures. High flow velocities can lead to high volumes of sediment in suspension, which then begin to verge on mass-flow conditions, which will be discussed below. Under high enough velocities, even gravel-sized material can be lifted into suspended transport, even if only intermittently.

Where the sediment particles are clay platelets, suspension transport can occur in even very slow laminar flow conditions. This is due to their small mass and their plate-like form which give them buoyancy, but also because of the electrostatic interaction between the water molecules and the charged clay platelets, which produces an electrostatic force support effect. Deposition of suspended clay particles will only occur under very still conditions in a standing body of water (lake, abandoned channel, swamp, deep ocean), by settling of clays out of the water column, producing massive clay layers (Fig. 10.20). These may periodically be interrupted by coarser layers introduced by fluvial crevasse splay events, or subaqueous turbidity current transport, or some other short-lived, high energy sedimentation event.

These epiclastic processes all operate in volcanic terrains in the rivers, lakes, shorelines and seas associated with different volcanic settings. The products will be tractional deposits such as those illustrated in a variety of situations in Figure 10.19,

Figure 10.20 Grey-to-black suspension-deposited muds, and thin associated volcanic sandstone turbidites deposited in a lake, intruded elsewhere by rhyolite cryptodomes. Late Devonian Boyd Volcanic Complex, Bunga Head, New South Wales, Australia.

Figure 10.21 (a) Floating pumice on a roadside pond (note splash from thrown stone) east of Lake Taupo, New Zealand. The pumice is derived from the deposits of the major AD 186 Taupo eruption, discussed further in Chapters 7 & 8. (b) Old lake deposits exposed in a 5 m terrace, shores of Lake Taupo, New Zealand. The deposits consist of muds, peat layers, phreatoplinian ashes and pumice falls that may have settled out of suspension from a quiet water body.

and suspension fall-out muds such as those illustrated in Figures 10.20 and 21.

However, in volcanic settings material very much coarser than clays, even gravel-sized material, may also result from very slow settling out of the water column under quiet water conditions. The origin of this very coarse material is neither tree rafting nor ice rafting, but as self-rafted floating pumice. The bulk density of pumice may be significantly less than that of water because of the dominant void space. It therefore has enormous flotational potential (Fig. 10.21a). There are many accounts in the literature of pumice which has floated around the world before coming to land after specific marine volcanic eruptions (e.g. Richards 1958). Pumice can, of course, also float in lakes (e.g. the giant pumice blocks discussed in Ch. 13 from La Primavera volcano, Mexico), and can be washed onshore as a shoreline deposit. However, because of its porosity pumice will soak up a lot of water and may eventually become sufficiently waterlogged to sink like waterlogged wood. In a significant study of the hydraulic behaviour of cold and hot pumice clasts, Whitham and Sparks (*in press*) have shown that the water-logging of cold

pumice can, predictably, take a long time, and is dependent on the degree of interconnection between vesicles, their diameter and surface tension effects. Surprisingly, hot pumice can draw in water very rapidly as a result of the cooling and contraction of hot gases in vesicles when the pumice falls into cold water, and the resultant suction effect that draws water into the pumice. The settling rate will vary, depending on the rate of water intake and the rate at which the bulk density of the pumice and soaked-up water increase relative to the density of water. Gravel-sized floating pumice may therefore be hydraulically equivalent to clay platelets in terms of settling velocity and, while floating, it has no equivalence! Deposits resulting from this process should therefore be essentially mud sequences containing isolated outsized pumice clasts and even concentrations of pumice in layers (e.g. Fig. 10.21b), perhaps representing a significant event (pyroclastic eruption, epiclastic (e.g. flood) event) that delivered an increased volume of pumice to the water body.

In solution

In all environments natural waters carry dissolved chemical components that have been weathered or leached out of ambient rock matter. This should be a particularly important process in volcanic settings with their abundance of metastable glass in lavas, intrusives, and pyroclastic and epiclastic deposits. In surface and subsurface waters these chemical components may flocculate as clays, or may precipitate as other mineral species as surface deposits and pore fillings, given the right chemical conditions. Flocculation of clays may occur in surface waters at the interface between fresh and saline waters. Precipitation of siliceous sinter (e.g. Fig. 10.22) and sulphide minerals, often under the influence of biogenic agents such as bacteria and algae, may occur from hydrothermal systems. Although these processes are not strictly epiclastic, they are included here briefly to illustrate that the waters so frequently responsible for surface epiclastic transport are simultaneously or penecontemporaneously also moving considerable volumes of rock matter in dissolved form, both at the surface and in recirculating subsurface systems. The volumes of water and rock matter involved over a significant geological time interval should not be underestimated.

Fluvial torrent flow, sheetflood

In the above, reference was made to large volumes of suspended granular sediment under conditions of high velocity surface flow. These occur during peak flow or flood events in river systems experiencing upper flow regimes, when the river system has enormous erosive capacity. Under these conditions huge volumes of sediment are carried in suspension, supported by the turbulent eddies in the water (Fig. 10.23). With large suspended sediment loads, flow conditions are similar to those in subaqueous granular mass flows (Section 10.3.3), i.e. grains are supported by fluid turbulence and grain-to-grain collisions are common. The essential difference lies in the nature of the force driving water–sediment movement: in fluvial systems movement is due to the gravitational potential of the water and the sediment is entrained; in granular mass flow, the

Figure 10.22 Siliceous sinter curtain precipated from inactive hot-spring pond, Whakarewarewa thermal area, Rotorua, New Zealand. The front-central part of the curtain is covered with a film of sulphur.

opposite is the case. Nevertheless, when flow velocities wane in high energy river systems, rapid rates of sedimentation occur, leading in places to the deposition of massive, structureless deposits of granular sediment (e.g. the upper layer in Figs 10.19d & 24), or plane bed, horizontally laminated deposits (e.g. the lower layer in Fig. 10.19d), both representing upper flow regime conditions. Such high energy conditions are called torrent-flow conditions and are most likely in high energy river systems such as braided rivers, and entrenched highland river valleys where all runoff is concentrated into narrow steep-sided gorges and valleys.

Vessell and Davies (1981) and D. K. Davies *et al.* (1978c) described the fluvial dynamics and

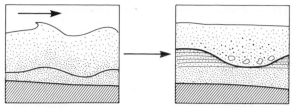

Figure 10.23 Torrent flow conditions in fluvial systems represent extremely high flow regime conditions corresponding with the antidune stage at the bed, whereas standing and upstream breaking waves characterise the state of the water surface. Resultant deposits are frequently massive to plane bed horizontally laminated, and significant erosional scour surfaces represent new flood cycles.

sedimentation history of the rivers draining from the Guatemalan stratovolcano Fuego following the 1974 eruption. Flow conditions are commonly in the upper flow regime. In the highlands part of the river systems, intermittent suspension transport occurs for all fine–medium sand, which constitutes 25% of the sediment load. Coarser sediment is moved as bed load. However, during peak flood events the granular sediment load in suspension would be even higher, and boulders 2–3 m in diameter are known to be moved as bed load. In the lower reaches, on lower gradients, the proportion of suspended granular sediment transport is lower.

The outwash rivers fed by melting glaciers may at times form meltwater floods and be very dynamic, experiencing torrent flow conditions. Under these circumstances they carry very high sediment loads, which may be instantaneously dumped upon rapid deceleration, producing massive, structureless deposits. In Iceland such floods, often triggered by sub-glacial eruptions, are called 'jökulhlaups' and these form extensive outwash fans of volcaniclastic sediment called 'sandurs' (Fig. 10.24b).

Sheetfloods commonly originate in areas of steep slope (>20°) and spread out onto relatively flat, lower slopes (<10°) after sudden torrential downpour. Water depths are a metre or more, and velocities reach 10 m s^{-1}. Flow conditions are therefore turbulent and highly erosive (Hogg 1982). Sheet floods dissipate quickly because they represent surges of water from a finite source. Sediment loads and sedimentation rates should be high, and upper flow regime conditions suggest that massive beds and horizontally laminated deposits should be common.

Subaqueous granular mass flow

Subaqueous granular mass flow refers to the downslope flow of a volume of cohesionless grains containing low viscosity, interstitial water, in a standing body of water (lakes, oceans). The mechanics of transport have been discussed by various authors including Middleton and Hampton (1976), Carter (1975), Lowe (1979, 1982), Nardin *et al.* (1979) and Allen (1982). Movement of the sediment mass is maintained by its own gravitational potential, and may begin initially by creep, sliding or more dramatic gravitational collapse. However, once in motion ambient water may be mixed into

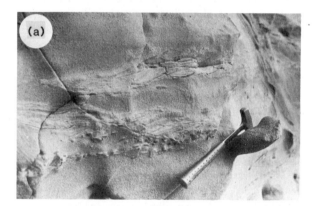

Figure 10.24 (a) Massive to very diffusely layered fluvial volcanic sandstone containing soft-sediment deformed bedded sandstone–mudstone intraclasts. See Figure 10.19d for setting. (b) Massive deposits of basaltic sediment dumped by flash floods of glacial meltwater or 'jökulhlaups'. Sandur near Vik, Iceland.

the sediment mass at its top, sides or be ingested along clefts in the flow-front. As a consequence, the volume of the sediment mass expands and fluid turbulence is initiated in the interstitial spaces within the flow. If the degree of expansion is low, grain-support in the flowing mass is due to both fluid turbulence and grain-to-grain collisions. If the flow is highly expanded with grains widely dispersed in the fluid medium, then they can only be supported by turbulence in the fluid. Although turbulence in the interstitial water is providing support for the grains, the water plays a passive role in the sense that it is not driving sediment motion, because its velocity and momentum is the same as that of the sediment, if not less. The water is thus entrained by the sediment. In fact, Middleton (1966) suggested that sediment movement in the body and head of such flows was faster than the speed at which the whole flow was advancing, leading to upward recirculation of the sediment in the flow-head.

Subaqueous granular mass flows like these, in which turbulence plays a major grain-support role, are called *turbidity currents* and their deposits are called *turbidites*. In turbidity currents with *low sediment concentrations* and high degrees of expansion, particle interaction is minor, so particles can be sorted according to their hydraulic properties. When sedimentation begins, the heaviest particles settle first. If all particles are of uniform density, the first particles to settle will also be the heaviest and largest, so producing a well defined upward fining or size-grading. Where particle densities are variable, for example due to the inclusion of variably vesicular grain types, size grading of the dense components may be accompanied by a density grading unrelated to size, perhaps producing reverse grading in the pumice population.

Other features of low sediment concentration, expanded turbidity currents include the common development of an upward sequence of structures reflecting the changing flow regime in the current as it decelerates. Bouma (1962) first recognised the regularity of the succession of structures (now known as the Bouma sequence, Fig. 10.25), but the significance of this in terms of declining flow regime

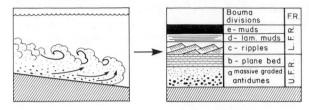

(a) Low concentration turbidity currents

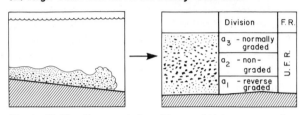

(b) High concentration turbidity currents

Figure 10.25 Characteristics of low and high concentration turbidity currents and their deposits. The resultant sedimentary structure divisions are interpreted in terms of the flow regime (F.R.) at the right (U.F.R. = upper flow regime, L.F.R. = lower flow regime).

conditions was first recognised by Harms and Fahnestock (1965) and R. G. Walker (1965) (Fig. 10.25). Complete Bouma sequences are not always present in outcrop, depending on distance from initiation point, flow distance, sediment load at that point, velocity at that point and degree of expansion. Relatively close to the initiation point the upper divisions are most frequently lacking (divisions c, d & e), and such turbidites are described as being proximal in character. Distal turbidites most frequently lack the basal divisions (a & b) and are found in settings distant from the source point, or in near-source areas, in overbank settings, between subaqueous distributary channels. When successive granular turbidite beds lack an intervening mudstone layer (divisions d and/or e), they are said to be amalgamated. These frequently occur as channel-fill deposits in proximal to medial distances from source point, or as the deposits of channel mouth lobes in medial to distal settings, associated with submarine fan systems (e.g. R. G. Walker 1984).

In turbidity currents with *high sediment concentrations* and low degrees of expansion, particle freedom is inhibited and size grading is only poorly developed, or is not developed at all, or is only represented by the coarsest and densest and heaviest grains (called coarse-tail grading). Large clasts can be supported by such a dense grain dispersion. Reverse-grading may also be developed in the basal part of such deposits, due to intense shearing in the basal part of the highly concentrated depositing grain mass. Internal shearing may also produce diffuse layering. These variations in the nature of the basal division of the Bouma sequence (division a) led Cas (1979) to suggest that this division could be subdivided into a basal reverse graded division a_1, a middle non-graded division a_2, and an upper normally graded division a_3, which is equivalent to division a for low concentration turbidity currents (Figs 10.25, 26a & b). Allen (1970b) also recognised that Bouma's (1962) sequence did not take account of very coarse sand to pebbly sand turbidites, which sometimes developed a lower cross-bedded division as well as an extra division of horizontal lamination (Fig. 10.26b). All of these variations were encompassed into an extended Bouma sequence for fine- and coarse-grained turbidites by Cas (1979) (Figs 10.25a & b), based on studies of volcaniclastic turbidites. R. G. Walker (1975) developed generalised, Bouma-like facies models for mass-flow conglomerate turbidites. The common development of reverse grading, no grading and normal grading in these suggests that these are also the deposits of highly concentrated turbidity currents, in which

clast suspension plays an important part. Importantly, these coarse-grained conglomeratic turbidites testify to the competency of turbidity currents to transport and support gravel-sized clasts by fluid turbulence as suggested by Komar (1970). The variations in the character of conglomeratic turbidite deposits suggest that a composite facies model along the lines of the extended Bouma sequence can also be *tentatively* made for these (Fig. 10.26c; R. G. Walker 1984) although such a model is not as confidently constructed as for sandy turbidites. Again, in any deposit not all divisions need be present. Lowe (1982) produced a comprehensive breakdown of the variations in the deposits of subaqueous mass-flow processes, with particular emphasis on those of high density, high particle concentration granular mass flows. In his scheme, Lowe distinguishes the effects of traction, traction carpet and suspension fall-out sedimentation, the first two developing in coarse sand to gravel deposits. Lowe's facies subdivisions from bottom to top in a theoretically complete sedimentation unit are: a basal gravelly, plane laminated to cross-stratified division (S_1-traction), a division of shear layers with reverse grading (S_2-traction carpet, equivalent to our a_1 division), a division of massive non-graded to normally graded, sometimes with water escape structures (S_3-suspension, equivalent to our a_2 and/or a_3 divisions), followed perhaps by

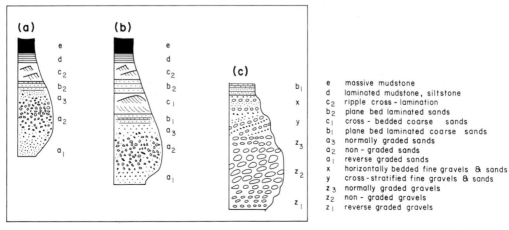

Figure 10.26 Three generalised, complete facies models for turbidity currents. (a) Silt to coarse sand turbidites (after Cas 1979). (b) Coarse sand to fine gravel turbidites (after Cas 1979, Allen 1970b). (c) Gravel turbidites, in which imbrication may be present in the gravels (after R. G. Walker 1975).

normal divisions of the Bouma sequence. Where a coarse, true gravel is involved, Lowe recognises a basal traction division (R_1), followed by a reverse graded layer (R_2, equivalent to our z_1 and z_2 divisions) and an upper normally graded division (R_3, equivalent to our z_2 and/or z_3 divisions), which could be followed by the S divisions as defined above.

It should also be mentioned here that short-lived turbidity currents can be generated on storm-dominated continental shelves. Onshore directed storm surges are balanced by offshore directed turbulent surge-like underflows. These carry shore-line and near-shore sediment offshore, depositing massive graded sands with sharp bases into areas of shelf mud sedimentation. Hummocky cross-stratification may occur, and the setting should be identifiable from fossils and trace fossils in rocks of the appropriate age. Gravels may also be trans-ported in this way. Storm-generated shelf-turbidites do not necessarily develop as much of the Bouma sequence as normal turbidites, and may be particu-larly deficient in divisions c and d. This is thought to be due to their relatively low degrees of expansion, the low slopes over which flow occurs and the low initial potential energy, all of which lead to rapid deceleration and rapid rates of sedimentation. (See R. G. Walker 1984 for further review; see Fig. 10.28f, below, for a volcaniclastic shelf turbidite.)

There has also been some debate in the literature as to the significance and likelihood of *liquefied* and *fluidised* granular mass flows (e.g. Middleton & Southard 1978, Allen 1982, Lowe 1982). Lique-faction is the process of the breaking of grain contacts in a loosely packed aggregate of cohesion-less grains without there being any volume increase in the aggregate. This can only occur in a very inefficiently packed aggregate. After breaking of contacts the grains are suspended in the interstitial fluid, supported by high excess pore pressures (viscous dispersive pressure), and the whole aggre-gate is in a 'quick', fluidal state. How far such liquefied aggregates can flow as such is uncertain. If unconfined, the interstitial water can escape. This process of escape itself gives grain support, because as the fluid streams out and up it exerts a lift force on the grains through which it streams. The streaming of a fluid through a clastic aggregate to provide grain support is called fluidisation, and has been discussed with respect to support processes in pyroclastic flows and surges in Chapter 7. The simple settling of dispersed grains causes upward displacement of fluid and very short-term self-fluidisation support, but it cannot be maintained for very long unless there is an external source for the fluidising fluid (Allen 1982), for example by ingestion of ambient water at the head of a granular mass flow. However, Allen (1971, 1982) suggests that relatively little fluidisation of this type actually occurs. Nevertheless, coarse granular mass-flow deposits do show evidence of liquefaction and fluidisation processes (e.g. dish-structures, pillar structures, which reflect water escape; Lowe 1975) which, if nothing else, reflect excess pore pressures *at the time of sedimentation*. However, it is doubtful that wholly liquefied and fluidised flows can travel very far without assistance from interstitial fluid turbulence, in which case they translate into highly concentrated turbidity currents.

Grain flows have also been assessed in the subaqueous realm (Middleton & Hampton 1976), but strictly grain flow, as discussed in Section 10.3.1 in subaerial settings, faces the same con-straints subaqueously. It involves very significant frictional interaction with the substrate and will therefore only operate on slopes near or at angle of repose. It does operate subaqueously (e.g. lee-side grain-flow forming cross-stratification in migrating ripples, sandwaves and dunes; subaqueous scree slopes), but once initiated subaqueously on long steep slopes it may evolve into a turbulent turbidity current. True grain-flow layers will be thin and reverse-graded (Lowe 1982).

Finally, *slurry flows* are essentially turbidity currents in which the interstitial fluid is not pure water, but contains a significant component of mixed clays (5–10+%), giving that interstitial fluid significant viscosity and strength. As such grain support is also provided by the strength of the fluid and the added buoyancy caused by the lessened density contrast between the fluid and the grains (e.g. Hampton 1972, 1975). However, with in-creasing viscosity, turbulence levels in the fluid are damped according to the Reynolds Number criterion (Ch. 2). Resultant deposits should be sandy mudstones or sandstones with abundant mud matrix. Intraclasts may be important throughout the bed, and tractional structures may be notably lacking due to suppressed turbulence (Hiscott & Middleton 1979; see Carter 1975 for more details). Slurry flows are clearly transitional between viscous subaqueous debris flows and low viscosity turbidity currents.

This lengthy discussion of subaqueous granular mass-flow processes is justified by virtue of the significance of such deposits in the sedimentary record. Volumetrically, they are dominant. In volcanic terrains they are also important. They operate in relatively deep lakes, especially caldera lakes supplied by steep, erodible margins or streams and deltas, and they operate in seas marginal to or hosting volcanic provinces. Accounts of the influences of subaqueous mass flows in marine volcanic settings are plentiful. They have been documented from modern oceanic settings, where they have been cored (e.g. Sigurdsson *et al.* 1980, Sparks & C. J. N. Wilson 1983, Klein 1975, Klein & Lee 1984, Carey & Sigurdsson 1984), and amongst the successions of ancient volcanic terrains ranging from the Tertiary to the Archaean (e.g. J. G. Jones 1967b, Mitchell 1970, Fiske 1963, Fiske & Matsuda 1964, J. V. Wright & Mutti 1981, Cas 1978b, 1979, 1983a, Cas *et al.* 1981, Tassé *et al.* 1978, Lajoie 1984, Ricketts *et al.* 1982). In all of these modern and ancient successions normal turbidite deposits of volcanic detritus have been commonly identified. In many the detritus is very juvenile in character, including unabraded cuspate glass shards (e.g. Cas 1979, J. V. Wright & Mutti 1981). This is important from two points of view. First, pyroclastically fragmented detritus is being redeposited by epiclastic mass-flow processes and, secondly, subaqueous mass-flow processes are capable of transporting fragile, pyroclastically fragmented detritus tens to hundreds of kilometres without significant abrasion. So two points raised initially in Chapter 1 – that modes of fragmentation and final deposition may be quite independent and have to be evaluated independently, and that subaqueous mass-flow processes can be non-abrasive and can transport detritus long distances without significant textural modification – are reinforced.

The nature of the supply process of the volcanic detritus may be variable and is difficult to determine in some cases. Sometimes the origin can be essentially epiclastic (epiclastic reworking, gravitational collapse and epiclastic mass-flow redeposition; e.g. J. G. Jones, 1967b, Mitchell 1970). In other cases the supply originates from subaerial pyroclastic activity. Pyroclastic air-fall and pyroclastic flow processes then transport the debris into water and it is then redeposited subaqueously by epiclastic mass-flow processes (e.g. Ch. 9; Sigurdsson *et al.* 1980, Cas 1979, 1983a, Cas *et al.* 1981). For example, Cas (1979, 1983a) has suggested that anomalously thick individual sedimentation units (metres to ten of metres thick) of crystal rich, juvenile volcanic detritus with rare preserved cuspate shards, were initiated when subaerial pyroclastic flows flowed into the sea, perhaps interacted explosively with it and were then reconstituted into huge, highly concentrated turbidity currents. This seems to have occurred several times, producing a coherent stratigraphic interval made up of multiple massive, graded to non-graded, very thick beds with all the characteristics of normal highly concentrated turbidites (Fig. 10.27), except for their huge thickness which is explained by the nature of the supply (see Ch. 11 for further discussion). Similar features were documented in coeval deposits by Cas *et al.* (1981) (Fig. 14.12b). In other cases the supply could be directly and indirectly from shallow subaqueous eruptions (e.g. Fiske 1963, Fiske & Matsuda 1964; Fig. 9.12), but nevertheless most of the deposition appears to involve redeposition by subaqueous, epiclastic, water-supported mass-flow processes rather than pyroclastic flows (Ch. 9).

The distinction between what is pyroclastic flow and what is epiclastic mass flow in such situations may seem academic, and has been discussed in considerable detail in Chapter 9. From the point of view of establishing criteria for palaeoenvironmental interpretation, it is an extremely important distinction. We will iterate here our ideas from Chapter 9. Except in very shallow settings, it has not been convincingly demonstrated that gas-supported

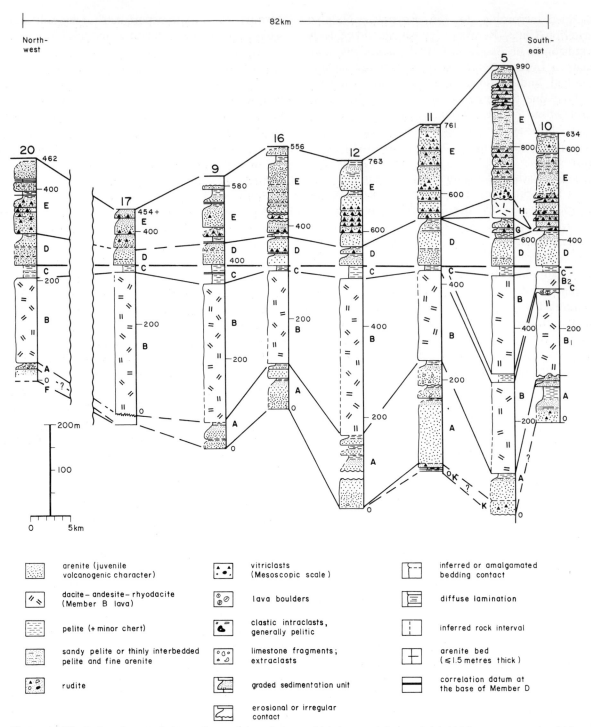

Figure 10.27 Facies characteristics and correlations for very thick (see scale) crystal-rich highly concentrated turbidity current flow deposits and intercalated lavas, Lower Devonian Merrions Tuff Formation, southeastern Australia. (After Cas 1978b.) Numbers at the tops of columns are section numbers, letters identify informal stratigraphic members.

pyroclastic flows can maintain their integrity sub-aqueously. We therefore favour epiclastic flow and grain-support mechanisms for relatively deep-water, coarse granular volcaniclastic deposits, unless they can be demonstrated to be of air-fall origin.

Unfortunately, there are numerous references to subaqueous and submarine pyroclastic flow deposits and events in the literature, when the deposits clearly bear the imprint of epiclastic mass-flow facies characteristics (e.g. Lajoie 1984). Perhaps the intention was to identify (epiclastic) flows of pyroclastic debris (i.e. debris originally fragmented pyroclastically, but redeposited by epiclastic mass-flow processes. Lajoie (1984) for example, illustrates numerous deposit types which are identical to the facies models described here in Figures 10.25 and 26 (see R. G. Walker 1984), but clearly labels them to be subaqueous pyroclastic flow deposits implying, perhaps unintentionally, gas support flow mechanisms for which there is no evidence. Similarly Niem (1977) and Yamada (1984) discuss subaqueous pyroclastic flow deposits and imply a pyroclastic flow mechanism. There is no doubt that fragmentation was by pyroclastic explosive activity, as shown by well defined shards in the case of Niem's successions. However, final transportation and deposition appear to have been from water-supported epiclastic mass-flow processes, as was the case for the shard bearing Merrions Tuff units of Cas (1979, 1983a) and the Dali Ash of J. V. Wright and Mutti (1981). Niem (1977) even documents tabular cross-stratification high in his depositional units, and flute marks at the bases of his units, suggesting aqueous turbulent flow conditions in the flows. Yamada (1984) describes the deposition of his units as being by turbidity currents, yet still calls them pyroclastic flow deposits. They would appear to be epiclastic turbidites transporting pyroclastic debris! That the detritus was originally fragmented pyroclastically is incidental to the fact that epiclastic mechanisms were responsible for final deposition, and we believe that this distinction should be clearly made. Both the Niem and Yamada accounts are also valuable in demonstrating the flotational capacity of pumice, which in places forms high level, reverse-graded pumice zones in depositional units.

The question of subaqueous air-fall deposits also brings into focus just how much of a subaqueous volcaniclastic succession might be air-fall and how much is epiclastic mass-flow deposits. The answer is that there is no consistent answer. For example, Sigurdsson et al. (1980) have shown that around the Lesser Antilles arc system, prevailing westerly wind systems disperse nearly all of the air-fall material eastwards, whereas westwards, behind the arc in the back-arc basin, a huge apron of redeposited mass-flow volcaniclastic sediment almost completely devoid of air-fall deposits is forming.

Finally, we provide illustrative documentation of this spectrum of subaqueous granular mass-flow epiclastic processes and facies from a variety of volcanic settings (Fig. 10.28). The foregoing has suggested that in any subaqueous granular mass-flow, a number of different grain-support mechanisms (Fig. 10.29a) may operate in combination, or at different times during the initiation and flow history (Fig. 10.29b). These include interstitial fluid turbulence, grain-to-grain collisions, fluidisation and liquefaction effects. Where interstitial fluids are viscous, fluid strength and buoyancy effects also operate (Fig. 10.29a). As a result, and also dependent on flow velocity and competence with distance travelled, a wide diversity of facies may result (Fig. 10.29b & c).

Mud flows, debris flows, lahars

These three flow types differ from the previous mass-flow types in being viscous. *Cohesive mud flow or debris flow* generally involves a cohesive mud fluid with a strength which is capable of supporting large clasts if available (Hampton 1975, 1979). The second element of support for large clasts is a buoyancy effect resulting from the low density contrast between clasts and high density mud fluid (cf. Archimedes' principle). A third source of support comes from the viscous dispersive pressure that builds up in the matrix mud fluid during near approaches of large clasts. Hampton (1979) suggests that, on this last count alone, the higher the proportion of large clasts, the greater the mobility of the viscous flow and the greater the competence to carry large clasts.

Viscous mass flows have been called mud flows

where coarse debris is absent, debris flows where very coarse to fine sediment is involved, or lahars where the detritus consists of contemporaneous volcanic debris. Appreciable granular sediment can be mixed with the mud fluid without affecting its viscous properties. As long as the granular sediment particles are not in mutual contact and locking, the overall internal friction of the mass is low and viscous flow can be maintained, slopes permitting. This has led to the recognition of debris flows with as little as 5% interstitial mud-water fluid, but apparently enough to lubricate the movement of the granular sediment component (Lowe 1979, 1982, Rodine & Johnson 1976). These have been called grain modified debris flows (Lowe 1979), a better term perhaps being *grain dominant debris flow*. According to Rodine and Johnson (1976), the support of large clasts in such debris flows occurs because of the poor sorting, and the resultant hierarchy of grainsizes. Each grainsize level, together with the finer grainsize levels and the interstitial fluid, provide support for the next coarser grainsize level, and so on.

A. M. Johnson (1970) has shown that debris flows usually move by combined laminar and plug flow (Ch. 2). The marginal laminar flow zone represents the zone of maximum shear at the interface between the substrate and the body of the debris flow. The inner part of the debris flow moves as an inert plug on the laminar zone, and can be viewed as an extreme laminar flow layer. Lateral levées and surface mound topography are not uncommon, indicating a high yield strength. Middleton and Southard (1978) suggested that thick debris flows may develop turbulent parts, and Lowe (1982) suggested possible facies characteristics for such deposits. Because of the cohesive nature of debris flows, tractional structures are unlikely. Subaqueous debris flows can transform into slurry flows and perhaps turbidity currents, provided sufficient ambient water can be mixed into the interstitial mud-fluid to reduce the fluid strength and viscosity sufficiently to allow interstitial fluid turbulence to develop (Hampton 1972). Debris flows occur subaerially and subaqueously in any setting where slopes are relatively high, slope instability prevails and water-saturated, fine-grained cohesive sediments, with or without associated coarse debris, are available. Debris flows are initiated by slumping–sliding on relatively steep slopes (Fig. 10.30) and are therefore commonly associated with alluvial fans, steep slopes with expandible clayey soils, marine slopes, delta fronts, upper parts of submarine fans and, of course, the slopes of both subaerial and subaqueous volcanoes.

Their mobility and competence should not be underestimated. They are known to flow several tens of kilometres, carrying very large boulders well away from the source. Speeds of 40 m s^{-1} were calculated for the 1980 Mt St Helens lahars.

Many documented accounts of debris-flows carrying volcanic debris in volcanic terrains are known, the majority being associated with stratovolcano terrains (e.g. Mullineaux & Crandell 1962, Mt St Helens; Fiske *et al.* 1963, Crandell 1971, Mt Rainier; Schmincke 1967b, eastern margin of Cascades arc; Janda *et al.* 1981, Mt St Helens). Again, the most widely published debris flows are those from the 18 May 1980 eruption of Mt St Helens, which were derived directly from the slide and avalanche event, and flooded the major distributary systems draining the volcano with huge volumes of debris.

Debris-flow deposits should be very poorly sorted, usually with large clasts in open framework organisation in a finer-grained matrix (Figs 10.30 & 31) containing significant, although apparently not necessarily large, amounts of clays. They are there-

◀ **Figure 10.28** (a) Conglomeratic base of submarine crystal-rich volcaniclastic turbidite of the Lower Devonian Merrions Tuff discussed in Cas (1979). Note large flame structure. The bed is several metres thick and shows coarse-tail grading. (b) Steeply dipping, white weathering pumiceous ash turbidites, Obispo Tuff, Twichell Dam, California. (c) Amalgamated basaltic volcaniclastic turbidites, Eocene–Oligocene Waiareka–Deborah volcanics, Oamaru, New Zealand. (d) Thinly bedded, amalgamated, pumiceous lacustrine turbidites, Quaternary Huka beds, Huka Falls, New Zealand. (e) Normally graded, 25 cm thick, lacustrine volcaniclastic turbidite, Late Devonian Boyd Volcanic Complex, Bunga Head, New South Wales, Australia. (f) Storm-generated shelf turbidites derived from a latite source. Tops of turbidites are highly bioturbated. Units are internally massive and graded. Permian 'Gerringong Volcanics', Broughton Formation, Kiama, New South Wales, Australia.

(a) Grain support mechanisms and flow types

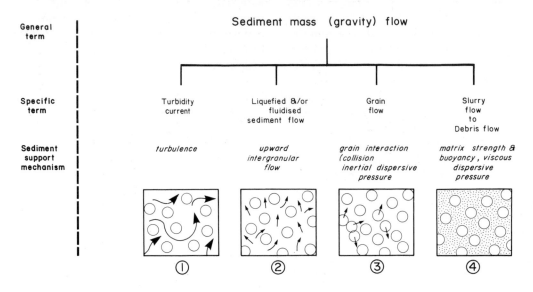

(b) Flow initiation, possible evolution and deposits

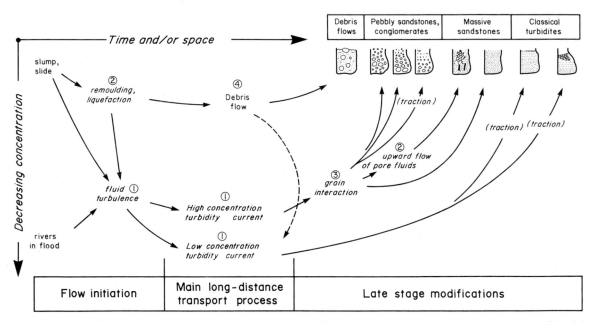

Figure 10.29 (a) Different grain-support mechanisms and related flow types in the realms of subaqueous mass flow (after Middleton & Southard 1978). (b) Modes of initiating subaqueous mass flows, possible evolution paths and generalised characteristics of deposits (after R. G. Walker 1978). The circled numbers correspond to the principal grain-support process operating, depicted in (a).

(c) Subaqueous mass-flow facies and process spectrum

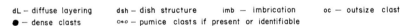

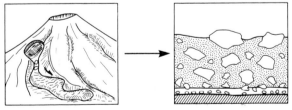

dL – diffuse layering   dsh – dish structure   imb – imbrication   oc – outsize clast

● – dense clasts   o●o – pumice clasts if present or identifiable

Figure 10.29 (continued)   (c) A generalised process and facies spectrum for subaqueous mass flow, depicting the general facies characteristics, the flow types, the physical properties of these, the grain-support processes operating and the general physical conditions during flow (after Cas 1977).

Figure 10.30 Initiation and general internal texture of debris flows.

fore a type of diamictite as defined in Section 10.3.2. Water seems to be a major lubricant in the whole system. Thicknesses of deposits may be less than a metre to several tens of metres. A basal shear zone layering may be developed (Fig. 10.31c), out of which most large clasts will have moved because of the effects of dispersive pressure in thin, shearing layers. Lahars are thought to accompany eruptions frequently, and as such they may carry hot debris, evidence of which may be found in charred woody remains, although such material could easily be picked up in transit by a cold debris flow. Lahars could easily have a generally similar texture to non-

Figure 10.31 (a) Volcanic debris flow (lahar) mantled by air-fall or surge deposits, Mt Egmont, New Zealand. Note open framework organisation of clasts in the matrix which contains a high proportion of sand-sized volcanic detritus. (b) Tangiwai railway bridge over the Whangaehu River, draining off Mt Ruapehu, New Zealand. On Christmas Eve 1954 the railway bridge was wiped out by a debris flow from Ruapehu, only minutes before a crowded train passed by and crashed into the river, with major loss of life. Boulders in the river may have been transported by debris flow, but the river has washed finer matrix away. (c) Semicircular erosion channel gouged out into underlying bedded volcaniclastic deposits by a marine grain-dominant debris flow, Bridge Point, Kakanui, New Zealand, Eocene–Oligocene Waiareka–Deborah volcanics. The base of the debris flow has a diffuse layering parallel to the outline of the channel representing the basal laminar shear flow zone. Large clasts are concentrated in the central massive plug zone. (d) Same debris flow deposit as in (c), showing open framework texture of basaltic blocks, including a prominent breadcrusted block. (From Cas & Landis in prep.)

welded ignimbrites, the major distinguishing criteria to look for are gas escape pipes (Ch. 7) and evidence of sintering and vapour-phase crystallisation in ignimbrites.

Slumps

Slumps are coherent masses of aggregate that creep, slide or flow down slope leaving an erosion scar behind, to which the tail of the slump is still attached. They are therefore local in significance and represent incipient slides, with which they form a continuum. Internally soft-sediment deformation may be apparent, but original layering is still preserved. The role of water in this case is to lubricate movement. Slumping can occur subaerially or subaqueously; in the former case slumping is not unusual after heavy rainfall.

Soil creep

Imperceptible, downslope creep of soil or surface aggregate, lubricated by interstitial water is common on all landscapes with steep slopes.

10.3.4 SEDIMENT TRANSPORT IN WHICH AIR IS AN ESSENTIAL INTERSTITIAL MEDIUM

The role of air in this case is either to drive sediment movement through the effects of wind, or perhaps to lubricate movement of rock avalanches. In particulate form wind moves sediment by:

traction and
suspension,

whereas in mass-movement it may be involved in:

air-lubricated rock avalanches.

Traction and suspension

These two modes will be discussed together since the basic principles have already been touched on in discussing the same modes subaqueously. The main difference is that because particles are not immersed in water their transportable weight is greater, and because the density and viscosity of air

Figure 10.32 (a) Coastal titanomagnetite-rich beach dunes, Bethel's Beach, Auckland, New Zealand. Sediment is eroded from Tertiary intermediate volcanics exposed in coastal cliffs. (b) The Taupo ignimbrite ponded against a palaeovalley margin made of eroded Oruanui ignimbrite (right) draped by a thick 1 m deposit of loess.

is less, it cannot exert as high a shear stress at equivalent velocities. As a result wind cannot move coarse granular material (except during hurricanes, etc.), the optimum sizes being fine–medium sand, although once eroded finer grainsizes are easily suspended. As with all of the other modes discussed, this is based on dense grainsize equivalents, and not on the hydraulic properties of low density pumiceous material, for which no information is known. The processes of bed load and saltation transport again constitute tractional transport, and are responsible for producing structures such as ripples and large-scale wind dune systems, the characteristics of which are adequately described by Brookfield (1984).

In volcanic terrains, wind erosion and transport can be major influences, particularly when newly erupted pyroclastic ash mantles the surface in an unconsolidated state. Richards (1965) mentions the extreme cloudiness of the atmosphere around Bárcena due to wind-borne ash, and records the development of wind-generated barchan dunes. In most cases these may only be ephemeral features, but in some circumstances they could be preserved. Again, angle of repose cross-stratification will be distinctive, and cross-stratification sets may be highly irregular in geometry with well defined bounding surfaces representing breaks in movement. Coastal erosion of volcanic successions could produce significant coastal beach dune successions of volcanic sands (e.g. Fig. 10.32a).

One major consequence of significant wind erosion in volcanic terrains may be to mantle the landscape with a blanket of ash, superficially similar to air-fall ash. Figure 10.32b depicts prehistoric loess mantling the New Zealand volcanic landscape. The loess was apparently mobilised during the influence of the recent ice age and associated strong winds. The loess shows mantle bedding and is relatively thick, and to the unwary could be misinterpreted as a blanket of pyroclastic airfall heralding a major eruption! Heaven help the geologist confronted with such a deposit in an ancient volcanic terrain!

Air-lubricated rock avalanches

In the discussion of rock avalanches in Section 10.3.1 mention was made of ideas that rock avalanches in some circumstances could glide on a trapped, compressed cushion of air, trapped at the hydraulic jump or break in slope where the rock mass changed from a fall trajectory to a lateral flow trajectory. In such a situation the air would also fluidise the flowing rock mass. While some support this mechanism (e.g. Shreve 1968), others discount it (e.g. Hsü 1975, 1978, Howard 1978) but the concept should not be totally rejected (Allen 1982).

10.4 Further reading

This chapter has attempted to provide a general overview of principles involved in epiclastic transport and deposition and we have shown by way of example that almost every process can be found represented by deposits in the record of volcanic terrains. However, the content of the chapter should only be considered as an introduction to the subject, and should be followed by more-detailed reading where necessary. In particular, no discussion has been possible of facies models for the overall sediment successions of specific depositional environments, for which the reader is referred to excellent sources such as R. G. Walker (1984) and Reading (1978). For discussion of principles and mechanics of sediment movement Allen (1970a, 1982), Leeder (1982), Middleton and Southard (1978) and Harms *et al.* (1982) are recommended. Few reviews of epiclastic processes and influences in volcanic settings are known to us but Fisher (1984) has provided a useful evaluation of the interaction between sedimentation and volcanism in submarine environments, as has Lajoie (1984). However, in both these reviews the term 'subaqueous pyroclastic flow' has to be interpreted with care.

Plate 11 Photomicrograph of crystal-rich aggregate from the Lower Devonian Merrions Tuff, central western New South Wales, Australia. Note the juvenile grains of volcanic quartz, microperthitic orthoclase and albitised plagioclase in closed framework organisation.

CHAPTER ELEVEN

Crystal-rich volcaniclastics– pyroclastic or epiclastic?

Initial statement

Crystal-rich volcaniclastics are not uncommon in volcanic terrains or in the surrounding basins that derive their sediment from volcanic terrains. In this chapter we show that their origins may be diverse, involving the influence of pyroclastic processes, or epiclastic processes, or combinations of both. Having dealt with pyroclastic and epiclastic processes in preceding chapters, we can thus use this type of deposit to illustrate the relationships between, and the possible interface between these two groups of processes, and the importance of evaluating the role of each group of processes in volcanic terrains. Crystal-rich volcaniclastics are frequently called 'crystal tuffs' without careful evaluation of the possible diverse origins, or without consideration of the genetic implications of such terminology. In this consideration of crystal-rich volcaniclastics, and especially the use of the term 'crystal tuff', we can consider some of the problems of classification and terminology as an introduction to a comprehensive discussion on terminology in Chapter 12.

11.1 Introduction

Having dealt with pyroclastic and epiclastic processes and products in previous chapters, we now look at one particular lithofacies, crystal-rich volcaniclastics, as a way of showing that particular lithofacies can have diverse origins. Like volcaniclastics rich in glassy or lithic components, crystal-rich volcaniclastics can be produced by either pyroclastic processes or epiclastic processes, or by a combination of the two. They are not uncommon in volcanic terrains and in the basins peripheral to volcanic terrains, so evaluation of the origins of specific deposits requires familiarity with the pyroclastic and epiclastic processes that produce crystal-rich volcaniclastics.

Given that crystal-rich volcaniclastics can have diverse origins, there also exists a problem of nomenclature, in that the name applied should clearly reflect both the mode of fragmentation and the mode of transport and deposition. Crystal-rich volcaniclastics have frequently been called 'crystal tuffs' (or 'crystal-lithic tuffs' where lithic components are also prominent), in the stratigraphic literature, and whether intended or not, this has specific genetic implications, indicating both a pyroclastic mode of fragmentation and a pyroclastic mode of transportation and deposition. It is therefore also appropriate here to evaluate the usage of the term 'crystal tuff', as a prelude to a comprehensive discussion of classification and nomenclature in Chapter 12.

11.2 Three types of ash and tuff

Ashes and their lithified equivalents, tuffs, have for a long time been subdivided into three types: vitric, lithic and crystal ashes and tuffs (Pirrson 1915, Holmes 1920, H. Williams *et al.* 1954, Carozzi

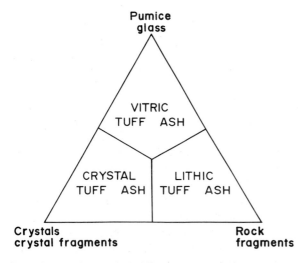

Figure 11.1 Standard classification of ashes and tuffs into vitric, lithic and crystal types. (After Schmid 1981.)

1960, MacDonald 1972, Pettijohn 1975, H. Williams & McBirney 1979), which serve as end members in the formal classification of ashes and tuffs (Pettijohn 1975, Schmid 1981; Fig. 11.1).

By definition, ashes and tuffs are pyroclastic deposits with an average grainsize of 2 mm or less (Ch. 12). They are deposited by pyroclastic processes (Chs 5–9, inclusive) and are not reworked or redeposited by epiclastic processes (Ch. 10), in which case they would essentially be tuffaceous sediments (Ch. 12). The three principal components of explosive eruptions – cognate glassy clasts, crystals and lithic fragments can be present in varying combinations in ash deposits (Heiken 1972, 1974). Their abundance depends first on the make-up of the magma at the time of eruption, including its degree of crystallisation, secondly on the type of explosive activity (Ch. 3) and thirdly on the effects of sorting processes during pyroclastic eruptions.

Vitric ashes and tuffs are dominated by largely uncrystallised glassy fragments with an average grainsize less than 2 mm. They are the products of the explosive eruption of poorly crystallised magmas, and/or they are very fine distal ashes deposited far from the vent, owing their vitric character to sorting in the eruption column and plume which has left behind coarser and denser crystals and lithic fragments closer to the vent, or to elutriation of fines from pyroclastic flows, resulting in crystal enriched ignimbrites and vitric co-ignimbrite ashes (Chs 5 & 8). In explosive eruptions, in which magmatic explosions are dominant, the fragments will be highly vesiculated pumice or scoria, or cuspate shards which are the fragmented walls of vesicles (Ch. 3; Figs 3.23a & 24; Heiken 1972, 1974). In eruptions where phreatomagmatic explosions are dominant the fragments may be less vesiculated, blocky, glassy fragments (Ch. 3; Fig. 3.18; Heiken 1972, 1974, Self & Sparks 1978, Sheridan & Marshall 1983).

Lithic ashes and tuffs are dominated by variably crystallised volcanic rock fragments of cognate character and, frequently, also by fragments of older lithified igneous (or other) rocks, eroded from the walls of the vent as accessory inclusions during explosive eruptions of magmatic, phreatomagmatic or phreatic origin (Heiken 1972, 1974). The influ-

Figure 11.2 'Crystal tuff' from the Carboniferous Tulcumba Sandstone, Keepit Dam, New South Wales, Australia. Note the resorption embayments in some quartz and feldspar grains. Crystals are juvenile and the matrix is still glassy.

ence of magmatic explosions should be reflected by highly vesiculated cognate fragments, whereas phreatomagmatic origins should be reflected by less vesiculated cognate fragments and often also by increasing proportions of accessory clasts.

Crystal ashes and tuffs, by definition, should contain a dominance of crystal grains over vitric and lithic fragments (Figs 11.1 & 2; Plate 11). The crystals should be discrete fragments with little or no attached groundmass selvedge. The influence of magmatic explosions should be reflected by shards and highly vesiculated cognate fragments. Crystal-rich ashes and tuffs could be produced by both magmatic and phreatomagmatic eruptions, but special conditions attend, particularly in the formation of extremely crystal-rich pyroclastic deposits, as discussed further below.

Of the three, it is acknowledged by early workers that crystal tuffs are rarest and probably require relatively special conditions of formation. Holmes (1920) credits Cohen in 1871 with the introduction of the term 'crystal tuff', and it is explicit in early definitions and discussions of the characteristics and origins of tuffs, that crystal tuffs originate as near-vent ash-fall deposits derived from the explosive disruption of crystal-rich magmas or crystallised igneous country rock, or both (Pirrson 1915).

Such origins are demonstrable for some modern crystal-rich ashes which mantle the flanks of modern active volcanoes, their close spatial relationship to the vent being one of the most compelling reasons for invoking a near-vent ash-fall origin (e.g. the samples cited by H. Williams *et al.* 1954). The term crystal tuff pervades the general stratigraphic geological literature on ancient volcanic successions, often without clear qualification of the origins of the deposit in question. In some instances, the term has been clearly applied to rocks that are strictly redeposited tuffaceous sandstones, and even to very labile, turbidite greywackes. Unless qualified the term 'crystal tuff' can only be taken to mean that the relevant crystal-rich volcaniclastic aggregates are of ash-fall origin and that they are derived directly from the explosive eruption of crystal-rich magma. However, work in recent years is beginning to show that 'crystal tuffs' also occur as pyroclastic flow deposits. In addition, eruption-related pyroclastic processes may be responsible for causing the apparently crystal-rich nature of many 'crystal tuffs', and equally importantly, epiclastic processes involving reworking or redeposition (note the distinction), or both, may also be involved in the genesis of so-called 'crystal tuffs'. However, with the exception of the recent paper by Cas (1983a), there has been little or no attempt to review the possible origins of the spectrum of crystal-rich volcaniclastic aggregates, and of the conditions needed to produce true crystal tuffs.

11.3 Possible fragmentation and transportation modes for crystal-rich volcaniclastic deposits

In the book so far, we have looked separately at the processes that produce pyroclastic deposits (Chs 5–9) and those that produce epiclastic deposits (Ch. 10). From this it should be apparent that different processes can produce superficially similar lithofacies. In particular, aggregates of pyroclastic debris that have not been significantly reworked, but have been redeposited by epiclastic processes shortly after eruption, may superficially resemble primary pyroclastic deposits (those that have both been fragmented, transported and deposited by pyroclastic processes). For example, epiclastic mass-flow processes can redeposit pyroclastic debris and produce deposits that are internally massive and structureless, and which resemble pyroclastic flow deposits (Ch. 9 & 10). Proximal, fine-grained ash turbidites could be mistaken for aqueously deposited air-fall deposits (Ch. 9). Cross-stratified epiclastics could be mistaken for pyroclastic surge deposits, and so on. It is therefore important to distinguish carefully the possible modes of fragmentation and the possible modes of transportation and deposition independently, before a final interpretation of the genesis is made (Ch. 1) and before a genetic name is given to the deposit (Ch. 12). In this section we briefly recapitulate the possible modes of fragmentation of volcaniclastics (Ch. 3), and their possible modes of transportation and deposition (Chs 5–10). The appropriate chapters should be referred to for more detail, and for appropriate facies characteristics.

A further point is that both pyroclastic and epiclastic transportation processes are capable of transporting their load tens of kilometres or more from the source, and so the identification of pyroclastic material, no matter how coarse, does not mean that the depositional site was close to the source vent. This can only be established on the basis of important associations of facies (Ch. 14).

Therefore, crystal-rich volcaniclastic aggregates are not necessarily pyroclastic in origin or deposited by pyroclastic processes such as fall, flow and surge, and they need not indicate proximity to the vent. This has to be demonstrated rather than assumed. Of the principal modes of producing volcaniclastic aggregates (Ch. 3; magmatic explosions, phreatomagmatic explosions, quench-fragmentation, autobrecciation and epiclastic erosional processes), only magmatic explosions, phreatomagmatic explosions and epiclastic erosional processes seem capable of producing large volumes of ash-sized volcaniclastic aggregates (Ch. 3). For crystal-rich volcaniclastic aggregates, it also seems necessary to be dealing with at least moderately crystallised magma or volcanic rock sources. However, the final aggregate may contain significantly different crystal contents relative to the source magma or rock as a result of efficient fractionation and sorting processes operating during eruption or transportation and deposition (see Section 11.4.2).

Pyroclastically fragmented, crystal-rich aggregates should be recognisable as an homogeneous, coherent assemblage of grain types of which the glassy clasts will be variably vesiculated and, for conclusive interpretation, should be dominated by abundant shards, particularly in the case of magmatically fragmented debris. Epiclastically fragmented debris should be characterised by relatively high degrees of sorting and rounding compared with pyroclastic deposits, and the grain population may be compositionally heterogeneous.

The mode of deposition may be by pyroclastic flow, pyroclastic fall, pyroclastic surge (Chs 5–9) or epiclastic processes (Ch. 10), each of which should be identifiable from **field facies characteristics**, particularly outcrop-scale textures and sedimentary structures, as long as metamorphic and deformational overprints are not too strong. Even the presence of abundant shards is *not* diagnostic of pyroclastic transportation *and deposition* processes although it is indicative of pyroclastic fragmentation. Subaqueous turbidity currents, for example, can transport even fragile shard fragments for long distances with little or no physical abrasion (e.g. Cas 1979, J. V. Wright & Mutti 1981) because the transportation time is short, the transport mechanism is turbulent suspension, so limiting abrasive particle interaction, and because the viscosity of the interstitial water cushions grain impacts. In each case, therefore, the depositional and transportational mode has to be assessed independently of the mode of fragmentation, and this must be done using field facies characteristics.

Perhaps the biggest problem in distinguishing between pyroclastic and epiclastic transportation influences relates to distinguishing pyroclastic flow deposits from redeposited pyroclastic deposits involving high concentration granular mass flow or debris flow, as already discussed (Chs 9 & 10). The former should be recognised by evidence for a hot state of emplacement, including *microscopic* evidence of welding (see Ch. 9), columnar jointing and gas segregation pipes (see Ch. 7). In particular, non-welded pyroclastic flow deposits may be difficult to distinguish from the mentioned types of epiclastic mass-flow deposits because they will lack evidence of welding and columnar jointing. In ancient deposits, epiclastic origins will have to be equally entertained with pyroclastic flow origins.

11.4 Factors influencing high crystal concentrations

The various factors that can effect a high crystal concentration in volcaniclastic aggregates are:

eruption of highly crystallised magmas
physical fractionation and sorting processes associated with pyroclastic eruption and transportation processes
epiclastic reworking and redeposition

11.4.1 ERUPTION OF HIGHLY CRYSTALLISED MAGMAS

At face value, the occurrence of volcaniclastic aggregates rich in crystal grains and fragments implies the need for highly crystallised magmas to erupt at the Earth's surface. However, van der Molen and Paterson (1979) suggested that crystal-rich magmas have difficulty in reaching the surface, because when the crystal content is about 60–65% a contact framework of crystals is produced. The bulk viscosity of the magma is increased (Marsh 1981), and so is the internal frictional resistance to flow, so preventing the magma from rising through the crust to the surface. Marsh (1981) suggests that this critical crystallinity limit is about 55% for basaltic magmas. At crystal abundances below this, basic magmas have the potential to be erupted as lava. At higher concentrations than 55% crystals, they are destined to become intrusive bodies (Marsh 1981). For such basaltic magmas with their low silica content, it is essentially the crystal abundance which exerts the dominant control on bulk viscosity and, hence, magma mobility (Ch. 2). However, for more-silicic magmas the inherently high silica content also controls the viscosity (Ch. 2). Therefore, for silicic magmas the critical limiting crystallinity necessary to render the magma immobile will be significantly less than for basaltic magmas (Marsh 1981).

Marsh's reasoning is supported by the rarity of lavas with 50% or more phenocrysts. For example, Ewart (1979) collated the modal phenocryst abundances for some 1650 volcanics with SiO_2 contents greater than 60% (Fig. 11.3). Only two specimens exceed 60% phenocryst content, and only 38 exceed 50% crystal content. However, it is unclear from Ewart's data whether the volcanics are lavas or pyroclastics, or what the proportions of each are.

From this, it is unlikely that highly crystallised magmas or their extrusive derivatives will have risen in a highly crystallised state from any significant depth in the subsurface (say, >3–4 km). It is therefore probable that moderately to highly crystallised magmas that are erupted will have crystallised wholly or at least partially in relatively *shallow* magma chambers (e.g. Clemens & Wall 1984) to

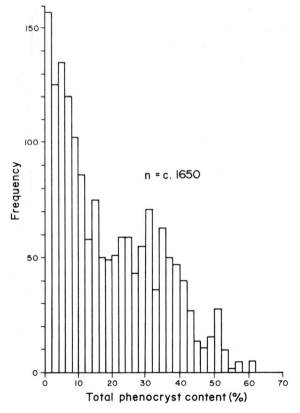

Figure 11.3 Total crystal contents for some 1650 Cainozoic volcanic rocks from the circum-Pacific region plotted from the data in Ewart (1979).

preclude the need to migrate far while in a rheologically unsuitable state. Although Marsh's findings preclude eruption of highly crystallised *lavas*, it is still possible that highly crystallised *magmas* may erupt explosively to produce crystal-rich pyroclastics (Marsh 1981) from the shallow near-surface magma chambers where they crystallised. The energy for explosive activity would come largely from the volatiles and latent heat released as a consequence of second boiling (crystallisation), and explosive eruption would occur when the pressure in the magma chamber equalled the confining pressure and the tensile strength of the country rock (Ch. 3; Burnham 1979, 1983). Collapse of the roof of the magma chamber, the cause of caldera collapse, could produce rapid explosive eruption of highly crystallised magma, producing crystal-rich air falls or crystal-rich ignimbrites, or both (e.g. Steven & Lipman 1976, Clemens & Wall 1981, Birch 1978, R. J. Roberts & Peterson 1961), because the release of pressure causes rapid boiling of the magmatic volatiles, leading to explosive expansion.

Even so, there are few ignimbrites with greater than 60 wt% phenocrysts. For most outflow ignimbrites at least, it can be argued that they have undergone crystal concentration processes during and after eruption (see below) so that the original magma crystal concentration at the time of eruption would have been less than that preserved in the ignimbrites.

Other implications can also be made about the modes of fragmentation of magmas with different crystal contents and compositions. Since crystal-rich magmas are likely to have crystallised over some time in relatively shallow but stable magma chambers, and because their high bulk viscosity prevents spontaneous rise towards the surface, phreatomagmatic causes for triggering explosive eruption are unlikely. Explosive eruptions of such highly crystallised magmas are therefore likely to be exceedingly rapid, and to be due to second boiling associated with crystallisation, and exsolution of magmatic volatiles from subsurface magma chambers as discussed above. Explosive eruption of poorly to moderately crystallised magmas (<40% crystals), on the other hand, may be triggered by either magmatic or phreatomagmatic explosions, depending on the circumstances. Being less crystal-rich they are mobile enough to rise through and reside in shallow crustal levels *en route* to the surface, and so interact with ground or surface waters, or both.

R. J. Roberts and Peterson (1961) noted that welded crystal-rich ignimbrites in the Western USA are very largely dacitic to rhyodacitic in composition, whereas welded crystal-poor ignimbrites are largely rhyolitic. This is apparently due to rhyolitic magmas being near-minimum melts for which the interval between the solidus and liquidus is narrow, particularly at high levels of water activity (Tuttle & Bowen 1958, Wyllie 1977). This means that the temperature range over which crystallisation can occur is narrow, implying that unless eruption of a rhyolitic magma occurs very soon after its formation, or the magma temperature is initially well above its liquidus, it may crystallise rapidly (depending on the size of the chamber), and it may become too crystal-rich to migrate upwards through the crust and erupt, especially if crystallisation occurs internally, rather than just along the walls of the chamber. On the other hand, more-calcic magmas such as dacites and rhyodacites have a wider solidus–liquidus interval, even over a wide range of water activities, and so are likely to experience a more protracted crystallisation history, allowing migration or eruption, or both, over a range of crystal abundances (V. J. Wall *pers. comm.*).

11.4.2 ERUPTION-RELATED CRYSTAL CONCENTRATION PROCESSES

Hay (1959), G. P. L. Walker (1972) and Sparks and Walker (1977) have shown that during pyroclastic eruptions much of the fine glassy ash (representing a significant part of the uncrystallised part of the magma at the time of eruption) can be lost by physical fractionation processes accompanying eruption and transportation of the tephra. The fractionation processes separate some of the fine glassy ash from the coarser, denser crystal and lithic fragments. This fractionation and sorting takes place first in the eruption column and, secondly, in pyroclastic flows, if these should be formed during eruption. It occurs because vesiculated glass fragments into much smaller, lighter particles than associated crystals which have greater strength.

Turbulence and convective circulation in the eruption column may winnow out a large proportion of fine glassy ash, carrying it into the upper part of the eruption column and then into a downwind eruption plume (Chs 5 & 6). Crystals and lithics therefore become more concentrated within the main body of the eruption column. From there they may either fall-out relatively close to the vent as a crystal-enriched ash, or alternatively they may be incorporated into pyroclastic flows (Chs 5, 6 & 8). During pyroclastic flow, elutriation of fine glassy ash from the fluidised head and body of the flow takes place, producing an overriding and trailing ash cloud fed by gas and ash escaping from the flow (Ch. 7).

Both column- and flow-related fractionation processes appear to have the capacity to remove huge volumes of glassy ash. G. P. L. Walker (1972) and Sparks and Walker (1977) have calculated that 50% of the erupted volume of a magma can be lost in these ways. This calculation comes from comparing the proportion of crystals in pumice clasts,

which represent whole or bulk samples of the erupting magma, with that in the body or matrix of the ignimbrite. This is calculated on a weight percentage basis to establish dense-weight equivalents for crystals and glass, and to overcome the problem of vesicularity or porosity. The studies of G. P. L. Walker (1972) and Sparks and Walker (1977) show that the matrix of ignimbrites, especially of outflow ignimbrites, is consistently enriched in crystals relative to the whole pumice fragments (Fig. 11.4). This difference seems to be minor in some caldera-fill ignimbrites (e.g. Birch 1978, Clemens & Wall 1981), suggesting that perhaps the eruption of these did not produce a large, well defined eruption column, implying that the ignimbrites were not generated by collapse of a high maintained column (Chs 6 & 8).

G. P. L. Walker (1979) suggested that a third-stage pyroclastic crystal concentration process could have been involved in producing the 42 000-year-old Rotoehu ash of the North Island of New Zealand (Nairn 1972). He suggested that some pyroclastic flows associated with the deposition of the ignimbrites of the coeval Rotoiti Breccia may have flowed into the Bay of Plenty along the northern margin of the North Island. By this stage

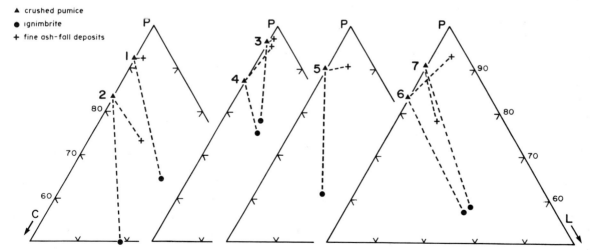

Figure 11.4 Graphical representation of the proportions of glass (P), crystals (C) and lithics (L) in samples of artificially crushed pumice (triangles) taken from within ignimbrites, and therefore representing whole rock samples of the erupting magma, ignimbrite (closed circles) and associated co-ignimbrite ashes (crosses). Samples 1–6 are from ignimbrites and ashes of Vulsini volcano, central Italy. Sample 7 is from the Minoan ignimbrite and its co-ignimbrite ash, Santorini. (After Sparks & Walker 1977.)

they would presumably already have undergone the fractionation processes within the eruption column, and subsequently in the pyroclastic flows. On flowing into the Bay of Plenty, Walker argues that they interacted explosively with sea water producing rootless vents (Ch. 3) and secondary eruption columns in which fine ash would again have been separated and from which the crystal-rich Rotoehu ashes were dispersed. These stages of progressive crystal-concentration are again reflected by the relative proportions of crystals in various products of this apparently complex eruption. Crystal contents are lowest in pumice fragments from Rotoiti Breccia ignimbrites and fall deposits, these pumices representing whole, unmodified samples of the erupting magma. Crystal abundances are higher in the matrix of the Rotoiti Breccia ignimbrites, which is the depositional residue of elutriation and sorting processes operating first in the eruption column and, secondly, during pyroclastic flow. Crystal contents are highest in the Rotoehu ash, the inferred product of secondary rootless vents and eruption columns (Fig. 11.5).

11.4.3 EPICLASTIC CRYSTAL CONCENTRATION PROCESSES

Epiclastic processes can concentrate crystal fragments in several ways. The two principal modes are normal reworking, especially of poorly consolidated and unwelded ignimbrites and fall deposits, and redeposition by relatively non-viscous regime massflow processes such as turbidity currents and fluidised sediment flows (Ch. 10).

Reworking has two principal effects in this regard. First, epiclastic reworking can sort grains of different size or hydraulic equivalence, or both (Ch. 1). A large amount of fine ash and highly vesicular glassy fragments should therefore be readily separated from denser crystal and lithic fragments. Secondly, the abrasion effects associated with reworking will abrade fragile, irregularly shaped (especially vesicular) glassy fragments more readily than crystals. Any degree of abrasive reworking should therefore lead to physical breakdown and removal of vesicular glassy fragments and concentration of crystals and dense lithics (e.g.

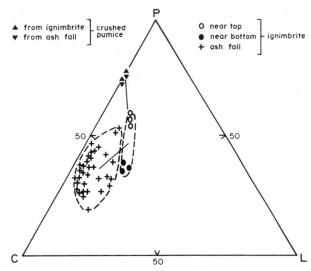

Figure 11.5 Graphical representation of the proportions of glass (P), crystals (C) and lithics (L) in the Rotoehu ash (unit G) and Rotoiti Breccia ignimbrite, and pumice in both for the sum of grainsize classes 0.25 mm and coarser. (After G. P. L. Walker 1979.)

Roobol 1976). The effects of this should also be reflected by rounding of the corners and edges of crystals. However, with volcanic detritus, not all rounding is necessarily due to epiclastic reworking. Magmatic resorption may also cause smoothing of the corners and edges of a whole array of mineral crystals before eruption (Fig. 11.2).

It is also possible for crystal-rich clastic aggregates to be produced by erosion of plutonic sources. In such cases identification of the plutonic origins would entail establishing: the presence of plutonic quartz (irregular shape, fluid vacuoles, semicomposite character, slight undulose extinction; see Folk 1980, Blatt *et al.* 1980 for details) rather than clear, unstrained bipyramidal volcanic quartz, which frequently has marked resorption embayments (Fig. 11.2); the presence of feldspar types such as microcline or coarse perthite, which can only develop in plutonic bodies; a high proportion of subhedral to anhedral crystals due to mutual interference between crystals during growth in the plutonic situation; and the occurrence of plutonic rock fragments. The epiclastic origins should also be identifiable from relatively good sorting and rounding where reworking has been significant.

Subaqueous *redeposition* of volcaniclastic aggregates by relatively non-viscous mass-flow processes such as turbidity currents and fluidised sediment flows (Ch. 10) can also produce significant separation of fine ash from denser crystals and lithics. Water ingested at the head of the flow streams backwards from the head through the body and carries fine ashy sediment into the dilute and turbulent trailing sediment cloud (Sparks & C. J. N. Wilson 1983). The coarser, denser fraction, such as crystals and lithics, is deposited from the head and body as a massive, variably graded layer, grading upward into the fine suspension deposits of the trailing cloud. Such deposits are volcaniclastic turbidites and may contain several or more of the Bouma succession of sedimentary structural zones (Bouma 1962, R. G. Walker 1984; Ch. 10). Ash turbidites could be several centimetres thick (Sparks & C. J. N. Wilson 1983) or tens of metres thick (Cas 1979, J. V. Wright & Mutti 1981; Ch. 9) and could be deposited up to hundreds of kilometres from source (e.g. Sigurdsson et al. 1980, Sparks & C. J. N. Wilson 1983).

11.5 Several 'crystal tuff' deposits and their interpretation

On the basis of the previous discussion it seems desirable to consider several 'crystal tuff' deposits with diverse origins, these being discussed according to whether the genesis was essentially pyroclastic, mixed pyroclastic and epiclastic, or essentially epiclastic.

11.5.1 CRYSTAL TUFFS OF PYROCLASTIC ORIGINS

In this category, deposits whose modes of fragmentation, transportation and deposition were purely pyroclastic are included. Deposits such as crystal-rich air-fall ashes and tuffs, crystal-rich surge deposits and crystal-rich pyroclastic flow deposits (ignimbrites) are also included.

G. P. L. Walker (1971) and G. P. L. Walker and Croasdale (1971) have described crystal-enrichment processes associated with *air-fall de-*

posits of the Fogo A Member in the Azores (also see Chs 5 & 6). The air-fall deposit is essentially a pumice-fall deposit but, in the middle distances of its dispersal area, it shows distinctive crystal enrichment (up to 46 wt% crystals) associated with sorting processes in the upper, spreading, umbrella-like part of the eruption plume. The explanation is that large pumices drop out of the plume close to the vent, while smaller particles, with lower terminal fall velocities (Ch. 6) are carried further, leading to increased proportions of crystals further from the vent. In a similar way, pumice that is lighter than crystals, and especially fine, glassy, shard-rich ash, will occur as an enriched component in the ashes most distant from source because the terminal fall velocity of these elements is less than that of the bulk of larger, heavier crystals. The Fogo A deposit is essentially a massive, structureless deposit, interrupted only by thin pyroclastic flow units. Although its characteristics have been described only from onshore exposures, a significant part must have fallen offshore as water-lain ash.

G. P. L. Walker (1979) also interpreted the very crystal-rich deposits of the Rotoehu Ash of New Zealand as air-fall deposits resulting from secondary, rootless vent eruption columns generated when pyroclastic flows flowed into the Bay of Plenty. The several phases of crystal concentration, as described above, led to concentrations of crystals to as high as 60 wt% and more, in the coarsest of several beds in the Rotoehu Ash, especially Bed G. Bed G is described as 'conspicuously rich in crystals, resembling a waterlain sand. The relative uniformity of grainsize and thickness over a great area, the perfect mantle bedding, and the lack of rounding of the grains, however, leave no doubt as to its shower origin' (G. P. L. Walker 1979).

Crystal-rich *surge deposits* have also been documented. Fisher (1979), for example, described ash-cloud surge deposits (Chs 5 & 7) from the Upper Bandelier Tuff (Ch. 8), in which distinct crystal-rich laminae occurred as parts of planar to cross-stratified bed sets. The surge deposit occurs as a variably thin (2–35 cm), discontinuous, lensoidal horizon between two ignimbrite cooling units. The surge deposit is gradational into the underlying cooling unit, interdigating with it, and also contains gas segregation pipes, continuous from the underlying ignimbrite, and so also indicating a contemporaneous relationship. The crystal enrichment in these surge deposits can thus be viewed as a product of elutriation of fines, including some crystals, from pyroclastic flows, their concentration and separation into the trailing ash-cloud, and then hydraulic sorting within the turbulent ash-cloud leading to cells enriched in, and depleted in crystals, which are represented by the alternating crystal-rich and crystal-poor laminae in the surge deposit.

Self (1976) also records crystal-rich (up to 44 wt%) surge deposits at the base of the Angra ignimbrite on Terceira, Azores, although detailed field descriptions are lacking and there is no consideration of whether they are base-surge deposits or ground-surge deposits.

Crystal-rich ignimbrites or *pyroclastic flow deposits* with crystal contents in excess of 40% are documented from both relatively young and old volcanic successions. From the Western USA, for example, Lipman (1975) and Byers *et al.* (1976) have described the successions associated with mid- to late-Tertiary, largely acidic volcanism. Lipman's account is of the volcanic history associated with the 29–30 Ma old Platoro Caldera Complex of the economically important and very large San Juan volcanic province of Colorado and New Mexico. The Platoro caldera formed during eruption of the Treasure Mountain Tuff, which consists of three members: the La Jara Canyon member (oldest), the Ojito Creek member and the Ra Jadero member. All are described as being quartz latites, but vary markedly in crystal contents. Only the La Jara Canyon member is particularly crystal-rich, containing approximately 40–50% crystals in the intracaldera fill units and 20–35% crystals in the outflow sheet units. This discrepancy is not clearly explained, but may be due to tapping of different parts of the magma chamber, suggesting that the outflow sheet ignimbrites and the intracaldera fill are of different ages and did not originate from the same eruption phase (cf. Fig. 8.18). The overall crystal-rich nature is presumably due to eruption of an originally crystal-rich magma. No comparison of the crystal content of the ignimbrite and pumice

fragments was carried out, so it is difficult to evaluate how much loss of glassy ash or crystal enrichment took place during eruption. For the other two, less crystal-rich members, crystal contents in pumice appear to be up to 10% lower (by volume) than in the ignimbrite matrix, suggesting some loss of fine vitric ash during eruption or flow, or both (Lipman 1975).

The emplacement of the Treasure Mountain Tuff and associated lavas and intrusives was followed by a period of erosion and reworking of volcanic sediments, and then by a succession of regionally extensive ignimbrites originating from other eruptive centres. These include the two most extensive ignimbrites in the San Juan province, the Fish Canyon Tuff and the Carpenter Ridge Tuff (Lipman 1975; Ch. 8; Fig. 8.1). The Fish Canyon Tuff is credited with being the largest ignimbrite in the world (3000 km^3 distributed over 15 000 km^2) and is a crystal-rich, quartz–latite ignimbrite whereas the Carpenter Ridge ignimbrite is a crystal-poor rhyolitic ignimbrite sheet. As a caldera fill, the Fish Canyon Tuff is over 1 km thick. As an outflow sheet it varies from 20 to 500 m thick. Crystal contents recorded by Lipman (1975) vary from 34–51% for the intracaldera fill to 34–46% for the outflow sheet. Again, there is little information on crystal contents of pumice fragments for use in evaluating ash loss during eruption, but Lipman (1975, p. 49) alludes to the latter having occurred (see also Section 8.4.3).

Crystal-rich ignimbrites are also well represented in the Upper Devonian volcanic successions of southeastern Australia (Birch 1978, Clemens & Wall 1984). Those that have been studied in detail are thick intracaldera fill ignimbrite piles. Crystal contents vary from relatively low (<20%) to high (65%), and there is in many instances a systematic change in the quantity and types of crystals through the succession, suggesting progressive eruption from a zoned magma chamber (Birch 1978, Clemens & Wall 1984). The high crystal content again implies eruption of highly crystallised magma. However, Birch (1978) suggests that juvenile magma clots (called 'schlieren') are chemically and mineralogically similar to the surrounding magma, implying minimal loss of fine ash during eruption.

This observation is also supported by Clemens and Wall (1984), who imply that collapse of large eruption columns was therefore not likely in generating these ignimbrites, and also suggest that much of the crystallisation in the Violet Town Volcanics took place in high level magma chambers, after a crystal-poor magma rose from deeper crustal melting source areas.

It should also be noted that local crystal- or lithic-rich pods, or both, may occur in ignimbrites. The pods may be diffuse or well-defined pipe-like structures representing gas segregation pipes, from which fines have been removed (Ch. 7). Sparks and C. J. N. Wilson (1983) have also noted that the flow-head deposits of violent pyroclastic flows (layer 1, or the ground layer) may also be enriched in crystals or lithics or both, resulting from the extreme turbulence in the head of such pyroclastic flows and the loss of fines because of this.

The correct terminology to use for all of these deposits, which are all wholly pyroclastic in origin, is the appropriate genetic terminology for primary pyroclastics as outlined in Chapter 12, once the wholly pyroclastic origin has been established. Deposits such as these are true ashes and tuffs, and where crystal-enriched, are crystal-rich ashes or tuffs.

Summary

Crystal-rich primary pyroclastics are not abnormal, but their crystal-rich origin is not just simply due to the eruption of a crystal-rich magma. Crystal-enrichment processes during eruption are just as important. Crystal-rich ashes and tuffs have been documented from each of the main pyroclastic deposit types: fall, surge and flow. Ignimbrites can vary markedly in their crystal contents. This, in large measure, reflects the degree of crystallisation before eruption, and for crystal-rich ignimbrites it probably implies significant crystallisation in high level or shallow crustal chambers. During eruption the crystal content may be modified by separation of fine ash. However, it seems that some ignimbrites, especially intracaldera fill ones, may be erupted without significant loss of glassy ash, which has implications for the eruption mechanism (e.g. Fig. 8.18). Clearly, more attention needs to be paid

to the crystal contents of pumice fragments and the enclosing ignimbrite matrix to get a clearer picture of eruption mechanisms and sorting processes. To establish a pyroclastic flow origin for crystal-rich volcaniclastics, the field facies characteristics (Chs 5, 7 & 8) need to be established.

11.5.2 'CRYSTAL TUFFS' WITH MIXED PYROCLASTIC AND EPICLASTIC ORIGINS

There are several well documented cases where the genesis of so-called crystal tuffs has been influenced by both pyroclastic and epiclastic processes. Cas (1978b, 1979, 1983a) described submarine crystal-rich tuffaceous sediments from an ensialic deep marine Siluro-Devonian basin, the Hill End Trough, in southeastern Australia. These tuffaceous sediments had previously been called crystal tuffs (Packham 1968), and form part of the so-called Merrions Tuff formation. The Lower Devonian Merrions Tuff consists of nearly equal volumes of andesitic/dacitic and minor rhyodacitic lavas and crystal-rich volcaniclastics. The volcaniclastics are composed of crystals and fragments of volcanic quartz, albitised plagioclase, orthoclase (see Plate 11) and minor vitriclasts. These occur in medium- to coarse-grained sedimentation units up to tens of metres thick, which in places grade into siliceous pelitic tops containing well preserved shard pseudomorphs. In extreme instances, crystal contents are in excess of 80% but average 64.2%, much higher than the average 37.6% phenocryst content of the associated lavas and glassy intraclasts. The volcaniclastics are therefore largely closed-frame-work grain aggregates with a subordinate matrix component. They are also chemically different from the associated lavas and are therefore not just unmodified pyroclastic equivalents of the lavas (Cas 1983a).

Sedimentation units are largely massive or internally structureless, with sharp, frequently erosive bases, and they are non- to normally-graded. They are akin to very thick, voluminous, high particle concentration turbidity current deposits (Cas 1979; Ch. 10) and show no evidence of having been emplaced by hot phase pyroclastic transportation processes (they lack welding and columnar jointing). Yet, clearly, their fragmentation mode was pyroclastic, as evidenced by shards. They are juvenile and have not been reworked by epiclastic surface processes, but they have been transported and redeposited by epiclastic processes (Cas 1979, 1983a). Their juvenile character and the fact that they were used together with the lavas to define a tight whole rock Rb/Sr isochron (Cas *et al.* 1976) suggests that they were deposited shortly after eruption. If they had been reworked and then deposited long after eruption, they would have mixed with older detritus, and would probably not have constituted part of a coherent isotopic suite of rocks.

However, the crystal contents are in many instances much higher than in ignimbrites. Cas (1983a) therefore suggested that they have experienced very efficient fractionation processes leading to very high concentrations of crystals. Based on concepts discussed in Section 11.3, Cas (1983a) suggested that the genesis of the Merrions Tuff crystal-rich volcaniclastics could have involved explosive eruption of a highly crystallised magma, and one or more very efficient stages of separation of glassy ash. This could have occurred in the eruption column, and subsequently from resulting pyroclastic flows and secondary eruption columns, perhaps generated when the initial pyroclastic flows entered the sea and interacted explosively with it after the model of G. P. L. Walker (1979). This led to contemporaneous submarine slumping and submarine mass-flow redeposition by turbidity currents, during which more fines were elutriated from the head and body of such flows (Sparks & C. J. N. Wilson 1983).

Given that there are many sedimentation units and that these occur in compositionally discrete stratigraphic packets (Cas 1978b), with minor interstratified background basinal sediments (hemipelagic pelites, greywackes), it appears that such a process was repeated many times in quick succession. Cas *et al.* (1981) suggested a similar origin for part of the Kowmung Volcaniclastics of the southeastern Hill End Trough, and it is clear from the stratigraphic account of Packham (1968) that many more such events are recorded in the stratigraphic record of the Hill End Trough.

Sparks and C. J. N. Wilson (1983), Sparks *et al.* (1983) and J. V. Wright and Mutti (1981) also describe crystal-rich ash turbidite layers, which are, however, much smaller in scale than those described above from the Merrions Tuff. The deposits of Sparks and C. J. N. Wilson (1983) and Sparks *et al.* (1983) were by-products of the historic Minoan eruptions of Santorini in the Mediterranean. Their origin is attributed to slumping of very extensive marine ash-fall deposits. These slumped ashes were then transported by deep-marine turbidity currents, and show many features of turbidites. Generally, they are rich in glassy ash, but several of these ash turbidites have a well defined 5–10 mm thick crystal-rich to pumice-rich basal layer which Sparks and Wilson attribute to sorting processes in the head of the transporting turbidity currents. They suggest that ingestion of ambient water into the head of the turbidity current leads to fluid streaming through the head and backwards (fluidisation), leading to elutriation of fines and concentration of denser and larger particles in the head from which they are deposited. This head layer deposit is then overridden by the body of the flow from which the bulk of the vitric-rich ash is deposited. Again, the initial fragmentation mode of these deposits was pyroclastic. They were initially transported and deposited by pyroclastic processes, and then finally retransported and redeposited, perhaps appreciable distances, by epiclastic mass-flow processes which were apparently entirely responsible for producing crystal-rich layers. A similar origin was proposed earlier for crystal-rich layers in the Dali Ash by J. V. Wright and Mutti (1981; Ch. 9).

The terminology for this class of deposit is perhaps the most problematic. The terminology used should reflect both the initial pyroclastic influence during eruption, fragmentation and initial pyroclastic transport, and the influence of the final epiclastic mode of transportation, given that this type of deposit is not an *in situ* pyroclastic deposit. Terms such as 'crystal tuff' and 'crystal-lithic tuff' are inappropriate. These deposits are tuffaceous or pyroclastic *sediments*, and therefore qualified terms such as tuffaceous sandstone, arenite, siltstone, lutite, etc., are more appropriate.

Summary

These cases have highlighted the intimate inter-action between pyroclastic and epiclastic processes. It is clear that although a pyroclastic fragmentation mode is demonstrable, final transportation and deposition involved epiclastic processes that appear to have been capable of causing significant crystal concentration. Identification of this epiclastic influence has been dependent on familiarity with the sedimentary structures, textures and general facies produced by epiclastic processes (Ch. 10). It is important to note that mass-flow transportation and deposition has not caused any abrasion or rounding of even delicate shard fragments. This is because the transportation time is too short and abrasive impacts are buffered by the viscous interstitial water. Furthermore, subaqueous re-sedimentation and mass-flow transport can re-deposit juvenile volcanic debris long distances away from source.

11.5.3 CRYSTAL-RICH VOLCANICLASTICS OF LARGELY EPICLASTIC ORIGIN

In this situation we are dealing with volcanic sediments which have been reworked considerably, imparting a distinctive sedimentary imprint and overprinting any original primary volcanic features. Comprehensive accounts of crystal-rich reworked volcanic sediments and their origins are not common. The first example chosen lacks detailed sedimentological data, but serves to illustrate the effects of reworking. Roobol (1976) documents the existence of well stratified, well sorted, crystal-rich, greenish-grey sands and grits interbedded with shallow-marine limestone horizons from the Maastrichtian Summerfield Formation in Jamaica. Overlying this basal marine interval is a succession of less well sorted, pink coloured sandstones and conglomerates and two ignimbrites. The sandstones and conglomerates are poorly stratified, and in-distinct lenses of boulders are often present. Roobol interprets this upper sediment section as being of continental fluviatile origin, with all the rocks having an hornblende–andesite parent-age. Roobol suggests that most of the volcanic detritus had a

pyroclastic origin. Crystal concentration processes during pyroclastic flow, as well as during epiclastic transportation and reworking, especially in the marine part of the section, led to removal of glassy ash and pumice, some of which is concentrated in mudstone interbeds. Comparison of the crystal abundance in the resultant sediments with the crystal contents of lava boulders in the succession suggests that up to 60% of glassy ash could have been separated from crystal grains and fragments (Roobol 1976).

Sigurdsson *et al.* (1980) describe redeposited, rounded beach sands from the deep-sea cores of the Grenada Basin in the Caribbean (Fig. 9.4d). Although they do not discuss crystal contents, R. S. J. Sparks (*pers. comm.*) indicates that some are crystal-rich, essentially crystal beach sands derived from the erosion, reworking and mass-flow redeposition of debris derived from ignimbrites exposed onshore. During reworking, significant crystal concentration has occurred as in the case described by Roobol (1976). The authors have sampled similar crystal-rich beach sands on the island of Kos in the Dodecanese Islands, Greece, which were derived from Quaternary rhyolitic pyroclastics exposed on Kos (Fig. 13.44). Such crystal-enriched beach sands have, without doubt, also been redeposited offshore into surrounding basins.

Similar crystal-rich clastics of volcanogenic-derivation, involving erosion, variable degrees of reworking, and final redeposition, abound in the rock record and have frequently been called 'greywackes', which is appropriate to their almost completely sedimentary origin. The use of the term 'crystal tuff' in such cases would be totally inap-propriate in spite of the volcanic provenance of many of the grain types. An alternative terminology would include terms such as feldspathic or crystal-rich volcanic sandstone, arenite, conglomerate, rudite, etc., depending on grainsize (Ch. 12). Such deposits can be found in any sedimentary environ-ment, and the transportation mode could be any of the appropriate ones discussed in Chapter 10. For example, Tanner *et al.* (1981) and Storey and Macdonald (1984) describe volcaniclastics of the Lower Tuff Member of the Annenkov Island

(a) Explosive eruption of crystal-rich magma

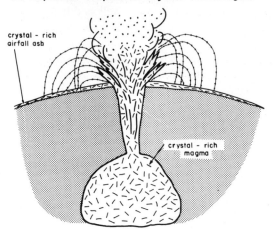

crystal - rich
airfall ash

crystal - rich
magma

(b) Convective transport of fine ash into upper column and downwind transport in plume

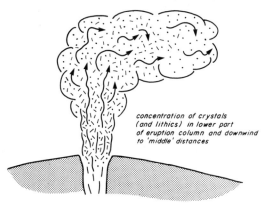

concentration of crystals
(and lithics) in lower part
of eruption column and downwind
to 'middle' distances

(c) Collapse of eruption column and elutriation of fine glass into trailing ash cloud of pyroclastic flow

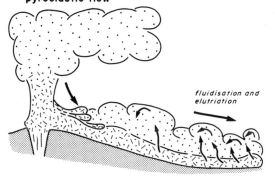

fluidisation and
elutriation

(d) Passage of pyroclastic flow into a water body explosive interaction producing secondary rootless vents

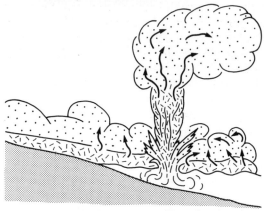

(e) Reworking of crystal-rich ignimbrite (beach or river)

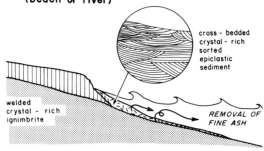

cross - bedded
crystal - rich
sorted
epiclastic
sediment

welded
crystal - rich
ignimbrite

REMOVAL OF
FINE ASH

(f) Slumping and subaqueous turbidity current transport of crystal-rich pyroclastic detritus from (d) or reworked crystal-rich detritus from (e)

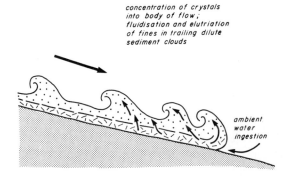

concentration of crystals
into body of flow;
fluidisation and elutriation
of fines in trailing dilute
sediment clouds

ambient
water
ingestion

Figure 11.6 (a)–(f) Schematic representation of the factors controlling the formation of crystal-rich pyroclastic and epiclastic volcaniclastic deposits.

Formation of the South Georgia back arc basin as 'crystal-lithic tuffs'. In places this nomenclature is qualified as 'turbiditic', indicating that the mode of transportation and deposition was by epiclastic turbidity currents. However, in spite of this it is not convincingly demonstrated that the volcaniclastics had a pyroclastic fragmentation mode, although it seems likely. The use of 'turbiditic' is a valuable qualification in this case, but given the epiclastic influences, 'tuffs' could be dropped in favour of 'tuffaceous sandstones'. Unless the inferred pyroclastic fragmentation origin is demonstrable, a preferable term may be 'crystal-lithic volcanic sandstone'.

Summary

Epiclastic derivation of crystal-rich volcaniclastics is also a common way of producing crystal-rich volcaniclastics. Given the major role of epiclastic processes, the final deposit should bear the imprint of reworking or redeposition, or both, in terms of textures and field facies characteristics (Ch. 10).

11.6 Overview

The foregoing has clearly demonstrated that the genesis of crystal-rich volcaniclastics may be complex (Fig. 11.6). In many cases they probably owe their origin to explosive eruption of a relatively highly crystallised magma. Thereafter they may be modified by efficient fractionation processes involved with pyroclastic or epiclastic transportation and deposition processes, or both. The latter could transport such aggregates well away from source.

The analysis of such deposits must involve determination of the modes of fragmentation *and* the modes of transportation and deposition independently of each other. The determination of the full history of these deposits (and others) is best attempted by an integrated approach involving understanding of sedimentological, volcanological, petrological and chemical concepts as outlined by Cas (1983a), and herein.

11.7 Further reading

To our knowledge, this topic has not been adequately covered elsewhere.

Plate 12 Poorly sorted, matrix-supported volcanic breccia on the upper slopes of Mt Ruapehu on Pinnicle Ridge. Is this volcanic breccia an explosion breccia (i.e. agglomerate) or a redeposited epiclastic breccia? The open framework character, the sandy–muddy matrix, and the thick massive nature suggest the latter, which is consistent with the well–known frequency of debris flow and lahar events on Mt Ruapehu.

CHAPTER TWELVE

Classification of modern and ancient volcaniclastic rocks of pyroclastic and epiclastic origins

Initial statement

In this chapter we first provide classification schemes for modern pyroclastic deposits. We then describe an approach to the classification of lithified, indurated and metamorphosed volcaniclastic rocks of the rock record, stressing the problems, and the caution needed in giving names, even with basic genetic meanings, to such rocks. This is followed by a discussion of descriptive lithological criteria and their relevance in determining the genesis of ancient volcaniclastic rocks. The problems associated with the use of the terms 'agglomerate' and 'vulcanian' for ancient volcaniclastic rocks are

discussed and, finally, we comment on the nomenclature of quench-fragmented and autobrecciated (flow brecciated) volcaniclastics.

12.1 Introduction

From the foregoing chapters it should be apparent that recent work on young volcaniclastic deposits has shown that genetic nomenclature is useful in representing our understanding of fragmentation, transportational and depositional mechanisms, especially for pyroclastic deposits. It should also be clear that genetic terminology cannot always be immediately applied, especially to ancient volcani-clastic rocks, because of the problems caused by poor exposure, lack of exposure of contact relation-ships, weathering, alteration, metamorphism and deformation (Ch. 14). In both modern and ancient volcanic terrains two different approaches to nomenclature of deposits are needed. An initial non-genetic approach is required, especially for those working on ancient successions. Application of genetic terminology should be a final step after all lithological and field facies characteristics have been evaluated, including consideration of both the modes of fragmentation and final deposition.

However, this approach is not obviously em-bodied in the recommendations of the IUGS Subcommission on the Systematics of Igneous Rocks in their publication on the 'Descriptive nomenclature and classification of pyroclastic de-posits and fragments' (Schmid 1981). Although Schmid (1981) professes that the approach used is descriptive rather than genetic, this is not borne out by the use of terms such as 'tuff' and 'agglomerate'. The nomenclature scheme proposed by Schmid *only has application to hand specimen and petrographic characteristics* being based largely on grainsize, and to some degree on grain types and morphology and fails in not comprehensively reviewing necessary field facies characteristics. It is a scheme that merely duplicates the valuable contributions of Fisher (1961, 1966b), does not contribute anything significantly new to Fisher's summaries and does not attempt to go that one step further in the process of genetic interpretation – that of identifying the transportational and depositional origins of volcaniclastics. Without considering these aspects erroneous use of names can easily occur. We hope that by reading this book, particularly this chapter, and Appendix II, a clear approach to objective description and nomenclature, and then to genetic interpretation and nomenclature, will become apparent.

12.2 Modern pyroclastic deposits

There is no unique approach to classifying modern pyroclastic deposits. Many schemes and approaches have been proposed (e.g. Fisher 1960, 1961, 1966b,

A. E. Wright & Bowes 1963, Parsons 1969, J. V. Wright *et al.* 1980, Schmid 1981). J. V. Wright *et al.* (1980) indicated that at least two systems are required:

genetic classification and
lithological classification

The purpose of a genetic classification is to interpret the genesis of deposits which can then be related to the history, eruptive pattern and mechan-isms of a volcano or volcanic terrain. The purpose of a lithological classification is primarily des-criptive, describing and documenting the major characteristics of a deposit, such as grainsize and constituent fragments. However, these features themselves may indicate a particular process and allow some conclusions to be made about the genesis of a deposit.

12.2.1 GENETIC CLASSIFICATION

The genesis of a pyroclastic deposit is partly deduced from its lithology, but also from geometry and field relations. Because of this, a genetic classification can only be rigorously applied in the field to very young, well exposed Quaternary deposits. However, even then initial description and consideration of the total facies characteristics should be made before a genetic name can be applied. As discussed in Chapter 1, the term 'pyroclastic', or rock names pertaining to a pyro-clastic origin (e.g. ash or tuff, agglomerate) should only be applied to deposits or rocks with a demonstrated pyroclastic mode of fragmentation *and* a pyroclastic depositional origin. We have already discussed in Chapter 5 how pyroclastic deposits can be divided into those derived from:

pyroclastic falls,
pyroclastic flows and
pyroclastic surges,

and this will be the basis of the following classifi-cation schemes.

Pyroclastic fall deposits

For pyroclastic fall deposits the classification scheme of G. P. L. Walker (1973b) (Ch. 6) is the most practical approach yet proposed for working on

Table 12.1 Genetic classification of pyroclastic falls and their deposits.

Eruptive mechanism		Pyroclastic fall	Dispersal	Fragmentation	Essential fragment
magmatic	open vent	hawaiian	small	low	scoria
		strombolian			scoria
		sub-plinian			pumice† (scoria)
		plinian			pumice† (scoria)
		ultraplinian	v. wide	high	pumice
phreatomagmatic	closed vent	vulcanian	wide to moderate*	v. high to moderate*	
		surtseyan	moderate	v. high	basic‡
		phreatoplinian	v. wide	v. high	silicic ash‡

* Dispersal and fragmentation of vulcanian fall deposits depends on whether they are generated by short-lived explosions or periods of intense maintained explosions (Ch. 6).

† Most sub-plinian and nearly all plinian deposits that have been documented are pumice falls. However, scoria-fall deposits that are sub-plinian and even plinian in their dispersal characteristics are known (Ch. 6).

‡ The type examples used to define these terms were basaltic for surtseyan and rhyolitic for phreatoplinian but, given the right conditions, it may be possible to find rhyolitic surtseyan and even basaltic phreatoplinian ash falls (Ch. 6). Further studies are needed to characterise air-fall deposits produced in different types of hydrovolcanic eruption, i.e. between true surtseyan (open vent in a surface body of water – lake, sea) and those formed by interaction of magma and ground water, but having surtseyan dispersal characteristics.

modern pyroclastic fall deposits. As already discussed, this quantitative scheme relies on the accurate mapping of a deposit and detailed granulometric analysis (see Figs 6.15, 16 & 21). Enough data for plinian and sub-plinian deposits are also now available to use thickness or maximum clast size to characterise them on 'area plots' (Figs 6.18 & 19). These plots have certain advantages because if, for reasons of time or geography, it is not possible to map a deposit to the 0.01 isopach, or to carry out an extensive programme of sieving, the deposit can still be accurately classified. A summary of this genetic scheme for pyroclastic fall deposits is given in Table 12.1.

Pyroclastic flow deposits

Pyroclastic flow deposits have been described by a lexicon of terms and classified in numerous ways. A genetic scheme for the three main types of pyroclastic flow deposit (Chs 5, 7 & 8) is given in Table 12.2. This is based on the eruptive mechanism and characteristics of the deposit, and is adapted from J. V. Wright *et al.* (1980). Some of the other names used for these types of flow deposit are shown in Table 12.3, which attempts to compare various classifications. In our scheme we have also included deposits of hot, dry volcaniclastic debris flows of

the type generated at Mt St Helens in 1980 and Bandai-san in 1888 (Ch. 5).

One term that needs further comment is 'nuée ardente', meaning glowing cloud. This was first used by La Croix (1903, 1904) to describe pyroclastic flows he observed during the 1902 Mt Pelée eruption. In a later publication La Croix (1930) expanded the use of the term to include all types of pyroclastic flows. 'Nuée ardente' has since become widely used, but in the past few years it has become somewhat unfashionable (e.g. see Nairn & Self 1978, J. V. Wright *et al.* 1980). This is because the term was used too generally to describe eruptions from many volcanoes, as well as to describe the mechanism producing widespread ignimbrite sheets. More-meaningful words should be used to describe the processes and deposits, as it is quite obvious that La Croix meant 'glowing cloud' to describe the complete volcanic phenomenon (both overriding ash cloud and basal underflow or pyroclastic flow proper), and we now know these behave differently during transport, and produce very different types of deposit.

Although 'glowing cloud' is perhaps a very good description of how some small volume pyroclastic flows appear when moving down the sides of large volcanic cones, we would suggest that their deposits

Table 12.2 Genetic classification of pyroclastic flows and their deposits.

Eruptive mechanism	Pyroclastic flow	Deposit	Essential fragment	
eruption column (fountain) collapse	pumice flow, ash-flow	ignimbrite, pumice flow deposit, ash-flow tuff*	pumice	↑
	scoria flow	scoria flow deposit	scoria	*decreasing density of juvenile clasts*
lava, dome collapse (explosive and gravitational)	block and ash flow (nuée ardente)	block and ash flow deposit	dense lava	
explosive cryptodome release	hot, dry volcaniclastic debris flow	volcaniclastic debris flow	accessory lithics (± juvenile fragments)	

* In a strict definition, 'ash-flow tuff' should only refer to deposits with >50 wt% finer than 2 mm. 'Ignimbrite' and 'pumice-flow deposit' can be used more loosely irrespective of grainsize, although 'pumice-flow deposit' is sometimes used to emphasise those flow deposits with a large proportion of bomb-sized pumice fragments.

Table 12.3 Comparison of various classifications of pyroclastic flows.

This book: based on eruptive mechanism and characteristics of the deposits		Murai (1961): based on type eruptions and characteristics of the deposits	Williams and McBirney (1979): based on site of and type eruptions	Smith (1979): based on volume of the deposits
Pyroclastic flow	Deposit			
pumice flow	ignimbrite (large volume)		Valles-type	large-volume ash-flow tuffs (100–1000 km³)
		VTTS-type Krakatoa-type	VTTS-type Krakatoa-type	intermediate-volume ash-flow tuffs (1–100 km³)
	(small volume)	St Vincent type		
scoria flow	scoria-flow deposit		St Vincent-type	small-volume ash-flow tuffs (0.001–1.0 km³)
block and ash flow*	block- and ash-flow deposit	Sakurajima-type	Peléan-type	
block and ash flow†		Pelée-type		
block and ash flow‡		Merapi-type	Merapi-type	

* Produced by collapse of vertical eruptive column accompanying lava or dome collapse.
† Produced by explosive collapse of an actively growing lava flow or dome.
‡ Produced by gravitational collapse of an actively growing lava flow or dome.
VTTS = Valley of Ten Thousand Smokes.

Table 12.4 Genetic classification of pyroclastic surges and their deposits.

Eruptive mechanism	Pyroclastic surge type	Temperature, water content	Essential fragment
phreatomagmatic (outward moving radial collar and column collapse)	base surge	cold, wet (hot, dry)	juvenile (vesiculated to non-vesiculated); accessory lithics
accompanying pyroclastic flows	ground surge	hot, dry	juvenile (vesiculated to non-vesiculated)
	ash-cloud surge	hot, dry	juvenile (vesiculated to non-vesiculated)
accompanying pyroclastic fall eruptions (but without generation of a pyroclastic flow)	ground surge	hot, dry	juvenile (vesiculated)

can be better categorised in the manner indicated here. Alternatively, as suggested by J. V. Wright *et al.* (1980), we should restrict 'nuée ardente' to its original definition and only use it for pyroclastic flows produced by collapse of an actively growing lava flow or dome. This is indicated in Table 12.2.

Pyroclastic surge deposits

A classification of pyroclastic surges is given in Table 12.4. Ground and ash cloud surges accompanying pyroclastic flows are generated by the whole spectrum of mechanisms indicated in Table 12.2.

12.2.2 LITHOLOGICAL CLASSIFICATION

The main bases of a lithological classification are:

the grainsize limits and overall size distribution of the deposit,
the constituent fragments of the deposit and
the degree and type of welding.

Grainsize

Table 12.5 summarises the grainsize limits of pyroclastic fragments from the schemes of Fisher

(1961, 1966b). Concerning the overall grainsize distribution, granulometric analysis of non-welded and unlithified pyroclastic deposits can be an important discriminant in distinguishing different types of deposit. We have seen that a convenient way of representing grainsize data is an Md_ϕ/σ_ϕ plot (Ch. 5; App. I). Pyroclastic flow deposits usually have σ_ϕ values greater than 2.0, whereas in general, pyroclastic fall deposits have values less than 2.0. Md_ϕ/σ_ϕ plots of pyroclastic surge deposits tend to overlap both fields. Individual laminae of surge deposits can be well sorted, while channel samples through several laminae are often poorly sorted. Ground and ash-cloud surges may be better sorted than base surge deposits, and this may be due to the cohesion of water-saturated fine ash in the base-surge clouds. However, an Md_ϕ/σ_ϕ plot should be viewed with caution, and not used alone to interpret the origin of a deposit. Pyroclastic deposits are formed by complex processes. For example, some concentration zones or segregation structures in pyroclastic flow deposits may be as well sorted as pyroclastic fall deposits or, conversely, some phreatomagmatic ash-fall deposits may be as poorly sorted as flow deposits. In the more-recent publications other types of grainsize plots are being used,

Table 12.5 Grainsize limits for proven **pyroclastic fragments** and **pyroclastic aggregates** (after R. V. Fisher 1961, 1966b).

Grainsize (mm)	Pyroclastic fragments			Name of unconsolidated aggregate	Lithified equivalent
	round and fluidally shaped		angular		
coarse 256 —————— fine		bombs	blocks	agglomerate (bombs) or pyroclastic breccia	agglomerate (bombs) or pyroclastic breccia
64 ——————————————————					
		lapilli		lapilli deposit	lapillistone
2 ——————————————————					
coarse 1/16 —————— fine		ash		ash deposit	tuff

Table 12.6 Summary of the components in pyroclastic deposits (after J. V. Wright *et al.* 1980).

(a) Pyroclastic flows and surges

Type of flow or surge	Essential components		Other components
	Vesicular	Non-vesicular	
pumice flow and surge	pumice	crystals	accessory and accidental lithics
scoria flow and surge	scoria	crystals	cognate, accessory and accidental lithics
block and ash flow and surge. (nuée ardente)	poor to moderately vesicular juvenile clasts	cognate lithics and crystals	accidental lithics

(b) Pyroclastic falls

Predominant grainsize	Type of fall	Essential components*		Other components
		Vesicular	Non-vesicular	
>64 mm	agglomerate	pumice or scoria		cognate and accessory lithics
	breccia		cognate or accessory lithics, or both	
>2 mm	lapilli deposit	pumice or scoria	cognate or accessory lithics, or both	crystals
<2 mm	ash deposit	pumice or scoria	crystals and/or cognate and/or accessory lithics	

* Depending on type of deposit.

and again these should be treated with the same caution.

Constituent fragments

A summary of the dominant components in a pyroclastic deposit provides a qualitative lithological description as well as information on the genesis (Table 12.6).

Welding

We have already discussed the process of welding and the lithological variations it can produce in Chapters 6 and 8.

12.3 Classification of lithified, indurated and metamorphosed volcaniclastic rocks

The foregoing review of the classification of *modern* pyroclastics is an essential prelude to the consideration of ancient volcanic successions because it gives an appreciation of the great range of products produced by a diversity of *observed* (or confidently inferred) processes in, around and stemming from modern volcanic vents and centres. Such an awareness of the possible range of origins is critical for the geologist working in ancient successions because, in most instances, the definitive context and spatial relationship to vent or volcanic centre, that are often (but not always) observable or inferable for modern successions, are lacking. The approach therefore initially has to be more objective and less genetic, with the overall context, extent and characteristics of the lithological unit(s) having to be established.

Other factors which complicate the genetic interpretation and classification of lithified volcaniclastics include: devitrification, recrystallisation, new mineral growth during diagenesis and low grade metamorphism, and deformation, all of which lead to modification of original textures and mineralogy to varying degrees (Ch. 14). Add to this that epiclastic volcaniclastics can also be exceedingly abundant, and equally modified, and that modern weathering also takes its toll in producing confusion. It is then a brave person who walks up to an outcrop and applies a genetic classification or terminology. For example, devitrification of an originally glassy lava can produce an equigranular mosaic or spherulitic texture, so giving a pseudo-granularity in hand specimen. Metamorphism and weathering can further accentuate this, so producing a granular texture which may be very difficult to distinguish from a truly fragmental texture. Thin section examination may be no more helpful than the hand specimen, because the original glassy character of the rock may have been overprinted, with the distinction between a coherent glassy groundmass and vitriclastic (e.g. shards) or epiclastic textures being difficult to identify.

Application of genetic terminology to lithified volcaniclastics should therefore be avoided until thorough evaluation of the *complete* set of characteristics of the lithology (or lithologies) has been made, including:

(a) hand specimen characteristics (textural, compositional),
(b) outcrop characteristics (bedded as opposed to massive; structures and fabrics that are essentially contemporaneous with emplacement),
(c) contact relationships (sharp or continuous, gradational),
(d) geometry (three-dimensional form and thickness),
(e) associated facies and
(f) context and palaeogeographic setting.

Rarely will any one of these, or any single outcrop, be definitive enough to allow an unequivocal interpretation of the genesis. Furthermore, it cannot be assumed, as is commonly done, that because a rock is volcaniclastic it had a pyroclastic origin. Even having demonstrated that a rock is of probable pyroclastic origin, it cannot be assumed that it was deposited close to the vent, as is also often done.

Finally, it cannot be assumed that the imprint and character produced by a particular mode of eruption will reflect the final transportational and depositional mode. For example, pyroclastically fragmented detritus (crystals, pumice, shards) can be transported and deposited by means other than pyroclastic ones (e.g. surface reworking, lahars,

Table 12.7 Non-genetic classification of volcaniclastic rocks (modified from R. V. Fisher 1961).

Volcanic breccia
    closed framework
    open framework
        non-cohesive, granular matrix
        cohesive mud-sized matrix

Volcanic conglomerate
    closed framework
    open framework
        non-cohesive, granular matrix
        cohesive mud-sized matrix

2 mm— — — — — — — — — — — — — — — —2 mm

Volcanic sandstone

0.0625 mm— — — — — — — — — — — —0.0625 mm

Volcanic mudstone
    volcanic siltstone

                if sufficiently well sorted
                and volcanic origin is clear
    volcanic claystone

subaqueous mass flows; Ch. 10) and in environments away from the near vent area. *Consideration of genesis therefore involves consideration of both the fragmentation mode and final depositional mode.*

A useful starting place with ancient volcaniclastic rocks before attempting any kind of genetic classification, as presented in Section 12.2, is Fisher's (1961) suggested non-genetic nomenclature. Following Fisher's lead, a suggested initial non-genetic terminology is given in Table 12.7. If a wholly pyroclastic, rather than epiclastic, origin can be established, then the nomenclature of Table 12.5 can be used, and if beyond that the pyroclastic transportation and depositional origins can be established, then the appropriate nomenclature of Tables 12.1–12.4 can be used.

12.4 Descriptive lithological aspects of ancient volcaniclastic rocks relevant to determining their genesis

Few of the textural or compositional characteristics of volcaniclastics are by themselves indicative of one particular mode of fragmentation or deposition.

For example, the term 'agglomerate' (which has distinct genetic connotations, see Section 12.5), is frequently used for any volcanic breccia but, as seen in Table 12.8, there are over twenty ways of producing volcanic breccias!

Another example is accretionary lapilli, which are generally assumed to be diagnostic of air-fall deposits, formed by rainfall through a downwind ash cloud or from a moisture-laden eruption column, especially a phreatomagmatic column. It is now known that accretionary lapilli are perhaps more commonly generated within the explosive eruption column, particularly within those of phreatomagmatic eruptions (Ch. 5). They are therefore common in base surge deposits and some ignimbrites, as well as surtseyan and phreatoplinian ash-fall deposits (Fig. 5.24). They have also been noted in fossil fumarole pipes in ignimbrites, and could form in the secondary explosion columns of rootless vents formed where pyroclastic flows or lavas interact explosively with a body of water (Ch. 3) or ice.

The following is a brief listing of important descriptive properties of volcaniclastics and some qualifying comments on their usefulness or limitations. The assemblage of properties should be used together with the larger-scaled outcrop and field properties and relationships to evaluate the genesis. The majority of the listed properties are sedimentological in origin, and we feel that such an approach is a useful one.

Textural:
    coherent crystalline igneous texture versus fragmental texture
    welding
    grainsize
    sorting
    shape
    angularity or rounding, and
    framework type

Compositional:
    compositional affinities
    compositional homogeneity, and
    clastic components.

Table 12.8 Grainsize–textural classes of volcaniclastic rocks and some possible origins (see App. II for suggested diagnostic characteristics).

	Grainsize–textural class		Origin
A	conglomerate – closed framework (rounded clasts essential)	1	epiclastic reworking (fluvial, shoreline)
		2	mass-flow redeposition (subaqueous)
		3	pumice and scoria concentration zones in ignimbrites and scoria-flow deposits
		4	fines-depleted ignimbrite
B	conglomerate – open framework (rounded clasts essential)	5	epiclastic reworking and mass-flow redeposition (deposits with granular matrix)
		6	cohesive pebbly mudflows and lahars
		7	non-welded (uncollapsed pumice) ignimbrite and scoria-flow deposits
C	breccia – closed framework (angular clasts essential)	8	epiclastic redeposition and mass-wastage (includes gravitational collapse, including caldera margin collapse breccias)
		9	aa lavas
		10	block lavas
		11	lava dome and flow-front talus deposits
		12	agglutinates
		13	agglomerates
		14	quench-fragmented lavas, cryptodomes and shallow intrusives (hyaloclastites)
		15	hydrothermal explosion breccias
		16	hydraulic fracture breccias
		17	pumice-fall deposits
		18	scoria-fall deposits
		19	lithic concentration zones (base of *layer 2b*) and ground layers of violent ignimbrites
		20	co-ignimbrite breccias (lag breccias and ground breccias)
		21	fines-depleted ignimbrite
D	breccia – open framework (angular clasts essential)	22	glacial till and moraines (diamictites)
		23	glacial dropstone deposits
		24	epiclastic reworking and mass-flow redeposition with granular matrix
		25	cohesive debris flows and lahars
		26	ignimbrite (*layer 2b*) and other (denser clast) pyroclastic flow deposits (block and ash flows, scoria flows)
		27	co-ignimbrite breccias and proximal ignimbrites
		28	near-vent base surges
		29	ground or ash-cloud surge
		30	giant pumice beds
E	sandstone (sand-sized framework grains essential)	31	epiclastic reworking
		32	epiclastic mass-flow redeposition
		33	weathered and/or devitrified lavas or dykes
		34	fine-grained ignimbrite
		35	air-fall ashes or tuffs
		36	base surge deposits
		37	ground or ash-cloud surges
F	mudstone (mud-sized grade predominant)	38	epiclastic
		39	fine-grained ignimbrite
		40	air-fall ashes or tuffs
		41	surge deposits

12.4.1 TEXTURAL

Coherent igneous texture versus fragmental texture

Intrusive or extrusive porphyritic rocks should be characterised by an even distribution of euhedral to subhedral crystals in a fine glassy or variably devitrified and recrystallised groundmass. Apart from *occasional* xenoliths, the rocks should lack evidence of clastic or vitriclastic texture. If a clastic character is demonstrable, for example broken and shattered crystals and abundant lithic fragments, then any of the fragmentation origins discussed in Chapter 3 are possible, and have to be evaluated.

Welding

If textural evidence of welding (eutaxitic texture, pumice lenticle foliation) can be found, it is important because this is indicative of the rock having a pyroclastic origin. Welded rocks are likely to be preserved in the record because of their resistance to contemporaneous erosion compared with non-welded equivalents. Welding is not uniquely indicative of a pyroclastic flow origin. Welded air-fall tuffs are a common rock and, if recognised, can be important in determining vent proximity (Ch. 6). They have been identified in ancient volcanic successions.

However, in modern volcanic successions, many ignimbrites are totally non-welded, and few welded air-fall deposits are known compared with non-welded ones. Alignment of pumice and other juvenile clasts can be caused by sedimentary processes, and flattening can be caused during tectonic deformation associated with the strain effects of the formation of a tectonic cleavage. As a result the origins of apparent welding textures should be investigated. Welding is only unequivocally demonstrable at the microscopic scale.

Grainsize

This is the most obvious characteristic of any rock, and it is the first observed in any specimen or outcrop (Ch. 1). However, grainsize, especially average grainsize, is not indicative of particular deposits or origins (Ch. 11). Table 12.8 and Appendix II summarise the possible origins of the major non-genetic size classes listed in Table 12.7, and emphasise the diverse range of possibilities for each textural category. Most importantly, *breccias do not necessarily indicate proximity to the vent*. Ignimbrites, for example, can have coarse, basal, lithic concentration zones many kilometres from the vent (Chs 7 & 8). Volcanic mudflows or lahars are also capable of carrying exceedingly large blocks tens of kilometres from the source point of the flow, which is usually high on the volcanic edifice. Many kilometres from source this coarse volcaniclastic material may then be reworked by river systems, or in the marine environment into other types of epiclastic breccias and related deposits (Ch. 10).

Sorting

Sorting, combined with various grainsize statistical parameters, has been used to distinguish different types of young pyroclastic deposits. For lithified, indurated and even metamorphosed aggregates, grainsize parameters and quantitative values of sorting are not accurately determinable. The usefulness of sorting is also dubious, except in a qualitative way, and only then when combined with other properties (see Ch. 1). Most deposit types, irrespective of their clastic origins, can be well sorted or poorly sorted. The distinction between simple size sorting and hydraulic sorting is important (Ch. 1).

Shape

Grain shape can be distinctive for explosively ejected detritus (Chs 2 & 3). Glass shards are diagnostic of explosive fragmentation during eruption (Ch. 3), but could thereafter be deposited in ignimbrites, air-fall and surge deposits, or could be reworked to minor degrees, or could be redeposited by mass-flow processes (Ch. 10), particularly subaqueously without being destroyed (Chs 9 & 11). Shaped bombs, breadcrust bombs and irregular spatter fragments are typical of the explosive eruption of fluidal (basaltic) magma. These components should be the only criteria for the use of the term 'agglomerate'. Large irregular blocks, irrespective of composition, could have diverse fragmentation and depositional origins.

Flattened, attenuated pumice fragments (fiamme) and in thin section, plastically deformed shards, are indicative of welding. The distinction between euhedral–subhedral and angular crystals *may* be indicative of volcaniclastics as opposed to coherent, un-fragmented igneous rocks, but not always so.

Angularity or rounding

Angularity or rounding of clasts should be interpreted with care. Angularity by itself does not imply proximity to the vent or a primary volcanic fragmentation or depositional origin (Ch. 1). Rounded clasts imply post-eruptive reworking by surface processes, but exceptions occur. Accretionary lapilli are spherical and rounded, but clearly have a pyroclastic origin. Rounded fluvial clasts can be picked up by ignimbrites as accidental lithics. Pumice clasts become rounded in transport within an ignimbrite. Rounded edges and corners on crystals (especially volcanic quartz) may be due to magmatic resorption before eruption, rather than to post-eruptive reworking.

Framework-type

This is not indicative of any particular origin (see Table 12.8), but is nevertheless a useful indicator of transport conditions.

12.4.2 COMPOSITIONAL

Compositional affinities

The actual compositional affinities of a volcanic rock are of little use in determining the genesis, except for the fact that volumetrically the majority of basaltic products are lavas, while the majority of silicic products are pyroclastic.

Compositional homogeneity

Compositional homogeneity of the clastic aggregate may be useful in identifying the degree of reworking, if any, of the aggregate. Homogeneity is probably best evaluated from lithic components, including partially crystallised glassy fragments. Their compositional, and for that matter textural homogeneity, should reflect the degree of mixing of different source materials. Note, however, that mixtures of crystals, glassy fragments and cognate lithics may be a compositionally homogeneous assemblage of components because they can all be derived from the one source, reflecting the fragmentation of a partially crystallised magma.

Clastic components

Abundance of glass shards, pumice or scoria, certainly indicates a pyroclastic eruption mode and possibly, *but not necessarily*, deposition by pyroclastic processes. As the crystal or lithic content, or the content of both, increases, and if compositional homogeneity decreases, a greater diversity of fragmentation and depositional origins, including epiclastic origins, becomes more probable.

12.5 Use of the terms 'agglomerate', 'vulcanian breccia' and 'tuff' in ancient successions

'Agglomerate' and 'vulcanian' are the two terms that seem to have caused more confusion and problems than any others in discussion of old volcanic successions. Both can be highly misleading terms, and we would try to avoid use of either for ancient volcanic rocks.

An agglomerate is a coarse pyroclastic deposit composed of a large proportion of rounded, fluidally shaped, volcanic bombs (predominant grainsize is >64 mm). It is essential that evidence of true volcanic bombs be present (e.g. shaped or breadcrust types or bomb sags). In the strict sense the term implies a fall deposit (Table 12.6), and is a very good indicator of proximity to the vent. It is perhaps best applied to the scoria deposits that build strombolian cones (Figs 6.5 & 8).

The main problem is that geologists working in ancient volcanic successions have used the term 'agglomerate' indiscriminately for any volcanic breccia, and then, even worse, suggested connections with distance from vent or volcanic centre, which in many cases could be totally wrong. Caution is necessary when using this term or interpreting other geologists' work and maps in which it is used, because of the obvious genetic implications of the term.

'Vulcanian' is often applied to (bedded) sequences of volcanic breccias, and as with 'agglomerate', suggests proximity to the vent. However, as described in Chapter 5, vulcanian air-fall deposits from individual eruptions are thin, small-volume stratified ash deposits, although near vent they contain large ballistic blocks and bombs. Vulcanian deposits have a low preservation potential. To identify a breccia as vulcanian in an ancient succession, evidence of ballistic blocks and bombs with bomb sags must be present and, again, as with agglomerates, most rocks identified as such in the record are in fact other types of volcanic breccia.

Although denser clast types of pyroclastic flows are commonly generated during vulcanian eruptions (Chs 5 & 6), such volcanic breccias are better described as scoria-flow deposits or block- and ash-flow deposits. This avoids the implication that 'vulcanian breccias' are necessarily near vent.

'Tuff' is the lithified equivalent of an ash deposit (Table 12.5). Both terms are genetic and should only be applied to aggregates that have been demonstrably fragmented *and deposited by pyroclastic* processes, and in which the grainsize of the pyroclasts is <2 mm. Where significant proportions of lapilli also occur (>10%) the terms lapilli-ash and lapilli-tuff can be used. The pyroclastic fragmentation mode is interpreted from the grain morphology (Ch. 3) and the pyroclastic depositional mode from field facies characteristics (Chs 5–9). Where more-comprehensive indications of the transport and depositional origins are required, terms such as 'ignimbrite', 'air-fall tuff' and 'base-surge tuff' can be used.

12.6 The consequences of redeposition on nomenclature

In Chapter 1, Section 1.3.2 pyroclastic and epiclastic deposits were distinguished based on their modes of fragmentation *and* final deposition. Pyroclastic deposits are those which have a demonstrated pyroclastic mode of fragmentation *and* a demonstrated pyro-clastic mode of deposition (fall, flow, surge). Epiclastic deposits, as defined in Chapter 1, are clastic deposits which were fragmented by normal surface processes (weathering, physical abrasion, gravitational collapse) *or* were deposited by normal surface processes, irrespective of the mode of fragmentation, or both. Therefore, where pyroclastic deposits are reworked or redeposited, or both, *to any degree* by epiclastic processes (Ch. 10; interpretation comes again from facies characteristics) genetic terms such as 'agglomerate' and 'tuff' can no longer be used (Ch. 11). The deposit is then a volcanic gravel (? breccia), sand or mud or, where lithified, a volcanic breccia, sandstone or mudstone. If such a deposit still clearly contains evidence of its pyroclastic fragmentation mode, then qualifying terms may be used, e.g. 'tuffaceous sandstone' or tuffaceous 'mudstone'. In this way, the general textural character and the combined influences of pyroclastic and epiclastic processes in the formation of the aggregates are clearly reflected. In some countries the term 'tuffite' is used for such redeposited pyroclastic deposits.

12.7 Nomenclature of quench-fragmented and autobrecciated volcaniclastics

In the foregoing sections we have essentially addressed the nomenclature of pyroclastics and epiclastics. Quench-fragmented and autobrecciated volcaniclastics do not fit into either category, and this point cannot be stressed strongly enough. Many workers equate aggregates of angular volcanic fragments with explosive, pyroclastic origins. However, in subaqueous, and especially in deep subaqueous, settings many such aggregates are the result of quench fragmentation (chilling and shattering) or autobrecciation, or both (Ch. 3).

Quench-fragmentation can produce aggregates that are coarse and angular or finely granulated. Such deposits are called hyaloclastites (Ch. 3). However, in the first instance such an origin may not be clear, especially in ancient volcanic successions. The initial approach should be non-genetic – i.e. the aggregates are either volcanic breccias or volcanic sandstones according to the grainsize (Table 12.7). If the quench origin can be established, the genetic names to use are 'hyaloclastite breccia' and 'hyaloclastite sand-stone'. For fluidised mixtures of hyaloclastite and sediment, formed either when magma is intruded through, or lava flows over wet, unconsolidated sediment (Ch. 3), the term 'peperite' or 'peperitic hyaloclastite' may be used. However, as discussed in Section 3.6, the term 'peperite' should be used with care and in a non-genetic, descriptive sense only, because of the diverse possible origins of deposits with peperitic texture. 'Peperitic breccia' may, in fact, be a preferable less genetic term than 'peperite'.

A similar approach should be used towards autobreccias (Ch. 3). In the first instance they are volcanic breccias. Once their origins have been determined confidently from the other possible origins of breccias (Table 12.8; App. II), the genetic term 'autobreccia' can be used.

12.8 Further reading

The paper by J. V. Wright *et al.* (1980), much of which has been summarised here, is essential reading on the classification of pyroclastic deposits. The earlier literature, especially the papers by Fisher (1960, 1961, 1966b) and Parsons (1969) are also useful.

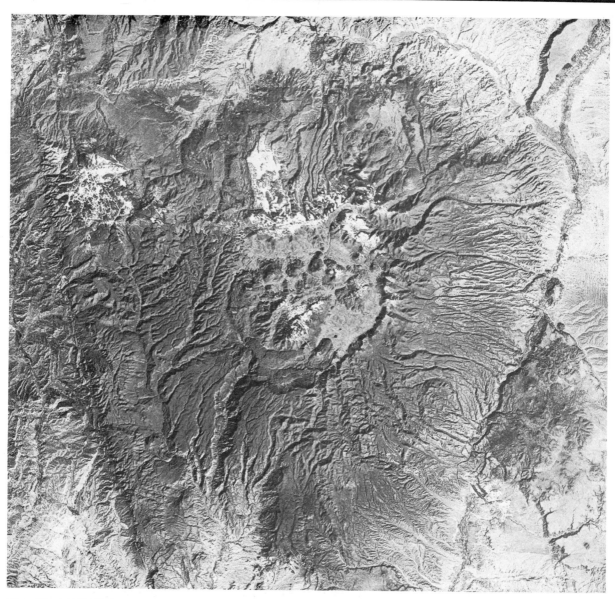

Plate 13 LANDSAT image of the Valles Caldera, New Mexico, the source of the Bandelier Tuff. The caldera has a diameter of nearly 24 km, and is surrounded by the wide, shallow dipping ignimbrite shield. Note that the arcuate alignment of the dome complex within the caldera. See text for further discussion of characteristics. (After Sheffield 1983.)

CHAPTER THIRTEEN

Modern volcanoes and volcanic centres

Initial statement

This subject warrants much fuller attention than we are able to give within the scope of this book. We therefore document the important physical and stratigraphic characteristics of modern volcanoes and volcanic centres, as a preliminary account before considering them within the context of general facies models for modern and ancient successions in Chapter 14. Our attention is focused here on individual volcanoes rather than on provinces, an important distinction, given the relatively short life-span of individual volcanoes compared with the longer lived, larger volcanic–sedimentary provinces of which they are part. Initially we concentrate on the volcanic forms commonly built by the subaerial eruption of basaltic magmas: basaltic shield volcanoes, flood basalt plateaus,

scoria cones, maar-type craters, and also pseudo-craters and littoral cones. Pumice-forming eruptions can also build steep-sided cones somewhat akin to scoria cones, and these are briefly mentioned. Maar-type craters of virtually every composition are known. We go on to describe stratovolcanoes, which are the most abundant type of large volcano on the Earth's surface, and are built from repeated eruptions of either calc-alkaline or alkaline magmas. Multi-vent intermediate-silicic centres are then briefly covered, and include some large calderas and volcano-tectonic depressions. The large rhyolitic volcanoes are discussed next. The remaining two sections deal first with submarine spreading centres and seamounts and, lastly, with intra- or subglacial volcanoes. From this chapter the duration of the active life of different types of volcanoes and of the duration of volcanic versus epiclastic processes should be appreciated before considering ancient volcanic successions in Chapter 14.

13.1 Monogenetic and polygenetic volcanoes

Volcanoes can be subdivided into two types: **monogenetic** and **polygenetic**. Monogenetic volcanoes are built up by the products of one eruption or one eruptive phase, while polygenetic volcanoes are those resulting from many eruptions, separated by relatively long periods and often involving different magmas. An eruption from a monogenetic volcano could last several years, but is essentially one prolonged eruption involving one magma type. For example, the relatively recent eruptive phases of Parícutin, Surtsey and Heimey all represent the activity of monogenetic centres, even though the activity varied from months, in the case of the 1973 activity on Heimey, to nine years, in the case of the Parícutin activity. By contrast, the Hawaiian

volcanic islands are frequently polygenetic, involving not only repeated eruptive phases separated by significant repose periods from the one vent, but also being marked by multiple vents or eruptive centres. The island of Hawaii hosts several major volcanoes in their own right: Mauna Loa, Mauna Kea, Kohala, Hualalai and Kilauea (Fig. 13.1b), not to mention numerous other smaller ones. The whole complex is clearly a major polygenetic centre.

A significant distinction between monogenetic and polygenetic volcanoes appears to be that monogenetic volcanoes have a simple magma conduit system used only during one eruption or one prolonged eruptive phase. However, polygenetic volcanoes may have complex 'plumbing systems', with an intricate, complicated conduit system that is used many times, or different parts of which are used, during spaced eruptive phases.

(a)

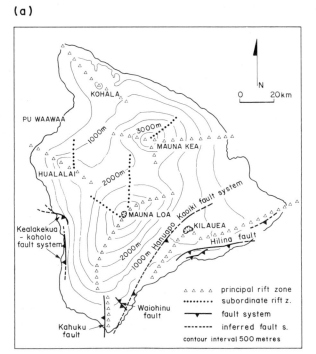

(b)

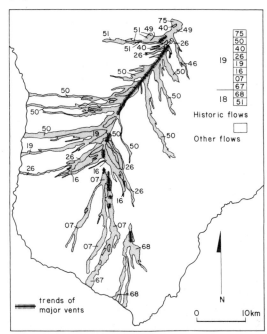

Figure 13.1 (a) The island of Hawaii and shield volcanoes. (b) Distribution of lava flows and associated vents of documented historic eruptions along Mauna Loa's southwest rift zone. (After Lipman 1980.)

13.2 Basaltic shield volcanoes

In plan view, basaltic shield volcanoes are symmetrical and circular to elliptical in shape. In profile they are convex-up piles of basaltic lava, with gentle slopes of $<10°$. They are built up by frequent eruptions of fluidal basaltic lava from central vents or flank eruptions, or both. The smallest shields have basal diameters (W_s) of a few kilometres (e.g. Mauna Ulu on Hawaii), whereas for the largest shields found in the Hawaiian Islands W_s is over 100 km (e.g. Mauna Loa; Fig. 13.1; Pike & Clow 1981). Shield heights (H_s) are on average about $\frac{1}{20} W_s$. The flanks of basaltic shields are dominated by well defined overlapping lava flows (Fig. 13.1b). Lithologically, basaltic shield volcanoes are almost entirely composed of lava flows. Both pahoehoe and aa types occur; most lavas are pahoehoe near the vent and may change to aa with distance (Ch. 4).

Other than basalt lavas in the pile, there may be minor ($<1\%$) pyroclastic deposits, including scoria fall deposits of limited dispersal (hawaiian, strombolian), and deposits produced by phreatomagmatic and phreatic explosions. Oxidised soil horizons (**red boles**) and epiclastic layers which may be of diverse origins (e.g. glacial as on Mauna Kea; Porter 1979) may also be interbedded. From Figure 13.1b it is obvious that in detail the lava pile will have a complex stratigraphy, and it is certainly not of layer-cake character.

In the literature basaltic shield volcanoes are sometimes divided into different types. The two most important are:

Hawaiian or large shields **and**
Icelandic or small shields

and a third type we shall describe is:

Galápagos shields.

13.2.1 HAWAIIAN SHIELDS

The Hawaiian volcanoes are well known through the studies of Gordon MacDonald and the reader should refer to MacDonald (1972) and MacDonald and Abbott (1979). The largest polygenetic basaltic shields (Figs 13.1 & 2) have summit calderas, and major rift zones marked by spatter cones, spatter ramparts, collapse craters (pit craters), scoria cones and smaller superimposed (Icelandic) monogenetic shields, e.g. Mauna Ulu on Kilauea. Eruptions within the calderas occur slightly more frequently than on the rifts, but it is the eruptions from the lateral rifts that give these shields their elongate form. Mauna Loa and Kilauea have the largest calderas, both slightly over 5 km in diameter, on the island of Hawaii, but older shields on other islands in the chain have calderas between 10 and 20 km in diameter. Vents for eruptions within the calderas are generally fissures that cut across their floors. The Halemaumau crater within Kileaua caldera (Fig. 13.2) is somewhat of an exception. It may have a pipe-like vent, although eruptions still tend to occur through fissures at the bottom of the crater, and it has been the site of a lava lake many times. As well as lavas and minor pyroclastics, the shields are also built by high level intrusions which may be present in the walls of the summit calderas (Fig. 13.3).

The two most active shields on Hawaii are Kilauea and Mauna Loa. Mauna Loa is the world's largest active volcano, rising nearly 9 km from the Pacific Ocean floor to its summit at 4169 m above sea level. The total volume of Mauna Loa is around 40 000 km^3. At their present combined growth rates (including intrusions) of ~ 0.1 km^3 $year^{-1}$ (Swanson 1972) both Mauna Loa and Kilauea could have been built in less than 1 Ma. However, much of Mauna Loa and Kilauea is likely to consist of pillow lava formed by submarine extrusions as the volcanoes rose from the sea bottom, in the same way as seamounts grow (Section 13.10; Fig. 13.50; Ch. 15). These shields therefore closely interface with the marine environment. Some of the important structural features of the shields such as rifts can be traced offshore (e.g. Fornari et al. 1979). Gravity sliding and slumping of their flanks into the ocean basin has also occurred, e.g. at Kilauea (Duffield et al. 1982). The southern flank of Kilauea is constantly being displaced towards the sea along normal faults (pali) as a result of oversteepening caused by addition of lava flows and the intrusion of magma into the summit and rift

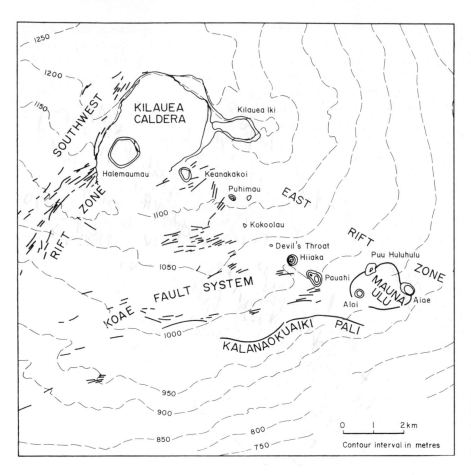

Figure 13.2 Map of Kilauea shield volcano showing eruption centres, principal rift zones, fault systems and pit craters.

zones (Figs 13.1 & 2). The area south of the Hilina Fault system may be part of a large gravity slide extending to the base of the volcano at abyssal depths (e.g. J. G. Moore 1964).

Volcanologists have recognised a series of stages in the growth of the Hawaiian shields, and these are sometimes named after the volcano showing those characteristics, e.g. Mauna Kea stage. Although we do not wish to go into details here, it is important to note that compositional differences occur between the different stages, and lavas and other erupted products change from tholeiitic, becoming more alkalic with time, although some of the earliest submarine products may be alkaline. More explosive activity accompanies the eruptions of alkaline magmas, and over 300 scoria cones have been built on the sides of Mauna Kea during the last phase of its eruptive activity (Porter 1972, 1973).

These are concentrated along the rift zones of the volcano and reach a density of eight per square kilometre (Fig. 13.4). The explosive eruption of a rare highly differentiated trachytic magma on Hualalai formed the Puu Waawaa pumice cone (located on Fig. 13.1). A high aspect ratio lava flow was also extruded during this eruption and this is morphologically very different from the basaltic lavas of this shield or elsewhere on the island. Frequency of eruptions also decreases with time, but the precise nature of the transition between stages in terms of compositional and output variations remains unclear. This is discussed by Feigensen and Spera (1981), who presented a dynamic model for temporal variation in magma type and eruption interval at Kohala volcano.

Figure 13.3 (a) Wall of Kilauea caldera showing part of the succession of thin lava flows that have built the shield. Irregular body (bottom centre) is an intrusion called the Uwekahuna lacolith, and this is seen in (b).

13.2.2 ICELANDIC SHIELDS

Icelandic shields are smaller ($W_s < 15$ km), almost symmetrical and entirely, or nearly entirely, built up by effusive eruptions from a central summit vent. The summit craters of Icelandic shields are usually <1 km across, and often have raised rims of spatter. There are generally few radial fissures, or lines of parasitic cones. These shields are composed of large numbers of thin pahoehoe flows. One of the best known volcanoes of this type is Skjaldbreidur in Iceland (Fig. 13.5), which has a W_s of about

10 km and H_s of 600 m. Most Icelandic shields are probably monogenetic, and even the larger ones such as Skjaldbreidur may be constructed in less than 10 years.

Although no historic eruptions producing Icelandic shield volcanoes are known, the eruptive activity of Surtsey in the period 1963–7 is perhaps an example of the duration of the eruptive event required to produce Icelandic shield volcanoes. Although the subaqueous, explosive phase of the Surtsey eruption is somewhat anomalous, the later subaerial effusive phase of the activity of Surtsey is

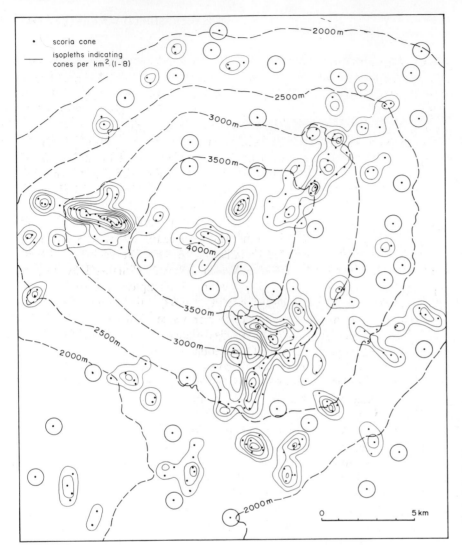

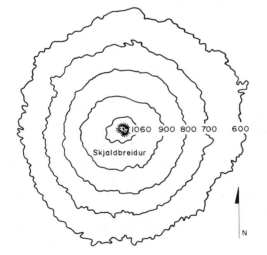

Figure 13.4 Map showing distribution and density of late-stage alkaline scoria cones on the upper slopes of Mauna Kea. Their distribution clearly defines the rift zones (see Fig. 13.1a). (After Porter 1972.)

Figure 13.5 The nearly perfectly symmetrical shield of Skjaldbreidur formed by lavas erupted from a central summit vent (cf. the Hawaiian shields where eruptions from the rift zones determine their shape; note scale difference).

perhaps indicative of the style of activity involved in the building of Icelandic shield volcanoes. The Surtsey eruption involved the eruption of over 2 km³ over the four years of activity. The activity of Mauna Ulu on the flanks of Kilauea in 1969 to 1974 can also be cited as an example of the duration and style of eruption producing Icelandic shield volcanoes.

13.2.3 GALÁPAGOS SHIELDS

The Galápagos Islands of Isabela and Fernandia have several major polygenetic shield volcanoes. These are very regularly spaced occurring about 35–40 km apart at the intersections of orthogonal tectonic lineaments. W_s at sea level varies from 45 to 80 km, and H_s is between 3000 and 3500 m above sea level (Pike & Clow 1981). These shields have gentle lower slopes which rise to steeper central slopes that flatten off around spectacular summit calderas. The summit calderas are between 3.0 and 8.7 km in diameter (the largest being on Sierra Negra), and are surrounded by well defined concentric fissures which are marked by small scoria and spatter cones. Lavas have been erupted from these, some pouring back into calderas, and lavas have also issued from radial fissures on the flanks of the shields. Alkaline basalt is more important than in the Hawaiian shields. It has been suggested the high level intrusion of dykes and sills may account for their characteristic shape.

13.3 The source vents in flood basalt plateau and plains basalt provinces

So far we have looked at volcanic centres dominated by effusive, lava-forming eruptions. In this section we look at the nature of vent systems with high magma discharge rates that are the source of extensive plateau basalts and plains basalts, as discussed in Section 4.5. This is done to show that not all vents are necessarily parts of central or point-source volcanoes. In a strict sense, these vent systems are source vents to entire provinces rather than single localised volcanoes, but because of their importance, we consider them here.

Flood basalts are some of the largest single eruptive units known, and are believed to have flowed great distances from source (Ch. 4). Repeated eruptions build up a flood basalt pile, forming a vast lava plateau which may cover areas of $>10^6$ km², and which generally has slopes <2–$3°$. These are typified by the Columbia River Plateau in Washington, Oregon and Idaho which is of mid-Miocene age (Fig. 4.1b). The Columbia River Plateau basalts cover an area of about 220 000 km², have an estimated volume of 195 000 km³, and were erupted during a very short interval of only 2 or 3 Ma (Swanson et al. 1975, Swanson & Wright 1981, H. Williams & McBirney 1979). Other younger examples occur in Iceland (G. P. L. Walker 1963), the Tertiary North Atlantic volcanic province, and the mid-Tertiary Ethiopian–Yemen plateaux. Older examples include the Cretaceous Deccan Traps in northwestern India, which cover an area of more than 500 000 km² and have a volume of more than 1 million km³ (Choubey 1973, Subbarao & Sukheswala 1981), the Cretaceous Paraná plateau of southern Brazil–Uruguay, the Jurassic Karoo in South Africa, the central sector of the Siberian platform, which is early Mesozoic in age, and the Proterozoic Keweenawan district of north-central America.

Many flood basalt plateaux are closely associated with the initiation and early development of rifted continental margins (Ch. 15). Some of the above examples became active during the formation of new continental margins born with the Mesozoic–Cenozoic fragmentation of Pangaea. The Ethiopian flood basalt province, the evolution of which has been recently reviewed by Mohr (1983), is shown in Figure 13.6. The area covered by Tertiary flood basalts in Ethiopia (even excluding the once contiguous Yemen province) is about 600 000 km², and must have been close to 750 000 km² before plateau uplift and ensuing erosion occurred during Pleistocene time.

The characteristics of flood basalt lavas and their eruptions are described in Chapter 4. Plateaux are built of simple lava flows which cover vast areas because they may be erupted at discharge rates as high as 10^6 m³ s⁻¹. (Compare this with eruptions that produce Icelandic shields which may take

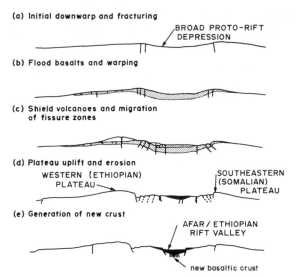

(a) Initial downwarp and fracturing

BROAD PROTO-RIFT DEPRESSION

(b) Flood basalts and warping

(c) Shield volcanoes and migration of fissure zones

(d) Plateau uplift and erosion

WESTERN (ETHIOPIAN) PLATEAU

SOUTHEASTERN (SOMALIAN) PLATEAU

(e) Generation of new crust

AFAR / ETHIOPIAN RIFT VALLEY

new basaltic crust

Figure 13.6 Schematic evolution of the Ethiopian flood basalt province from about 30 Ma BP to the present. Length of sections is about 1000 km. Note with (b), tholeiitic basalts were erupted in the northern sectors of the province during the late Oligocene–Miocene with the development of the Afar Margin, but in the southern sectors further into continental Africa coeval flood basalts were mildly alkaline. (c) Later in the Miocene eruptions of alkaline basalts built shield volcanoes over the present plateau surface. (After Mohr 1983.)

several months. For a medium-sized shield, the volume may be about 10 km^3 and, assuming an eruption duration of six months, this leads to an average discharge rate of 6×10^2 m^3 s^{-1}.)

Although flood basalts were generally believed to be erupted from fissures, recognising the vent systems has been difficult, and Swanson *et al.* (1975) were probably the first to document vent systems for a large flood basalt province. Part of the problem is that these vents tend to be covered by younger flood basalts. Also, features that may mark the vents, such as spatter cones and ramparts, are ephemeral features which are easily eroded. Swanson *et al.* (1975) documented vent systems for the Roza Member (Ch. 4; Fig. 4.1b) and the Ice Harbour flows in the Columbia River Plateau. They are both associated with one linear dyke system at least 200 km wide and 450 km long. The vent systems are confined to linear zones within this dyke system, are tens of kilometres long and a few kilometres wide, and have the same trend. The vent

and near-vent areas (Fig. 13.7, Table 13.1) are defined by remnants of spatter cones and ramparts, agglutinates and welded scoria which are the typical near-vent products of lava-fountaining. Lava flow units are also more abundant in these areas.

Vent systems within flood basalt plateaux are therefore likely to be large and to have complex stratigraphies, with overlapping spatter cones and interdigitating agglutinated layers, clastogenic lavas and welded scoriar fall deposits (Chs 4 & 6). Downwind, fall deposits may have formed when scoria was not incorporated into moving lava, and because of the very high discharge rates these might have wide dispersals of perhaps plinian proportions.

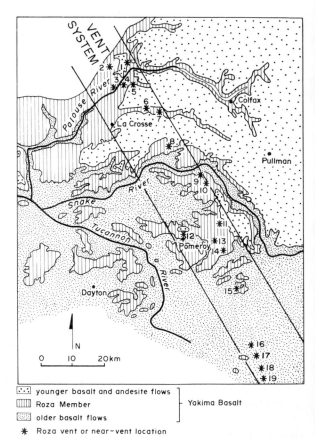

younger basalt and andesite flows
Roza Member — Yakima Basalt
older basalt flows

* Roza vent or near-vent location

Figure 13.7 Map locating linear-vent system for the Roza flood basalt flows in the Columbia River Plateau (total areal extent of the Roza Member is shown in Fig. 4.1b). Features found that indicate a vent or near-vent position for the numbered locations are listed in Table 13.1. (After Swanson *et al.* 1975.)

Table 13.1 Vent and near-vent areas for the Roza flood basalt flows (after Swanson *et al.* 1975).

Locality	Elevation (m)	Evidence
1	500	bed of scoria and spatter, welded at base, between two Roza lava flows
2	490	thick beds of spatter and scoria, in places welded, partly define cones and ramparts
3	490	spatter and thin flows
4	470	spatter and scoria ramparts and cones overlain and bulldozed by Roza flows
5	470	spatter and scoria, in places welded, interbedded with thin flows; small dykes; relics of cones and ramparts
6	475	bed of spatter and scoria overlain by a Roza flow
7	490	bed of spatter and scoria overlain by a Roza flow
8	490	poorly exposed spatter and scoria beneath a Roza flow
9	500	dyke of Roza chemistry and lithology high in section
10	645	ash, spatter, and scoria between two Roza flows
11	760	four thin Roza flow units with interbedded ash and spatter
12	645	thick beds of spatter and scoria, dykes and thin flow units associated with Roza flows
13	790	spatter and cinder
14	770	welded spatter between two Roza flows
15	1035	large cone of spatter and pumice cut by Roza dykes and associated with thin, dense Roza flows
16	1360–1385	thick piles of dense, viscous Roza flows; platy joints and ramp structures
17	1465–1525	thick piles of dense, viscous Roza flows; platy joints and ramp structures
18	680	thin dyke of Roza lithology
19	610	at least seven dykes (some wider than 2 m) of Roza chemistry and lithology

As discussed in Chapter 4, the vents in plains basalt provinces are both of the central vent type, largely within low profile shields, as well as fissure systems (Greely 1982). As such, these provinces share characteristics of both flood basalt provinces and shield volcanoes. According to Greeley (1982) some lava fields consist of coalescing very low profile shields, others of moderate-sized, fissure-fed lavas. The lavas are frequently multiple or compound types, their volumes are small to moderate (cf. flood lavas) and the inferred magma discharge rates are considered to be considerably less than those for flood basalt provinces.

The lava shields have slopes less than $\frac{1}{2}°$, except at the summit where slopes may rise to 5°. The summit region may also be marked by one or more irregular craters. The shields are up to 15 km in diameter, and have volumes of less than 7 km^3. These low profile shields have also been described on Hawaii and Iceland (Greeley 1982). In the Snake River Plain, many contemporaneous sets of shields appear to be aligned in distinctive rift zones, representing fissures along which localised eruption points formed.

In the type example of plains basaltic provinces, the Snake River Plain, scoria cones, spatter cones, tuff cones and maars are not common.

13.4 Scoria cones (and pumice cones)

Scoria or cinder cones (Figs 13.8 & 9) are small volcanic landforms built typically during subaerial strombolian eruptions (Ch. 6) of basaltic and

Figure 13.8 The almost fresh basaltic scoria cones in the San Francisco volcanic field, Arizona. (a) Merriam Crater. (b) Cone 173.

Figure 13.9 The partially degraded cone of Mt Elephant in western Victoria, Australia.

also found to be characterised by relatively wide craters for any given basal diameter of the cone, and crater width, $W_{cr} = 0.40W_{co}$ (Fig. 13.10c). This is because most of the cone is built of larger fragments, which are ejected on short ballistic trajectories and so fall only around the vent.

Scoria cones are very susceptible to weathering and erosion, which changes their morphology (cf. Figs 13.8 & 9). In the Massif Central of France, Kieffer (1971) showed that 10 000-year-old cones

basaltic andesite magmas. Wood (1980a) gives an excellent review of their characteristics and evolution, and indicates that they are the most common type of volcano. In another paper, C. A. Wood (1980b) also studied the degradation of scoria cones and modelled their changes in shape with time.

The cones are often approximately circular in plan. Elongate forms are built when eruptions continue along a large part of a fissure which does not localise to a single point-source vent. They have central bowl-shaped craters, although these tend to become infilled by subsequent mass-wastage. In a morphometric analysis of 910 scoria cones, C. A. Wood (1980a) found that the basal diameter, W_{co}, ranges between 0.25 and 2.5 km (mean 0.9 km), and for 83 fresh cinder cones, cone height, $H_{co} = 0.18W_{co}$ (Figs 13.10a & b). These volcanoes were

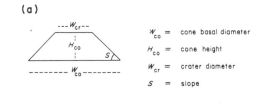

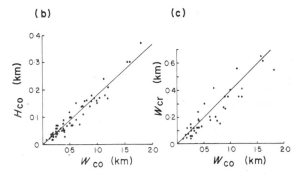

Figure 13.10 Cone dimensions for 83 relatively fresh basaltic scoria cones from various parts of the world. (After C. A. Wood 1980a.)

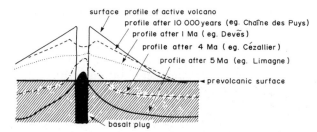

Figure 13.11 Changes in cone morphology with increasing age in the Massif Central, France. (After Kieffer 1971.)

have slopes several degrees shallower than the most recent cones, and that slopes of cones 1 Ma old are as low as 15° (Fig. 13.11). Cones 4 Ma old are marked only by residual necks, the scoria having been completely removed. With time and increasing degradation, cones show decreases in H_{co}, $H_{co} : W_{co}$ ratio and slope, but the ratio of $W_{cr} : W_{co}$ does not appear to change (C. A. Wood 1980b). Variation in H_{co} and W_{co} for 38 scoria cones of different ages from the San Francisco volcanic field are shown in Figure 13.12a (after C. A. Wood 1980b). The two young cones from this field shown in Figure 13.8 are of Merriam age. In Figure 13.12b the future degradation history of Sunset Crater, which is the youngest cone in the San Francisco field, is predicted. This has been derived from past degradation trends of other cones in this field.

Fresh scoria cones have steep angle of repose slopes of about 33°. Avalanching and rolling of coarse, mainly ballistic scoria down the outer and inner slopes and back into the crater, are important depositional processes occurring on the cone during an eruption. Many of the scoria layers found in the cones are therefore of mass-flow origin (inertial grainflow; Chs 6 & 10). Deposits of the cones also include bombs of lava spatter (e.g. cow-dung bombs and spindle bombs), and sometimes these form agglutinated layers (Chs 3 & 6). Clastogenic lava flows can form with the rapid accumulation of spatter (Chs 4 & 6). Frequently, the gas content of a magma will decrease towards the end of an eruption, so the proportion of lava spatter ejected increases. Many cones are therefore capped by a collar or ruff of spatter. Sometimes dissected scoria cones show welded interiors with columnar jointing and compaction profiles broadly similar to welded tuffs

(Fig. 6.8d). The different deposits found in the interiors of scoria cones have been fully discussed and illustrated in Chapter 6. A large proportion of the explosively ejected scoria is transported downwind to form a mantling fall deposit of usually limited dispersal, although more widely dispersed (sub-plinian and plinian) examples are known (Ch. 6). Most cones have accompanying lava flows which form during cone-building, or when degassed magma continues to be extruded after explosive activity has ceased. Scoria cones generally have smaller volumes than their associated lavas, and

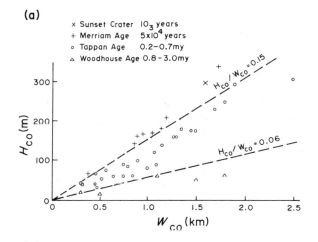

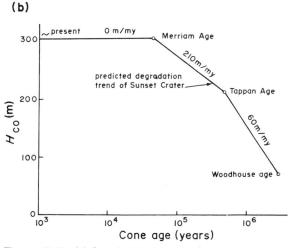

Figure 13.12 (a) Cone height versus basal diameter for 38 cones of different stratigraphic ages in the San Francisco volcanic field. (b) Possible future degradation history for Sunset Crater derived from past degradation trends in the San Francisco volcanic field. (After C. A. Wood 1980b.)

therefore the cone only represents a small part of the total magma erupted.

Scoria cones are typically monogenetic. They most commonly occur as isolated cones in large basaltic volcanic fields, but they also occur in nested clusters within complex tuff ring and maar centres (Section 13.5) and as parasitic cones on polygenetic shield and stratovolcanoes, on which they are also essentially monogenetic elements. Eruptions range in duration from a few days to a few years (e.g. the nine year eruption of Parícutin in Mexico between 1943 and 1952). Of observed scoria cone eruptions, 50% have lasted for less than 30 days, and 95% stopped within one year (C. A. Wood 1980a). Later eruptions produce new cones at separation distances of a few kilometres, and it has been speculated (McGetchin & Settle 1975) that these distances may be approximately equal to the depths of the magma reservoirs. Repose periods in these basaltic fields are in the order of 10^2–10^3 years,

and such fields may be active for 10^6 years. There are many examples of this type of basaltic volcanic field which are today potentially active. One example is the Newer Volcanics in southeastern Australia (Fig. 13.13), where volcanic centres are dated from <5 Ma BP to as young as 4000–6000 years BP (Aziz-Ur-Rahman & McDougall 1972, G. Blackburn et al. 1982, Joyce 1975). Basaltic volcanism may be active for much longer periods (5×10^7 years) in the same region when a number of separate magmatic provinces (10^6 years) are relatively closely related in space and time. The Older Volcanics of Victoria are an example of this, and basaltic volcanism persisted from 95 to 17 Ma BP (Wellman 1974, Day 1983).

Pumice-forming eruptions can also build steep-sided cones at vent, although **pumice cones** are not well documented in the literature. An example previously mentioned in this chapter is the trachytic cone Puu Waawaa on the side of the Hualalai shield

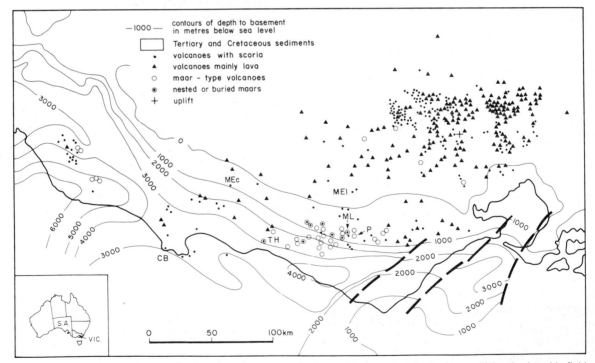

Figure 13.13 Map showing distribution of scoria cones and other volcanic centres in the Newer Volcanics basaltic field in southeastern Australia, and relationship to basement contours and sedimentary cover (after Joyce 1975). Mt Elephant (ME1; Fig. 13.9) is identified, and the maar-type volcanoes (some with associated scoria cones), Lake Purrumbete (P), Tower Hill (TH) and Mt Leura (ML) and a scoria cone, Mt Eccles (MEc) discussed in text and Figures 13.15 & 22, are also highlighted.

(Fig. 13.1). Some very large examples with basal diameters much greater than those found for scoria cones occur within La Primavera volcano (e.g. Figs 4.21a, 4.22e & 13.37; Section 13.9). Larger cones have complex stratigraphies and may be polygenetic. Many smaller ones seem quite similar internally to scoria cones, and are constructed predominantly of steeply dipping, reversely graded mass-flow layers, e.g. as at Puu Waawaa. However, lava spatter, agglutination and welding are probably not found except in peralkaline examples (see Section 6.11). Effusion of lava produces short coulées (Ch. 4).

Figure 13.14 Maars, tuff rings and tuff cones. (a) Crater Elegante, Pinacate Volcanic Field, Sonora, Mexico (after C. A. Wood 1979). (b) East Ukinrek Maar, Alaska. View into crater during eruption on 3 April 1977 (after Kienle *et al.* 1980). (c) Fort Rock tuff ring, Oregon, USA, with spectacular wave cut cliffs cut into palagonite. The volcano is an isolated feature in a flat lake basin; the south rim was breached by waves of the former lake. (d) Koko Crater tuff cone, Oahu, Hawaii. (e) Diamond Head tuff cone, Oahu, Hawaii.

(a)

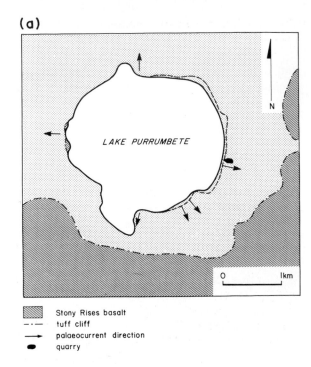

Stony Rises basalt
-·-   tuff cliff
→   palaeocurrent direction
●   quarry

(b)

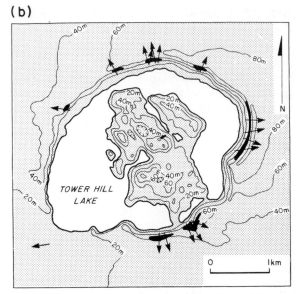

Figure 13.15   Maps of two maars in the Newer Volcanics basaltic field. (a) Lake Purrumbete. (b) Tower Hill which is a very large nested maar–scoria cone complex. The stratigraphies of these two centres are discussed later in the text.

13.5 Maars, tuff rings and tuff cones

These volcanic craters (Figs 13.14 & 15) are usually monogenetic and are produced by phreatomagmatic and phreatic eruptions (Chs 3, 5 & 6). They are second only to scoria cones in abundance (C. A. Wood *in press*), and all of these landforms are often associated with each other in large basaltic volcanic fields, e.g. the Fort Rock, Christmas Lake Valley basin, Oregon (Heiken 1971) and the Newer Volcanics, southeastern Australia (Fig. 13.13). There is usually some geographical control on the distribution of the different volcano types, determined essentially by hydrological factors. In Oregon, phreatomagmatic craters are found in an area once occupied by a lake, while scoria cones occur only outside the former lake basin (Fig. 13.14c). In southeastern Australia, phreatomagmatic craters are concentrated where a cover of aquifer-bearing Cainozoic sediments overlies the Palaeozoic basement (Figs 13.13 & 16). Examples of other magma compositions are also common, e.g. rhyolitic (Sheridan & Updike 1975), dacitic (Sillitoe *et al.* 1984), phonolitic (Schmincke *et al.*

1973) and, more exotic, carbonatitic (Dawson & Powell 1969). Their classification and terminology has been a subject of debate for the past ten years. 'Maar' has been used as the general term for broad, low rimmed volcanic craters (including tuff rings and tuff cones), but it also has a more restricted usage (and below we follow the distinctions of C. A. Wood *in press* as summarised in Table 13.2)

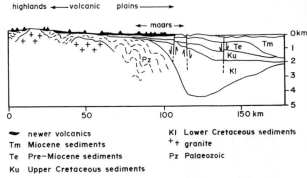

↗ newer volcanics
Tm   Miocene sediments
Te   Pre–Miocene sediments
Ku   Upper Cretaceous sediments

Kl   Lower Cretaceous sediments
++   granite
Pz   Palaeozoic

Figure 13.16   North–south cross section through western Victoria (see Fig. 13.13) showing the distribution of maar-type craters and other volcanoes (scoria cones and lava shields) in relation to the basement of Palaeozoic rocks and trough of Cretaceous and Tertiary sediments. (After Joyce 1975.)

Table 13.2 Distinguishing characteristics of maar-type volcanoes (after C. A. Wood, *in* press).

	Maar	Tuff ring	Tuff cone
rim dips	outward	quaquaversal	quaquaversal
floor level	below surroundings	higher than surroundings	higher than surroundings
magmatic materials (%)	0–100	90–100	90–100
inner slope	steep or vertical	steep	steep
outer slope	gentle	steep	steep
crater or cone diameter	large	large	small
ejecta or crater volume	small	small	large

(a) Maars

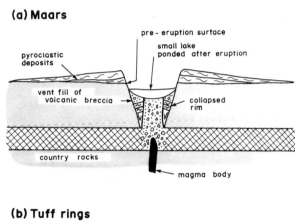

(b) Tuff rings

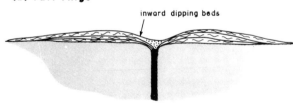

(c) Tuff cones

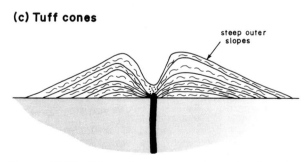

Figure 13.17 Schematic cross sections showing differences between the three types of crater formed by phreatomagmatic eruptions and for maars also by phreatic activity.

for phreatomagmatic craters with steep to vertical inner slopes, with floor level below the level of the surrounding terrain and with shallow, outward dipping slopes on the surrounding tuff ring. Other important reviews describing these volcanoes are given by Ollier (1967a), Lorenz *et al.* (1970), Lorenz (1973) and Wohletz and Sheridan (1983).

Maars, when new and little eroded, have craters which are cut into the surrounding country rock (Fig. 13.17a). They generally lack or have only minor inward dipping beds, and frequently exhibit near-vertical scarps below rim beds. Maar craters range in diameter from a few hundred metres to about 3 km. W_{cr} for 116 maars measured by C. A. Wood (*in press*) from throughout the world are given in Figure 13.18. The youngest maars have crater depth-to-diameter ratios of 1 : 5, but this ratio increases with age as the craters become infilled with epiclastic sediment, often deposited in ponded shallow lakes, and as erosion increases their diameter. They form when rising magma explosively interacts with ground water or surface-derived water below the original topographic surface (Figs 13.17a & 19), and the deposits may contain very little or, in the case of phreatic eruptions, no juvenile magma. Note that Wohletz and Sheridan (1983) use 'maar' somewhat differently, applying it to the crater of either a tuff ring or tuff cone occurring below the pre-existing ground surface. According to these authors tuff rings are more commonly associated with maars than tuff cones.

Maars may lie above **diatremes**, which is the term used for a pipe-like volcanic conduit filled with

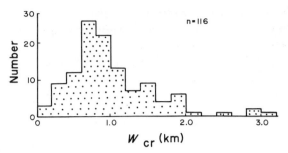

Figure 13.18 Histogram showing measurements of crater diameter for maars. (After C. A. Wood in press.)

volcaniclastic debris (Fig. 13.17a). Diatremes are only exposed by deep erosion and are therefore generally only seen in older volcanic successions. They can be very important economically (e.g. Sillitoe *et al.* 1984).

Tuff rings have craters that lie on, or above, the pre-eruption surface, and have relatively steep rims which dip both inwards and outwards with approximately equal slopes (Fig. 13.17b). Less information is available on the dimensions of tuff rings (although this may be partly because of the confusion in terminology), but they seem to have dimensions comparable with maars (C. A. Wood *in press*). Tuff rings form when rising magma interacts explosively with abundant water close to or at the ground surface. They usually contain a much

higher proportion of juvenile material than do maars. Tuff rings are also usually basaltic, but more-acidic ones are also common (e.g. Phlegrean Fields, Italy).

Tuff cones (Fig. 3.14) differ from tuff rings by having smaller craters, and having larger height to width ratios (Fig. 13.17c). They are formed in areas where surface water (lake or shallow-marine) is located above the vent. They show bedding angles of 20–25° near rim crests.

The pyroclastic deposits forming the constructional parts of maars, tuff rings and tuff cones consist of stratified and cross-stratified ash. Cross-stratified units can be readily identified as base surge deposits and should show radial palaeocurrent distributions (e.g. Fig. 13.15). However, it may be difficult to distinguish between flat-bedded layers which are base surge plane beds and those which are surtseyan air-fall layers (Ch. 7). From the number of examples of tuff rings and maars we have visited, the conclusion is that most of the thinly stratified flat beds found in or near the crater walls are base surge rather than air-fall ash deposits. Realistically, base surge and air-fall material would often have been deposited simultaneously. The layers often show abundant evidence that they were wet and sticky when deposited (Chs 5 & 7). Debris-flow deposits may also occur.

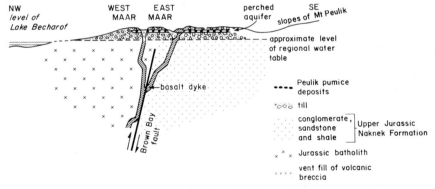

Figure 13.19 Schematic cross section showing the setting of the Ukinrek maars in Alaska, formed during the April 1977 eruption (also see Figs 13.14b & 5.18c). Subsurface geology is inferred and shows the proposed basaltic dyke that triggered the explosive eruption, an evacuated feeder under West Maar which led to collapse of the crater rim along arcuate fractures, and a dyke feeding a lava dome which was extruded on the crater floor of East Maar. The ground water contained in the underlying till and silicic volcaniclastic deposits from nearby Peulik volcano controlled the dominantly phreatomagmatic course of the eruption. (After Self *et al.* 1980.)

Figure 13.20 Large ballistic block of country rock with bomb sag at Tower Hill (Fig. 13.15b). The local stratigraphy suggests this fragment of limestone was brought to the surface from a depth of 600 m (W. Edney *pers. comm.*) and this was part of the aquifer that triggered the hydrovolcanic eruption.

Near-vent ballistic blocks and bombs are often common, and are usually fragments of country rock. If the local stratigraphy is known, then the ballistics can be used to estimate at what depths explosions were taking place, and to identify subsurface aquifers (Fig. 13.20).

Penecontemporaneous collapse of parts of crater rims, including country rock, often occurs during an eruption, and structural and stratigraphic complexities in the ring succession may be found. Erosion surfaces and unconformities within the deposits of crater rims record changes in erosional and constructional phases around the rim during the eruption. On a small scale these may represent changes in the direction of movement and velocity of surge blasts. On a large scale they could result from a change in the position of the vent, or renewed activity from the same vent after a major break in the eruption.

The evolution and eruptive histories of these volcanoes have been discussed by Wohletz and Sheridan (1983) and Heiken (1971). Generalised stratigraphic sections of the deposits found in maars, tuff rings and tuff cones are shown in Figure 13.21. Typically the first deposit to be ejected is a volcanic breccia consisting of coarse angular fragments of broken country rock. This is essentially the product of initial phreatic explosions. Most of

the fragments are deposited by ballistic fall-out, but these may be interbedded with less-coarse fall deposits and fine-grained surge layers. Small maars are predominantly composed of such deposits, e.g. the Ukinrek maars (Figs 13.14b & 19). During the main construction phase of larger maars and tuff-rings, thinly bedded deposits are mainly emplaced by highly inflated and energetic surges (Ch. 7). Tuff cones follow the same initial eruptive pattern but continue into a third stage, characterised by deposition from poorly inflated surges and falls. This forms massive, crudely bedded deposits which construct the major portion of the cone.

A number of centres show a common progression from phreatomagmatic to strombolian or hawaiian activity, and a scoria cone may be built or a lava lake pond inside the maar crater or tuff ring may be produced by the earlier activity. This reflects a decrease in the degree of water–magma interaction or the water–magma ratio during the eruption (Ch. 3). Either the ground water or surface water source was gradually exhausted, or rising magma was effectively cut off from the water supply.

To illustrate the effects of availability of water on the volcanic stratigraphy, the successions found in four centres in the Newer Volcanics basaltic field in southeastern Australia are shown in Figure 13.22. At Mt Eccles the eruption that built a scoria cone

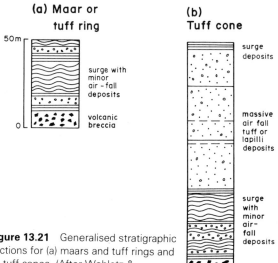

Figure 13.21 Generalised stratigraphic sections for (a) maars and tuff rings and (b) tuff cones. (After Wohletz & Sheridan 1983.)

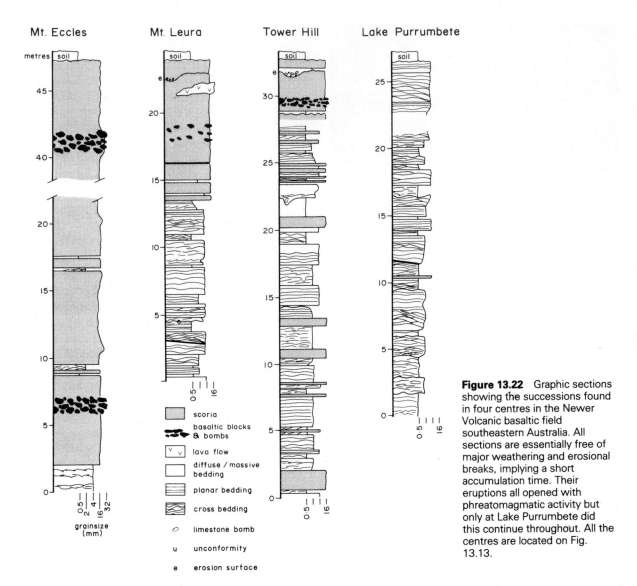

Figure 13.22 Graphic sections showing the successions found in four centres in the Newer Volcanic basaltic field southeastern Australia. All sections are essentially free of major weathering and erosional breaks, implying a short accumulation time. Their eruptions all opened with phreatomagmatic activity but only at Lake Purrumbete did this continue throughout. All the centres are located on Fig. 13.13.

was initially phreatomagmatic, but there was insufficient water for this phase to continue long enough to form a tuff ring. At Lake Purrumbete (Fig. 13.15a) abundant ground water was available throughout the eruption which was solely phreatomagmatic. Both Mt Leura and Tower Hill (Fig. 13.15b) are major 'nested' tuff ring–or maar-scoria cone complexes. The eruption at Mt Leura gradually progressed from a phreatomagmatic phase during which a tuff ring was formed to a strombolian phase when a scoria cone complex was built (Fig. 6.6a). Lava fountaining along a fissure also built a series of

spatter cones, and a lava flow also occurs in the tuff ring moat. Base surge layers and scoria-fall deposits are closely interbedded where the transition from phreatomagmatic to magmatic activity took place (Fig. 13.23a). This may have been due to pulsations in the magma discharge rate subtly controlling the water–magma ratio, and hence the type of eruptive activity. Tower Hill has a much more complex eruption history, and the maar crater wall shows a succession of interbedded phreatomagmatic and magmatic deposits (Figs 13.22 & 23b). Magmatic activity also built a considerable scoria cone complex

Figure 13.23 (a) Base surge and scoria-fall deposits near the top of the tuff ring at Mt Leura, Victoria, Australia, showing the change from phreatomagmatic to strombolian activity. (b) Interbedded phreatomagmatic and scoria-fall deposits (dark layers with arrows) near top of crater wall at Tower Hill, Victoria, Australia.

and eruption points were clearly aligned on fissures (Fig. 13.15b). Constantly alternating phases in the style of activity may be due to a very complex interrelationship between rate of supply of water, from a ruptured and fracturing subsurface aquifer, and magma supply rate (Edney 1984).

The original basaltic glass (sideromelane) in the phreatomagmatic deposits of some craters is quickly altered by hydration to palagonite (Hay & Iijima 1968; Ch. 14). This produces indurated and lithified tuffs, and primary textures become obscured, especially in layers composed of fine ash. Palagonitisation seems to be more commonly found in tuff cones rather than the other types of crater (Wohletz & Sheridan 1983), although many tuff rings are altered (e.g. Fort Rock, Fig. 13.14c). Wohletz and Sheridan (1983) suggest that this is because tuff cones are largely constructed by less-inflated surges (Ch. 7) containing a high proportion of water which is then trapped in the ash after emplacement. Alteration of the glass may also be caused by steam moving through the deposits with mild hydrothermal activity, as has occurred at Surtsey.

Eruptions of maars, tuff rings and tuff cones have, rarely, been witnessed. The two Ukinrek maars which erupted in the tundra of the Alaskan Peninsula between 30 March and 9 April 1977 are the best documented examples of the formation of this type of crater (Kienle *et al.* 1980, Self *et al.* 1980; Figs 13.14b & 19). Historical eruptions forming maars suggest durations from a few days to a few weeks. The eruption that produced Tower Hill deposits perhaps persisted for a few months (Fig. 13.22). Only one major erosional unconformity is found near the top of the succession and this does not necessarily indicate a long time break. The best known tuff cone is the marine volcano, Surtsey, which began to erupt in November 1963 and continued into the summer of 1965, although satellite vents remained active until 1967.

13.6 Pseudocraters and littoral cones

These are small secondary cones or **rootless vents**. Pseudocraters from Iceland have been described, and form where basalt lavas flow into lakes (Thor-arinsson 1953). Secondary explosions are caused by steam trapped under a lava, and small volcaniclastic cones are built. Littoral cones form in a similar way when basaltic lavas enter the sea, and are best described from Puu Hou, at the southern end of Hawaii (Fisher 1968). These examples formed in a five day period when two lava channels from the 1868 Mauna Loa flow entered the sea. A cone complex consisting of overlapping half-cones was constructed on land. They are built of poorly bedded hyaloclastite fragments and layers of agglutinated spatter which show crude welding profiles and form clastogenic lavas. Older littoral cones along this stretch of coastline show identical features (Fig. 13.24).

These types of volcanic cone are important palaeoenvironmental indicators, and it is important not to misinterpret examples as primary centres (e.g. see Cheshire & Bell 1977). The distinctive association of facies to look for is the combination of:

(a) pillow lavas,
(b) palagonitised hyaloclastite breccias which may show crude foreset bedding (flow-foot breccias) and define a lava delta (Ch. 4) and
(c) shallow marine and lacustrine sediments (and marine or freshwater fossils to distinguish between marine and lacustrine shorelines).

13.7 Stratovolcanoes

Stratovolcanoes or **composite volcanoes** are the characteristic volcanic landform found at subducting plate margins, and are the most abundant type of large volcano on the Earth's surface (Fig. 13.25). Stratovolcano morphology results from repeated eruptions of pyroclastics and relatively short lava flows from a central vent. The textbook diagram usually shows a symmetrical cone with steeply dipping pyroclastic deposits interbedded with lava flows. However, in detail the shapes of these volcanoes are more variable (e.g. there may be several eruptive points, or the presence of a caldera), and they rarely have a *layer-cake* geology. It is also important to stress that mass-wastage is

Figure 13.24 Puu Ki littoral cones on the southernmost coast of the island of Hawaii. (a) Crescent-shaped landward rim of one example. (b) Seaward section of a half-cone with steeply dipping layers of spatter and clastogenic lavas. (c) Layer of partially agglutinated spatter overlying a clastogenic lava.

just as important as, if not more important than, volcanic processes in determining their characteristics and lithologies (Ch. 10), although obviously an aggradational landform would not be present without the volcanic activity. Volcaniclastic deposits (pyroclastic and epiclastic) are usually very important volumetrically, and on some volcanoes can make up to 70–80% or more of the succession. However, the variations in the proportions of lavas to volcaniclastics can vary enormously from one stratovolcano to another. Some may be dominated by lava eruptions (e.g. Bultitude 1976a). However, even on volcanoes such as Ruapehu (Fig. 13.25d), where lavas are important in the edifice, they are surrounded by extensive **ring plains** of volcaniclastic debris.

At destructive plate margins, stratovolcanoes are built by eruptions of calc-alkaline magmas (Ch. 15). They are usually broadly andesitic or basaltic-andesite in composition, but their products may include basalts, dacites and rhyolites, and some are basaltic centres, e.g. Fuji in Japan. Alkaline magmas also generate stratovolcanoes which are on average larger than their calc-alkaline counterparts (Table 13.3). Examples built up of trachytic and phonolitic products are common on the Mid-Atlantic islands of the Azores and Tenerife (which are not associated with subduction margins; Ch. 6), and a basaltic example is Etna. Again it should be clear that chemical composition cannot be the sole control of volcano shape, and indeed the 'classic' stratocone, Fuji, is basaltic. Fuji has much lower average magma discharge rates than basaltic shield-forming volcanoes.

13.7.1 MORPHOMETRY

Pike and Clow (1981) collated morphometric data for over 200 stratovolcanoes. They divided stratovolcanoes into a number of classes according to

Figure 13.25   Examples of stratovolcanoes. (a) Mt Shasta in the Cascades, western USA. (b) Crater Lake, western USA. (c) Mt Egmont, North Island, New Zealand. (d) Ruapehu, North Island, New Zealand.

whether the magma type is calc-alkaline or alkaline, and according to the presence of a summit crater, sector collapse scar (Ch. 5) or caldera. Those volcanoes with calderas were further subdivided according to the proportions of silicic and pyroclastic products. Data for these different classes are summarised in Table 13.3. Some stratovolcanoes are topographically very impressive, rising steeply as high as 5 km above their bases. The tallest seems to be Queen Mary's Peak on Tristan da Cunha. Average slopes of stratovolcanoes range from about 15° to 33°.

Another term we need to note in connection with the morphology of stratovolcanoes is 'somma'. This is used to describe a high, circular or crescent-shaped ridge with steep inner walls, which may be the rim of an older crater or caldera surrounding a central volcanic cone. The name is taken from Monte Somma (Table 13.3) which forms a low ridge encircling the active cone of Vesuvius on its northern and eastern sides.

13.7.2   OUTPUT RATES, REPOSE PERIODS AND LIFE EXPECTANCY

Little information on the output rates of stratovolcanoes is available because the activity of very few centres has been monitored over a sufficient time interval. Output of Cerro Negro in Guatemala has averaged 0.15 km^3 per 100 years for the first 121 years of its lifetime (C. A. Wood 1978), and Fuego has averaged 0.38 km^3 per 100 years for the past 450 years (Martin & Rose 1981). The three active stratovolcanoes on São Miguel, Sete Cidades, Agua de Pau and Furnas, have a combined average output rate of 0.1 km^3 per 100 years (dense rock

Table 13.3   Summary of dimensions of different classes of stratovolcano (after Pike & Clow 1981).

Class	Sample	Basal diameter (km)	Height* (m)	Summit crater, collapse scar or caldera diameter (km)	Average volume (km^3)	Examples
1 calk-alkaline with summit crater	59	6–55	625–3700	0.1–1.75	60	Fuégo, Popocatepetl, Mt Shasta, Mayon, Fuji, Ngauruhoe, Mt Egmont
2 alkaline with summit crater	18	1.5–60.5	425–4900	0.15–0.9	80	Etna, Vesuvius, Stromboli, Hekla, Queen Mary's Peak, Mt Erebus
3 calc-alkaline with sector collapse scar	7	11–30	1020–2700	1.2–4.5	100	Bezymianny, Shiveluch, Mt St Helens, Lassen Peak
4 calc-alkaline with caldera	44	7.5–68.5	500–3350	1.75–11	160	Coseguina, Toluca, Aniakchak, Batur, Ruapehu
5 calc-alkaline with caldera, more silicic and more pyroclastics than 4	32	8–59	200–1700	2.2–18	80	Crater Lake, Santorini, Krakatau, Rabaul, Deception Island
6 alkaline with caldera	57	6.5–87	350–4700	0.75–9	250	Sete Cidades, Agua de Pau, Furnas, Kilimanjaro, Tambora
7 alkaline with caldera, more silicic and more pyroclastics than 6	19	7.5–64.5	100–1250	1.8–18.5	200	Lake Bolsena, Monte Somma (Vesuvius), Pantelleria, Fantale

* Average height of crater or caldera rim crest above pre-volcano topographic datum.

Table 13.4   Average lifetime output rates for some stratovolcanoes (after Wadge 1982).

Volcano	Volume (km^3)	Age (10^3 years)	Average output rate (km^3 per 100 years)
Oshima	45	20–30	0.18
Sakurajima	40	13–15	0.28
Akan	75	31	0.24
Shikotsu	140	32	0.44
Asama	60	30	0.20
Fuji	400	80	0.50
Hakone	200	400	0.05
Aso	45	25	0.18
Kirishima	50	17	0.29
Fuego	50	100	0.05
Santá Maria	20	30	0.07
Kluchevskoy	250–300	10	2.75
Shiveluch	1000	200	0.50
Avachinsky	100	60	0.17

equivalent, DRE) over the past 5000 years (Booth *et al.* 1978), and this figure includes an estimate of the large volume of trachytic pumice dispersed far away from the cones as air-falls.

In Table 13.4 a compilation is given of estimates of the volumes of some large active stratovolcanoes and the time taken to build them. Average lifetime output rates can be calculated from these, but note that the volumes in general represent less than the total volume of magma erupted at each volcano because dispersed pyroclastics are not accounted for. Wadge (1982) indicated that periods of dormancy for stratovolcanoes may be twice as long as periods of activity, for which average output rates are three times higher than the total construction rate. Wadge (1980, 1982) suggested that during active periods average output rates are fairly constant over periods of many years, even though magma may appear in a complex series of eruptions. According to Wadge, periods of steady-state volcanism may be maintained for up to a few hundred years.

Repose periods between eruptions seem to vary considerably. Some stratovolcanoes are almost persistently active, and we can think here of Fuego, Soufrière on St Vincent, the island volcano Stromboli, and Ruapehu. Other stratovolcanoes seem to have repose periods as long as 10^3 and 10^4 years, and these are sometimes erroneously termed dormant, and even extinct, although they may be very much alive.

Most active stratovolcanoes appear to be less than 10^5 years old (e.g. Table 13.4). Some are older, Mt Pelée having been K–Ar dated at least as old as 0.4 ± 0.2 Ma BP (400 000 years) (Briden *et al.* 1979). The oldest volcanic rocks on Santorini are dated at about 1 Ma BP. However, it is perhaps more important to know the length of time over which overlapping and closely spaced stratovolcanoes of a polygenetic complex can be active in an area. It is unlikely that within ancient terrains (Palaeozoic and older) all the individual centres could be distinguished, so with what time periods could we be dealing? Martinique, as an example, has at least six (probably more) centres, and K–Ar dating (Briden *et al.* 1979) indicates their development over at least 20 Ma. We are therefore considering time periods perhaps as large as 10^7 years or more, in many cases reflecting the age of the volcanic arc of which stratovolcanoes are characteristic elements.

13.7.3 ERUPTIONS, CHARACTERISTICS AND DEPOSITS

Eruptions of stratovolcanoes are very variable in style, duration and frequency. As a result even modern stratovolcanoes present a myriad of problems in the stratigraphic analysis of their deposits.

Several modern cones exhibit nearly continuous mild strombolian activity, e.g. Stromboli. Vulcanian eruptions are common and these may generate small denser-clast types of pyroclastic flow, e.g. Ngauruhoe and Fuego (Chs 5 & 6). Phreatomagmatic eruptions are also common where volcanoes have, or had, a crater lake, e.g. Soufrière St Vincent (Shepherd & Sigurdsson 1982; Ch. 6), and Ruapehu (Healy *et al.* 1978). More cataclysmic

Figure 13.26 Wizard Island, which is a scoria cone formed within the Crater Lake caldera.

are peléan eruptions, involving the explosive collapse of viscous domes as exemplified by the 1902 eruption of Mt Pelée (Ch. 5). Less frequently, plinian and ignimbrite-forming eruptions occur from these cones, producing much larger (>10 km^3) types of pumiceous deposit. Welded tuffs of both air-fall and pyroclastic flow origin can be formed during this type of activity. These eruptions also cause the more important changes in cone morphology, e.g. Mt St Helens 1980, and caldera collapse may result, as happened during or after the ignimbrite-forming eruptions of Crater Lake (Figs 13.25b & 26), Santorini 1470 BC and Krakatau AD 1883. Effusive eruptions are frequently observed, and lavas and domes may be important in the pile of some stratovolcanoes, although these may vary considerably in composition, even on the same volcano. Other eruptions can occur from parasitic centres, and scoria cones (Fig. 13.26), spatter cones and phreatomagmatic craters may form, perhaps aligned along a fissure. Magma is also intruded into the piles, and dykes and shallow intrusive bodies are seen in the crater or caldera walls of stratovolcanoes (Fig. 13.27).

Stratovolcanoes can therefore show a wide variety of primary volcanic products, and these may be distributed quite differently around the volcanic centre. Viscous lava flows and welded air-fall tuffs may be confined to near-vent areas. Pyroclastic flows will be channelled down canyons and valleys,

Figure 13.27 Near-vertical dykes which form ridges cutting a succession of basaltic lavas in caldera wall of Ceboruco volcano, Mexico.

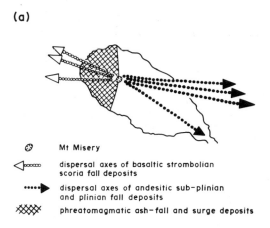

(a)

⊗   Mt Misery

◁∘∘∘∘   dispersal axes of basaltic strombolian scoria fall deposits

•••••▶   dispersal axes of andesitic sub-plinian and plinian fall deposits

⨯⨯⨯⨯   phreatomagmatic ash-fall and surge deposits

(b) Eruption of basaltic strombolian scoria falls

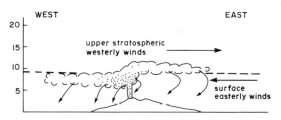

(c) Eruption of andesitic sub-plinian and plinian falls

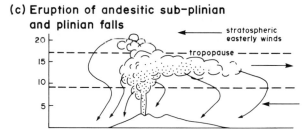

(d) Eruption of phreatomagmatic ashes

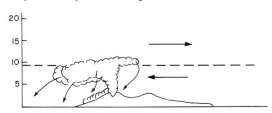

that are often repeatedly re-excavated and infilled (Figs 1.2 & 8.7b) and may be radially distributed. Parasitic cones and dykes may also have radial distributions. The dispersal of pyroclastic fall deposits will be controlled by the height of the eruption column and the prevailing wind direction at the time of the eruption. This means that on many stratovolcanoes the distribution of fall deposits may be quite asymmetric. This is well illustrated by many volcanoes in the Lesser Antilles because of a layered atmospheric system with contrasted bipolar wind directions at different altitudes (Sigurdsson *et al.* 1980, Roobol *et al.* 1985). Around Mt Misery on St Kitts (Fig. 13.28), basaltic strombolian scoria falls and phreato-magmatic ash falls (Fig. 6.32) have been dispersed mainly to the north-west and west by lower surface tropospheric winds. In contrast, andesitic sub-plinian and plinian falls with higher eruption columns have been dispersed to the south-east and east by upper tropospheric winds.

In addition to these types of 'primary' strati-graphic complexity, further problems arise because mass-wastage and epiclastic processes are very important on stratovolcanoes, and their deposits form major parts of the successions (Ch. 10, Sections 14.8.2 & 6).

The volcanic successions of two modern strato-volcanoes are illustrated in Figures 13.29–31. Stratigraphies show complex and rapid changes in

Figure 13.28 The effect of a layered wind system and height of eruption columns on the dispersal of air-fall deposits erupted from Mt Misery, St Kitts. The lower tropospheric easterlies are today confined to altitudes of ~5–9 km depending on season. We have depicted the boundary during the wet summer season. The tropopause occurs at ~17–18 km. Trajectories are schematic transport paths of air-fall ejecta. (After Roobol *et al.* 1985.)

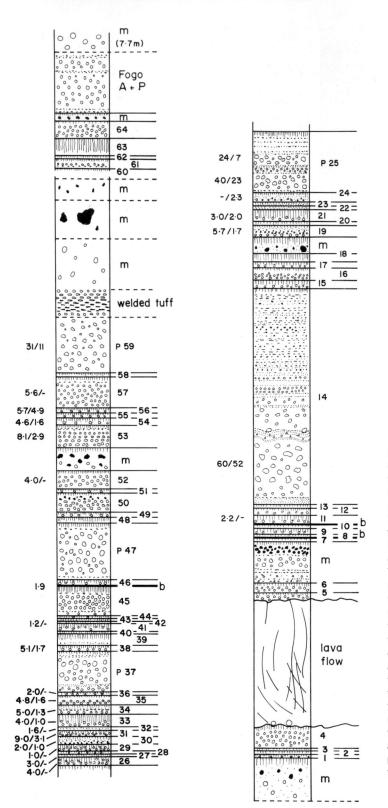

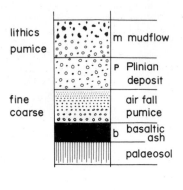

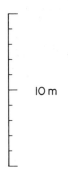

Figure 13.29 Composite section through part of the volcanic succession of the Agua de Pau volcano on São Miguel. The section comprises 65 separate trachytic pumice-fall deposits, numbered 1–65 in sequence upwards, and includes several plinian (P) deposits, together with three thin basaltic ashes, one welded ignimbrite, a number of mud flows (m), some of which have been omitted, and a lava flow. The thick, in part, very coarse fall deposit numbered 14, has a local source. The figures to the left of the column give the average maximum diameters in centimetres of, respectively, the three largest pumice clasts and the three largest lithic clasts found. The Fogo A eruption has been ^{14}C dated at 4550 years BP, and the volcano has erupted four younger trachytic pumice-fall deposits not shown here, the last being in 1563 (Ch. 6). Probably several tens of thousands of years of activity are represented by the section. Carbonised wood found in deposit number 14 has given a ^{14}C age of >34 200 years BP. It is also quite probable that some of the falls were erupted from other vents. (After Booth et al. 1978.)

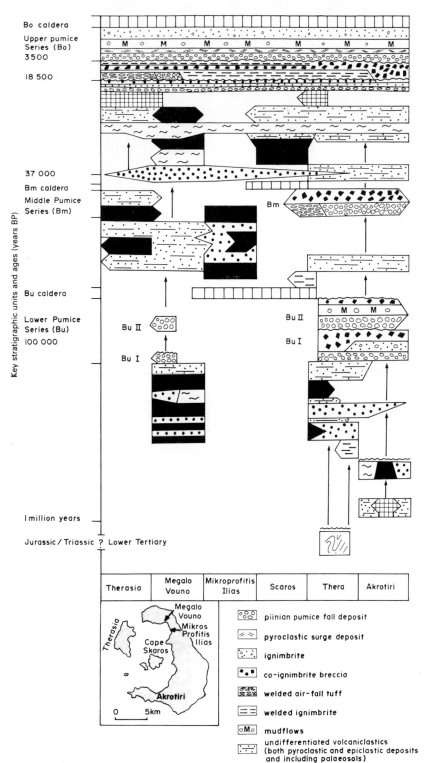

Figure 13.30 Interactive stratigraphic section illustrating the volcanic succession of Santorini volcano. Note section is not to scale in thickness or time. Bu, Bm and Bo are abbreviations of the German '*Bimsstein, unterer, mittler* and *oberer*', stemming from Reck's classic work, and are key stratigraphic markers. Ages are based on ^{14}C and fission-track dating, and the application of oxygen isotope stratigraphy to deep-sea cores containing correlated distal Santorini ash layers. (Based on Pichler & Kussmaul 1980, Druitt & Sparks 1982 and J. V. Wright *unpub. data*.)

lithologies, both vertically and laterally. Volcaniclastic deposits are the much more voluminous products of both examples. Santorini also shows complex compositional variations, and products range from basaltic through to rhyodactic. Many of the pumice deposits also result from mixed magma eruptions, e.g. the Lower Pumice (Fig. 6.13c), the Middle Pumice (Ch. 6) and the Minoan deposits (Bond & Sparks 1976). Obviously, the generation, evolution and type of magma erupted from these volcanoes can be complex. This could involve magma chambers on different levels and a complex plumbing system, as well as different batches of *primary* basaltic magma periodically rising into the system. Marine stratovolcanoes could consist of an initial stratigraphy much like basaltic seamounts (Section 13.10), i.e. an initial pile of pillow lavas passing upwards into hyaloclastites, pyroclastics and epiclastics, and then even a subaerial stratigraphic succession like the one just described.

It has also been proposed that many porphyry copper–molybdenum–gold deposits are formed during final consolidation of high-level magma chambers beneath andesitic–dacitic stratovolcanoes (Sillitoe 1973). The porphyry-copper bearing stocks may intrude the basal parts of the volcanic pile, although more commonly they appear to be in the subvolcanic basement (Ch. 14). Stratovolcanoes found in island arc settings, especially those with submarine calderas (e.g. Santorini, Krakatau (Fig. 9.10), Rabaul and Deception Island), are thought to be economically important because they are considered to be favourable sites for the formation of a whole selection of sea-floor polymetallic massive sulphide deposits, which are generally grouped together under the general heading of 'Kuroko-type deposits' (e.g. Colley 1976). Stratiform massive sulphide deposits can therefore be a very important part of the successions of these volcanic centres.

13.7.4 MASS-WASTAGE AND EPICLASTIC PROCESSES

Unfortunately volcanologists, igneous petrologists and geologists in general have largely ignored or underestimated the influence of sedimentary processes on modern volcanoes (Ch. 10). Yet if we observe modern volcanic regions, then it is quite evident that reworked volcanic rock is volumetrically important and must be significant in the geological record, and we have tried to emphasise this in this chapter and especially in Chapter 10. Stratovolcanoes are very prone to mass-wastage because they are high topographic features, although all modern volcanoes are continually being degraded.

The most comprehensive study to date on epiclastic processes is by Vessell and Davies (1981) on the cone of Fuego. They divided the deposits of Fuego into four facies associations (Fig. 13.32):

(a) Volcanic core facies of lavas, air-fall deposits and colluvium breccias.
(b) Proximal volcaniclastic facies of volcanic breccias (block and ash-flow deposits, colluvium breccias) and air-fall deposits.
(c) Medial volcaniclastic facies of debris-flow deposits (lahars) and fluvial conglomerates with some air-fall deposits.
(d) Distal volcaniclastic facies dominated by fluvial sands, breccias and conglomerates, which interfaces with the coastline near Fuego.

The succession examined by Vessell and Davies (1981) was deposited in the past 20 000–30 000 years, and repeated cycles of deposits and sedimentary processes have been recognised. Three cycles, each triggered by an eruption, could be divided into four phases, as discussed in Section 10.2 and summarised in Table 13.5. From Table 13.5 it is apparent that there is much more happening on volcanoes than simply volcanic

---

◀ **Figure 13.31** Three consecutive views of the caldera wall of Santorini illustrating part of the volcanic succession shown opposite. Views move north-south from south of Cape Skaros to south of Thera town, and cover a distance of about 2 km. Height of caldera wall is up to 300 m. Key stratigraphic markers: Lower Pumice Series (deposits of younger Bu II eruption) is prominent light coloured layer in (c); Middle Pumice Series (Thera welded tuff, Ch. 6) is prominent dark layer in lower part of (b), this mantles the Lower Pumice to south which is cut-out by the Bu caldera; Upper Pumice Series (Minoan deposits) is white layer seen at the very top of the succession in (c) which can be traced northwards on top of cliffs.

(a)

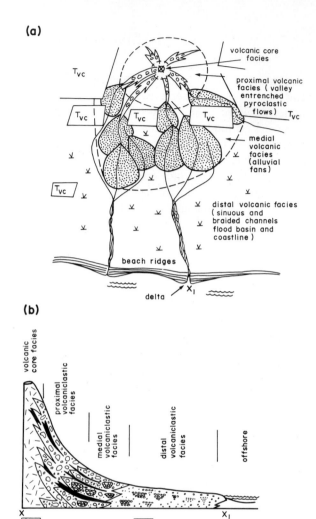

(b)

X — lavas

air fall deposits

pyroclastic debris flow deposits

fluvial channels

Figure 13.32 Facies model of a stratovolcano based on studies of Fuego, Guatemala. In (a), T_{vc} are massifs of Tertiary volcaniclastics which separate elongate troughs filled with modern volcaniclastic sediment. The active cone is flanked by numerous alluvial fans (dotted areas) which extend onto the edge of the coastal plain. X to X' is the cross section shown in (b). (After Vessell & Davies 1981.)

eruptions. On Fuego, only eruptions producing greater than 6×10^7 m^3 of ejecta were found to be capable of triggering large-scale sedimentary events, and the repose period between these eruptions is 80–125 years. Minor eruptions with a shorter repose period do not significantly affect the sedimentary system, which proceeds as a series of

Table 13.5 Sedimentary cycles triggered by larger eruptions of Fuego volcano (after Vessell & Davies 1981).

Phase	Duration (years)	Sedimentation process
1	80–125	stream incision and delta reworking
2	1	air-falls and pyroclastic flows
3	~2	debris flows
4	~20–30	braided fluvial transport and delta construction

short pulses. Since major eruptions between 1972 and 1974, at least 6×10^6 tonnes of volcaniclastic debris have been removed from the cone of Fuego, which is at present in a Phase 4 stage (Table 13.5; Section 10.2). More data describing fluviatile processes on Fuego are given in D. K. Davies *et al.* (1978c).

Kuenzi *et al.* (1979) also described fluvial and deltaic sedimentation on the Guatemalan coastal plain in front of Santá Maria volcano. They document effects of the dramatic increase in sediment supply after the 1902 plinian eruption (Chs 6 & 10). The bed of the Samalá River was raised 10–15 m, and between 1902 and 1922 a deltaic platform prograded approximately 6.4 km seaward and deposited a prism of sediment having a volume of about 4 km^3 (Figs 13.33 & 10.1). However, with waning sediment supply, the delta was destroyed and sands were redistributed laterally into prograding shoreface and beach environments. This developed the present arcuate shoreline, which has essentially remained unchanged since before 1947.

Figure 13.34 shows facies patterns resulting from long-term progradation along the entire length of the coastal plain supplied by Santá Maria and nearby stratovolcanoes of the Guatemalan volcanic front (also see Fig. 10.1). In this setting, progradation would not be constant along the length of the volcanic arc, but would vary with time, and with magnitude and frequency of volcanic eruptions. Sporadic episodes of delta formation and subsequent reworking of deltaic sands along sedimentary strike would produce a complex sand body showing a variety of vertical sedimentary sequences

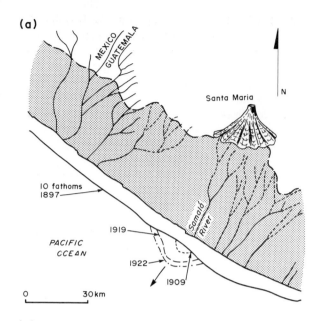

(a)

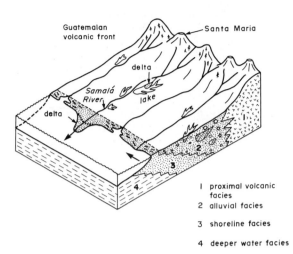

Figure 13.34 Facies relationships resulting from long-term progradation of coastline parallel to Guatemalan volcanic front. (After Kuenzi *et al.* 1979.)

(b)

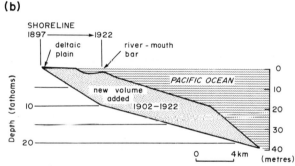

Figure 13.33 (a) Map showing progressive seaward displacement of the 10 fathom depth contour from the mouth of Samalá River, as shown by hydrographic charts for 1897, 1909, 1919 and 1922. (b) Cross section of body of volcaniclastic sediment deposited by deltaic progradation at mouth of Samalá River between 1902 and 1922. (After Kuenzi *et al.* 1979.)

generated by a number of shoreline processes. This would overlie deeper water facies (e.g. prodelta turbidites) and be overlain by an alluvial complex which would intertongue landward with proximal volcanic facies.

Similarly, marine stratovolcanoes are also subject to major mass-wastage, involving landslides, debris flows, and to high energy, fluvial gullying, and torrent flow, on the subaerial flanks, to subaqueous slides, debris flows and turbidity current processes

(Ch. 10) on the subaqueous flanks. The subaqueous flanks would expand by the radial growth of a subaqueous volcaniclastic apron fed by the mass-flow processes mentioned above.

In addition to these degradational processes, stratovolcanoes are particularly prone to explosively or non-explosively induced sector collapse whereby a large segment of the edifice is mobilised into a dynamic slide and avalanche (Ch. 10, Ui 1983, Siebert 1984, Francis *et al.* 1985).

13.8 Intermediate–silicic multivent centres

Some other andesitic–dacitic and alkaline volcanoes seem to be multivent complexes that lack a central cone.

The highlands of St Lucia form a volcanic core from which radiate valley-filling dacitic pumice flow fans (Fig. 13.35). Studies of the volcanic succession indicate that eruptions have occurred from a number of vents located over a wide area, involving different dacitic magma batches at various times (J. V. Wright *et al.* 1984). Eruption of volatile- rich magma led to highly explosive pumice-forming activity. Degassed magma was extruded later, from the same vents or from the attenuated

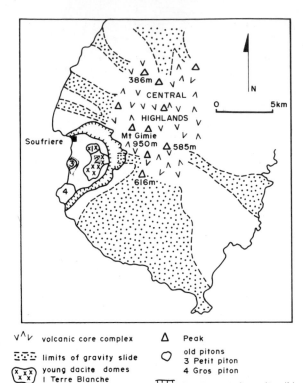

v^v volcanic core complex   △ Peak

⌐⌐⌐ limits of gravity slide   ○ old pitons
⌐⌐⌐   3 Petit piton
(x x x) young dacite domes   4 Gros piton
 x x x   1 Terre Blanche
         2 Belfond Complex   ⊤⊤⊤ head scarp of gravity slide

Figure 13.35 Map showing the multi-peaked central volcanic core of St Lucia from which fans of late Quaternary dacitic pumice-flow deposits radiate. Some of the peaks are the suspected source vents, but others, including Mt Gimie, are older andesitic lavas and volcaniclastics. (After J. V. Wright *et al.* 1984.)

flank of the Qualibou depression, which is thought to have formed by a large gravity slide on the side of the volcanic complex (Ch. 9). This activity produced lava domes with associated block and ash-flow deposits. The island of Dominica has a similar central volcanic core complex, and vents for pumice flows are plugged by lava domes with block- and ash-flow aprons (Sigurdsson 1972, Carey & Sigurdsson 1980; Fig. 9.3).

Much larger intermediate–silicic multivent centres are known which are associated with large calderas, and some come under the heading of **volcano-tectonic depressions**. Like rhyolitic volcanoes (Section 13.9), these are surrounded by large ignimbrite sheets. Andesitic examples would include the Laguna de Bay and Taal volcano-tectonic depressions in the Philippines (Wolfe &

Self 1982; Fig. 13.36). Laguna de Bay has the appearance of a caldera with dimensions of 25 × 12 km (300 km²), nested in a large collapse structure of uncertain origin. This latter structure is 35 × 45 km in diameter (1500 km² in area), which would make it one of the largest volcano-tectonic structures yet described. It may be compared with, for example, the Taupo volcanic centre covering an area of 1100 km² (see below). Major Pleistocene eruptions of alkali-rich andesitic ignimbrites from Laguna de Bay and Taal have joined together several Tertiary island volcanoes to form the present landmass of southwestern Luzon (Fig. 13.36). The Phlegrean Fields–Bay of Naples

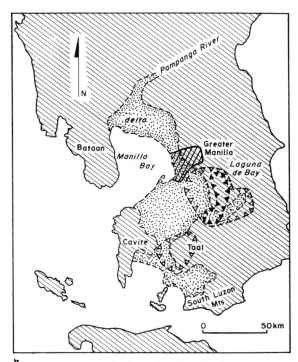

▷◁△ structural margin of volcano–tectonic depression

▼◁▲ caldera margin

▓▓ new land

Figure 13.36 Location of the Laguna de Bay and Taal volcanic centres, and palaeogeographic reconstruction of southwestern Luzon before the main ignimbrite-forming eruptions from these centres about 1 Ma BP. New land was largely added by deposition of ignimbrites, and has joined islands together to form the present landmass. The Taal ignimbrites may, in part, post-date eustatic change and have substantially added to the land area. (After Wolfe & Self 1982.)

volcano-tectonic depression is a large multivent alkaline centre surrounded by the Campanian ignimbrite, which is compositionally a potassic trachyte (Barberi *et al.* 1978, Rosi *et al.* 1983; Ch. 8). Caldera collapse in the Bay of Naples undoubtedly accompanied the eruption. More-silicic examples would have to include the very large dacitic ignimbrite-forming centres like Cerro Galan (Fig. 8.2), and perhaps some of those in the San Juan volcanic field which erupted ignimbrites that are not truly rhyolitic (Ch. 8; Fig. 8.1). The large calderas of Kyushu, Aso, Aira and Ata, all of which are surrounded by extensive andesitic to rhyodacitic ignimbrite sheets, fall into this group. The Aira and Ata calderas have collapsed beneath Kagoshima Bay, which is a large volcano-tectonic graben open to the sea (Fig. 8.3). Again, this would be a contemporary setting for the generation of Kuroko-type massive sulphide deposits.

13.9 Rhyolitic volcanoes or centres

Rhyolitic volcanic centres are some of the largest volcanic landforms on the Earth's surface. They are polygenetic and invariably consist of multiple eruption points or volcanoes. They are found in extensional tectonic settings in rifts, grabens and marginal basins of continents and microcontinents (Ch. 15). Their form is in marked contrast to stratovolcanoes, and in many cases from a distance, it would be difficult to convince the layman that they were even volcanoes! Typically they lack a topographically impressive cone, and in extreme cases they form large, broad volcano-tectonic depressions which G. P. L. Walker (1981e) termed **inverse volcanoes**, the type example being Lake Taupo.

They are multivent complexes, and the volcanic centres are a collection of low rhyolitic hills which

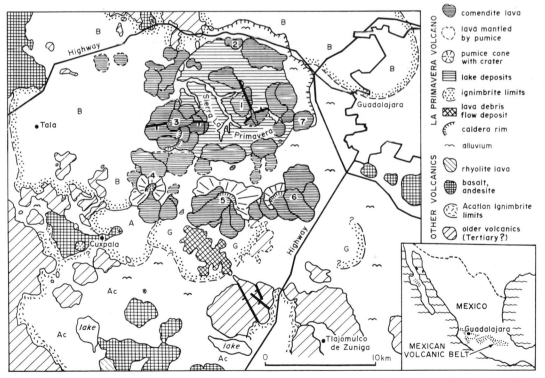

Figure 13.37 Geological map of La Primavera volcano and adjacent areas. A–C refer to different ignimbrites erupted separately; B is the Rio Caliente ignimbrite. Numbers are rhyolite hills: 1, El Majahuate; 2, Pinar de la Venta; 3, El Pedernal; 4, San Miguel; 5, Cerro Las Planillas; 6, Cerro Pelon; 7, El Colli. The lava debris flow deposit results from the collapse of an obsidian flow on Las Planillas. (After G. P. L. Walker *et al.* 1981d.)

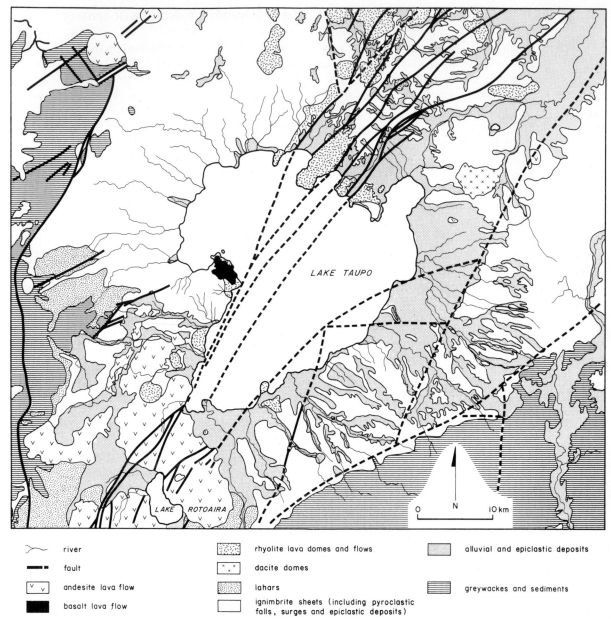

river

fault

andesite lava flow

basalt lava flow

rhyolite lava domes and flows

dacite domes

lahars

ignimbrite sheets (including pyroclastic falls, surges and epiclastic deposits)

alluvial and epiclastic deposits

greywackes and sediments

Figure 13.38 Map of Taupo volcanic centre, showing the major inward drainage system developed in this low profile volcano, the diversity and distribution of volcanic and epiclastic products, and older basement. (After Grindley 1960.)

may be scattered over a few hundred square kilometres, and situated within a shallow caldera (Figs 13.37 & 38). These rhyolite hills are composed of rhyolite domes, coulées and pumice cones. The hills rise from gently sloping ignimbrite sheets, often containing more than one ignimbrite, and interbedded pumice-fall and epiclastic deposits. The whole form is therefore of a broad shield, and C. A. Wood (1977) has drawn an analogy between their shape and basaltic shield volcanoes.

13.9.1 MORPHOMETRY

Detailed morphometric data are not available, but we have made some approximations based on examples from the USA, Mexico and New Zealand. W_s (diameter of shield) varies from about 50 to >200 km, and W_{vc}, which is the width of the volcanic centre defined by the rim of the caldera (Figs 13.37 & 38) is 10–60 km. Slope varies from <1° to 5° and H_s (height of shield) would generally be a few hundred metres, unless resurgence has occurred (see below). Total volume of these shields varies from 10^2 to >10^3 km^3. These volume estimates ignore the large amounts of ash lost far beyond the limits of the volcanoes as co-ignimbrite and distal plinian ash-fall deposits. It should be noted that larger rhyolitic calderas do exist. The largest recognised is Lake Toba in Sumatra, having rim dimensions of 100 × 35 km (Ch. 8).

13.9.2 OUTPUT RATES, REPOSE PERIODS AND LIFE EXPECTANCY

Again, we can make some order of magnitude estimates of the volumetric output of this type of volcano. For La Primavera volcano, which has been active for the past 100 000 years, the average output (as dense rock equivalent, DRE) is 0.06 km^3 per 100 years (G. P. L. Walker *et al.* 1981d). This compares with an average of about 0.4 km^3 per 100 years (DRE) for each of Taupo and Okataina, which are the two most active rhyolitic centres in New Zealand and probably the most productive rhyolitic volcanoes known.

A striking difference between rhyolitic volcanoes and stratovolcanoes and their types of activity is the repose periods between eruptions. Although stratovolcanoes have been persistently active throughout historic time, only one rhyolitic eruption of any kind has been observed by volcanologists. For La Primavera there has been an eruption on average about every 2000 years for the past 100 000 years. For Taupo the average repose period for the past 50 000 years is ~3000 years, but has varied from a few hundred years to 16 000 years, and for Okataina for the same period it is ~2500 years (Froggatt 1982).

Repose periods between eruptions of very large ignimbrites may be of the order of 10^5 years, and perhaps even 10^6 years. Such intervals suggest that some of the rhyolitic centres in the western USA, especially Long Valley and Yellowstone, are certainly still capable of erupting a large-volume ignimbrite.

The life expectancy of some individual rhyolitic centres could therefore be greater than 10^6 years. The Yellowstone centre is at least 2 Ma old (Christiansen 1979). As for large rhyolitic volcanic fields, the large San Juan calderas and associated ignimbrites, were all formed within a period of ~7 Ma, between 29 and 22 Ma BP (Steven & Lipman 1976). The Taupo volcanic zone has been active for 0.6 Ma (Cole 1981, 1984).

13.9.3 ERUPTIONS, CHARACTERISTICS AND DEPOSITS

From the discussion above it is apparent that eruptions from rhyolitic volcanoes are relatively infrequent or there is a relatively long repose period. The primary eruptive products which characterise rhyolitic volcanoes are large explosive eruptions of rhyolitic pumice as plinian fall deposits, voluminous ignimbrites and small-volume rhyolite lavas (Fig. 13.39).

Plinian pumice fall deposits will vary from <1 m to >10 m in thickness in and around a rhyolitic centre, and we have already discussed in detail the controls of their dispersal and distribution (Ch. 6). Measured sections showing the pumice fall stratigraphy of La Primavera volcano are illustrated in Figure 13.40. It is important to note the number of fall deposits, their *individual* thicknesses and, from the foregoing discussion, to consider the time framework.

Ignimbrites are the dominant components of the low-angle shield flanking a rhyolitic centre. We have already discussed their many depositional and welding facies, and a complex variety of lithologies may be found in the same ignimbrite (Chs 7 & 8). Stratigraphic relationships between different ignimbrites are also likely to be complex, as each one fills an erosional surface cut into the older succession (e.g. Fig. 13.41; Ch. 1). In addition, ignimbrites from nearby centres may overlap.

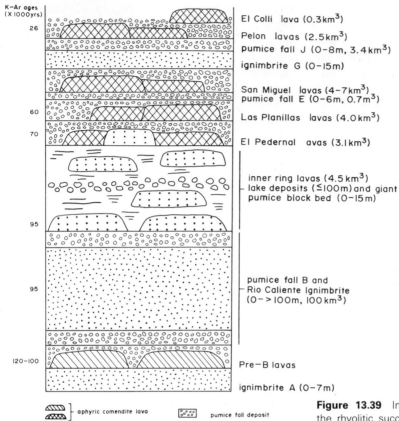

K–Ar ages
(x 1000yrs)

26 — El Colli lava (0.3km³)
Pelon lavas (2.5km³)
pumice fall J (0-8m, 3.4km³)
ignimbrite G (0-15m)

San Miguel lavas (4-7km³)
pumice fall E (0-6m, 0.7m³)

60 — Las Planillas lavas (4.0km³)

70 — El Pedernal avas (3.1km³)

inner ring lavas (4.5km³)
lake deposits (≤100m) and giant
pumice block bed (0-15m)

95 —

pumice fall B and
Rio Caliente Ignimbrite
95 — (0->100m, 100km³)

120-100 — Pre-B lavas

ignimbrite A (0-7m)

aphyric comendite lava
porphyritic comendite lava
ignimbrite
pumice fall deposit
lake deposits
giant pumice bed

Figure 13.39 Interactive stratigraphic section illustrating the rhyolitic succession of La Primavera volcano (see Fig. 13.38). Note all volumes are DRE. (Based on G. P. L. Walker *et al.* 1981d and Mahood 1980.)

The ignimbrite-forming eruptions are generally associated with major structural changes of the volcano and with caldera collapse (Ch. 8). Caldera collapse occurs after or during the eruption, around a circular ring fracture formed above the drained or draining magma chamber. At this time, **caldera-collapse breccias** of the type envisaged by Lipman (1976) may form by the caving or gravitational collapse of the caldera walls, producing rock slides and falls, which may be interbedded with ignimbrite (cf. co-ignimbrite breccias, Ch. 8).

Later volcanic activity is then concentrated on this ring fracture, perhaps for 10^5 years. Explosive phases producing plinian and sub-plinian air-fall deposits and small ignimbrites precede the eruption of rhyolite domes and flows. Individual rhyolite lavas do not travel far from the vent (Ch. 4), but in time they may coalesce on some volcanoes to produce a nearly complete ring of rhyolite lava. La Primavera (Fig. 13.37) and Valles (Fig. 13.42) volcanoes illustrate these features particularly well. At Valles all the rhyolite lavas are located on the ring-fracture of the Valles caldera, or part of an incomplete ring related to the earlier Toledo caldera. Each of these domes is probably monogenetic with a repose period of perhaps 10^3 years between eruptions. La Primavera is somewhat more complicated. Some of the lavas are located on an almost complete ring, with some lavas on a transverse line which crosses it. Most of these domes are again monogenetic, with a vent-to-lava ratio of 1 : 1. However, there are three larger, polygenetic structures which could be called independent volcanoes in their own right. These are younger, petrologically different (Fig. 13.39), and may lie on an outer second ring structure. Rhyolite

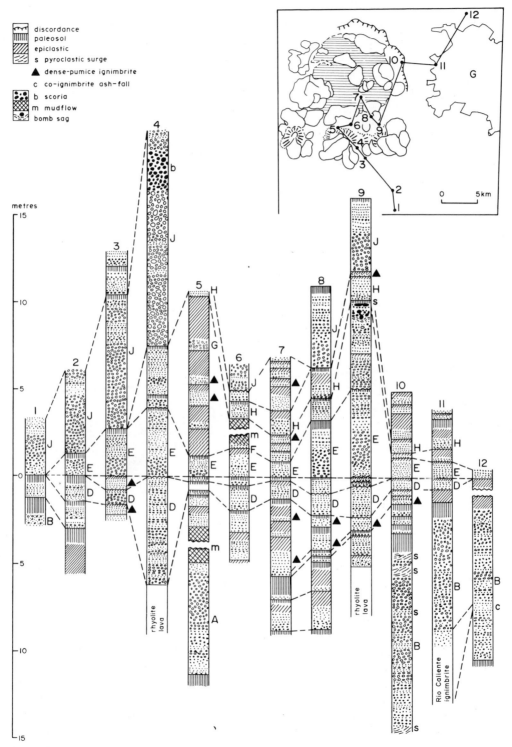

Figure 13.40 Twelve measured stratigraphic sections showing lateral variations and established correlations for the rhyolitic pumice-fall succession erupted from La Primavera volcano. (After G. P. L. Walker *et al.* 1981d.)

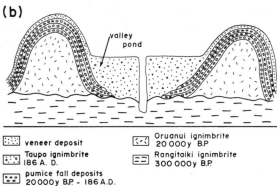

veneer deposit

Taupo ignimbrite
186 A. D.

pumice fall deposits
20 000 y B.P. - 186 A.D.

Oruanui ignimbrite
20 000 y B.P

Rangitaiki ignimbrite
300 000 y B.P.

Figure 13.41 (a) View to Lake Taupo from east showing rounded hills cut into Oruanui ignimbrite and valley pond of Taupo ignimbrite (light-coloured vegetation). (b) Schematic diagram showing geological relations at the above location. Valley pond of Taupo ignimbrite is up to about 40 m thick. The Rangitaiki ignimbrite is welded but both the Oruanui and Taupo ignimbrites are non-welded.

volcanoes without a well defined ring structure also occur. The rhyolite lavas at Okataina have a more linear distribution, and many of the vents seem to have been located along fissures and to have fed nests of domes (Nairn 1981, Cole 1970; Chs 6 & 8). Regional extension seems likely to have been a much more important structural control on the distribution of eruptive points than magmatic processes.

The central parts of some rhyolitic calderas are updomed giving rise to **resurgent domes** (R. L. Smith & Bailey 1968; Ch. 8). This is thought to occur when new magma surges back into a magma chamber following a climactic ignimbrite eruption. R. L. Smith and Bailey (1968) suggested that these

events were part of an evolutionary cycle shown by rhyolite volcanoes. The cycle was divided into seven stages:

(a) regional tumescence and generation of ring fractures,
(b) ignimbrite eruption,
(c) caldera collapse,
(d) pre-resurgence volcanism and intra-caldera sedimentation,
(e) resurgent doming,
(f) major ring-fracture volcanism and
(g) terminal fumarolic and hot spring activity.

The type example is the Valles centre (Plate 13) where two successive cycles are recognised. The resurgent dome of Redondo Peak (Fig. 13.42), formed after the second cycle, is associated with the eruption of the Upper Bandelier Tuff and formation of Valles caldera (Ch. 8). At La Primavera the updoming is much later than the ignimbrite event and the possibility is that this might be related to

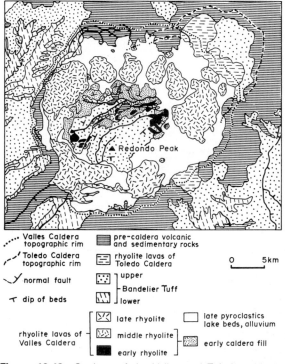

Valles Caldera topographic rim

Toledo Caldera topographic rim

normal fault

dip of beds

pre-caldera volcanic and sedimentary rocks

rhyolite lavas of Toledo Caldera

upper
Bandelier Tuff
lower

rhyolite lavas of Valles Caldera
late rhyolite
middle rhyolite
early rhyolite

late pyroclastics lake beds, alluvium

early caldera fill

0   5km

Figure 13.42 Geology of the Valles and Toledo calderas. Jemez Mountains, New Mexico. (After R. L. Smith & Bailey 1968.)

Figure 13.43 Delta at mouth of Tongariro River prograding into Lake Taupo. This transports reworked Taupo rhyolitic pumice and andesitic epiclastic debris from the Tongariro centres (Tongariro, Ngauruhoe and Ruapehu) into the caldera lake.

the buoyant uprise of a new pluton (G. P. L. Walker *et al.* 1981d). Updoming, resurgent or otherwise, has significant structural effects, particularly on deposits within the caldera.

Although rhyolitic volcanoes generally contain little other than rhyolite, there is usually a small proportion of basaltic volcanic rock, and characteristically there is a so-called **bimodal association** (e.g. Cole 1970). Basaltic rock may be present as scoria or mixed pumice in rhyolitic pyroclastic deposits (e.g. in J at location 4, Fig. 13.40), as cognate xenoliths in rhyolite lavas, or as small scoria cones (e.g. near Cuxpala, Fig. 13.37). The evidence of mixed magma deposits suggests that basaltic magma may participate in at least some of the eruption of rhyolitic magma. Rhyolite magma chambers are widely thought to be generated and stoked up by basalt from a mantle source. Occasionally basaltic dykes do reach the surface, but generally only outside the rhyolitic centre (defined by the caldera rim) which may be a **shadow zone** above the magma chamber through which higher density, but lower viscosity, basaltic magma cannot pass.

Epiclastic processes and deposits are also significant, and it is important to stress the repose periods between eruptions, and the possible effect on the hydrologic and geomorphic system of a large ignimbrite eruption. Rhyolitic volcanoes are characterised by the rapid production of vast volumes of loose, easily erodible sediment over irregular intervals. This sediment can be transported out of the volcano by rivers, or back into the system to be deposited in the caldera (Fig. 13.43). Caldera fills are complex sequences of fluvial and lacustrine epiclastic sediments, pyroclastics and lavas. Also very significant are the products of

hydrothermal alteration and precipitation. The near-surface intrusion of rhyolite domes and magma at depth below these long-lived centres can promote very active geothermal systems. Hot springs associated with rhyolitic volcanoes are important environments for the formation of epithermal gold–silver deposits (Henley & Ellis 1983; Ch. 14).

Submarine rhyolitic calderas are thought to be very important sites for the formation of Kuroko-

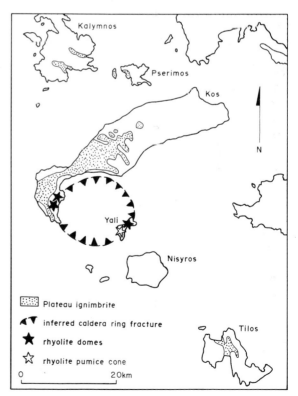

Figure 13.44 Distribution of the Plateau ignimbrite and outline of a submarine rhyolitic caldera between Kos and Yali in the Dodecanese Islands.

type massive sulphide deposits (Ohmoto 1978, Ohmoto & Takahashi 1983). In the Kuroko area in Japan, the Miocene mineralisation is associated with rhyolitic lava domes and volcaniclastic rocks believed to have been emplaced at minimum water depths of 1000 m (Ohmoto & Takahashi 1983). There are no well known Quaternary submarine rhyolitic calderas for comparison. One good example exists in the eastern Aegean Sea between Kos and Yali in the Dodecanese Islands (Fig. 13.44). Caldera collapse was associated with the eruption of the rhyolitic Plateau ignimbrite which is found on Kos, and in valleys and bays facing Kos on some of the other Dodecanese Islands (Fig. 13.44). It would be very enlightening to explore this centre further with deep-sea submersibles to try to locate hydrothermal vents and recent polymetallic sulphide deposits. Other submarine rhyolitic calderas may exist in the northern offshore extension of the Taupo volcanic zone (Ch. 15), where submarine geothermal activity has been recorded (Duncan & Pantin 1969, Pantin & Duncan 1969), but these centres would have to be completely submerged.

13.9.4 CALDERA SEDIMENTS AND DOMES: LA PRIMAVERA

La Primavera lacks a well defined structural expression of a caldera because subsequent uplift of the central parts of the volcano has masked original relationships. What is significant is a circular area of lake sediments 10 km in diameter associated with rhyolite lavas erupted on a ring fracture (Clough *et al.* 1981, 1982; Fig. 13.45). For this reason the sediments are believed to have been deposited in a caldera lake. The caldera was formed by the biggest explosive eruption of the volcano, i.e. the event which produced the Rio Caliente ignimbrite (Figs 13.37 & 39; see Ch. 8). A 30 m scarp around the northeastern edge of the centre may be the original embryo rim (continuous with the outer ring or rhyolite extrusives mentioned above).

Updoming and inversion of topography (Fig. 13.45b) have provided an opportunity to examine the caldera fill of a rhyolitic volcano. The lake deposits have a maximum exposed thickness of ~100 m, and they rest on top of the ignimbrite

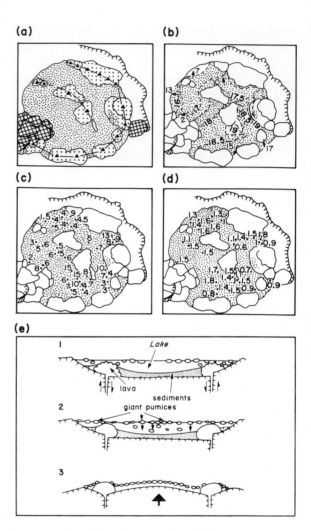

Figure 13.45 The giant pumice block bed and lake sediments, and their relationships to the rhyolite lavas of La Primavera volcano. (a) Distribution of lake sediments and rhyolite lavas: fine stipple is lake sediments; coarse stipple is porphyritic comendite lavas lying on an 'inner ring' with a transverse zone; triangles represent inferred vent positions; cross-hatch is aphyric comendite lavas lying on an 'outer ring'; hachured line is a low escarpment, possibly an embryo caldera rim. (b) Altitude contours of the top of the giant pumice block bed in hundreds of metres above sea level. Dip of lake sediments is also shown (arrows). (c) Thickness of giant pumice block bed in metres. (d) Median diameter of giant pumice blocks in metres. (e) Three stages in the envisaged formation of the giant pumice block bed. (After Clough *et al.* 1981.)

which was eroded before their accumulation. The sedimentary succession consists largely of thinly bedded white ashy sediments, ash turbidites, some beds of diatomite and an unusual bed of giant pumice (Figs 13.45 & 46). This bed contains blocks of pumice up to 8 m in diameter, which are thought to have formed when rhyolite lava erupted underwater, as was observed to happen during the submarine eruption of rhyolite lava at Tuluman Islands (M. A. Reynolds *et al.* 1980; Ch. 4). Rhyolite lavas commonly contain a pumiceous carapace which at La Primavera is thought to have spalled off and floated away (Fig. 13.45e). Eventually the pumice blocks became waterlogged and were deposited. Another giant pumice deposit which formed at the end of the AD 186 Taupo

Figure 13.47 Silica sinter deposit within the La Primavera caldera.

eruption has been identified by C. J. N. Wilson and Walker (1985), and it therefore seems that this facies may not be particularly unusual in rhyolitic caldera fills.

The associated porphyritic rhyolite lavas at La Primavera (Figs 13.39 & 45a) also show features that suggest they were erupted underwater or emplaced as shallow intrusions or cryptodomes into the lake sediments (Clough *et al.* 1982). Many of these lavas have thick, glassy, quench-fragmented and lobate margins, and at some exposures lava can be demonstrated to have intruded lake sediments. Reverse faults and folds are seen in some of the lake sediments, suggesting forcible intrusion. An outward dip is commonly seen in the sediments which mantle the domes, but one could also argue that this was due to sedimentary draping or post-depositional compaction.

Numerous hot springs are active within the caldera, and sinter deposits have formed (Fig. 13.47). These point to the existence of hot rock or a magma body below the caldera.

13.9.5 OTHER CRATERS

Two other types of explosive crater that might be found associated with rhyolitic centres are:

rhyolitic tuff rings and
hydrothermal explosion craters.

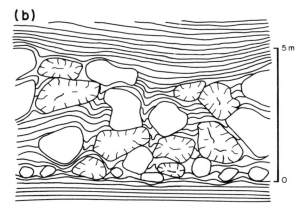

Figure 13.46 (a) Giant pumice blocks enclosed in bedded lacustrine sediment. (b) Field sketch showing disturbed and contorted lake sediments with blocks of giant pumice. Some blocks show cooling joints normal to their margins. (After Clough *et al.* 1981.)

Rhyolitic tuff rings

The early explosive phase accompanying the eruption of domes and coulées on the ring-fracture may be initially phreatomagmatic and build a tuff ring (Section 13.5). Often these are not seen because they are destroyed or overwhelmed by later lava flow.

Hydrothermal explosion craters

Hydrothermal explosions occur when superheated water at shallow depths flashes to steam, disrupting the confining rocks and ejecting solid debris, water and steam. In these explosions magma is not involved (cf. *phreatomagmatic*, where magma is involved, Ch. 3). The explosions form small craters ($W_{cr} < 1$ km) but they may also occur as multiple structures. Deposits are bedded, and consist of angular fragments of old ignimbrite, rhyolite lava and other volcanic, epiclastic and local rocks within a matrix of more-finely granulated fragments and hydrothermal clays. Some of the layers may be deposited as base surges. Good descriptions of these rock types in the Quaternary Kawerau geothermal field, New Zealand, are given by Nairn and Wiradiradja (1980). Hydrothermal explosions have occurred in most boiling spring areas of New Zealand during the late Quaternary, including several historic events. At depth hydrothermal explosion craters should root in breccia pipes or diatremes, as may be the case for many maars (Section 13.5). Again they may be economically very important, concentrating gold-bearing, deep system fluids into the near-surface epithermal depositional environments (Henley & Ellis 1983, Sillitoe *et al.* 1984).

13.10 Submarine spreading ridges and seamounts

Over the past ten years detailed investigation of the ocean floor has provided a wealth of information about submarine basaltic volcanic processes at constructive plate margins and in the ocean basins. Although fissure-fed basaltic lavas create most of the newly accreted upper oceanic crust, large-scale volcanic cones that are the products of central volcanism are also important components in both axial rift systems as well as oceanic plate interiors.

Submarine oceanic plate volcanism is important because it produces the largest volume of volcanic rocks on Earth, especially at active spreading ridges. However, because of subduction, the vast majority of these volcanic rocks are destined to be subducted, so in terms of their preservation potential in the rock record they represent a much less significant group of volcanic products than their present extent suggests. Where preserved, the volcanics occur as stratigraphic elements of oceanic crustal slices, which are usually tectonically emplaced as *ophiolite* slices in forearc, subduction related accretionary prisms, or obducted onto continental margins. Ophiolite slices may also be preserved in ancient greenstone belts and as the basement to island arc complexes. More will be said about ophiolites in Chapter 15.

Although much has been written on the stratigraphy of oceanic crust, it is only recently that insight has been gained on the nature of the volcanoes that erupt oceanic crustal volcanic rocks. In this section we consider the nature of the volcanoes, rather than the stratigraphy of oceanic crust.

13.10.1 SPREADING RIDGES

Macdonald (1982) has reviewed the controls on the nature of the basaltic volcanoes, the morphology and tectonic form of major oceanic spreading ridges. His 1982 work is a review of recent advances through the use of submersibles, deep-tow instrumentation, side-scan sonar and high resolution bathymetric surveying (e.g. Ballard & van Andel 1977, Luyendyk & Macdonald 1977, Macdonald & Luyendyk 1977, Searle & Laughton 1981).

Mid-ocean ridge (MOR) volcanism issues from fissures within a median rift valley. This is dominated by ridge-parallel fractures and faults, producing a very rough terrain with horsts and grabens (Fig. 13.48). Small volcanic mounds with a topography made of overlapping flow-fronts are prominent. Pillow lavas can be expected to be abundant,

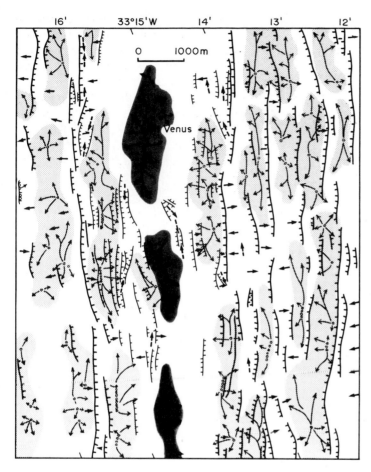

Figure 13.48 Map of fault patterns (hachured lines) and volcanic mounds (shaded areas in the inner rift valley of part of the Mid-Atlantic Ridge. Darkest shading indicates youngest volcanism on present rift axis. Heavy fault lines indicate wall faults; light fault lines, inner-floor faults. Black dots (most recent) and open circles (older) indicate vents and crest lines of volcanoes; thinner arrows show flow lobes. Dip and plunge of fault blocks on walls shown with thick short arrows. Fault lines are placed near the base of fault scarps and throw on faults varies from <0.1 to 25 m. Mount Venus rises about 200 m from its base. Data are from bathymetric map, dive observations, bottom photography and deep-towed geophysical studies. (After Ballard & van Andel 1977.)

but non-pillowed lavas are also common (Ch. 4). At the foots of fault scarps talus fans of brecciated lava are found (e.g. Ballard & van Andel 1977). Hydrothermal vents and fissures are also very important features. One of the most exciting recent discoveries in volcanic and economic geology has been the direct observation of active sulphide-depositing vents, or black 'smokers', on the East Pacific Rise (Francheteau *et al.* 1979, Hekinian *et al.* 1980, Ballard *et al.* 1984). There are close similarities between these active hydrothermal systems and the fossil ore depositing systems that have formed Cyprus-type massive sulphide deposits (Rona 1984; Ch. 14).

Macdonald (1982) pointed out that spreading rates not only affect the morphology of the ridge crest, but also of the volcanoes from which the basalts of layer 2 of the oceanic crust are erupted.

The neovolcanic zone (the zone of recent and on-going volcanism *at the spreading centre,* cf. on the ridge flanks) is only 1–2 km wide. At low spreading rates the neovolcanic zone lies within a well defined rift valley (Fig. 13.49) and is dominated by a discontinuous line of central volcanoes. These are elongate parallel to the spreading axis, and typically have dimensions of 1–4 km in basal diameter and heights of about 250 m. They are dominated by pillow lava (Ballard & van Andel 1977, Luyendyk & Macdonald 1977, Macdonald 1982).

At intermediate spreading rates the medial rift valley is less well defined, and the volcanoes are more elongate along strike, appear to be fissure-fed, and may be interrupted by *en echelon* offsets (Fig. 13.49). Heights are only a maximum of 50 m and pahoehoe-like sheet flow lavas are more common. At high spreading rates there is no definable axial

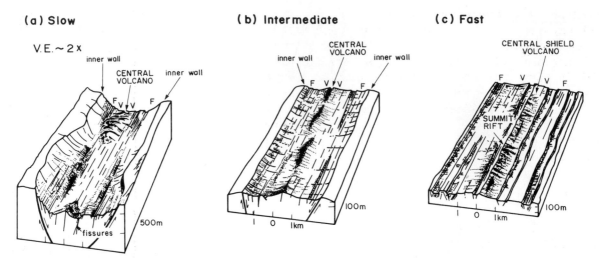

Figure 13.49 Schematic illustration of characteristics of volcanoes and the central rift of mid-oceanic spreading ridges with different spreading rates. (After Macdonald 1982.)

rift valley. The central volcano resembles a very elongate hawaiian-type shield volcano (see Section 13.2) with gentle slopes and a summit rift (Fig. 13.49c). The 'volcano' is only 1–2 km wide (cf. Hawaiian shields), and may be 100 km long, so is clearly fissure-fed. Both pillow lavas and massive sheet lavas have been observed (Macdonald 1982). The only sediment deposited is from pelagic sources. A pelagic sediment mantle only becomes prominent at the edges of the median rift valleys, and on the ridge flanks and abyssal plains as the lithosphere spreads with time away from the MORs. The pelagic sediment layer and the underlying basaltic crustal layer make up layers 1 and 2, respectively, of the oceanic crust. These should be closely associated with an underlying mafic sheeted dyke complex, and gabbroic and mafic cumulate igneous rocks. Such crustal profiles coincide with ophiolite profiles, as discussed in Chapter 15.

Marginal seafloor spreading centres (Ch. 15) also produce a similar crustal profile, but in addition the basaltic crust should be overlain by a mixed volcaniclastic–carbonate clastic mass-flow sediment apron derived from associated island arcs.

It is uncertain, however, what the nature of the volcanic centres is, since little or no work has been done on these. Although the axial spreading systems are usually topographically elevated ridge regions of high heat flow, the axial rift morphology that is relatively easily definable for mid-oceanic ridge spreading systems is not easily definable in marginal basins. In addition, magnetic anomaly patterns are also difficult to define clearly, suggesting that the spreading process may be more diffuse and the eruptive centres less clearly identifiable.

13.10.2 SEAMOUNTS

This is a poorly known group of submarine volcanoes which, because they have not been readily observable, have not received much attention until recently. However, advances with submersibles, high resolution side-scan sonar and narrow-beam bathymetric surveys have greatly increased our understanding of the morphologies of these volcanoes. Recently Searle (1983, 1984) and Batiza and Vanko (1983) have added considerable insight to the nature of submarine seamounts and guyots documented by earlier workers such as Menard (1964, 1969). These central volcanoes are variable in size and shape. Volcanoes that are near circular in plan-view have basal diameters ranging from the minimum limits of resolution (<1 km) to 25 km (although usually less than 10 km), and their heights are usually several hundred metres, but may exceed 2 km. They have dominantly concave-up or sometimes convex-up slopes, which at their maximum are inclined at up to 40°, and on average about 20°. The summits are usually flat-topped with diameters up to 10 km, and frequently these have well defined craters or caldera collapse structures less than a kilometre in diameter (Hollister *et al.* 1978, Searle 1983, 1984, Batiza & Vanko 1983).

Although seamounts are probably mostly relatively small monogenetic centres, some may evolve

(d) Terraced and subsided

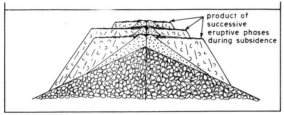

(c) Truncated and subsided

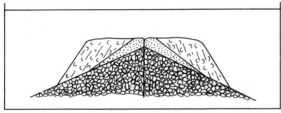

(b) Emerged

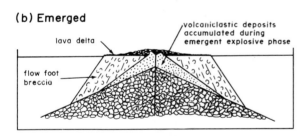

(a) Submerged

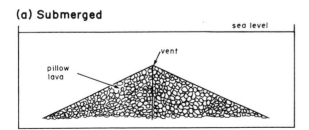

Figure 13.50 Growth of oceanic seamounts, as first postulated by Jones from his studies on intraglacial volcanoes (see Fig. 13.51). (After J. G. Jones 1966.) Recent work suggests that even in stage (a) a flat-topped morphology may be developed (see text).

into very large, complex polygenetic centres, such as Hawaii. In this case the subaerial part of the volcano has evolved into a major shield volcano (Section 13.2.1).

Variations in shape and form stem from variations in conduit geometry (Batiza & Vanko 1983). Submarine central volcanoes are frequently located along fracture systems of the oceanic lithosphere, and may have associated lava fields.

Oceanic seamounts are cones and domes that grow upwards from the abyssal plains near spreading ridges, or away from them, in which case they are thought to be derived from mantle hot spots. They are therefore essentially basaltic, and pillow lavas appear to be prominent in the basal parts where, due to the great water depth of the vent, the hydrostatic pressure of the water column is sufficient to prevent explosive eruption of the basalt (Fig. 13.50), although hyaloclastites could form. If the (pillow) lava pile grows closer to sea level, where the hydrostatic pressure is less, exsolving magmatic volatiles or hydrovolcanic interaction, or both, may be capable of producing explosive eruptions, so that the basal pillow pile becomes mantled by pyroclastics and/or hyaloclastites of variable grainsize (e.g. Fig. 4.15). If the seamount emerges above sea level and the vent is no longer intruded by sea water, more-passive lava eruptions will produce a lava cap over the volcaniclastic layer. Lavas flowing into the sea may transform into pillow lavas or, more commonly, will quench-fragment and build out coarse hyaloclastite deltas and flow-foot breccias (Ch. 4; J. G. Jones 1966, J. G. Moore & Fiske 1969). Complex relative sea level changes associated with large tidal ranges or tectonic movements could produce complex stratigraphic–structural relationships in the upper part of the stratigraphy, including several levels of coherent lava-flow-foot breccia transition, as discussed by J. G. Jones and Nelson (1970) (Fig. 4.17). Surtseyan volcanoes (e.g. Thorarinsson 1967) are examples of basaltic seamounts whose eruptions were initiated in relatively shallow seas. As a result the initial products are phreatomagmatic pyroclastics (Jakobsson & Moore 1980) the usual early pillow lava and hyaloclastite stage of deep water seamounts being by-passed. After Surtsey emerged above sea level and its vent was protected from access by the sea, a passive lava cap formed.

However, some seamounts with flat tops never reached sea level (Searle 1983). Searle's explanation for their morphology is that the steep slopes are built of pillow breccias and hyaloclastites (Lonsdale & Batiza 1980) tumbling over the edge of the flat-topped summit, which is built of sheet flows with high effusion rates, and pillow lavas. Flat tops are therefore not necessarily produced by wave erosion and truncation, as was previously thought.

Older seamounts are stabilised by precipitation of volcanogenic ferromanganese crusts and elements. They may become colonised by fringing, framework-building biota, which will then contribute carbonate clastic debris to the submarine flanks of the volcanic pedestal. Prolonged growth of the seamount could produce an apron of coarse volcaniclastic and carbonate clastic detritus, transported and deposited by mass-flow processes fed through a well developed canyon system. The facies of such a seamount should merge laterally into the pelagic sediment facies mentioned above. Isostatic subsidence of the seamount and cessation of volcanism could lead to the development of an atoll and eventually complete submergence of the seamount.

13.11 Intra- or subglacial volcanoes

The best-documented intra-or subglacial volcanoes are from Iceland. Compositionally, all types occur: basaltic, andesitic, dacitic and rhyolitic (Furnes *et al.* 1980). Typically they form steep-sided ridges called **tindas**, or steep circular table mountains called **tuyas**, which were first described from British Columbia by Mathews (1947).

Basaltic subglacial volcanoes consist principally of masses of pillow lavas, palagonitised hyaloclastite breccias and sideromelane fragments. J. G. Jones (1969) envisaged that a subglacial basaltic eruption first builds up a steep-sided pillow pile, as lava from a fissure is extruded into a meltwater vault created under the ice sheet (Fig. 13.51). Much of the hyaloclastites may form later as lava deltas build out into a meltwater lake as the growing pile emerges. Flow-foot breccias develop as the deltas capped by

lava grow out into the meltwater, eventually forming an equidimensional table-mountain. Intraglacial basaltic eruptions in Iceland have formed very substantial deposits, often referred to as '*moberg*'. Overlapping centres form compounded

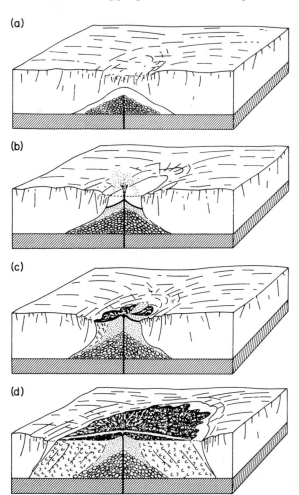

Figure 13.51 Growth of a tuya. (a) Aquatic effusive phase. Magmatic heat causes melting of ice sheet above eruptive fissure forming meltwater vault. Within this vault erupting lava builds a steep-sided pillow lava pile. (b) As pile builds, roof of vault collapses forming intraglacial lake. Effusion gives way to explosive phase on emergence and resulting volcaniclastic deposits accumulate between walls of ice on top of pillow lava pile. (c) Emergent explosive phase gives way to aerial effusive phase. Lava issues from vent and pushes out into meltwater lake on deltas of flow-foot breccia. (d) Advanced stage of aerial effusive phase. Products of earlier eruptive phases overwhelmed and buried by flow-foot breccia. (After J. G. Jones 1969.)

FURTHER READING   409

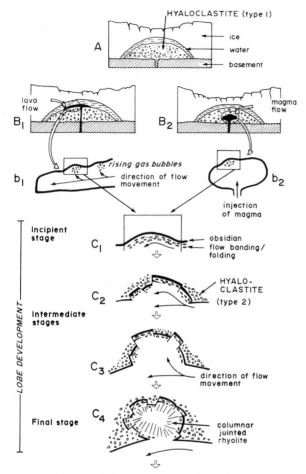

Figure 13.52 Evolutionary model for formation of the two types of silicic hyaloclastite and lava lobes during a subglacial eruption. (After Furnes *et al.* 1980.)

deposits extending tens of kilometres and forming mountains up to 1500 m in height, and resulted from eruptions during the latter parts of the late Pleistocene. Older moberg masses (late Pliocene–early Pleistocene) are also known and are buried by younger volcanic rocks.

Several rhyolitic subglacial accumulations of smaller volume (0.01–0.1 km³) also occur in Iceland (Furnes *et al.* 1980). These consist of two main components (Fig. 13.52):

(a) hyaloclastites and
(b) lava lobes averaging a few metres (<10 m) in diameter.

Two types of breccia are found, and these are believed to represent major changes in the style of activity during subglacial eruptions. Type 1, by far the most important type, is brecciated pumice and glass, and is thought to result from explosive activity (stage A, Fig. 13.52). By analogy, with subaerial silicic eruptions the initial phase of the eruption of rhyolite magma beneath ice is also likely to have been explosive (plinian or phreatoplinian) if the confining pressure is not too high. Type 2 is characterised by fragments of obsidian and stony rhyolite, and is genetically related to the lava lobes (stages B & C, Fig. 13.52). The lobes generally consist of (from the margin inwards), a chilled obsidian rind, a zone of flow-banded and flow-folded pumiceous rhyolite, and a central zone of radially joined rhyolite. These lobes are thought to be the remnants of disintegrated subglacial lava flows or lava intrusions into the waterlogged pumiceous hyaloclastite. Type 2 breccias are hyaloclastites formed by the brecciation and quench-fragmentation of the lobes. The lava flows or intrusive bodies are emplaced by a continuous process of alternating construction and fragmentation of lobes. Rhyolite lavas which have been emplaced into wet sediments show similar lava lobes at their margins (Clough *et al.* 1982; Section 13.9), and these, as well as extensive quench-fragmentation, seem to be an important feature of rhyolites intruded into water-saturated hosts.

Also associated with subglacial eruptions are huge meltwater floods, called '*jökulhlaups*' in Iceland. These have deposited large debris fans (*sandurs*; Fig. 10.24b), and some of the larger flows are believed to have continued into the North Atlantic Ocean and deposited volcaniclastic turbidites (Laughton, Berggren *et al.* 1972).

13.12 Further reading

Several data sources are now available which list active and potentially active volcanoes. The International Association of Volcanology and Chemistry of the Earth's Interior (IAVCEI) has published the invaluable *Catalogue of the active volcanoes of the world including solfatara fields.*

This, at present, consists of 22 volumes, which are listed on the back of every issue of *Bulletin Volcanologique* which now continues from 1986 as *Bulletin of Volcanology*. Descriptions of the different centres vary in quality, but for many a geological map, short description, summary of historic activity and some petrological information are found. Simkin *et al.* (1981) of the Smithsonian Institute have compiled *Volcanoes of the world*, which presents a very large amount of data in digital format. When eruptions do occur, the location and type of activity are reported in the *Scientific event alert network (SEAN) bulletin* published by the Smithsonian Institute. Also, the *Bulletin of volcanic eruptions* is an annual report of world volcanic eruptions published by the Volcanological Society of Japan and the IAVCEI, and this appears in *Bulletin volcanologique*, and will continue to do so in the new *Bulletin of Volcanology* which will also release condensed reports from the *SEAN Bulletin*.

In addition, more-general literature sources on the characteristics of different types of volcanoes include MacDonald (1972), Bullard (1976) and H. Williams and McBirney (1979), on basaltic volcanoes, C. A. Wood (1980a, b and *in press*), Greeley and King (1977) and *Basaltic volcanism on the terrestrial planets* (Basaltic Volcanism Study Project 1981), and on calderas, the special issue of the Journal of Geophysical Research (volume 10, B10 edited by Lipman *et al.* (1984).

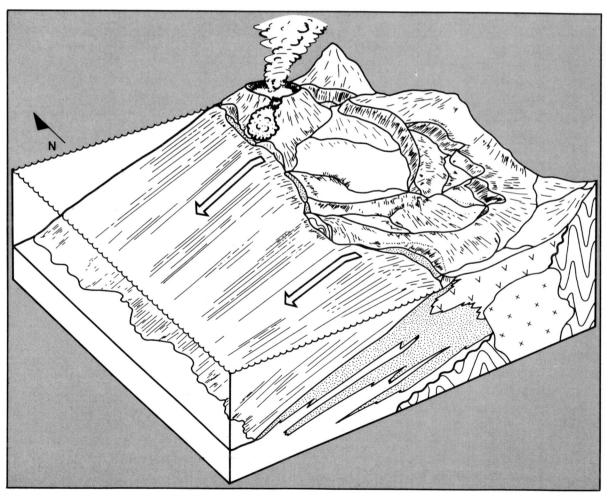

Plate 14 A schematic reconstruction of the palaeogeography and facies relationships in the lower Devonian Kowmung Volcaniclastics, southeastern Australia. (After Cas *et al*. 1981.)

CHAPTER FOURTEEN

Facies models for ancient volcanic successions

Initial statement

The preceding chapters have laid the groundwork for this chapter, by emphasising processes and products in modern volcanic terrains, which are extremely important in understanding equivalent elements in ancient terrains. Any study in modern terrains is only half complete if the relevance to the rock record is ignored. The majority of workers in volcanic terrains are not working in modern successions, but are geologists (most of them exploration or survey geologists) trying to make sense of discontinuously outcropping, variably deformed, metamorphosed and hydrothermally altered successions. Therefore, in this chapter we try to bring together what has been presented in previous chapters to construct general but workable facies models. These serve to represent significant associations of facies and, if possible, their spatial and genetic relationships, and to summarise, however schematically, the palaeoenvironmental–palaeogeographic context. If this can be done, then such facies models may have predictive value for ancient successions and in basin analysis studies involving volcanic successions. We also consider the influences that modify original facies characteristics in ancient successions, including the effects of erosion, alteration, metamorphism and diagenesis.

14.1 Introduction

The consideration of facies descriptors in Chapter 1, and the subsequent documentation of the facies characteristics of volcanic successions and of the processes involved (Chs 3–13), have paved the way for discussing the analysis of facies, depositional setting and general palaeogeographic context of ancient successions. However, in ancient successions all of the facies descriptors (*geometry, lithology, sedimentary structures, sediment movement patterns* and *fossils*; Ch. 1) are subject to varying degrees of modification by processes contemporaneous with deposition or post-dating it, or both. The former processes are represented largely by erosion and hydrothermal alteration, and the latter by protracted hydrothermal alteration, diagenesis, metamorphism and deformation. These are considered before proposing an approach to facies analysis and the construction of general facies models.

The facies models we will develop represent generalised summaries of many of the volcano types discussed in Chapter 13, and the descriptions of the various facies models in this chapter should be read in close association with the relevant parts of Chapter 13. However, in this chapter we take the discussions of Chapter 13 one step further, by placing the various volcano types and their successions in a broader geographic context, i.e. within the context of the basins within which volcanism is occurring. As such the facies models that are developed should have significance in basin analysis studies involving volcanic successions.

14.2 Facies geometry and facies–stratigraphic relationships: factors affecting them in ancient successions

Ancient volcanic successions are, on the whole, the erosional relics of complex volcanic centres and very few ancient centres will look the same as presently active volcanoes. In the first instance, the preservation of the constituent facies and their geometry is dependent on the interplay between aggradation (deposition) and degradation (erosion; Chs 1, 10 & 13). Because of the relatively high slopes and the generally high availability of loose debris in proximal and near-vent settings, erosion rates in such settings are usually high (Ch. 10). Unless a particular unit is resistant (lavas, welded deposits) or is rapidly buried under a resistant unit, or both, it will be eroded. The record of such rock units may be

lost or their original extent and geometry may be severely modified. A consequence of this is that, in many volcanic successions, much of the original volcanic and sedimentation record is lost, resistant units may be disproportionately preserved in the rock record relative to original abundance, and original depositional geometries are frequently not preserved. Lateral facies and age relationships may therefore be complex, change abruptly (Fig. 1.2), and are not necessarily related to fault contacts. Furthermore, unconformities may be common, relatively local in significance, not related to regional deformation, and the erosion that unconformities represent may have been initiated by an aggradational (eruptive) event rather than tectonic uplift.

Nevertheless, volcanic terrains are also the sites of contemporaneous crustal movements and, given that many are associated with orogenic belts, they may also be the sites of subsequent penetrative deformation. Deformation may also therefore severely change original stratigraphic relationships and, where strain is high, the original geometry of rock units. However, until evidence one way or another is found, deformational influences should be demonstrated rather than assumed, and be given only as much weighting as erosional contacts in assessing lateral discontinuities.

Furthermore, although deformation may cause steep dips, these are not necessarily the consequence of tectonic deformation in volcanic terrains. Volcanoes may have steep initial slopes upon which deposition occurs. Fall deposits may mantle an irregular, steep topography, producing steep depositional attitudes (Fig. 5.2), which may be at least partly preserved in the rock record.

It is also clear from the preceding that discordant stratigraphic relationships (e.g. high angle unconformable contacts, Figs 8.55 & 10.32b) may be common, and need not represent major, regionally extensive deformational events or time breaks. It may be difficult and unrealistic in some instances to construct conventional 'layer-cake' stratigraphies or columns. This is particularly so in proximal and near-vent settings, with the irregularities in topography and the diversity and frequency of events (eruptive, depositional, erosional, tectonic). Unless

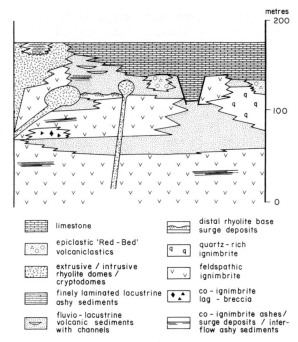

	metres
	200
	100
	0

limestone

epiclastic 'Red - Bed' volcaniclastics

extrusive / intrusive rhyolite domes / cryptodomes

finely laminated lacustrine ashy sediments

fluvio - lacustrine volcanic sediments with channels

distal rhyolite base surge deposits

quartz - rich ignimbrite

feldspathic ignimbrite

co - ignimbrite lag - breccia

co - ignimbrite ashes / surge deposits / inter-flow ashy sediments

Figure 14.1 Example interactive stratigraphic column, based on the mapped geology of the uppermost Lower Devonian Snowy River Volcanics, Murrindal – W–Tree area, eastern Victoria, Australia. Also see Figures 13.30 and 13.39 for other examples.

critical contacts are well exposed, it may be difficult to ascertain the exact age relationships between facies. In such cases **interactive stratigraphic–facies diagrams** (Fig. 14.1; also see Figs 13.30 & 39) may be preferable as a means of schematically representing spatial facies–stratigraphic relationships, the possible age relationships and the preserved geometries. In the distal parts of basins away from vents, slopes may be lower, facies should be more laterally continuous and more-consistent stratigraphic relationships should occur (e.g. Cas 1978b, McPhie 1983, Mathisen & Vondra 1983).

14.3 Factors affecting original lithological characteristics and depositional structures

The preserved outcrop, handspecimen and petrographic characteristics of volcanic facies may be significantly influenced by post-depositional

processes. These can generally be grouped into those that are contemporaneous with emplacement, and those that clearly post-date emplacement. Contemporaneous processes include:

polyphase hydrothermal alteration
devitrification
palagonitisation (in basaltic glasses)
hydraulic fracturing

Processes post-dating emplacement include:

prolonged hydrothermal alteration
diagenesis
metamorphism
deformation

Where ancient successions have been subjected to prolonged polyphase hydrothermal alteration, diagenesis, metamorphism and deformation, the effects of each may be difficult to distinguish.

These processes not only have the ability to overprint original textures, but in some instances to obliterate them totally and produce new textures (even apparent clastic textures) that can lead to misconceptions of the original rock-types. Original textures therefore have to be proven rather than assumed, and unusual textures have to be explained.

Each of the alteration and modification processes will now be briefly introduced. However, the coverage is not intended to be comprehensive. The interested reader is referred to other sources, and some useful references on these topics are provided in the following discussion.

14.3.1 POLYPHASE HYDROTHERMAL ALTERATION

'Hydrothermal alteration is a general term embracing the mineralogical, textural and chemical response of rocks to a changing thermal and chemical environment in the presence of hot water, steam or gas' (Henley & Ellis 1983, p. 10). The upper limits of this process are ill-defined and are gradational into low grade burial metamorphism. According to Henley and Ellis (1983), hydrothermal alteration involves ion-exchange reactions, mineral phase transformations, mineral dissolution and new mineral growth. Although the original rock composition has some effect on the secondary alteration mineralogy, the main influences are the permeability of the rock pile, temperature, and the composition of the fluids moving through the rock pile.

Hydrothermal alteration can take many forms. Some of the alteration minerals produced include amorphous silica, quartz, K-feldspar, albite, calcite, montmorillonite, montmorillonite–illite, illite, kaolinite, alunite, chlorite and a wide array of zeolites and low grade metamorphic minerals. Needless to say, bulk rock compositions can be very significantly altered. Stages of alteration or distinctive alteration mineral assemblages, or both, are identified by a variety of terms, which are by no means standardly defined and used (e.g. chloritic, sericitic, potassic, phyllic, argillic and propylitic alteration, silicification, etc.). For a discussion of

alteration types, processes and assemblages the interested reader is referred to Ellis (1979), Franklin *et al.* (1981), Beane (1982), Titley (1982), Urabe *et al.* (1983), Henley and Ellis (1983) and R. W. Hutchinson (1984). One general comment that should be made about alteration is that many of the 'type' alteration minerals (e.g. chlorite, silica, sericite) can also be produced by low grade metamorphism. Silicification can also be produced simply from circulating meteoric waters passing through a glassy pile. It may also be difficult to distinguish the effects of devitrification from true hydrothermal silicification.

Hydrothermal alteration and processes are, of course, an extremely important adjunct to epithermal precious metal and base-metal sulphide mineralisation (see preceding references as well as Rona 1984). From our point of view it is also

Figure 14.2 Gradational to sharp textural contacts in Cambro-Ordovician chlorite–quartz–sericite altered submarine rhyolite lava, eastern Australia. (a) Apparent clasts are sharply to diffusely defined, and irregular in shape. (b) Close-up of sericite alteration (light) superimposed on flow banding, and note uniform phenocryst texture throughout. Drill core diameter is 3.5 cm.

extremely important because of the modifying effects it can have on original rock textures, and the problems this can lead to in terms of recognising original rock types. Rocks may be subjected to more than one alteration event or type, and these may occur in close spatial and temporal relationship. Where this occurs the resultant textures and mineralogy may be complex.

The alteration takes place as a result of ion diffusion and fluid migration. Neither is necessarily pervasive in its coverage, leading to variation in the intensity of the alteration from barely perceptible, to patchy, and in the extreme to pervasive. The second of these stages may cause the most significant complications. Patchy alteration may juxtapose areas of original texture and mineralogy, or of an earlier alteration event, with later superimposed alteration effects. The overall effect of this may be to produce an apparent clastic texture (Fig. 14.2) wherein the relic patches appear as clasts in a differently coloured, mineralogically different matrix. The 'clasts' may even resemble fiamme and are sometimes called 'pseudo-fiamme'. R. L. Allen (*pers. comm.*) has shown convincingly that Palaeozoic rocks in southeastern Australia that were previously interpreted as acidic pyroclastic and epiclastic volcaniclastics are mostly variably devitrified, polyphase altered acidic lavas, cryptodomes and quench-fragmented equivalents. Allen suggests that the apparent clastic texture represents intensely seriticially altered patches in less sericitic, silica–sericite material, or intensely chloritically altered patches in less chloritic, more siliceous material. In both cases, original glass has been initially altered and then incompletely silicified. Recognition of such effects depends on recognising relic primary (pre-alteration) textures, even if only ghosts of the original texture. Relic primary textures which were important to the correct identification of lithology in Allen's study were faint, alteration-modified flow banding, evidence of quench fragmentation, autobrecciation, and continuity of homogeneous porphyritic texture from the apparent clasts into the apparent matrix. In both 'matrix' and 'clasts', similar crystals (or their ghosts) occurred in similar proportions, sizes and distribution. Apart from these criteria, it is also the gradational nature of the

contact between the zones or patches of different alteration which is the key to recognising the complication of polyphase alteration. Contacts between stratigraphically distinctive and adjacent lithological units should be relatively sharp, and not gradational in terms of colour, alteration and intensity of alteration (Fig. 14.3) as frequently seems to be the case in polyphase altered volcanics. Where such polyphase alteration is overprinted by a penetrative tectonic cleavage, the apparent clastic texture can resemble eutaxitic texture because the apparent clasts are streaked out within cleavage planes. Irregularly smeared out, highly strained crystals can also produce an apparent or pseudo-eutaxitic texture.

Figure 14.3 Effects of variable alteration type or intensity, or both, on rock type. (a) Six core samples through the *same* dacite unit. In order from left to right the principal alteration phases are: 1, quartz–kaolinite–sericite; 2, sericite–chlorite; 3, chlorite–pyrite; 4, sericite–quartz–chlorite; 5, chlorite–sericite; 6, sericite (pale)–chlorite (dark) resulting in an *apparent* clastic, breccia texture. (b) Variable visibility of phenocrysts in a dacite as a function of varying intensity of sericite–quartz alteration. Phenocrysts are easily visible in the core on the left, moderately visible in cores 2 and 3 and barely visible or imperceptible in the other cores.

Although the emphasis here has been largely on textural changes, it is inherent from the types and intensity of processes operating that significant chemical changes should also occur, causing significant changes to the original igneous bulk-rock compositions.

14.3.2 DEVITRIFICATION

Since glasses are thermodynamically unstable they can undergo post-eruption devitrification, which involves the nucleation and growth of fibrous crystallites of largely quartz (after cristobalite and rarely tridymite) and both sodium-rich and potassium-rich alkali feldspar (Lofgren 1970, 1971a & b). Lofgren's studies involved the experimental devitrification of rhyolitic obsidian glass under varying conditions of temperature, pressure, time and fluid compositions. He was able to simulate many of the devitrification textures found in natural glasses, including spherulites, bow-tie aggregates, axiolites, orb texture and devitrification fronts (cf. Ross & Smith 1961). Devitrification can occur in both lavas and pyroclastic rocks. Given the scarcity of natural glasses in rocks older than Tertiary, it must be assumed that glasses either devitrify penecontemporaneously or are altered because of their metastable state, as discussed above, under diagenetic to low grade metamorphic conditions.

The stages defined by Lofgren (1971b) are an initial hydration stage, a glassy stage, a spherulitic stage and a granophyric stage. The hydration stage is characterised by a polygonal mosaic of fractures in glass enclosed by a sharp curviplanar fracture,

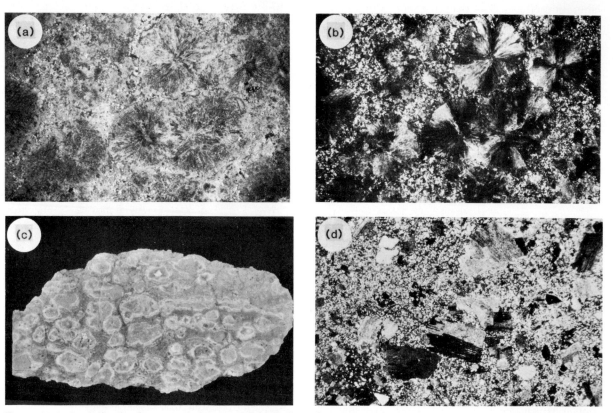

Figure 14.4   Devitrification features of glassy rocks. (a) Radiate, fibrous devitrification spherulites from Carboniferous rhyolite dyke, Bathurst, New South Wales, Australia; plane-polarised light; width of field is 10 mm. (b) As in (a), crossed nicols. (c) Devitrification spherulites, sometimes amalgamated and arranged in trains, Upper Devonian Boyd Volcanic Complex, New South Wales, Australia; slab is 8 cm long. (d) Granophyrically recrystallised, originally glassy dacite lava, Lower Devonian Merrions Tuff Formation, New South Wales, Australia.

called the hydration front. In addition, a strain-induced birefringence, a change in colour from pale yellow-green to pale green and pale reddish brown, and the development of micrometre-sized bubbles were noted. The glassy stage is marked by felsitic texture and minor spherulites, whereas the spherulitic stage is marked by abundant spherulites and micro-poikilitic quartz. Spherulites consist of radiating clusters of fibres, and vary in size from 100 mm to several centimetres (Figs 14.4a & b; Section 4.10.2). When large, they are generally spherical with smooth surfaces, but some irregularities are found (Fig. 14.4c). Sometimes spherulites can amalgamate to produce an elongate train of connected, overlapping spheroids (Fig. 14.4c), often concentrated and aligned along flow layering. Large spherulites with internal cavities or vughs are called lithophysae (Fig. 14.4c). Lofgren (1971a) found that the morphology of spherulites varied according to the temperature of the run. At runs below 400°C their outlines were nearly circular, between 400°C and 600°C bow-tie shaped aggregates occurred and at 700°C lath-like fibres, or open framework circular clusters of spaced fibres occurred. The granophyric stage, although not noted in Lofgren's experiments, is common in old, devitrified glassy rocks and is dominated by fine-grained, roughly equidimensional, recrystallised aggregates of quartz and feldspar, rather than elongate fibres (Fig. 14.4d).

Devitrification can produce an apparent granularity through the development of spherulites and orbs, which at times may give rocks an apparent clastic character and overprint original textures. Patchy devitrification can also give rocks a patchy domainal texture, the boundaries between devitrified and non-devitrified zones (glassy, altered) being called devitrification fronts.

Devitrification and hydration occur through the depolymerisation effects of water in breaking Si–O–Si bridges and hydrolysing tetrahedral oxygen to OH^-. The liberated SiO_4 tetrahedra are then free to reorganise and to nucleate quartz and feldspar crystal structures in the presence of alkalis (network modifiers), which diffuse through the glass network (Lofgren 1970; see also Ch. 2). Lofgren found that adding alkalis in solution increased the devitrifi-

cation rate by four to five orders of magnitude, and the hydration rate by one to two orders of magnitude.

The original textures of all glassy rocks are therefore inevitably modified by devitrification and ensuing alteration. However, in addition Lofgren (1970), following the observations of Lipman (1965) and Noble (1967), found that there were significant changes in the bulk rock chemistry accompanying hydration and devitrification, especially variations in SiO_2, H_2O, Fe_2O_3/FeO ratio, Na_2O and K_2O contents (up to 2% for individual element oxides). Ancient glassy volcanics are unlikely to reflect original chemistry on this count alone, not to mention the effects of alteration, diagenesis and metamorphism.

Devitrification in ignimbrites can be more complex. It can have a zoned distribution through ignimbrites, and be accompanied by vapour-phase crystallisation and hydrothermal or fumarolic alteration. The essential devitrification stages (hydration, glassy, spherulitic and granophyric) can all occur. However, because of the variation in welding and zonation in welding (Ch. 8), different degrees of devitrification can occur at different levels (R. L. Smith 1960b, Lipman & Christiansen 1964, Lipman et al. 1966, Scott 1971, Briggs 1976, Carr 1981; Ch. 8). Carr (1981) also suggested that devitrified pumice seemed to host spherulite growth more than the glassy welded matrix. In addition, in the porous, partially and non-welded zones of ignimbrites, especially the upper zones, vapour-phase crystallisation can occur. Vapour-phase crystallisation involves the growth of tridymite, less commonly cristobalite (both of which readily invert to stable quartz), alkali feldspar (usually sanidine), haematite and, less commonly, biotite, amphiboles and zeolites in open pore spaces (e.g. between uncompacted shards and pumice, in the vesicles of uncollapsed pumice). It occurs contemporaneously with or after devitrification, and at least the early vapours from which these secondary mineral species are precipitated, are derived from trapped volatiles, volatiles which continue to exsolve or diffuse, or both, from juvenile glassy fragments, and heated ground water, which percolates through the ignimbrite shortly after its emplacement and during cooling (Gilbert 1938, R. L. Smith 1960b, Ross & Smith 1961; Ch. 8). These fluids escape and move upwards during compaction and welding. After emplacement of an ignimbrite, downward percolation of rain water leaches elements out of the porous, glassy top of the ignimbrite, and may also lead to secondary mineral precipitation in open pore spaces.

In basic rocks, spherulite-like radiating aggregates of largely feathery, needle-like crystals of plagioclase and pyroxene are called variolites and the texture variolitic texture. Variolites may in part result from devitrification, but where they occur in submarine glassy rocks, especially in the margins of pillows, they may be a product of quench-induced crystallisation.

14.3.3 PALAGONITISATION

When basaltic magma is chilled or quenched by contact with water it forms a pale brown to reddish-brown glass called sideromelane or a black glass called tachylite (MacDonald 1972). Tachylite was thought to be black due to very fine, dispersed grains of magnetite (Peacock & Fuller 1928, R. E. Fenner 1932, MacDonald 1972, Kawachi *et al.* 1983). However, Kawachi *et al.* (1983) have established that the black grains are not magnetite but quenched crystallites of pyroxene. In pillow lavas the two glass types may occur together in layers at the margins of pillows (J. G. Moore 1966, Kawachi *et al.* 1983). These glasses, especially sideromelane, are susceptible to alteration.

When sideromelane alters it changes to a pale-yellow to yellow-brown hydrated altered glass called palagonite (Peacock & Fuller 1928, J. G. Moore 1966, Hay & Iijima 1968, MacDonald 1972, Kawachi *et al.* 1983), which is either finely fibrous or gel-like, and isotropic. The formation of palagonite involves hydration and ion-exchange (J. G. Moore 1966). Moore found that during the formation of palagonite in submarine hawaiian basaltic glasses, sodium, calcium and manganese were lost, whereas potassium, iron and titanium were gained relative to the unaltered glass, presumably due to exchange with sea water. He also found that the palagonite layers at the margins of

pillows became thicker with age. A manganese crust precipitated from sea water was observed on the surfaces of pillows, and was also noted to be thicker as the age of the pillows increased. Moore also suggested that tachylite could be replaced by palagonite but Kawachi *et al.* (1983) found that this had not happened with their Eocene pillow lava succession. Kawachi *et al.* (1983) indicated that palagonite was a mixed-layer montmorillonite–illite mineral and that palagonitisation involved considerable element mobility. In general palagonitic layers were depleted in Si, Mn, Ca and Na, and enriched in K and H_2O. Fe was either depleted or enriched, as were Ti, Mg and Al. *Ti, contrary to its reputation as a stable element, was found to be considerably mobile.*

Palagonitic alteration of basaltic glass is thus clearly pre-metamorphic, is essentially a weathering effect due to the interaction of water with metastable basaltic glass (J. G. Moore 1966, Hay & Iijima 1968, MacDonald 1972), and produces significant compositional changes in affected rocks. Jakobsson (1978) has summarised his significant research on the palagonitic alteration of the tephra erupted on Surtsey during 1963. He indicates that palagonitisation can occur at normal surface temperatures and pressures over the years, but is speeded up by the influence of temperature. In the area of a significant thermal anomaly, palagonitisation appears to have occurred within 1–1½ years, under the influence of temperatures between 40 and 100°C. In association with concurrent precipitation of secondary minerals such as opaline silica, zeolites and calcite (Jakobsson 1978, Jakobsson & Moore 1980) palagonitisation has the potential to stabilise and cement Surtseyan volcanic piles within several years after cessation of volcanism.

14.3.4 HYDRAULIC FRACTURING

If a fluid system in the subsurface develops a pressure which exceeds the tensile strength of the enclosing rock and the minimum principal stress component, then that fluid has the capacity to propagate a fracture or fracture system, and to open it in a tensile manner (see Chs 3 & 15; Secor 1969, Phillips 1972, Shaw 1980). Such fracture propa-

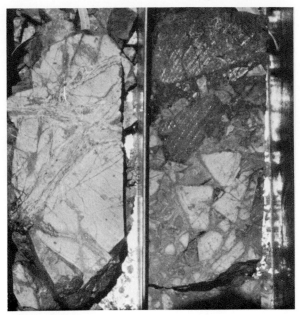

Figure 14.5 Hydraulic fracturing and brecciation, and copper–gold mineralisation in Permo-Carboniferous rhyolite intrusive, North Queensland, Australia. Core diameter is 3.5 cm. (Photograph by R. Allen.)

gation and opening may lead to brecciation of the wall rock associated with the fracture, or system of fractures. The rock is essentially shattered in tensile fashion by the overpressured fluids working through the rock (Fig. 14.5). The resultant fracture need not be involved in shear motion, and so the breccia fragments and the associated wall rock to the fracture need not be pervasively foliated or sheared. This fluid induced fracturing and brecciation is called *hydraulic fracturing* (see Secor 1969 and Phillips 1972 for further discussion of the mechanics and principles).

In volcanic successions, hydrothermal fluid systems have enormous potential for causing hydraulic fracturing and brecciation. The breccia fragments are generally angular, blocky to splintery in shape (Fig. 14.5) and, out of context, could be confused with quench-fragmented hyaloclastite deposits. Where observable in outcrop, hydraulic fracture breccias should be distinguishable by their cross-cutting character and confinement to a fracture zone. The width of such hydraulic fracture zones may vary from several centimetres to many metres.

In small, incipiently brecciated zones, breccia clasts may have jigsaw puzzle like fit relationships. In well developed breccia zones, clasts may be moved relative to each other, significantly rotated, and even transported through the system, leading to a heterogeneous mixture of clast lithologies. Such breccia zones may also obviously be the sites for significant hydrothermal alteration and mineralisation (e.g. Phillips 1972). Also, where through these alteration processes significant weakening of the rock has resulted, such zones may preferentially absorb strain and become highly deformed during subsequent tectonic movements.

14.3.5 DIAGENESIS

The concept of diagenesis encompasses the mineralogical and textural changes associated with lithification and the early stages of burial of any sediment or rock system. It is of fundamental importance in hydrocarbon exploration because it modifies original porosity, generating secondary porosity and causing hydrocarbon maturation. Significant textural and mineralogical changes can be produced by dissolution of original components, precipitation of and replacement by new mineral phases, and compaction. These processes occur under the influence of increasing pressure, temperature and significant fluid flux during burial. Diagenesis can be viewed as the low grade, initial stages of metamorphism.

In volcanic rock systems, diagenesis is just as significant if not more so than in sedimentary systems, given the usual abundance of metastable glassy material and the presence of labile mineral components, especially in basaltic to intermediate rocks. Granophyric recrystallisation can be considered to be a diagenetic process. In addition to these, the growth and overprint by burial metamorphic mineral assemblages, typically zeolite facies assemblages, can be significant in causing major modification of original rock textures and components (e.g. Raam 1968, D. K. Davies *et al.* 1978b).

14.3.6 METAMORPHISM

Metamorphism, whether it be regional or contact, is a higher grade extension of diagenesis, but should not be confused with hydrothermal alteration, which is usually more local in effect. Metamorphism produces pervasive mineralogical and textural changes and, where allied with the strain effects of deformation, can cause major problems in the recognition of original textures and rock types. Nevertheless, perseverance is usually rewarded, because frequently enclaves of rock, still preserving original textures or ghosted relics of such, can be found, particularly in greenschist or lower grade rocks. For example, relic shards, welded textures and perlitic textures are all documented in greenschist grade volcanics. Weathering frequently etches out original components (e.g. fiamme), and even in high grade metamorphics or deformed rock successions, or both, original features may be etched out on weathered surfaces of rocks which internally (e.g. in fresh core) are totally recrystallised and contain only obvious metamorphic and/or structural features. For example, in amphibolite and granulite facies rocks it is sometimes possible to recognise sedimentary structures, pillow lavas and even spinifex textures on weathered surfaces.

14.3.7 DEFORMATION

Deformation is important, first from the point of view of its regional and outcrop effects; for example, in modifying geometry. Secondly, on a mesoscopic and microscopic scale, deformational fabrics and strain create major problems in recognising original textures and rock types, particularly where, as suggested above, pervasive deformational and metamorphic effects are both well developed. However, even in such successions there will be significant local variations in the degree of strain, and therefore the effects on original textures (Fig. 14.6). Such variations may misleadingly give the impression of different juxtaposed lithologies. However, original textures are frequently preserved in only mildly affected enclaves, which are well worth the time spent in looking for them (e.g. S. F. Cox 1981; Fig. 14.6).

14.3.8 RELATIONSHIP BETWEEN DEFORMATION AND ALTERATION

Many highly altered volcanic rock successions are penetratively deformed and cleaved (e.g. S. F. Cox 1981). The question frequently arises as to whether the alteration preceded the deformation or whether deformation fabrics produced permeability for alteration fluids. There is obviously no simple answer, but it should be clear that if a rock

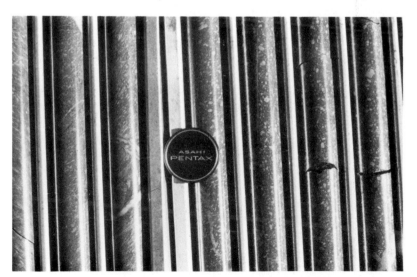

Figure 14.6 Variable strain and its effect on original porphyritic dacite. At right, undeformed porphyritic dacite with well defined feldspar phenocrysts; in centre cleaved dacite with strained, smeared feldspars; left, highly cleaved (and more pervasively altered) chloritic schist after dacite.

succession is significantly altered, either in discrete zones or at large, the affected areas will be mechanically weakened and will preferentially absorb more strain than unaltered zones will. As suggested above, there may thus be significant variations in the degree of deformation.

14.4 Recognition of pumice in the rock record

Although pumice is one of the most abundant components in pyroclastics and epiclastic volcanic sediments in modern silicic volcanic terrains, its recognition can be very difficult in the rock record. Not that pumice should be any less abundant in the rock record than it is today. It is, however, one of the components most susceptible to post-depositional modification. It is porous and so readily allows passage of circulating fluids; it is glassy and metastable, so readily alters or is replaced; and it is mechanically weak, so may take up significant strain during burial compaction or deformation, or both. It also hosts vapour phase crystallisation (in ignimbrites) and diagenetic minerals, and it readily devitrifies and may preferentially develop de-vitrification spherulites.

In ancient volcanic successions pumice is most readily identified in welded ignimbrites since its flattening produces the characteristic eutaxitic tex-ture (Ch. 8; Fig. 8.38). Even in metamorphosed, weathered ignimbrites the flattened pumice lenticle foliation, or fiamme may be distinctive and etched out by weathering. In non-welded ignimbrites and sediments the pumice fraction could be especially susceptible to weathering. It could also be flattened by mechanical, brittle collapse during burial and compaction, or it could be 'densified' by the precipitation of secondary minerals in vesicles, and so retain an original unflattened shape, perhaps making distinction from lithics difficult.

Fiske (1969) addressed the problem of recog-nising pumice in ancient marine volcaniclastic deposits, and he noted the problems outlined above. However, he cited the preservation of pumice in rocks as old as Precambrian. Essentially, pumice may be preserved in nearly pristine state or

be beyond recognition, depending on its post-depositional history. Fiske (1969) suggests a com-bination of both careful hand specimen and thin-section examination to verify the existence of pumice.

14.5 Facies as diagnostic indicators of palaeoenvironments and palaeoen-vironmental conditions

The point of carefully documenting the character-istics of facies is to identify those characteristics which give clues to the mode of deposition or environmental conditions, or both, at the time of emplacement. For example, angle of repose cross-bedding in epiclastics suggests tractional sediment transport and deposition; structureless to faintly laminated epiclastic claystones or mudstones sug-gest settling out of suspension and very quiet aqueous conditions; well defined eutaxitic texture in a thick, massive, undeformed volcaniclastic deposit suggests welding and deposition by pyroclastic flow.

However, there are few facies or facies character-istics which by themselves unequivocally identify the host depositional environment or the spatial context of a facies within a broader palaeogeogra-phy. This applies in both sedimentary and volcanic terrains. For example, eutaxitic texture by itself may indicate a near vent, welded air-fall deposit (Ch. 6) or ignimbrite (Ch. 8), in both cases suggesting a subaerial setting. Only more detailed analysis, and consideration of related facies will determine which it is and, if it is ignimbrite, whether it is proximal or distal. Pillow lavas (Ch. 4) indicate a subaqueous depositional environment, but the environment could have been a lake or the sea. The water depth could be anywhere from only several metres to kilometres, and the vent could have been subaqueous or subaerial. Also, sub-aerially erupted basaltic lavas can flow into water and transform into pillow-lavas (Section 4.6). Accretionary lapilli (Ch. 6) indicate the existence of subaerial conditions, but final deposition could have been in a subaerial or aqueous environment and may have occurred close to or far away from vent (Ch. 5). Ballistic bomb-sag structures are one of the few diagnostic facies features of volcanic terrains, suggesting a near-vent (within hundreds of metres) setting (Ch. 5). However, impact sag structures can also be produced by epiclastic processes (Section 10.3).

Therefore, to arrive at a sound interpretation and reconstruction of the palaeoenvironment and its conditions, the geologist working in ancient terrains has to establish **significant associations of facies**. The context of any single facies is most likely to be revealed by its relationship to and association with other facies. For example, a lake-deposited facies in a subaerial volcanic terrain could be part of an eruptive centre (crater or caldera lake) or could have formed in the surrounding landscape far from an eruptive centre. The facies by itself may not suggest which setting, but the associated facies, e.g. perhaps thick, areally restricted rhyolite intrusives and flows (Ch. 4), proximal ignimbrite facies (Ch. 8), base surge deposits (Chs 5 & 7), critical epiclastic facies (Ch. 10), should help to identify the relevant setting.

Similarly, a succession of ignimbrites could represent an intracaldera fill succession, or a series of outflow ignimbrites deposited tens of kilometres from the vent. For example, McPhie (1983) has identified a succession of Carboniferous ignimbrites (both welded and non-welded) as an outflow succession. This was based on the association of moderately thick ignimbrites (cf. very thick intracaldera fill ignimbrites, Ch. 8) with valley fill morphologies, with fluvial volcaniclastic sedimentary rocks, *and the absence of* the expected association of caldera related facies as summarised in the previous example.

So, in summary, there are few facies in volcanic terrains that will be unique to a particular setting or environment, so that interpretation should be based on the association of facies. Associated sedimentary facies will be most valuable indicators of immediate sedimentary environments and the sedimentary processes and conditions operating. Readers are referred to current sources for the detailed facies characteristics of modern sedimentary environments and successions (R. G. Walker 1984, Reading 1978, Selley 1978, 1982, Friedman & Sanders 1978, Blatt *et al.* 1980).

14.6 A suggested approach to facies analysis

A thorough, careful analysis of facies and associations of facies in ancient volcanic successions will inevitably involve several stages.

Stage 1: where surface exposure or subsurface mine exposure occurs, produce *an outcrop map*, which is a map showing where each outcrop occurs, its size, its composition and any relevant structural information. Outcrop maps, being 'fact maps', are important because they constrain the degree of confidence that can be put on final interpretations, and serve as a basis for critically evaluating the possible spatial and age relationships between different facies. Where subsurface core is available, detailed logging will be necessary, and two- or three-dimensional correlation fence diagrams should be attempted.

Stage 2: determine the structure of the succession.

Stage 3: identify and describe all facies present based on field outcrop, core characteristics, hand-specimen characteristics, and *personal* thin-section observations, using the facies descriptors outlined in Chapter 1. Measure detailed sections or logs to represent facies characteristics. In successions that have been variably altered and deformed, facies analysis should first be done in the least altered areas or cores, and when confidence in recognition of original facies has been developed, only then work back into more altered areas and cores. Petrographic work should be done personally by geologists. Petrographic descriptions made by consultants who do not see the rocks in context should be avoided if possible.

Stage 4: work out the spatial and age relationships of facies (sharp, gradational, conformable, unconformable, faulted, intrusive) from outcrop patterns and cross sections, and represent these on interactive diagrams (Section 14.2).

Stage 5: assess the possible modes of fragmentation and/or formation, transportation and deposition.

Stage 6: consider the possible genetic relationships between *associated facies*, and then make genetic interpretations of each facies and the total association of facies in terms of origins (stage 5), environment of deposition and environmental conditions.

Stage 7: if a broader palaeogeographic context is required, then look at the identified significant associations of facies and their relationships to each other. Where outcrop permits, this could be supplemented by collecting data on transport and source directions such as palaeocurrent measurement, contouring of maximum clast sizes, and thickness variations.

14.7 Facies models – what they represent and their uses

Roger G. Walker, introducing the excellent volume he edited on facies models for sedimentary successions, writes:

> a facies model . . . [is] a general summary of a specific sedimentary environment. . . . The basis of the summary consists of many studies in both ancient rocks and recent sediments. . . . The increased need for models is due to the increasing amount of prediction that geologists are making from a limited local data base. This prediction may concern subsurface sandstone geometry in hydrocarbon reservoirs, the association of mineral deposits with specific sedimentary environments, . . . In all cases, a limited amount of local information *plus* the guidance of a well-understood facies model results in potentially important predictions about that local environment. (R. G. Walker 1984, pp. 3–4)

Walker goes on to emphasise that a facies model has four principle functions: as a norm for comparison, as a framework and guide for further observations, as an initial predictor in new geological situations and as a basis for hydrodynamic interpretation of the environment it represents. However, as discussed further in Section 14.8, although the facies model acts as a general guide, each case will have peculiar features that are slightly at variance to the norm. The identification of these differences is

just as important as the recognition of the similarities.

The construction of a facies model for a particular type setting involves recognition of the essential elements common to all the examples (modern and ancient) of that type setting and encapsulating these into the general facies model. It also involves ignoring the local, insignificant pecularities of each example, the guideline being that the omission of that idiosyncracy does not fundamentally affect the distinctive character of the facies model. R. G. Walker (1984) called this process of incorporating the essentials and omitting the insignificant idiosyncracies the distillation of a general facies model.

One common element in the approach to the interpretation of facies and the development of facies models in both non-volcanic and volcanic successions is the need to develop a framework of focal elements for reference, and to provide a scale perspective. For example, in fluvial successions the focal element is the channel; in deltas, alluvial fans and submarine fans it is the system of distributary channels; in near-shore marine successions it is the shoreline; and in volcanic terrains it is the eruptive centre. Irrespective of whether it is the actual eruption centre complex or the environments marginal to that complex that are of interest (e.g. for exploration purposes), it is necessary to locate the centre in order to predict where the marginal environments and facies will be found.

Another common approach in the analysis of facies in sedimentary successions and in the development of facies models is to identify distinctive sequences of facies, which may define trends unique to particular environments. For example, meandering fluvial facies models are dominated by upward fining channel-point bar packets of sediment; transgressive successions are upward fining; regressive successions are upward coarsening; mid-fan lobes of deep submarine fans are typically depicted as upward coarsening with beds thickening; associated channel fills are typically depicted as upward fining with beds thinning (see R. G. Walker 1984, Reading 1978). Such facies sequences are produced by a dynamic equilibrium in the sedimentary environments, leading to an essentially predictable progression of facies. However, in volcanic terrains, and in facies models for volcanic settings at large, such upsequence trends are not likely to be well defined because events are more random in terms of timing, intensity and volumes of material release. Although individual events may produce deposits for which a general facies model is applicable (e.g. ignimbrites, Section 8.7), the whole volcanic succession probably will not. Statistical facies sequence analysis (e.g. Markhov chain analysis; see R. G. Walker 1984) is unlikely to produce systematic patterns or trends. A possible exception to this rather negative assessment of regularity in facies successions, is in ophiolite and oceanic crust facies models (Sections 14.8.4, and 15.2 & 3). Although some volcanic successions show upsequence compositional changes, these are chemical changes rather than systematic, predictable changes in the physical facies, on which facies models are based.

Another approach frequently used in sedimentology, is the use of analogy, whereby the succession in question is compared with well documented, specific case studies of modern and ancient counterparts. This serves several functions. First it provides a scale perspective, particularly where a modern counterpart is being used for comparison. Secondly, it allows the similarities, but equally importantly the differences, to be identified, and their importance to be evaluated.

Finally, the ultimate aim of any facies analysis study is to produce a comprehensive, if at times a little imaginative, palaeoenvironmental–palaeogeographic reconstruction which serves as a working model (even if schematic) (e.g. Cas & Jones 1979, Fergusson *et al.* 1979, Cas *et al.* 1981) of the dynamics and products in the basin, or part thereof, in question. Such reconstructions can be in the form of interpretative two-dimensional cross sections, interactive stratigraphic diagrams (Fig. 14.1) or three-dimensional block diagrams (Plate 14), or as a combination of these. If the general facies models available are relevant, then the final reconstructions will be amended versions of these diagrams to suit the case in hand.

14.8 Facies models for volcanic successions

The distillation of general characteristics into a facies model, as outlined in Section 14.7, suggests that it should be possible to take the general features of modern volcanoes (Ch. 13), and those of documented ancient ones, together with a knowledge of the range of facies found (Chs 3–11) and an understanding of the eruptive, depositional and erosional processes (Chs 1–13) in modern volcanic terrains and associated basins, and use these to generate useful facies models for a wide range of volcanic settings. Such general facies models have predictive value to the degree that they summarise the range of associated facies to be expected in such settings. However, each case will have unique features, and the spatial relationships between facies will have to be determined in detail. The general facies model cannot be expected to be a consistent representation of scale and particular relationships between specific facies, but will serve as an initial framework to be modified to produce a more specifically relevant facies model for the case in hand. That specific, modified facies model will then have predictive value for the particular rock succession under consideration.

Mineralised zones and ore horizons and alteration zones can also be considered as facies within the context of our facies models, and their relative distribution can be shown. Syndepositional ores (e.g. exhalatives) should, in any case, be treated as a depositional facies and their critical relationships with other facies noted, and considered in terms of possible environmental significance.

The volcanic settings for which we will attempt to construct facies models are:

continental basaltic successions
continental stratovolcanoes
continental silicic volcanoes
submarine basaltic rift volcanism
oceanic basaltic seamounts
marine stratovolcanoes
marine felsic (silicic) volcanoes
silicic volcanic facies of ensialic marine basins

intraglacial basaltic and rhyolitic volcanoes
Precambrian volcanism

14.8.1 CONTINENTAL BASALTIC SUCCESSIONS

Selected references

Sections 13.3–6, Greeley and King (1977), Swanson *et al.* (1975), Swanson and T. Wright (1981), Waters (1961), Ollier (1967b), Ollier & Joyce (1964), Joyce (1975), Mohr (1983), Choubey (1973), Subbarao and Sukheswala (1981), Bultitude (1976b) White (1960) and Bristow and Saggerson (1983).

Description

The principal volcanic elements of continental basaltic provinces and the basins in which they occur will be flood and valley-fill lavas (Chs 4 & 13), cinder cones, maars or tuff rings and shield volcanoes. Important accessory sedimentary environments will include fluvial channels, which will be constantly relocated as flood lava ponds in valleys. The fluvial systems will be incisive and erosional, acting against the aggradational effects of the basaltic volcanism in building up the topography. Fluvial successions will therefore not be thick or extensive, and will only be preserved in the rock record through the capping effects of valley-fill lavas, producing 'deep lead' deposits which, in eastern Australia, are renowned for their precious metal and precious stone alluvial resources. Lacustrine and swamp deposits and environments may also be relatively short-lived. Their origin could be either within maars or tuff rings, or craters, or be due to damming up of valleys by lavas or ponding within the flat landscape dominated by flood lavas. They may be short-lived because contemporaneous erosion could breach their physical margins or because they become buried under a new lava flow. The basaltic succession may be underlain by alluvial–fluvial deposits and basement from which it may be separated by a regional unconformity, and upwards may pass again into alluvial–fluvial successions as volcanism wanes. Locally, where lavas have flowed over swampy areas, or into lakes or rivers, hyaloclastites, and even pillow lavas, may be found. If regional volcanism occurred in a continental edge setting, it may have been accompanied by regional updoming or tumescence, which is compensated by regional subsidence after volcanism ceases, perhaps leading to marine transgression. Throughout the section fossil soils should be found, reflecting the long erosion–weathering intervals between short-lived volcanic episodes. Magmas in these provinces may be alkaline to tholeiitic in composition.

Facies model

See Figure 14.7.

Economic significance

These settings may host concentrations of alluvial precious metals and stones (called 'deep lead' deposits) in river channel successions buried by flood basalt lavas. Diamonds may occur in kimberlite pipes, diatremes and breccia pipes in the root zones of maars and tuff rings.

14.8.2 CONTINENTAL STRATOVOLCANOES

Selected references

Section 13.7, Pike and Clow (1981), Vessell and Davies (1981), Roobol and Smith (1976), Fiske *et al.* (1963), Lipman and Mullineaux (1981), Thompson *et al.* (1965), Gregg (1960), Grindley (1960), Neall (1979), Gorshkov (1959), Aramaki (1963) H. Williams and McBirney (1979, pp. 312–3), Branch (1976) and R. W. Johnson (1976, 1981).

Description

Stratovolcanoes show complex primary variations in erupted products in time and space, and epiclastic processes and mass wastage of the steep-sided cones further complicates the stratigraphy of the pile (Ch. 13). We have already discussed the principal volcanic elements of stratovolcanoes in Chapter 13. Short lavas, domes and shallow intrusives (commonly basaltic andesite to dacite in composition) are the main cone-building elements, and armour the cone against erosion. These are flanked by, and interdigitate with, various types of pyroclastic and epiclastic rocks. Pyroclastics may

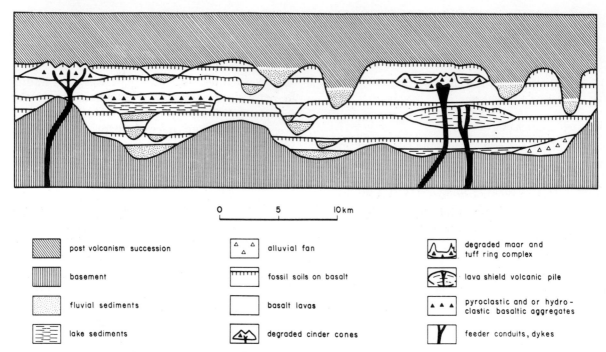

Figure 14.7 General facies model for continental basaltic successions.

show a wide range in composition and eruptive style. However, it will only be the deposits of large plinian and ignimbrite eruptions, especially welded tuffs, that tend to be preserved in the geological record. Other pyroclastic deposits will be for the most part eroded and redeposited as mass-flow deposits and reworked epiclastic sediments in basin areas marginal to the volcanic centre. Much of this probably occurs very shortly after their eruption (1–10 years) with much of the deposition taking place on the developing ring plain of these volcanoes.

Are there any generalisations that can be made about the volcanic successions of stratovolcanoes? It seems that their variability is their most important characteristic, and we can conclude that the stratigraphies of stratovolcanoes show:

(a) Rapid inconsistent lithological and compositional changes in the vertical succession;
(b) Rapid lateral lithological changes which, moving away from the source, may fit into a systematic facies pattern. Around the cone, facies will be discontinuous, controlled by

deep valleys channelling the coarse debris of pyroclastic flows and other types of volcaniclastic flows, and the presence of lavas will be characteristic;
(c) Near to the vent, a large proportion of volcaniclastic breccias, which may be pyroclastic flow deposits of various types and facies (Chs 7 & 8) or secondary epiclastic flows due to mass-wastage (Ch. 10); these will be closely spatially associated with lavas, domes and shallow intrusives, together constituting the 'core' of the volcano; and
(d) Away from source, thick successions of immature volcanic detritus in alluvial or marine settings, or both.

Facies model

See Figure 14.8. Note that this model is a very generalised one, depicting a relatively large proportion of volcaniclastics. However, as noted in Chapter 13, many stratovolcanoes consist of a much higher proportion of lavas and intrusives than shown here, especially in the near-vent 'core' area (e.g. Bultitude 1976a).

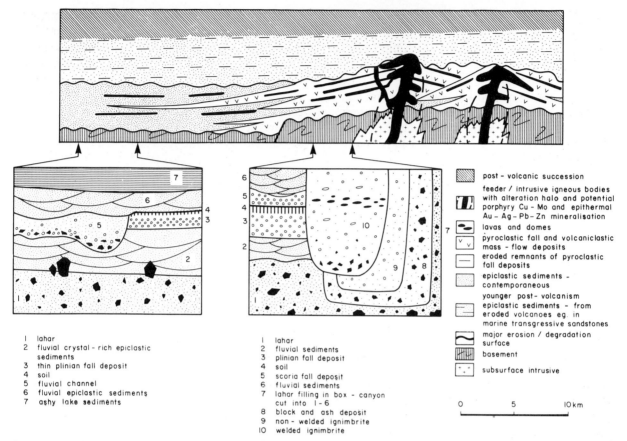

Figure 14.8 Generalised facies model for continental stratovolcanoes. Note the significant variations from the proximal near-vent cone facies to the distal ring plain environment where epiclastic volcaniclastics are dominant. Also see Figure 13.32.

Economic significance

Stratovolcanoes may host deep-level vein-stockwork porphyry copper–molybdenum deposits, higher level epithermal lead–zinc deposits and shallow level epithermal gold and silver deposits.

14.8.3 CONTINENTAL SILICIC VOLCANOES

Selected references

Section 13.9, G. P. L. Walker *et al.* (1981d), R. L. Smith and Bailey (1968), Byers *et al.* (1976), Lipman (1975, 1984), Bailey *et al.* (1976), Christiansen (1979), M. C. W. Baker (1981), Grindley (1960), Healy *et al.* (1964), McPhie (1983, ancient outflow ignimbrites) G. P. L. Walker (1984a), Lipman *et al.* (1984) and Nathan (1976).

Description

The principal volcanic elements of modern rhyolite volcanoes are topographically subdued clusters of rhyolite hills rising above gently sloping ignimbrite shields. The rhyolitic hills are composed of lava domes and short flows, with associated near-vent pyroclastic falls and other rhyolitic volcaniclastic deposits. The ignimbrite shields will contain the various ignimbrite facies discussed in earlier chapters (Chs 5, 7 & 8), and rhyolitic fall deposits erupted from the centre (e.g. subplinian, plinian, phreatoplinian, Ch. 6). The focal element of this association of primary volcanic facies in the shield, is the caldera structure, which will contain multiple lava and ignimbrite eruption points. The margins of the caldera may be abrupt, steep scarps, or gently inward dipping slopes and may have associated caldera margin collapse breccias (Ch. 13). The rhyolitic lavas and domes will lie within or at the margins of the caldera structure, although some may be erupted outside the caldera. The caldera subsidence basin will host not only lavas and domes (or cryptodomes), but also thick intracaldera ignimbrites (Ch. 8). These can be extremely thick (much thicker than ignimbrites of the shield outside the caldera), they may have associated near-vent co-ignimbrite breccias, they are frequently crystal-rich and often lack evidence of significant ash-loss associated with eruption. This caldera-fill succession will also contain intercalated epiclastic sediments. Contemporaneous basaltic scoria cones may occur around the margins or beyond the limits of caldera structures, but rarely within the caldera confines. Hydrothermal systems including sinter deposits and pools will be common within and outside the caldera, and these may be associated with hydrothermal explosion craters.

Epiclastic deposits may be volumetrically important caldera-fill elements, covering areas between 10^2 and $10^4 \, \mathrm{km}^2$. Such deposits could include fluvial sediments, subaerial mass flows, and caldera and valley-ponded lake deposits, subaqueous debris flows and turbidites of pumiceous and dense rhyolite debris, hemipelagic lacustrine muds, horizons of floated pumice, and even diatomaceous oozes. Where caldera centres are resurgent, caldera fill strata could be significantly tilted, faulted, and even folded.

The ignimbrite shield or plateau outside the caldera is dominated by relatively thin (tens of metres or more), sheet-like ignimbrites, known as outflow ignimbrites. Interspersed with these are pyroclastic fall deposits and epiclastic deposits (McPhie 1983). Erosion will be important between ignimbrite eruptions. Valleys, typically steep-sided box canyons, are quickly incised into freshly erupted ignimbrite as the drainage pattern tries to re-establish itself (e.g. Figs 8.7 & 42). If welded, the non-welded top will be stripped off in time (e.g. 10^2–10^5 years, depending on size and climate), otherwise rapid degradation of the whole sheet will continue unless covered by another ignimbrite, or other deposits. Epiclastic processes within the shield are largely erosional, and depositional processes are restricted to alluvial settings and to local, shallow lakes formed in valleys, or perhaps developed on the low slopes of the shield. The eroded volcaniclastic debris is largely removed from the volcano and is deposited in basins marginal to the centre, and the shields are therefore composed dominantly of ignimbrite with volumetrically small amounts of epiclastic interbeds. Large volumes of pumice, ash and crystals will be removed to be deposited in large fluvial outwash plains flanking the ignimbrites (perhaps more than 50% of most ignimbrites ends up in such epiclastic sequences), or are even transported into the marine environment. Many, especially small-volume, non-welded ignimbrites such as the Taupo ignimbrite (Chs 7 & 8) have a low probability of even being preserved in the geological record.

When a number of silicic centres are associated together in time, such as in the Taupo Volcanic Zone in New Zealand or the San Juan Volcanic Field in the western USA, ignimbrite shields will overlap. Any one stratigraphic section could therefore contain a number of ignimbrites from different centres, and in both of these fields ignimbrites from one centre are known to fill in older calderas of the other centres. Facies associations will be diverse, lateral facies relationships will be abrupt and stratigraphic relationships will be complex.

Facies model

See Figure 14.9.

Economic significance

Precious and other epithermal metal deposits are the most important resource associated with continental silicic volcanoes. Epithermal deposits of mercury, arsenic, antimony, gold, silver, lead and zinc may be found associated with shallow level hydrothermal systems both inside and outside the caldera margins, and may be associated with diatremes and breccia pipes (Sillitoe *et al.* 1984). In addition, porphyry copper–gold deposits may be associated with deeper-level intrusives. Bedded sulphides may occur in deep caldera lakes, but no substantial deposits of this type are yet known from continental silicic centres.

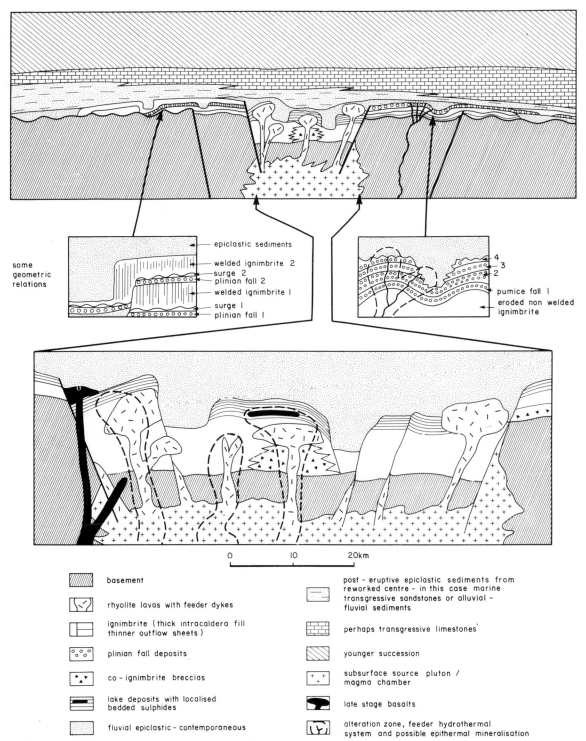

some
geometric
relations

epiclastic sediments
welded ignimbrite 2
surge 2
plinian fall 2
welded ignimbrite I
surge I
plinian fall I

4
3
2
pumice fall I
eroded non welded
ignimbrite

0   IO   20km

	basement
	rhyolite lavas with feeder dykes
	ignimbrite (thick intracaldera fill thinner outflow sheets)
	plinian fall deposits
	co-ignimbrite breccias
	lake deposits with localised bedded sulphides
	fluvial epiclastic - contemporaneous

post-eruptive epiclastic sediments from
reworked centre - in this case marine
transgressive sandstones or alluvial -
fluvial sediments

perhaps transgressive limestones

younger succession

subsurface source pluton /
magma chamber

late stage basalts

alteration zone, feeder hydrothermal
system and possible epithermal mineralisation

Figure 14.9 Facies model for continental silicic volcanoes depicting the variations in facies associations from the focal caldera complex to the ignimbrite plateau or shield outside the caldera.

14.8.4 SUBMARINE BASALTIC RIFT VOLCANISM

Selected references

Section 13.10, Macdonald (1982), Ballard and van Andel (1977), Luyendyk and Macdonald (1977), Macdonald and Luyendyk (1977), Searle and Laughton (1981), R. N. Anderson *et al.* (1982), Newmark *et al.* (1985), Moores (1982) and Coleman (1977).

Description

As summarised in Section 13.10, basaltic rift volcanism associated with mid-ocean spreading ridges (MORs) occurs within the median valley. The morphology of the volcanoes varies from the central type where spreading rates are low, to elongate, fissure-fed ones with intermediate spreading rates, to very elongate shield-like volcanoes at high spreading rates. Both pillowed and non-pillowed lavas are common in these volcanoes. However, ridge crests and oceanic lithosphere are dominated by ridge-parallel fractures and faults, which form in the environs of the ridge due to tectono-isostatic adjustments as the lithosphere ruptures and spreads away from the ridge crest. Because of this, the original volcanic mounds are likely to be highly dismembered. In the rock record their recognition as discrete volcanoes, or parts thereof may be nearly impossible. More commonly, their activity and original presence is preserved only as an almost imperceptible part of the layer 2 basaltic layer of oceanic crust and ophiolites.

A veneer of pelagic sediment and downward transition into a mafic sheeted dyke complex and gabbroic and mafic–ultramafic cumulate rocks, complete the stratigraphy associated with MOR volcanism (Ch. 15). This stratigraphy is similar to that of ophiolites, as discussed in Sections 15.2 and 3.

Basaltic rift volcanism may also take place in marine settings within submerged sialic crust (e.g. some Archaean greenstone belts). In this situation the nature of the volcanic pile will depend on the nature of the vent (whether a point-source or fissure), the rate of extension, the rate of magma discharge and the water depth. All transitions from small piles resembling seamount facies models (Sections 13.10 & 14.8.5; Fig. 13.50) to large piles similar in character to the median rift valley volcanoes of MORs (Section 13.10; Fig. 13.49) could be expected. Pillow lavas, massive lavas and hyaloclastites (quench-fragmented lavas) should be prominent. If eruption occurs in shallow enough water, and there is explosive interaction between magma and water (Ch. 3), then pyroclastics should also be present. Talus slopes and pelagic sediment or epiclastic sediments, or both, may also be prominent. Breaks in eruption activity will be recorded by pelagic–hemipelagic sediments and perhaps redeposited volcanic sediments, and hydrothermal activity and water depth permitting, bedded to massive sulphides.

Facies model

For a typical ophiolitic facies model see Figure 15.1. For ensialic rift basaltic facies model, see Figure 13.50.

Economic significance

Marine basaltic rift volcanism is significant for its potential as a source of Cyrus-type copper-pyrite mineralisation (and to a lesser extent zinc, lead, manganese, nickel, cobalt, silver and gold). Ore deposits occur as massive bodies at the sediment or seawater–basalt interface, and as subjacent, intensely hydrothermally altered stockworks. Manganese nodules with significant concentrations of Mn, Cr, Ni and Co may form post-depositionally at the sediment–basalt interface and within the sediment pile. Mineralisation at seafloor spreading centres has been reviewed by Rona (1984).

14.8.5 OCEANIC BASALTIC SEAMOUNTS

Selected references

Section 13.10, J. G. Jones (1966), J. G. Moore and Fiske (1969), Batiza and Vanko (1983), Searle (1983), Thorarinsson (1967) and Kokelaar and Durant (1983).

Description

See Section 13.10. Where basaltic seamounts develop on oceanic lithosphere, their preservation potential in the rock record must be poor. They, like their substrate, face being subducted. During this they may become detached from the substrate and are likely to be variably dismembered, depending on the tectonics of the subduction setting. So, like the volcanoes of MORs their recognition may be difficult.

Where basaltic seamounts develop on submerged sialic crust in a setting not necessarily subjected to subduction tectonics, they may be preserved in the record, and their recognition will be dependent on recognising the upsequence facies changes first documented within an overall facies model by J. G. Jones (1966) (Section 13.10; Fig. 13.50). After cessation of volcanism, the shallow tops of some centres may be sites for atoll and limestone cap formation.

Facies model

See Figure 13.50. Note: eruptions that begin in relatively shallow waters (e.g. Surtsey, Thorarinsson 1967), may consist only of the upper pyroclastic succession with or without the succeeding subaerial lava cap.

Economic significance

Basaltic seamounts are not noted for their resource potential, although development of ferromanganese crusts (see Section 14.3, discussion on palagonitisation) and nodules could produce significant manganese deposits, which in ancient successions have been called Lahn–Dill-type deposits.

14.8.6 MARINE STRATOVOLCANOES

Selected references

Section 13.7, Bryan *et al.* (1972), Bauer (1970), Black (1970), Ball and Johnson (1976), Pichler and Friedrich (1980), Pichler and Kussmaul (1980), Self and Rampino (1981), Sigurdsson *et al.* (1980), Mitchell (1970), J. G. Jones (1967b), Klein (1975), Klein and Lee (1984) and Ricketts *et al.* (1982).

Description

Marine stratovolcanoes are generally similar to their wholly subaerial counterparts: they have significant relief above their base, they are large and steep-sided, they may have summit calderas, and they have diverse magmatic compositions (Section 13.7). However, they are significantly different in that they have a foundation built in submarine conditions and a summit region which may be subaerial. If the initial eruptions are basaltic to basaltic andesite, then the lower submarine parts of the strato-volcanoes may develop similar stratigraphy and characteristics to volcanic seamounts, being dominated by lavas including pillow lavas and hyaloclastites. However, stratovolcanoes may be longer lived than individual oceanic seamount volcanoes, and they may build substantial subaerial cones dominated by lavas, pyroclastics and immature epiclastics intruded by feeder bodies. Such volcanoes are almost invariably associated with island arc systems (Ch. 15) built on prominent arc crustal blocks which consist of older, deformed arc volcanic piles, associated intrusives and younger carbonate cappings, upon a foundation of submerged oceanic or continental crust. Such settings are associated with significant epiclastic erosional, transportational and depositional processes (Ch. 10). High rates of subaerial mass-wastage and shoreline erosion will be accompanied by major submarine mass-flow processes, shedding debris into associated basins. These produce coarse, proximal forearc and back-arc sediment aprons built up largely by mass-flow deposits (debris flows, rubble avalanches, turbidity currents, slumps and slides). In this regard, emergent stratovolcanoes will be different from seamounts in that their initial lava-dominated foundation will be buried by, and flanked by a volumetrically, equally significant, volcaniclastic apron.

The back-arc apron, which is under the direct influence of the active volcanic arc, will contain much juvenile, contemporaneous volcanic debris, as well as older epiclastic, volcanic detritus. Sigurdsson *et al.* (1980) suggested that in the Lesser Antilles arc system the back arc apron is depleted in primary pyroclastic fall material because of the preferential transport of ashes eastwards by strong prevailing westerly wind systems at higher altitudes (Ch. 13; Fig. 13.28). Most of the back-arc apron therefore consists of mass-flow deposits (Klein 1975, Klein & Lee 1984). Both aprons may contain

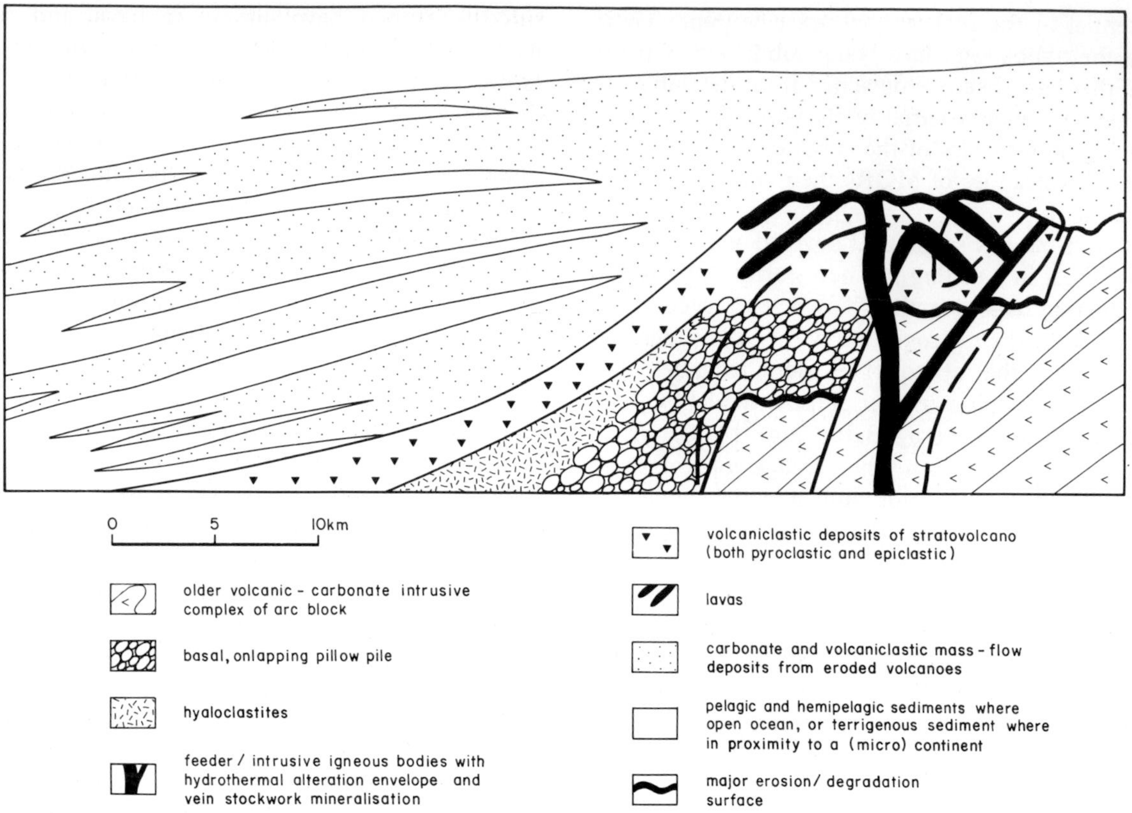
0
5
10km
older volcanic - carbonate intrusive complex of arc block
basal, onlapping pillow pile
hyaloclastites
feeder / intrusive igneous bodies with hydrothermal alteration envelope and vein stockwork mineralisation
volcaniclastic deposits of stratovolcano (both pyroclastic and epiclastic)
lavas
carbonate and volcaniclastic mass - flow deposits from eroded volcanoes
pelagic and hemipelagic sediments where open ocean, or terrigenous sediment where in proximity to a (micro) continent
major erosion/ degradation surface

Figure 14.10 Facies model for marine stratovolcanoes and environs. Compare with the model for continental stratovolcanoes. The significant differences are the presence of pillow lavas and submarine mass-flow volcanic sediments in the marine stratovolcano pile.

channel-form canyon fills as well as sheet-like sediment bodies. Intercalated pelagic intervals should be diagnostic facies of such intra-oceanic arc volcano terrains. The distal facies of such aprons will be dominated by relatively fine-grained volcaniclastic mass-flow deposits (largely turbidites), occasional ash horizons and more-abundant pelagic sediment horizons. Depending on the topography, they could show greater lateral continuity than the more proximal facies of volcaniclastic aprons. Where marine stratovolcanoes are associated with a significant, nearby landmass, there will be interface between the facies of the volcanic pedestal with terrigenous, continentally derived sediment, rather than with open ocean pelagic sediments.

Facies model

See Figure 14.10.

Economic significance

As for continental stratovolcanoes, marine andesite–dacite stratovolcanoes should have potential for porphyry copper–molybdenum–gold deposits and higher level epithermal vein zinc, lead, gold and silver mineralisation. In addition, where a substantial part of the stratovolcano is submerged, seafloor hydrothermal springs may have potential for producing massive, exhalative sulphides. These could occur in two sites – either on the flanks, associated with lateral hydrothermal springs, or where the summit is a collapsed caldera, in a summit basin fed directly by the vent hydrothermal system (e.g. Santorini, Krakatau). The steep flanks may be subject to considerable epiclastic erosional and mass-flow depositional processes long after eruption has ceased, so that ore precipitation could be interrupted or could also be redeposited down-slope by mass-flow processes, or both.

14.8.7 SUBMARINE FELSIC VOLCANOES AND VOLCANIC CENTRES

Selected references

M. A. Reynolds and Best (1976), M. A. Reynolds *et al.* (1980), Ohmoto (1978), Kouda and Koide (1978), Ohmoto and Takahashi (1983), Sato (1977), Sangster (1972), Spence and De Rosen-Spence (1975), De Rosen-Spence *et al.* (1980) and Kokelaar *et al.* (1984, 1985).

Description

Although acidic magmas can be erupted in intra-oceanic settings and result from differentiation and fractionation processes in basic–intermediate magma chambers, they are more common in marine environments floored by continental (sialic) basement (Ch. 13; Fig. 13.44). The popularity of submarine felsic and silicic centres as exploration targets stems from their association with Kuroko-type and Canadian Archaean base-metal ore deposits. Because of the difficulties associated with observing modern marine (especially submarine) volcanoes much of the current understanding comes from the rock record, especially from the closely studied Kuroko successions of Japan and those of the Archaean Abitibi greenstone belt of Canada. Although some workers have interpreted the Kuroko successions as submarine calderas (Kouda & Koide 1978, Ohmoto 1978, Ohmoto & Takahashi 1983), others (Cathles *et al.* 1983) have suggested that volcanism occurred in a narrow extensional rift-like basin, presumably from scattered point sources. The two ideas are, in fact, not mutually exclusive, and are similar in many regards to the ideas about the Canadian volcanic centres (e.g. Spence & De Rosen-Spence 1975; see Section 14.8.10).

The essential elements of the Kuroko successions seem to be subaqueous dome-like lavas, associated volcaniclastics, mudstones, the focal massive sulphides and their associated chemical sedimentary facies. The commonly preferred setting of the Kuroko deposits is in deep marine conditions, in a rift basin. The interpretation for water depth is based on both palaeontological grounds (Guber & Merill 1983), and by the need to prevent the boiling off of ore fluids (Ridge 1973, Franklin *et al.* 1981). The latest estimate of the depth of the depositional environment is about 3500 m (Guber & Merill 1983), although it may be a little less. Such depths have two effects on the eruption modes and products. First, because of the confining pressure of the water column at such depths, exsolution of volatiles may be considerably inhibited, so maintaining low magma viscosities. If the volume of magma that is erupted and the discharge rate is large, this could lead to areally extensive felsic lavas which flow anomalously large distances given the magma type (Ch. 4). Small volume, high viscosity lavas will be local in extent and produce small domes, as appears common in the Kuroko situation. Secondly, the confining pressure of the water column at depths of ~3000 m will preclude explosive fragmentation (Ch. 3). Although the volaniclastics associated with the Kuroko deposits have been described as tuff-breccias, lapilli-tuffs and pyroclastics, they are unlikely to be *in situ* pyroclastics, particularly given their low vesicularity of 17–34% (Ch. 3). At the water depths suggested above, they are more likely to be redeposited volcaniclastics derived from shallower depths (<500 m), or quench-fragmented hyaloclastites. Only at shallow eruptive depths are true, *in situ* pyroclastics likely to be significant, in which case significant sulphide precipitation may be precluded because of boiling off of host fluids. *In situ* hyaloclastites will also be important elements, associated with both surface domes and shallow cryptodomes intruding water-saturated sediments. With the former the domes may be mantled by aprons of hyaloclastite. The hyaloclastites associated with the cryptodomes may be dynamically mixed with the sediments they intrude through boiling-induced fluidisation of the interstitial pore fluids, producing peperitic textures (Ch. 3).

In summary, relatively deep marine silicic volcanic centres should be dominated by lavas, perhaps in overlapping nests, hyaloclastites, redeposited volcaniclastics derived from shallower eruption points, deep-marine pelagic–hemipelagic sediments representing hiatus in eruptive activity, possible associated sulphides, and variable proportions of terrigenous sediments, depending on proximity to land masses. Further consideration is given to the Canadian Archaean successions in Section 14.8.10.

Intermediate depth to shallow water centres will contain progressively increasing proportions of *in situ* pyroclastic deposits. Intermediate depth centres will contain local pyroclastic successions such as tuff cone and tuff ring deposits (e.g. Bunga outlier of the Boyd Volcanic Complex, southeastern Australia, Tuluman Volcano, Papua New Guinea; Reynolds & Best 1976, Reynolds *et al.* 1980). Shallow water centres may be highly explosive, and produce ignimbrites, that may be deposited in shallow waters or on exposed islands, as well as abundant water-lain tephra fall deposits, abundant redeposited pyroclastic debris, and domes. The geology of Kos in the Dodecanese islands of Greece is representative of subaerial accumulations associated with shallow submarine calderas (Fig. 13.44).

Facies model
See Figure 14.11.

Economic significance
Massive stratiform sulphides are the most significant resource likely in such settings. These should accumulate during quiescent times or in quiescent settings associated with pelagic–hemipelagic sediments, or both, may be in local trapped basins (between several domes), or perched (on top of domes) basins. There they will be protected from the influx of significant volumes of pyroclastics and epiclastics. Such settings should be fed by hydrothermal springs. It appears that many Kuroko-type massive sulphide deposits show clastic textures and are thought to have been mechanically transported. Hence, the difficulty lies in finding environments marginal to domes where these redeposited, slumped sulphide deposits have come to rest. A footwall stockwork system of sulphides and perhaps precious metals may also be prominent.

14.8.8 DEEP-MARINE FACIES DERIVED FROM SHALLOW-MARINE–SUBAERIAL SILICIC VOLCANIC CENTRES

Selected references
Cas (1978b, 1979, 1983a), Cas *et al.* (1981), Fiske and Matsuda (1964), Fiske (1963) and J. V. Wright and Mutti (1981).

Description
Shallow-marine to subaerial silicic volcanic centres produce abundant pyroclastic deposits, and the successions of the centres themselves will be similar to those of continental silicic volcanoes (see Section 14.8.3). However, where such centres interface directly with deep-marine basins, a significant proportion of the erupted products will be deposited in these basins. Much will be redeposited by normal subaqueous mass-flow processes, perhaps derived from shoreline settings. Slides, rubble avalanches, debris flows and turbidity currents (Ch. 10) will build an apron of volcaniclastics interdigitating with normal basinal facies. Other deposits will result from the interaction between pyroclastic flows and sea water (Ch. 11). Except for dense block and ash flows, there is considerable doubt that most pyroclastic flows will continue to flow *underwater* as gas-supported pyroclastic flows (Ch. 9). They may interact explosively with sea water, and the debris may then be redeposited as large-volume, very thick mass-flow deposits of juvenile volcanic debris, in associated basins, frequently well away from the original volcanic centres. In these settings stratigraphy may be continuous, and the volcaniclastic succession will interrupt the accumulation of normal basinal sediments and will interdigitate with them. These basin sediments will be largely sandy turbidites and interbedded hemipelagic mudstones.

Facies model
See Figure 14.12.

Economic significance
These successions will have little resource potential unless they are associated with distal hydrothermal

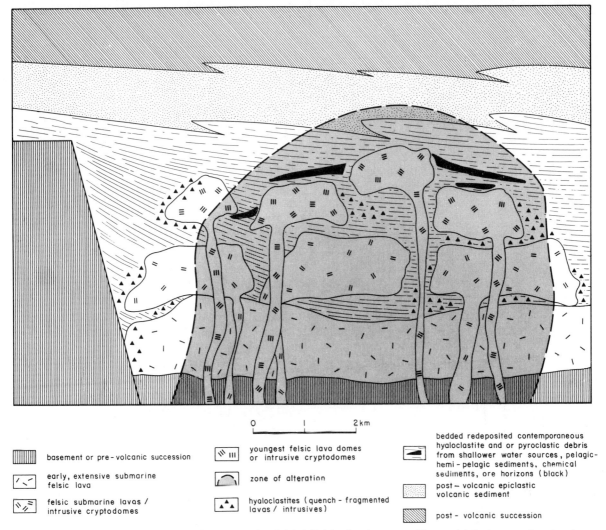

0   I   2 km

| | basement or pre-volcanic succession

| | early, extensive submarine felsic lava

| | felsic submarine lavas / intrusive cryptodomes

| | youngest felsic lava domes or intrusive cryptodomes

| | zone of alteration

| | hyaloclastites (quench-fragmented lavas / intrusives)

| | bedded redeposited contemporaneous hyaloclastite and or pyroclastic debris from shallower water sources, pelagic-hemi-pelagic sediments, chemical sediments, ore horizons (black)

| | post-volcanic epiclastic volcanic sediment

| | post-volcanic succession

Figure 14.11 Facies model for relatively deep marine felsic (silicic) volcanic centres. The model depicts areally extensive deep-marine felsic lavas, followed by smaller dome-like bodies. The mineralisation is depicted as being associated with the youngest, smallest domes, as appears to be the case in the Kuroko region. Intermediate depth volcanic centres may also contain local in situ accumulations of pyroclastic deposits, such as tuff cone and tuff ring successions.

spring systems on the slopes of the source volcanic centre, or unless they host significant redeposited, bedded clastic sulphide deposits.

14.8.9 INTRAGLACIAL BASALTIC AND RHYOLITIC VOLCANISM

Selected references

Mathews (1947), J. G. Jones (1966, 1969, 1970) and Furnes *et al.* (1980).

Description

The characteristics of the deposits and the processes involved have been described in Section 13.11.

Facies models

See Figures 13.51 & 52.

Economic significance

Being subglacial successions, these types of volcanics have no specific economic significance other

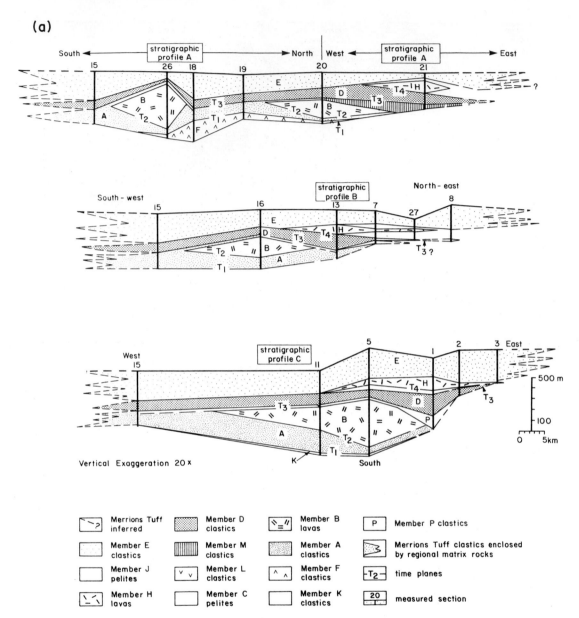

Figure 14.12 Facies model for deep-marine volcanic successions derived from shallow marine-subaerial silicic volcanic centres. (a) Basin floor facies fed by subaerial ignimbrites flowing into water and subaqueous fissure lava eruptions, interdigitating with normal basin terrigenous epiclastics, Lower Devonian Merrions Tuff, southeastern Australia. The three cross-sectional perspectives are arranged from northernmost (top) to southernmost (south). Each clastic interval consists of multiple thick, massive sedimentation units (after Cas 1978). Also see Figure 10.27 for greater detail. (b) (*opposite*) Volcanic apron, interdigitating with normal deep-marine basin, epiclastic terrigenous sediments, Lower Devonian Kowmung Volcaniclastics, southeastern Australia (after Cas *et al.* 1981).

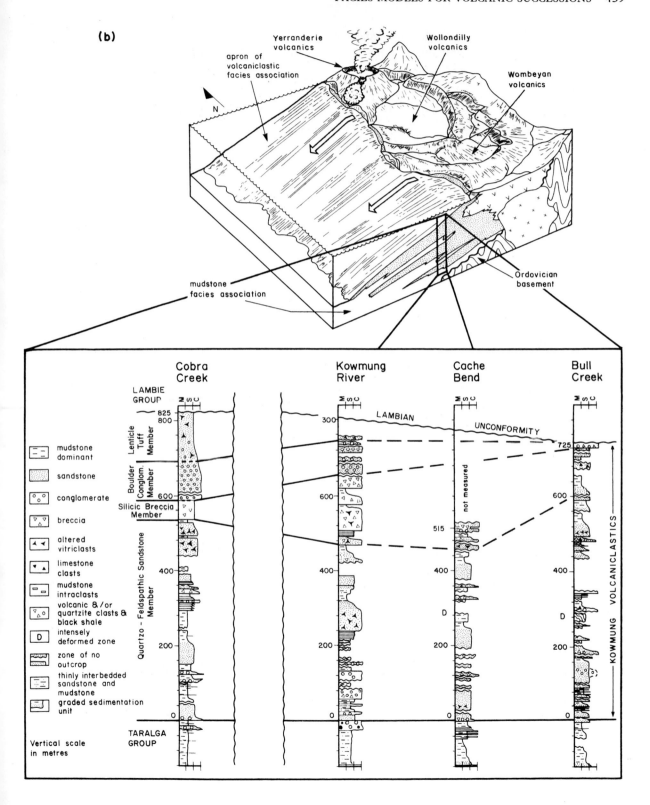

than as parts of basaltic and rhyolitic volcanic complexes, as discussed in previous sections.

14.8.10   PRECAMBRIAN VOLCANISM

Selected references

Arndt and Nisbet (1982), Huppert *et al.* (1984), R. W. Hutchinson (1973, 1980), Naldrett and Macdonald (1980), Ricketts *et al.* (1982), Dimroth *et al.* (1978), Sangster (1972), Spence and De Rosen-Spence (1975), De Rosen-Spence *et al.* (1980) and Thurston (1980).

Description

How different were physical volcanic processes in the Precambrian from those operating during the Phanerozoic? Tectonic controls, settings and conditions may have been different, and some of the erupted magmas may have been different from those erupted in more-recent times. In particular, during the Archaean, the atmosphere may have had a different composition and density, and the occurrence of komatiite lavas (Ch. 4) in Archaean greenstone terrains records the eruption of very high temperature ultramafic lavas quite different from any magmas erupted in modern settings (Huppert *et al.* 1984, Arndt & Nisbet 1982). The geothermal gradient may have been different and the tectonic regime was also almost certainly different. Nevertheless, the basic physical principles that we have reviewed in this book should be as applicable to Precambrian volcanic successions as they are to more-recent volcanics. Lavas of all types (basaltic, andesitic and rhyolitic) in all physical forms (pillowed, massive and dome-like) have been described in Precambrian successions (e.g. Ricketts *et al.* 1982, Dimroth *et al.* 1978, De Rosen-Spence *et al.* 1980). Similarly hyaloclastites, ignimbrites, air-fall deposits and redeposited epiclastic volcaniclastics have been recognised (Ricketts *et al.* 1982, Dimroth *et al.* 1978, De Rosen-Spence *et al.* 1980, Thurston 1980). It is clear that the physical volcanic processes were therefore similar to those operating in modern volcanic settings, although the larger-scale tectonic controls, processes and settings may have been different.

However, whether the volcanic centres were similar in character to modern ones is not clear. Perhaps their nature was affected by the different tectonic regime, influenced by the possibly higher geothermal gradient and a substantially thinner crust and lithosphere. In this regard, few comprehensive accounts of the structure and modern affinities of Precambrian volcanic piles have been ventured. Two distinctive types of volcanic pile have been recognised in Precambrian successions: mafic–ultramafic komatiite-bearing (Ch. 4) piles, and cyclical basic–intermediate–felsic piles (Naldrett & Macdonald 1980, R. W. Hutchinson 1973, 1980).

Mafic–ultramafic volcanic piles are dominated by high temperature tholeiitic to komatiitic lavas and intrusives, and volcaniclastics and sediments seem to be a minor component. Lava thicknesses are very variable, ranging from about a metre to several tens of metres. The features of komatiitic lavas have been briefly touched on in Section 4.12. These piles are thought to have been associated with early to mid-Archaean rifts (R. W. Hutchinson 1980) at a time when there was little or no crust and the near surface geothermal gradient may have been higher.

Basic–intermediate–felsic volcanic piles (Sangster 1972, R. W. Hutchinson 1973, 1980) appear to have become prominent from the mid-Archaean onwards, and to have continued into the Proterozoic. Sangster (1972), R. W. Hutchinson (1973, 1980) and Spence and De Rosen-Spence (1975) indicate that these centres were polygenetic, and erupted magmas ranging from basalts through to rhyolites: 'The volcanism is envisaged as the quiet outpourings of lava sheets on to the sea floor from magma rising through numerous fissures and small vents . . .' (Spence & De Rosen-Spence 1975, p. 94). The volcanic piles are often constructed of multiple cycles, commencing with basalt, followed by intermediate and then silicic lavas. Where mineralisation occurs it is most likely to be associated with the rhyolitic phase, which appears to be associated with domal masses, and has similarities to the successions described in Section 14.8.7. This style of volcanism appears to have developed at a time when the upper mantle and crust were becoming differentiated. Piles as thick as 10 000 m attest to significant subsidence contemporaneous with eruption, suggesting that in many cases relief of the volcanic piles above base level may never have been high. Nevertheless, volcaniclastics may constitute significant intervals, and include hyaloclastites and significant epiclastic sediments. Chemical sediments, including cherts and iron formations, are also common. Although pyroclastics have also been documented, some of these may, in fact, be misinterpreted hyaloclastites or epiclastics.

Ricketts *et al.* (1982) have recognised similarities between a Proterozoic basaltic pile, the Belcher Group, with modern basaltic seamounts in Icelandic rift-like settings (Section 14.8.4) or young island arc volcanoes (Section 14.8.6). However, many volcanic piles appear to be dominated by lavas and intrusives, and volcaniclastics are minor (e.g. Spence & De Rosen-Spence 1975).

Facies models

See Figure 14.13 for basic–intermediate–felsic piles. Also see Figure 14.11 for details of the upper parts of these piles.

Economic significance

Archaean mafic–ultramafic volcanic piles are hosts to many rich nickel–copper deposits, with accessory cobalt and gold. The basic–intermediate–felsic piles are known hosts to zinc–copper and lead–zinc–copper–silver massive sulphide deposits.

14.9 Summary

The aim of this chapter has been to provide those working in ancient volcanic successions with an awareness of the problems of rock identification,

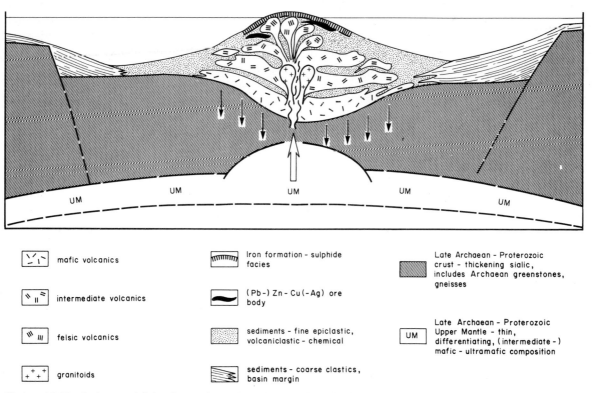

mafic volcanics	Iron formation - sulphide facies	Late Archaean - Proterozoic crust - thickening sialic, includes Archaean greenstones, gneisses
intermediate volcanics	(Pb-) Zn - Cu (-Ag) ore body	
felsic volcanics	sediments - fine epiclastic, volcaniclastic - chemical	Late Archaean - Proterozoic Upper Mantle - thin, differentiating, (intermediate-) mafic - ultramafic composition
granitoids	sediments - coarse clastics, basin margin	

Figure 14.13 Facies model for Precambrian basic–intermediate–felsic volcanic cycle, representing perhaps one of several such cycles in many volcanic piles. (After Hutchinson 1973.) Also see Figure 14.11, which can be taken to reflect details of the top of such piles.

with an approach to facies analysis, and with schematic facies models to be used as norms for comparison. In particular, we have emphasised the effects of various forms of post-emplacement alteration and deformation in causing significant problems in identifying original textures. This hinders identification of the original rock types and depositional facies. Even apparent clastic textures, with all kinds of potentially misleading implications for depositional processes, can result from polyphase alteration. Careful observation, description and interpretation are therefore necessary, and must be based on thorough understanding of Chapters 1–14. Recognition of significant associations of facies, and an understanding of the depositional context, may be very important guides in interpreting problematic rocks. We have emphasised again that volcaniclastic rocks may have diverse modes of formation. Some may be pyroclastic, but many may be quench-fragmented, or epiclastic, and the assumption must be avoided that volcaniclastics by and large are pyroclastic in origin. This must be proven rather than assumed.

The facies models are an initial attempt to summarise the essential characteristics of different volcanic successions. The models are based on understanding of both modern and ancient volcanoes and their products. Both the modern and the ancient have important strengths from a learning point of view. In modern terrains the processes, the way they interact, and their products can be directly observed. Workers in ancient terrains who have not looked first-hand at modern settings are at a major disadvantage. On the other hand, ancient terrains also have much to offer, especially with regard to subaqueous volcanic processes and products. In addition, the long-term development of stratigraphy of particular settings can be better evaluated and, most importantly, the context of volcanic hosted resources can be most easily assessed.

14.10 Further reading

Although Williams and McBirney (1979) and Fisher and Schmincke (1984) give some attention to analysis of ancient volcanic successions, there are no other comprehensive reviews known to us like the one in this chapter.

Plate 15 Aerial view of part of the East Africa Rift system on the border of Tanzania and Kenya. North is to the right. The volcanoes and lakes are characteristic of the landscape of the rift. The major volcano in the centre left is Ngorongoro. Immediately above (west of) Ngorongoro is Olduvai Gorge. (After Sheffield 1981.)

CHAPTER FIFTEEN

Volcanism and tectonic setting

Initial statement

The relationship between volcanism and tectonic setting is a diverse topic. One approach to this subject in recent years has been from a geochemical point of view, the approach being to correlate particular geochemical properties or trends in modern volcanics with their specific tectonic setting, and to use the patterns derived from modern successions in trying to identify the tectonic affinities of ancient volcanic successions. Ancient volcanic successions not only suffer from the effects of deformation in masking the original tectonic context, but also from the chemical overprints of metamorphism and alteration. In spite of advances in isotope and trace-element geochemistry, the geochemical approach offers little hope of unambiguously clarifying the original tectonic context by itself. The tectonic setting can only be evaluated by considering the overall regional geological framework of the volcanics in question, combined with a consideration of the original tectonic controls that allow volcanism to occur in the first place. In this chapter, we review the known tectonic settings in which volcanism occurs and the overall geological characteristics of these settings, and propose guidelines for the evaluation of tectonic settings in ancient terrains.

15.1 An introduction to volcanism in the modern global tectonic framework as a guide to the tectonic settings of ancient volcanic successions

The plate tectonic model for the modern global tectonic framework has produced a reasonably clear picture of the patterns and conditions of volcanism associated with different tectonic settings, and it is a useful starting place for developing an approach to interpreting the tectonic setting of ancient volcanic successions. The following is written with the aim of formulating guidelines for interpreting the tectonic context of *ancient* successions.

Although the approach of using modern global tectonic settings as analogues for past tectonic configurations is not new, many attempts at doing so suffer from two weaknesses: first, they frequently overestimate the scale of ancient configurations compared with the scale of supposed modern tectonic analogues and, secondly, they frequently lack critical evaluation of whether all the essential dynamically important tectonic elements of the modern analogue can be found in the ancient configuration. For example, modern plate margin, volcanic island arc systems are not small-scale tectonic systems. They are regionally extensive systems hundreds to thousands of kilometres long, usually hundreds of kilometres wide, and include at least the volcanic arc itself, the forearc region including the accretionary prism, perched basins and the trench, and they may include back-arc basins (see Section 15.7). The volcanic arc consists of a line of stratovolcanoes with basal diameters of at least 20 km. The spacing between individual stratovolcanoes in a modern plate margin/arc system is in the order of 10^1–10^2 km or more. The length of a volcanic arc is usually hundreds to thousands of kilometres. Therefore, in an ancient orogenic belt a linear volcanic belt only tens of kilometres to a hundred kilometres long does not constitute a plate margin/arc system. This is especially so without the proven coexistence of a contemporaneous forearc accretionary prism system, which is the physical evidence of the subduction process. The point of this digression is that reconstructions of ancient tectonic configurations have to be constrained carefully by the scale of supposed modern analogues, as well as by a clear understanding of all the dynamically important tectonic elements of each tectonic setting, their relationships and extent. However, having said this, care needs to be taken with Archaean terrains, because it is uncertain whether the geothermal gradient was the same as at present, and whether the scale of tectonic systems was the same as modern ones, or perhaps smaller.

Areas of modern volcanic activity can be classified into one of the following tectonic settings:

(a) open ocean (mid-ocean) spreading ridge volcanism,
(b) marginal sea–back-arc basin–interarc basin spreading volcanism behind oceanic island arcs (e.g. Lau-Havre Trough, Marianas Trough),
(c) intra-plate oceanic volcanism (e.g. Hawaiian chain and other oceanic volcanic seamounts),
(d) intra-plate continental (flood) volcanism (e.g. eastern Australia Cenozoic volcanism),
(e) continental rift volcanism:
narrow, linear zones (e.g. East Africa rift zone), including aulocogenes, and broad, wide zones (e.g. Basin and Range Province, western USA),
(f) young island arc volcanism associated with oceanic trench subduction zones (e.g. Tonga-Kermadec, Marianas, Aleutians, Hellenic, Scotia arcs),
(g) micro-continental arc volcanism associated with oceanic trench subduction zones (e.g. Japan, New Zealand, Indonesia) and
(h) continental margin arc volcanism associated with oceanic trench subduction zones (e.g. Andes, Cascades volcanic belt, western USA).

15.2 Mid-oceanic ridge volcanism and the geology of the crust and lithosphere

Mid-ocean ridge (MOR) volcanism (Sections 13.10.1 & 14.8.4) appears to issue from fissures which are arranged *en echelon* within a median valley (Searle & Laughton 1981, Macdonald 1982). Small volcanic mounds and a topography of overlapping flow-fronts are prominent in this median valley. The volcanics are overwhelmingly basaltic, and pillow basalts can be expected to be abundant. However, deep-sea drilling project results have shown that non-pillowed lavas are also common. Hydrothermal vents and fissures may also occur. The MORs are dominated by ridge-parallel fractures and faults, producing a very rough ridge and chasm terrain. The only source of sediment is through pelagic sources. A pelagic sediment mantle only becomes prominent at the edges of the median valleys and on the ridge flanks and abyssal plains as the lithosphere spreads away from the MORs (for details see Ch. 13; MacDonald 1982).

Although the full stratigraphy of the oceanic crust has not been penetrated by the Deep Sea Drilling Project, seismic, gravity and magnetic studies suggest that it closely resembles ophiolite stratigraphy (Coleman 1977, Moores 1982, R. N. Anderson *et al.* 1982, Newmark *et al.* 1985; Fig. 15.1a). Ophiolite stratigraphy was defined by a Penrose Conference (1972) and was expanded by Moores (1982) (Fig. 15.1b) to include, from bottom to top:

(a) a crystalline basement and shallow-water sedimentary sequence,

(b) a tectonic unit of thrust slices of continental margin, rise and abyssal sediments, and/or mélange,

(c) a metamorphic unit, as much as a few hundred metres thick, generally with higher grade rocks over lower grade ones,

(d) an ultramafic tectonite unit composed dominantly of multiply deformed peridotite, dunite and minor chromite,

(e) a cumulate complex, ultramafic at the base grading to mafic or intermediate at the top,

(f) a non-cumulus unit of variably textured gabbro and minor trondhjemite,

(g) a sheeted dyke complex,

(h) an extensive section of massive and pillowed flows and intercalated sediments,

(i) an abyssal or bathyal sediment sequence which may include radiolarian chert, red pelagic limestone, metalliferous sediments, breccias and/or pyroclastic deposits and

(j) post-emplacement deposits of laterite, reef limestone, or shallow-marine or subaerial sediments.

Ophiolites are usually tectonic slices that have been tectonically emplaced during orogenesis (Moores 1982). In this context, units (a)–(c) are basement onto which the oceanic slice, units (d)–(i) have been emplaced and (j) develops following emplacement.

Seismic stratigraphic studies of oceanic crust have defined a generally comparable stratigraphy based on seismic properties. The principal seismic layers identified based on seismic propagation velocities were layer 1, layers 2A, 2B, 2C and layer 3 (R. N. Anderson *et al.* 1982, and references therein). The most successful penetration through this stratigraphy has been in Hole 504B of the DSDP/IPOD programs, on the southern flanks of the Galápagos spreading centre (R. N. Anderson *et al.* 1982, Newmark *et al.* 1985). The stratigraphy and correlations with the seismically defined layers were (Fig. 15.1c):

layer 1, pelagic sediments (274 m),

layer 2A, permeable succession of pillow lavas, breccias and some massive flows (100 m),

layer 2B, more pillow lavas, breccias and minor massive flows; fractures filled with smectite (475 m),

transition layer, decreasing frequency of pillow lava; increasing frequency of dykes; includes a stockwork of Fe, Zn, Cu-sulphides in laumontite, chlorite, calcite and quartz veins in a fractured and brecciated pillow lava sequence (210 m) and

layer 3, sheeted dyke complex (300+ m).

A complete section of oceanic crust would have a thickness of 5–7 km (R. N. Anderson *et al.* 1982). The thickness and character of the layer 1 sediments is highly variable and depends on the age of the crust (thin for young crust, thick for old crust), biogenic productivity and proximity to areas of influx of continentally derived terrigenous clastic sediments. True *open* ocean oceanic crust will be mantled by calcareous or siliceous pelagic sediment, or both. Where oceanic crust comes under the influence of continental margin sedimentation it

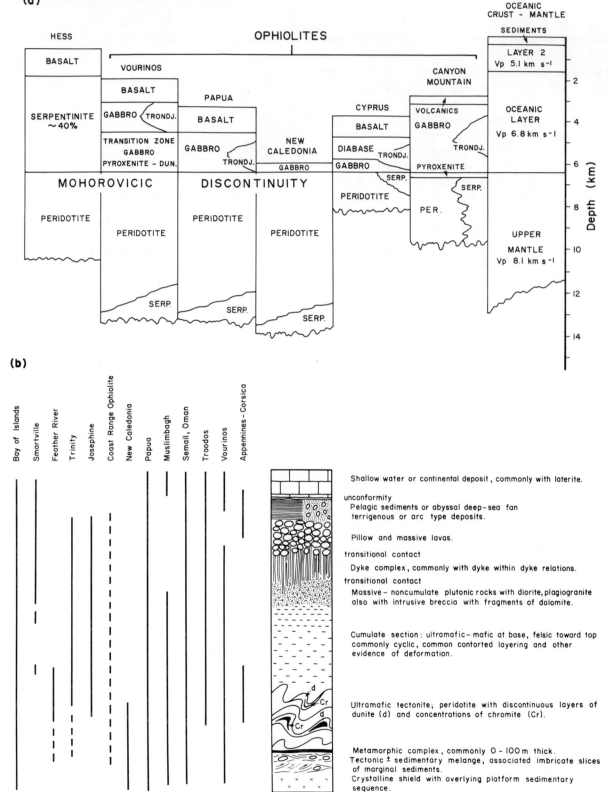

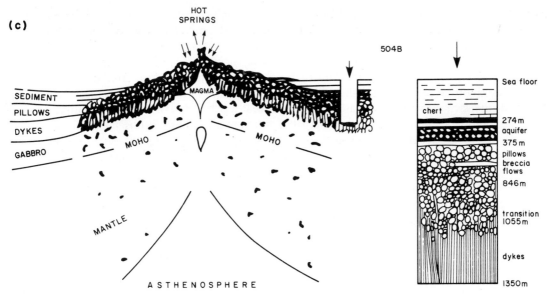

Figure 15.1 (a) Comparison of stratigraphy of several known ophiolites with the seismic stratigraphy of oceanic crust and mantle (after Moores 1982). (b) Expanded definition of a complete ophiolite (see text) (after Moores 1982). (c) Schematic representation of the section intersected in Deep-Sea Drilling Project (DSDP) Hole 504B, southern flanks of the Galápagos spreading centre, south of the Costa Rica Rift (after R. N. Anderson *et al*. 1982).

'vill also be marked by significant thicknesses of terrigenous sediments. Oceanic crust formed near island arcs will be mantled by pyroclastic and redeposited volcanic sediments (see Section 15.3).

The areal extent of oceanic crust during active ocean basin life is huge and measurable in terms of millions of square kilometres. However, because of the development of subduction systems and the subducted. In old orogenic belts preserved oceanic potential of oceanic crust is low, as most will be subducted. In old orogenic belt preserved oceanic crust is likely to occur as long fault slices caught up in accretionary prisms of subduction complexes, or as slabs obducted over the leading edge of the overriding plate and its volcanic arc, whether it be oceanic or continental (e.g. Coleman 1971, H. L. Davies & Smith 1971, Moores 1982).

Oceanic crust is pervasively normally faulted. This faulting occurs during the initial stages of rifting, as the oceanic crust moves from the initial volcanic rift valley beyond the flanks of the ridge (Macdonald 1982). It need not be the result of subsequent deformation.

The geochemistry of the basalts is typically described as being oceanic tholeiite in character (Basaltic Volcanism Study Project 1981), and the geochemical characteristics indicate derivation from poorly evolved magmas. However, although the magmatism is clearly mantle-derived, it need not be exclusively mafic. Silicic differentiates, including small-scale ignimbrites, are known on the rare subaerial exposures of mid-oceanic ridges such as in Iceland (G. P. L. Walker 1962). Very rare occurrences of subaerial lava, pyroclastic and epiclastic facies may be found. It should be noted here, however, that Iceland is very atypical of spreading ridge volcanism. More strictly it should be considered as a region of oceanic platform volcanism that happens to coincide with an active spreading ridge axis. In places alkaline volcanism may also occur, especially along transform faults.

Alteration may be variable in its grade, intensity and distribution, and varies from low grade sea floor weathering products to at least greenschist facies assemblage minerals. The variation is due to variable permeabilities controlling the throughflow of hydrothermal fluids (R. N. Anderson *et al*. 1982).

In ancient terrains, true oceanic crustal basaltic volcanism derived from mid-oceanic spreading centres will be distinguished not so much by geochemical characteristics as by overall, regional geological context. The basement to the volcanic succession should be clearly simatic; the layer 2 basalts should interface directly downwards with a sheeted dyke complex, which is the principal physical evidence of the tectonic extension associated with spreading; an essentially full oceanic lithosphere stratigraphy should be developed (layers 1–4); and the relict oceanic lithospheric slices should form an integral part of a complex orogenic terrain, irrespective of whether the final tectonic emplacement of the slices was by subduction-related accretionary prism offscraping and thrusting, or by obduction (Moores 1982). The scale of the whole complex should be large enough to be compatible with an original large oceanic basin setting and a subsequent major subduction setting.

15.3 Oceanic back-arc basin-interarc basin-marginal sea spreading volcanism and its geological context

Karig (1970, 1971, 1974) has shown that in the western Pacific many marginal basins are the product of the successive splitting and rifting of subduction-related volcanic arc blocks, and the oceanward migration of the frontal half of the rifted block, by the opening of a new back-arc or interarc basin containing a small-scale seafloor spreading centre (Fig. 15.2). This is suggested by irregular

topography, little or no sediment cover, high heat flow and limited magnetic anomaly patterns over elevated basin centre areas (e.g. Sclater *et al.* 1972). The successive splitting of the arc block and the seaward migration of the frontal arc should produce a sequence of progressively younger back-arc basins oceanwards, by successive 'small-scale' seafloor spreading in these basins (also see Section 15.7). The part of the rifted block that stays behind is called the remnant or third arc. The basins and ridges or rises east of Australia and east of the Chinese mainland have the ages predicted by this model, and are floored by oceanic crust (Burns, Andrews *et al.* 1973; Andrews, Packham *et al.* 1975; Karig, Ingle *et al.* 1975; Hussong, Uyeda *et al.* 1981). The currently active, youngest of these, the Lau-Havre Trough and the Marianas Trough, are marked by high axial heat flow and have a sufficiently linear, though not always symmetrical or well defined, magnetic anomaly pattern to suggest axial, if diffuse, spreading from a basin axial, irregular topographic rise (e.g. Karig 1970, 1971, 1974, Sclater *et al.* 1972, Bracey & Ogden 1972, Bibee *et al.* 1980, B. Taylor & Karner 1983, Eguchi 1984). The crustal thickness is thin (5–15 km, Brooks *et al.* 1984) and oceanic in nature, and the crust could also be expected to have an ophiolitic profile consisting of layers 1–3 as outlined in Section 15.2, overlying a layer 4 upper mantle basement.

However, not all arc systems necessarily have back-arc basins originating from the rifting of an arc block (Cooper *et al.* 1977, B. Taylor & Karner

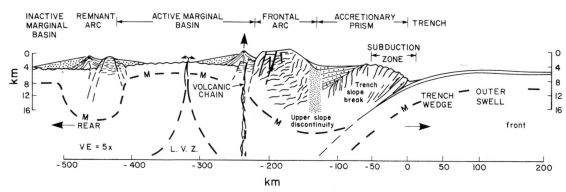

Figure 15.2 Essential elements of an active back-arc–interarc–marginal basin and its associated arc-subduction complex. (After Karig 1974.)

1983, Brooks *et al.* 1984). For example, the Aleutians arc appears to have formed upon old open ocean lithosphere and to have 'trapped' part of this in the associated back-arc basin.

Tamaki (1985) has suggested that some associations of marginal basins may have opened simultaneously. He cites the two areas of the Sea of Japan that are floored by oceanic crust and separated by the submerged block of continental crust of the Yamoto Rise. Tamaki (1985) suggests that these back-arc basins opened simultaneously as the microcontinental arc of Japan separated from the Asian mainland. He suggests that this dual basin system resulted because of a shallow dipping subduction zone and that single back-arc basins form where subduction zones are steep. Although Tamaki's suggestion may be correct, it is also possible that the refractory and thick nature of the continental basement under the Japan arc may have contributed to the opening of multiple basins.

Layer 1 sediments are largely pelagic where they are formed well away from the influence of sedimentation from the rifted arc blocks bounding the developing marginal sea. However, even within a hundred kilometres of the frontal arc and remnant arc blocks, redeposited volcaniclastic sediments transported by submarine mass-flow processes will form a significant pile of sediments above the layer 2 basalts (Klein 1975, Hussong, Uyeda *et al.* 1981, Klein & Lee 1984). Various amounts of redeposited carbonate debris may also be included. Ash-fall layers derived from the active frontal arc block may also occur. *The volcaniclastic character of this layer and sediment succession may help to distinguish marginal sea crust and lithosphere from large ocean basin oceanic crust*, the latter being mantled by only pelagic sediments or terrigenous, continentally derived clastics, or both, except where it is being subducted at an island arc plate margin, in which case it may also contain a volcaniclastic element in layer 1. These continentally derived clastics are likely to be more compositionally mature (quartz-rich) and diverse in their provenance than the sediment fill of an interarc basin. However, marginal basins directly adjacent to continental masses, may contain both provenance elements (e.g. Rodolfo 1969).

Although the geochemistry of modern marginal sea layer 1 basalts is generally similar to that of mid-oceanic ridge basalts (MORB), i.e. oceanic tholeiite (Sclater *et al.* 1972, Gill 1976, Hawkins *et al.* 1984, B. Taylor & Karner 1983), it shows sufficient geochemical differences to be considered transitional between MORB and island arc tholeiites (B. Taylor & Karner 1983).

In terms of recognising ancient marginal sea crust volcanism in the rock record, the most compelling criteria are the ophiolitic stratigraphy outlined above, the distinctive volcaniclastic character of layer 1 sediments, and the association with a regionally extensive, well documented contemporaneous arc succession and its forearc complex. The scale of the whole system should be compatible with suitable modern analogues (e.g. Cas *et al.* 1980, Cas 1983b). Modern marginal basins are hundreds of kilometres long and similarly wide at their widest parts, but may narrow right down to converging apices, such as in the Marianas system of the northwest Pacific Ocean. Deformational style of the preserved marginal sea succession could be variable depending on its tectonic history. Karig (1972) suggested that the active life of marginal seas may be terminated by the initiation of subduction zones within the marginal seas during complete polarity reversal of the arc system. Evidence of this should occur in the form of an intrabasinal subduction complex. Simpler deformational styles involving regional compressional deformation, perhaps associated with the jamming up of the subduction zone in front of the arc, could also occur.

Finally, it should be mentioned that back-arc and interarc basins can have on-land extensions or equivalents floored by continental crust. For example the wholly oceanic Tonga–Kermadec island arc extends into the microcontinental mass of New Zealand. The arc volcanoes in New Zealand (Section 15.8) are nested within the major Taupo graben system which Cole (1984) has described as an ensialic marginal basin.

15.4 Intraplate oceanic volcanism

Intraplate oceanic volcanism occurs upon a basement of older oceanic crust and lithosphere, the characteristics of which have been described in Section 15.2. It can be represented by a single oceanic volcano (Sections 13.10.2 & 14.8.5), or lines of volcanoes such as the Hawaiian–Emperor seamount chains. Individual seamounts may be <1 km to tens of kilometres in diameter and will have a volcanic pedestal stratigraphy up to several kilometres thick (e.g. J. G. Jones 1966, J. G. Moore & Fiske 1969; Fig. 13.50). This will consist of a basal pile of predominantly basaltic pillow lavas, hyaloclastites, and if the seamount approaches sea level, a veneer of pyroclastic–hyaloclastic layers, capped by coherent submarine lavas, perhaps all fringed and ultimately surmounted by limestones, organic colonies, etc., as atolls form. (Such a volcanic stratigraphy can also be produced subglacially, e.g. J. G. Jones (1969); Section 13.11; Fig. 13.51.) The basaltic rocks may be alkaline or tholeiitic and geochemically variable (MacDonald 1968, Frey & Clague 1984, Hawkins & Melchior 1984). Associated sediments are likely to be pelagic and volcaniclastic, and carbonate mass-flow deposits are likely to form aprons. Lines of oceanic islands with increasing age trends away from and perpendicular to MORs are interpreted in terms of passage of a lithospheric plate over resurgent sub-lithosphere mantle hot-spots (J. T. Wilson 1963, Menard 1964, McDougall 1974), and trace the direction of movement of the host lithospheric plate.

As for volcanism associated with mid-oceanic ridges and crust formation, the preservation potential of oceanic volcanic islands in the rock record is low and they, too, should get subducted along with the oceanic lithosphere. They could conceivably be preserved as parts of ocean crust slices in convergent margin accretionary complexes (Section 15.2), or as discrete off-scraped, sheared-off blocks in these settings (e.g. G. J. Macpherson 1983), in which case they may be highly deformed, even dismembered.

15.5 Intraplate continental volcanism

Intraplate continental volcanism (Section 14.8.1) is also essentially basaltic, although rarer, more acid centres are also known. Basaltic provinces are characterised by plains of valley-fill lavas (plains basalts) dotted with cinder cones and tuff rings (Section 13.5), such as occur in southeastern Australia (Ollier & Joyce 1964, Joyce 1975; Figs 13.13 & 16), and the Snake River Plain (Greeley & King 1977), or thick, regionally extensive flood lava sheets (plateau basalts; Section 13.3), such as the early flood basalts of Ethiopia (B. H. Baker et al. 1972, Mohr & Wood 1976, Mohr 1983), and the Columbia River Basalt Plateau (Swanson et al. 1975, Schmincke 1967c, Swanson & Wright 1981; Ch. 4). The effect of the outpouring of fluidal basaltic lavas into valley topography is to reorganise the drainage, in some cases damming up valleys, producing swamps and lakes. Fluvial, swamp and lake sediments should be interspersed with the basaltic volcanics (Fig. 14.7), but are probably best developed below and above the relatively short-lived volcanic succession.

Linear age trends in the eastern Australia Cainozoic province have also been interpreted in terms of the passage of a lithospheric plate over a mantle hot-spot (Wellman & McDougall 1974) and by intraplate extension (Pilger 1982). The geochemical character can range from alkaline to tholeiitic and more-differentiated rocks, including trachytic rocks, are also common. The predominance of basic volcanics in such settings suggests that the magmas are rising from mantle sources and passing through the crust rapidly enough to prevent partial melting of the crust, and the production of more-acidic magmas. Kimberlites could also be expected in these settings.

Intraplate continental volcanics should have a moderate preservation potential in the rock record, the major counteracting influence being long-term post-eruptive erosion in a tectonically, relatively stable continental setting. Volcanic successions of this type, erupted upon a stable continental rock record, should be recognisable by virtue of a regionally extensive basaltic suite erupted upon a demonstrable continental (sialic) basement.

Widespread continental basaltic volcanism probably suggests that the lithosphere is in a generally atectonic, or extensional state (see below), allowing large volumes of mantle magmas to reach the surface. Many major continental basaltic provinces can be related to major continental stretching, rifting and separation (Section 13.3) and to lithospheric adjustments (e.g. eastern Australia). These provinces frequently develop near rifted plate margins (e.g. Subbarao & Sukheswala 1981).

15.6 Continental rift volcanism

15.6.1 NARROW, LINEAR RIFT ZONES (e.g. EAST AFRICA RIFT ZONE)

A large body of literature is now available on the East Africa Rift Zone, so its morphology, sediments, sedimentary environments, geophysics, tectonics, petrology and geochemistry are well known (Di Paola 1972, B. H. Baker et al. 1972, Mohr & Wood 1976, Neumann & Ramberg 1977, Ramberg & Neumann 1978; Figs 15.3 & 4).

Several uplift-updoming phases in the Cainozoic were accompanied by areally widespread continental flood volcanism. In the late Cainozoic the areas of principal updoming were affected by initial rifting and extension, reflected by the formation of an axial graben system and by areally restricted axial graben volcanism (Fig. 15.4b). Igneous products are petrologically and geochemically diverse, ranging from mafic to silicic in character, and are alkaline to peralkaline rocks. Ignimbrites are common products, as are lavas, pyroclastics and epiclastics. The initial environments, at least, have subaerial settings and could include fluvial, alluvial fan and lake environments within a normal-fault-controlled topography (Fig. 15.3). In advanced rifting stages, continued extension and axial graben subsidence should produce marked marine incursions, and in time an overall transgressive succession should develop. The basal parts of the succession contain volcanics or thick volcaniclastics, or both, which slowly disappear upsequence as the rift widens into a narrow sea with a MOR, such as in the Red Sea (Fig. 15.4a) and the Gulf of California. During this, the volcanism should

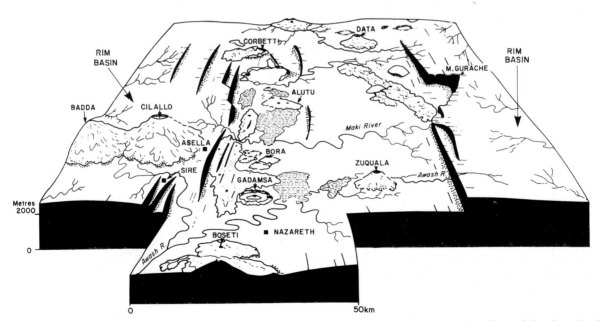

Figure 15.3 Geography of the Ethiopian Rift Valley between 7°00′ and 8°40′ North, showing the variation in volcanic landforms and centres and the interaction between volcanism, normal surface processes and environments, and tectonics. (After Di Paola 1972.)

(a)

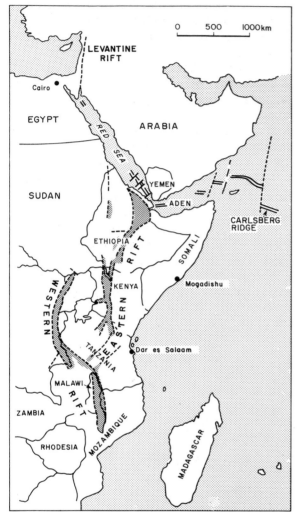

(b)

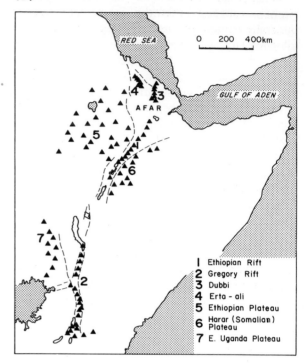

Figure 15.4   (a) Setting of the East Africa Rift System (after B. H. Baker *et al.* 1972). (b) Distribution of volcanic centres about the East Africa Rift System, but not including those of the southern Ethiopian Plateau (after Mohr & Wood 1976). Volcanoes of the graben–rift system are generally the youngest, their development coinciding with the initiation of active rift–graben formation. The widespread volcanism of the bordering plateaux preceded rift formation.

become more oceanic tholeiitic in character. The upsequence change from subaerial fluvial–alluvial–lacustrine successions to marine sedimentation should be accompanied by fundamental lithological changes upsequence. Mixed basaltic to silicic volcanics and volcaniclastics or basement-derived, immature, lithic sediments, or both, should become increasingly mature, more quartz-rich, and may give way to marine carbonates (also see Dickinson 1974). Failed rifts or aulacogenes could cease to develop at any stage (e.g. Burke 1978).

The basement in the initial stages is clearly continental explaining, in some cases, the abun-

dance of silicic volcanics, derived from subcrustal heating of the lower crust. As the extensional tectonic regime develops, giving access to more mantle magmas, the thinning continental crust becomes injected with mafic magmas, becomes transitional in character and, finally, when spreading proper commences, it becomes oceanic in character. Well defined rift systems are a hundred kilometres or more long and tens of kilometres wide, and may even consist of a dual rift system such as the East Africa Rift, the Western Australian rifted continental margin and others (C. A. Wood 1984, Veevers & Cotterill 1978). Veevers and

Cotterill (1978), in describing and modelling the Western Australia continental margin rift history in detail, have provided a useful general account of rift margin history.

15.6.2 BROAD CONTINENTAL RIFT ZONES (e.g. BASIN AND RANGE PROVINCE, WESTERN USA)

The Basin and Range Province is dominated by a broad terrain of graben formation, normal faulting and volcanism with a great diversity of petrological characteristics (Christiansen & Lipman 1972, R. B. Smith & Eaton 1978, and papers therein, Eaton 1982, 1984; Figs 15.5 & 6). The igneous rock types are frequently described as being a bimodal mafic–silicic association. Although both basaltic and acidic rocks are major rock types, intermediate rocks also occur. There is a complete spectrum of clan types from alkaline to tholeiitic to calc-alkaline

(Christiansen & Lipman 1972, Eaton 1982). The silicic rocks include large ignimbrites originating from large caldera structures. The *width* of the whole volcanic-graben terrain is *several hundred kilometres*. Magmatic activity may be concentrated along basement lineaments, and individual magmatically active graben basins may be similar in characteristics to the early stages of the linear rift zones described above. However, they are different in having many counterparts on a regional scale, so producing a regionally broad rift-basin terrain. Although the basement of the Basin and Range Province is clearly continental, it is thin (<30 km thick; Scholz et al. 1971, R. B. Smith 1978, Eaton 1984). The Basin and Range Province has anomalously high heat flow properties. Fault plane solutions suggest a predominance of normal faulting, with the present direction of extension being WNW–ESE, although in mid-Cainozoic times it was WSW–ENE. Basaltic dyke swarms of mid–late Miocene age are also prominent.

(a)

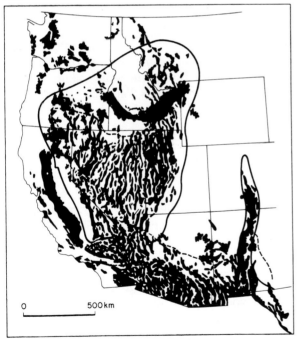

(b)

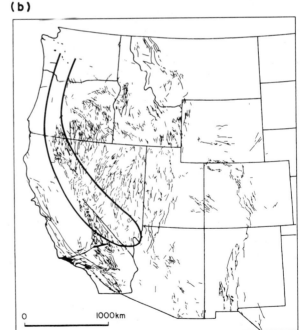

Figure 15.5 (a) Distribution of grabens and other structural basins in the western USA containing Miocene, Pliocene and Quaternary sedimentary and Quaternary volcanic rocks (after Eaton 1982). (b) Distribution of Miocene and younger normal faults with known or suspected Quaternary movement. The extent of the Miocene and younger continental volcanic arc is shown by the bold line, the modern arc being confined to the northern third of the area shown. Strike-slip faults in southern and coastal California are also shown. (After Eaton 1984.)

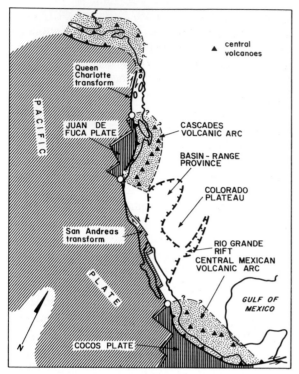

Figure 15.6 Present tectonic context of the Basin and Range Province of the western USA. (After Dickinson 1979.)

The origins of such a terrain are uncertain, but ideas include subduction of an active MOR under the western North American continent (Menard 1964) and the development of a back-arc continental basin system analogous to oceanic back-arc basins (Scholz *et al.* 1971, Eaton 1984). Zoback *et al.* (1981) and Eaton (1984) expanded on the latter hypothesis, suggesting that during the Oligocene, extensional tectonics occurred *within* a very wide calc-alkaline arc. This arc had narrowed by the mid-Miocene, and eastwards of it back-arc extension occurred in a WSW–ENE direction. This produced an ensialic back-arc basin with bimodal volcanism, behind a linear calc-alkaline arc that extended south along the trend of the present Cascade arc. During the late Miocene, the San Andreas transform system propagated northwards, the southern end of the arc retreated northwards, and under the influence of transcurrent faulting associated with the San Andreas system, extension became orientated WNW–ESE (Zoback *et al.* 1981; Fig. 15.5).

This extension is now described as 'back-transform extension' (Eaton 1984). Smaller-scale, more linear back-arc ensialic, verging on ensimatic, basins have also been proposed for the Mesozoic–Tertiary of Chile (e.g. Dalziel 1981, Bartholomew & Tarney 1984, Åberg *et al.* 1984), and the Quaternary Taupo Volcanic Zone has been described as an ensialic back-arc basin by Cole (1984).

In the Basin and Range, extensive sediment successions are associated and interstratified with the volcanics. These are typical continental sediments (fluvial, alluvial fan and lacustrine), frequently developed within graben basins, and often associated with volcanic landforms and successions. Significant extension and subsidence could produce marine incursions, transforming many of the graben basins into marine basins. Such a setting is envisaged for the Silurian–Early Devonian of southeastern Australia (Cas 1983b). Although the volcanics may still be similar to the subaerial situation, it is in deepwater basins associated with such extensional tectonic terrains that unique, large volume, regionally extensive, submarine silicic lavas might be expected (Cas 1978a). Their presence in association with regionally extensive, dominantly silicic volcanics and contemporaneous granitoid plutons clearly indicates the sialic nature of the basement upon which this Palaeozoic extensional province developed (Cas 1983b). Another, but subaerial, clearly bimodal Late Devonian extensional province then followed a mid-Devonian compressional event.

In the rock record, the principal criteria to look for are the widespread (hundreds of kilometres width) occurrence of bimodal volcanics, associated granitoids and multiple, localised, structurally controlled sedimentary basins (Cas 1983b).

15.7 Young island arc volcanism associated with oceanic trench subduction zones

The locus of volcanism is typically a narrow (no more than 50 km wide) line of volcanoes (some submerged, some emergent above sea level) lying parallel to an equally linear, deep-sea trench. The trench and line of volcanoes may have a straight-trace (Tonga–Kermadec Islands) or be curved (Marianas arc, Fig. 15.7). The arc system has a regular structure: trench; inner trench wall; forearc basin slope (or arc–trench gap), perhaps above an accretionary prism; arc block, including an outer non-volcanic line of islands; volcanic arc–back-arc basin-marginal sea-interarc basin; and, in some systems, a remnant or third arc (Figs 15.2 & 7), all developed above an arcward-dipping seismic (Benioff) zone. Karig (1970, 1971, 1974) has shown that in the western Pacific many marginal basins are the products of successive rifting and splitting of the arc block and the oceanward migration of the frontal half of the rifted block by the opening of a new interarc basin, as discussed in Section 15.3. The other half of the rifted arc block is called the remnant arc, or third arc. A new line of volcanoes forms

behind the frontal arc block which consists of an older volcanic–plutonic–volcaniclastic sediment complex mantled by younger volcaniclastics and perhaps carbonates. This sequential evolution indicates that the ages of basins will be successively younger in an oceanwards direction, and that the lifespan of the sediment fills of successive basins will be progressively shorter.

The volcanoes will probably have a similar stratigraphy to oceanic basaltic islands, as discussed in Sections 15.2 and 14.8.6. The sediment fills of the basins will be volcaniclastic mass-flow deposits laid down as aprons on the flanks of the arc volcanoes, as thinning mass-flow sheets in the near reaches of associated basins, and as wedges and sheets in the forearc basin. Carbonate debris may also be included in these deposits. Pelagic sediments will also be important. During the active life of the volcanic arc, pelagic sediments will be best represented furthest from the active arc apron, but are thin or non-existent in the basin centre. In volcanically inactive basins pelagic sediments will also mantle the previously active volcanic apron and will be interspersed with a declining proportion of epiclastic volcaniclastic sediments derived from the remnant arcs. Whereas some arcs are wholly intra-oceanic, others develop in relatively close proximity to continental masses. In old successions terrigenous clastics in basinal sequences can be used as an indication of proximity to continental crustal masses (e.g. Cas *et al.* 1980).

Most *young island arc* rocks are basalts or basaltic-andesites of 'island arc tholeiite' character derived from the subducting lithosphere and/or the overlying mantle and crust (Bryan *et al.* 1972, Ewart *et al.* 1973, Gill 1981, Hawkins *et al.* 1984, Gill *et al.* 1984). More recently, boninites (Mg, Cr, Ni enriched basaltic and andesitic rocks; see Hawkins *et al.* 1984, and references therein) have been considered to be important magmatic products emplaced in the forearc region of the frontal arc block, although these can also occur in continental settings (Wood 1980). Karig (1970) and Gill *et al.* (1984) also suggest that acidic rocks (low-K rhyolites, Gill *et al.* 1984) may be found as initial products of arc-block rifting that leads to formation of a new interarc basin. Sample and Karig (1982) noted that magma production rates

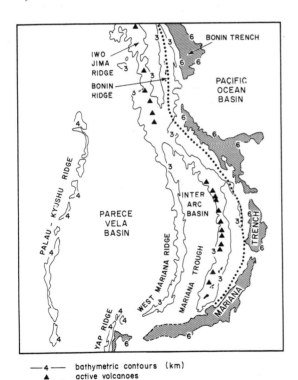

bathymetric contours (km)
active volcanoes
non - volcanic islands
topographic mid- slope basement high between
volcanic arc and trench

Figure 15.7 Principal elements of the Marianas island arc system (after Karig 1971). Note the arcuate character of the system and the apical terminations of Mariana Trough, the present active back–arc basin.

can vary *along* arcs, apparently as a function of variable along-arc subduction and back-arc spreading rates. Locally more-silicic differentiates may occur within the arc volcanoes (e.g. Bryan *et al.* 1972) and rare, small ignimbrites may occur. The basement in the basins is clearly oceanic lithosphere, whereas the basement in the arc block is initially oceanic lithosphere, succeeded upwards by a variably deformed volcanic– volcaniclastic–pelagic sediment–mafic intrusive complex, about 15 km thick (Karig 1970). The arc block itself contains an interesting array of *normal* faults, particularly towards the rear. The inner trench wall may be made up of an imbricate stack of faulted slices of off-scraped ophiolite slivers and deepwater sediments, both pelagic and redeposited volcaniclastic in origin, in which the faults dip towards the arc (Karig & Sharman 1975, Karig 1982; Fig. 15.2). This deformed, off-scraped sediment wedge or accretionary prism may be subducted some depth and subjected to high pressure, low temperature metamorphism, which should be reflected by a high pressure mineral assemblage. Tectonic mélanges may also be produced at this time. Although accretionary prisms should be sought-after evidence of subduction in the rock record, it is apparent that several modern island arc subduction systems lack accretionary prisms. This is apparently due to relatively low sediment influx rates and to a low degree of frictional coupling between the plates, so allowing what sediment there is to be subducted in step-like graben sediment traps in the down-going plate (Uyeda 1983, Von Huene 1984).

Arcs are at least several hundred kilometres long, the line of active volcanoes is less than 50 km wide, the distance between the active arc and trench may be tens of kilometres or more, and the width of the basin between the frontal and remnant arc varies from a few tens of kilometres to hundreds of kilometres wide.

Karig (1972) suggested that arc polarity may change in time, leading to the development of a subduction system in the previous back-arc basin and then to the migration and collision of the arc block back onto the nearby continental mass. The recognition of an ancient island arc system within an orogenic belt remains a problem, and some of the criteria considered necessary (scale, oceanic basement and subduction accretionary prism complex) have been alluded to in Section 15.1.

15.8 Microcontinental arc volcanism associated with oceanic trench subduction zones (e.g. Japan, New Zealand, Indonesia)

The arc-block structure is much the same as for the young island arc system discussed above. The main differences are that the arc block is substantially thicker (30–35 km, e.g. Katili 1973, Sugimura & Uyeda 1973, Cole 1984), there are considerable proportions of recycled, mature supracrustal sediments, and the arc block is much more sialic in character. This has profound effects on the magmatic products which are largely silicic and are in large part derived from the lower crust (Ewart & Stipp 1968, Cole 1979, 1984). However, rock types range from mafic to intermediate to silicic in character (Ewart *et al.* 1977, Whitford *et al.* 1979, Cole 1984). Calc-alkaline rocks are very prominent, but as in Japan and Indonesia, an across-arc trend from tholeiitic nearest the trench to calc-alkaline to alkaline and even shoshonitic volcanoes furthest from the trench, is known (e.g. Miyashiro 1972, 1974, Whitford *et al.* 1979, Hawkins *et al.* 1984). The active volcanic belt is again relatively narrow (several tens of kilometres), but may migrate towards or away from the trench with time (e.g. Sugimura & Uyeda 1973) or be widened by extensive outflow ignimbrite sheets. Along-arc variations also occur, with the more oceanic, juvenile parts of arc systems (e.g. east Indonesian arc, Tonga–Kermadec arc) being more mafic in character, with the more-mature parts (e.g. Java–Sumatra, New Zealand) having more silicic products (e.g. Ewart *et al.* 1977). Large silicic ignimbrites (Ch. 8), large calderas, some submarine in their setting, e.g. Krakatau (Self & Rampino 1981), and stratovolcanoes (e.g. Cole 1979, 1984; Ch. 13; Fig. 15.8) are common. Co-magmatic granitic intrusives may also be prominent, and testify to the existence of a sialic crust *at least* 20 km

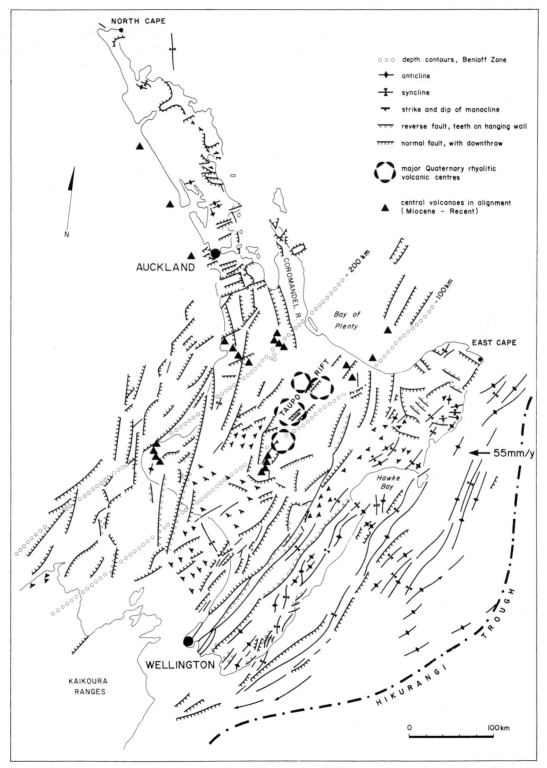

Figure 15.8 Extensional, graben-dominated basins and normal faulting associated with the Taupo Volcanic Zone of New Zealand. (After Sporli 1980.)

thick – thick enough for anatexis of the lower crust to occur. Basaltic volcanics and dykes testify to mantle influences in magma genesis (Cole 1979, 1984). This sialic crust is not only thick, but regionally extensive. For example, the marine straits between Malaysia, Sumatra, Java and Borneo are underlain by sialic crust, and the submarine Campbell Plateau off southeastern New Zealand is also sialic crust.

A forearc accretionary prism may also be developed (e.g. G. F. Moore & Karig 1980, Van der Lingen & Pettinga 1980, Sporli 1980, Von Huene & Arthur 1982, Shiki & Misawa 1982) and blue schist metamorphic assemblages should occur (Miyashiro 1972). A significant structural feature is the association of graben, subsidence basins with the volcanic belt (Sugimura & Uyeda 1973, Hamilton 1979, Healy 1962, Sporli 1980, Cole 1984; Fig. 15.8). In Indonesia and New Zealand there is a significant strike-slip component of movement associated with the extensional terrains, some basins being pull-apart basins (e.g. Ballance & Reading 1980). In New Zealand the Taupo Volcanic Zone can be viewed as an on-land extension of the Havre interarc basin (Karig 1970) and has been described by Cole (1984) as a back-arc ensialic marginal basin.

Sediments associated with the arc system will vary from continental (fluvial, alluvial fan, lacustrine) to marine. The forearc basin may contain all of continental, shallow-marine to deep-marine sediments. Pelagic sediments will only be prominent in far offshore environments, away from the dominating influence of near-arc terrigenous mass-flow sedimentation.

15.9 Continental margin arc volcanism associated with oceanic trench subduction zones (e.g. Andes, Cascades)

Magmatism in these settings takes place upon a wholly sialic, continental type crust, which in the Andes is up to 60 km thick (James 1971). Crustal involvement in magma genesis is evident from the high proportion of silicic volcanics, including huge ignimbrites (Ch. 8) and contemporaneous grani-

toids. Intermediate volcanics may also be common (e.g. Andes, Cascades), and the volcanics appear to be largely calc-alkaline. Forearc basin(s) and accretionary prisms should be expected (Kulm *et al.* 1982, Moberley *et al.* 1982, J. C. Moore *et al.* 1979; Fig. 15.9) and, as with microcontinental arc settings, extensional basins may be associated with the volcanic arc (Zeil 1979, H. Williams & McBirney 1979; Fig. 15.10). Sediments associated with the arc and back arc will be continental: fluvial, alluvial fan and lacustrine. Sediments in the forearc area will range from continental to shallow marine to deep-marine. The volcanic arc will again be hundreds of kilometres long, relatively narrow at *any one time* (tens of kilometres), but may migrate producing a time-transgressive, wider belt of volcanic rocks (e.g. Eaton 1982) and associated granitoids.

In the rock record such a setting should be recognisable by the association of a regionally extensive, linear, acidic–intermediate calc–alkaline magmatic belt with continental sediments, paralleled by a complexly deformed, forearc accretionary prism, all associated with a basement that was demonstrably sialic.

15.10 Igneous rock-types as indicators of basement

From Sections 15.2–9, several observations can be made about the usefulness of igneous rock types and their compositional characteristics in assessing

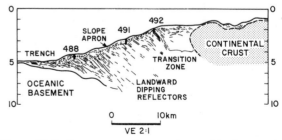

Figure 15.9 Forearc accretionary prism and basin complex of the Middle America trench and arc system as depicted in line drawings of a seismic profile. DSDP holes are numbered. Current slope basins and sediment apron lie above the dotted line. (After J. C. Moore *et al.* 1979.)

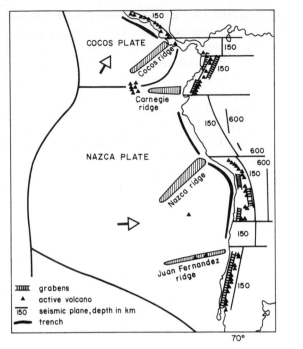

Figure 15.10 Association of regional extensional basins and Cainozoic volcanic activity in South America. (After Nur & Ben-Avraham 1983.)

the general tectonic setting and, more specifically, the nature of the basement at the time the volcanic succession was forming.

(a) Silicic magmas may be generated in areas where the basement is oceanic, as well as in areas where it is continental. Conversely, mafic magmas are not uncommon in continental settings, and can, in fact, be quite voluminous. However, large volumes of silicic magmas only occur wherever a significant sialic-type basement is present. Silicic ignimbrites also follow this pattern. Hildreth (1981) expressed the opinion that mantle derived magmatic activity is probably the ultimate source for all magmatic activity, whether oceanic or continental. In continental crustal settings it is subcrustal heat sources that cause the high geothermal gradient from which partial melting of the lower crust occurs, leading to large volumes of both volcanic and plutonic silicic magmas. In some cases, if the crust becomes sufficiently thickened that its base is depressed to depths of higher isotherms, partial melting could also occur, but this is speculative (R. S. J. Sparks *pers. comm.*).

(b) Continental basement will contain prominent granitoid complexes. In addition to indicating the existence of a well developed sialic-type crust at the time of cooling of the granitoids, the granitoids may give useful information about the nature of the crust at earlier times, according to whether they are S-, I- or A-type granitoids (Chappell & White 1974, Collins *et al.* 1982) and according to the isotopic characteristics of a regionally extensive suite of granitoids (Compston & Chappell 1979).

(c) Particular magmatic suites (alkaline, tholeiitic, calc-alkaline) are not unequivocally unique to any particular tectonic setting, although they may be more prevalent in some settings than in others. The geochemical characteristics of magmatic products are the result of many factors that are not unique to any particular tectonic setting. These include the degree of melting in the subsurface source area, the compositional character of the source area, its earlier history, and the local temperature and pressure regimes. These factors are also not going to be the same in different examples of the same tectonic setting.

(d) Stemming from this, the tectonic setting of old successions can only be reliably inferred *from the whole regional geological framework* including its volcanic and plutonic rock assemblages, its metamorphic state, the regional distribution and styles of diagnostic structural domains, the sedimentary facies, and the determined basement character. Geochemical fingerprinting techniques by themselves are unsatisfactory, although they may add support to the total picture.

(e) The nature of the basement is different in different tectonic settings and, most importantly, it can change in time (e.g. young island arc developing into a more mature arc system; rifting of continental blocks leading to the formation of new oceanic lithosphere; thin skin thrusting of lithosphere). Truly oceanic basement is likely to be ophiolitic and extensive. Even in old successions, it should be prominent in regionally extensive suture zones. Localised ophiolitic bodies could be allochthonous, could be subvolcanic complexes, or could be related to restricted crustal rupturing or rifting without wholesale spreading.

(f)   Several studies of modern volcanic provinces have tried to establish, with reasonable accuracy, based on geophysical information, a relationship between volcano spacing and crustal and/or lithospheric thickness, in both oceanic and continental settings (e.g. Vogt 1974, Mohr and Wood 1976). For old deformed terrains this is an almost impossible approach, but see Windley and Davies (1978). Recently Rickard and Ward (1981) and Rickard (1984) have used the spacing between granitoids in southeastern Australia and Baja, California, respectively, to estimate the thickness of the crust at the time of emplacement of the granitoids.

(g)   Finally, there are always exceptions to generalisations, even these!

15.11 Volcanism related to regional tectonic regimes and local stress field conditions

The plate tectonic-related settings discussed above can be regrouped according to whether the prevailing regional tectonic regime is divergent, 'passive' or convergent:

divergent	mid-oceanic ridges
	marginal seas or back-arc basins
	of spreading origin
	continental rifts
	   narrow African rift types
	   broad Basin and Range types

'passive'	intraplate oceanic volcanic islands intraplate continental volcanic provinces
convergent	young island arcs microcontinental arcs ('orogenic') continental margin arcs

The term 'convergent' was deliberately chosen, rather than 'compressional', to avoid the assumption that in convergent plate margin areas the stress field configuration is compressional in the sense that σ_1, the maximum *regional* principal stress component, is horizontal and directed parallel to the direction of convergence. The reasons for making this distinction are briefly outlined as follows.

The rise of magma to the surface is dependent on the existence of vertical or subvertical fractures, or pathways in the crust. Such fractures are most likely to form when σ_1, the maximum principal stress, is orientated vertically (or subvertically) or horizontally, *and* σ_3, the minimum principal stress, is orientated horizontally. Fractures will then propagate parallel or subparallel to σ_1 if it is high enough to exceed the tensile strength of the rock. Once the rock has fractured, the magma will move along these fractures only if the magma pressure is greater than σ_3 (Shaw 1980, M. A. Etheridge *pers. comm.*), which is orientated perpendicular to σ_1. In a situation where σ_1 is vertical and σ_3 is horizontal (equivalent to normal faulting), the magma will move through subvertical dyke-like fractures or fracture intersections upward to the surface (Fig. 15.11a). Where σ_1 is horizontal, the resultant fractures will be subhorizontal if σ_3 is vertical, and could lead to low-angle thrusting (Fig. 15.11b). The magma will move through subhorizontal sill-like fractures, and will not migrate significant distances vertically through the crust. Minor volumes may do so if there are already pre-existing vertical structural inhomogeneities in the crust. If σ_1 is horizontal and σ_3 is also horizontal, vertical fractures will be produced and strike-slip movement will occur (Fig. 15.11c). Such fractures could also allow the passage of magma to the surface. The failure mechanics in all three cases range from pure extensional fracture (rare) to extensional shear failure (more common) (Shaw 1980).

Stress-field configurations can be analysed at all scales. A single magma chamber or body generates its own local stress field, because it exerts a significant fluid pressure. If a magma chamber develops at a shallow crustal level it will be the local stress-field configuration in the wall-rock around the chamber, particularly the roof, that will control

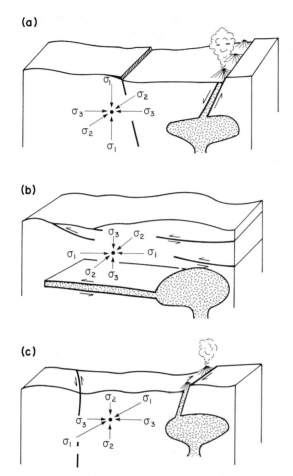

Figure 15.11 Schematic representation of crustal lithospheric stress-field configurations, orientations of resultant fractures and the controls of these on the passage of magma through the crust and lithosphere. (a) Normal faulting, (b) thrust faulting, (c) strike-slip faulting.

whether the magma will be erupted at the Earth's surface (also see Ch. 3). However, at a more regional scale the overall stress-field configuration in the lithosphere will also be important in controlling whether magmas, especially deep-seated mantle magmas, can rise through the lithosphere. If the lithosphere is in a largely compressional state (σ_1 horizontal, σ_3 vertical) this will be difficult. In addition, as Shaw (1980) points out, the overall lithospheric stress-field configuration will control magma ascent rates.

What, then, is the stress-field orientation in the lithosphere at convergent plate margins? Forearc

sedimentary accretionary prisms with their stack of imbricate, low-angle thrust faults (e.g. J. C. Moore *et al.* 1979), testify to a high sediment supply to the trench zone (Uyeda 1983, Von Huene 1984) and to a largely compressional tectonic regime at the leading edge of the overriding plate. However, many structures in the arc block of island arc systems suggest that *normal faulting* is prevalent over reverse faulting or thrusting, even in the forearc region (Uyeda 1983) (Fig. 15.12), and that *the tectonic regime is largely extensional* (σ_1 is therefore vertical and is equivalent to the lithostatic load). The location of the active volcanic arc at the rear edge of the arc block, at the oceanward edge of the extensional back-arc basins, suggests that island arc volcanism occurs in an extensional tectonic setting. Similarly, the occurrence of grabens with many active volcanic arcs on microcontinents (Japan, New Zealand, Indonesia) and continental margins (Andes, Cascades; H. Williams & Mc-Birney 1979) forces a similar conclusion. The physics of the situation outlined above requires that extensional, or strike-slip faulting tectonic regimes exist before large volumes of magmas reach the surface. Strike-slip faulting may well be significant in Sumatra (Hamilton 1979) and New Zealand.

For intraplate volcanism the stress field is neither compressional nor extensional, but perhaps passive or atectonic. It has been thought to be associated with vertical lithospheric fractures, including the inactive extensions of transform faults, which reflect at least an earlier strike-slip tectonic regime, presumably when the fractures were active ridge–ridge transforms (Searle 1984).

It is therefore suggested that voluminous volcanism will generally only occur in extensional, strike-slip or passive (neither compressional nor extensional but with associated pre-existing vertical crustal fractures) tectonic regimes (also alluded to by Hildreth 1981), but not highly compressional tectonic regimes. Under highly compressional regimes subsurface intrusive bodies are more likely, although large plutonic masses, which will create a large local extensional stress field in the roof region, may *in time* rise to sufficiently shallow levels of the crust to permit volcanic eruptions. Hildreth (1981) has suggested that in highly extensional settings,

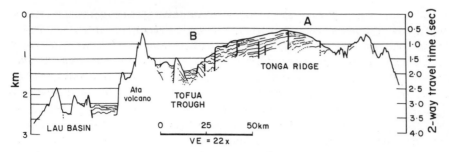

Figure 15.12 Normal faulting in the arc block of the Tonga–Kermadec arc system. (After Karig 1970.)

basaltic volcanics will be prominent and that in less extensional regimes, although basaltic volcanics may not erupt, basaltic magmas are probably the heat source for crustal magma-forming processes.

These findings pose interesting questions for the plate-tectonic model. How can significant extensional regimes exist in what is supposedly a dynamic, convergent (?compressional), tectonic regime? There are two possible answers:

(a) The overriding lithospheric plate is arched from the trench to the back-arc region. This would establish a compressional regime at depth, closest to the centre of curvature, and an extensional regime at shallow levels, where maximum stretching occurs. Magmas could reach the surface in this situation, unless they were derived from the mantle, irrespective of whether it was from the mantle at the base of the arched lithosphere or deeper. However, such mantle-derived magmas do exist in all convergent plate settings.

(b) The subduction process is not always as dynamic as is believed, only the leading edge of the overriding plate being a site of compressional tectonics for most of the life of the subduction system. This implies that the subducting plate is subducting passively, rather than thrusting forcefully under and against the overriding plate. This would suggest that the leading edge of the overriding plate is simply spreading laterally under its own gravitational potential (cf. van Bemmelen 1974), or that the downgoing plate is sinking passively (or in the jargon, has negative buoyancy), or the upper plate is experiencing backwards absolute motion (Cross & Pilger

1982), or any combination of these. This problem has been addressed by Uyeda and Kanamori (1979) and Uyeda (1983), who have introduced the concept of the degree of frictional coupling between two plates involved in subduction to account for the variation in development of extensional features associated with arc systems. They recognise two end-member coupling types: *Andean-type*, involving subduction of oceanic lithosphere under continental lithosphere and *Marianas-type*, involving subduction of oceanic lithosphere under oceanic lithosphere. Andean-type settings are supposed to involve high degrees of coupling and few, if any, extensional features; frictional interaction between the two plates is high. Marianas-type settings are supposed to involve low degrees of coupling or frictional interaction, and extensional features such as back-arc or interarc basins are well developed.

However, this classification seems a little simplistic, and even untenable. It has been pointed out above that Andean-type and microcontinental-type subduction settings can also have well developed extensional features associated with them, such as arc-associated grabens or back-arc ensialic basins (see Sections 15.6, 8 & 9). The splitting of continental fragments and their separation by back-arc spreading to produce marginal seas floored by oceanic lithosphere (e.g. Sea of Japan, Tasman Sea Basins) is evidence that Marianas-type dilatation can occur as successfully in Andean-type continental margin arc systems. Conversely, Marianas-type arcs also contain evidence of compressional, as well as extensional, tectonics as, in fact, do Andean-type settings. It is suggested that *both* types of settings may experience periods of high frictional coupling and low frictional coupling. The latter will be reflected by extensional features such as grabens, back-arc basin spreading and abundant magmatic activity. The former will be represented by widespread compressional deformation and limited or no *surface* magmatic activity. Extension and compression may be cyclical, i.e. alternating (Zeil 1979, Cas 1983b), and magmatic activity may also be cyclical (Cas & Jones 1979), reflecting alternately overall lithospheric compression (little or no magmatic activity), and lithospheric extension, or relaxation (magmatic activity).

The greater propensity for back-arc extension to occur with island arc systems could be a reflection of the lesser strength, the less refractory nature and thinner character of relatively young oceanic lithosphere compared with the greater strength, and the more refractory nature and greater thickness of continental crust. Arc systems are also periodically subjected to compressional deformation, perhaps when for whatever reason, the subduction system jams up (or in the jargon, experiences greater degrees of coupling). The older successions of the Tongan arc, for example, are deformed (Karig 1970, Bryan *et al.* 1972), and prominent unconformities occur, as well as breaks in volcanism (Cas & Jones 1979). Sample and Karig (1982) also noted synchronous fluctuations in volcanism in arc systems, over large regions and perhaps globally, but discount these being controlled by subduction rates because no corresponding changes in spreading rates occur. It is suggested that deformation, unconformity and a break in volcanism coincide with an overall compressional stress configuration in the overriding lithospheric plate, at least as far back as the volcanic arc, whereas normally, during relatively passive subduction, compression only occurs at the very leading edge of the accretionary prism, where the sediment supply exceeds the volume of small step-like sediment traps on the surface of the downgoing plate, so leading to off-scraping rather than subduction of the sediment (Uyeda 1983, Von Huene 1984).

15.12 Igneous rocks as palaeostress indicators in the crust and lithosphere

In tectonically active regions magmatic activity, or its absence, may be a potential indicator of the stress configuration of the lithosphere. The relationship between magmatic activity and intervals of crustal extension and compression has also been noted by other workers (e.g. Noble *et al.* 1974, Bussell 1983). Periods of extensional tectonics coincide with widespread magmatic activity, especially volcanism and mantle-derived dyke swarms, whereas lithospheric compression leads to a restriction of magmatic activity. However, compression may precipitate base of crust melting, as mantle-derived magmas are prevented from rising through the lithosphere because of its compressional stress state. As soon as relaxation of such lithospheric stresses occurs, for example immediately after the peak of orogenic compressional deformation, these crustal magmas may rise through the crust and be emplaced at shallower levels as granitoids, perhaps with co-magmatic eruptives. Throughout the 200 million year history of the Palaeozoic Lachlan Fold Belt of southeastern Australia, Cas (1983b) has noted alternations of magmatic activity, corresponding with associated extensional tectonics and diminished levels of, or cessation of, magmatic activity corresponding with compressional events.

Nur and Ben-Avraham (1983) and McGeary *et al.* (1985) have noted that gaps in magmatism ('magmatic gaps') occur along some arc systems where an oceanic ridge or plateau collides with, or is being subducted in, a trench. They suggest that the causes of such magmatic gaps could be diverse, including:

(a) complete jamming and cessation of subduction, so curtailing magma production,

(b) reduction in angle of subduction, so eliminating the asthenospheric wedge where magmas are generated between the arc and the subduction zone and

(c) continued subduction but the ridge or plateau changes the local stress field, suppressing a rise of fluids as suggested above, or it changes local chemical conditions because it has a different composition or hydrous sediments on its top have been scraped off.

Although McGeary *et al.* (1985) discount the stress-field effect because of lack of seismic evidence, it is peculiar that narrow magmatic gaps occur almost everywhere that there is such a collision. We suggest that stress field controls play a relatively significant part in terminating magmatism in such areas.

Igneous rocks can also be used in more-specific ways to reconstruct the palaeostress conditions in volcanic terrains. In particular, the orientation of planar or tabular igneous bodies, especially dykes, can be used to infer the palaeostress field configuration, given the conditions necessary for magma rise through the lithosphere and crust outlined in Section 15.11. Widespread, uniformly orientated vertical dykes suggest that σ_3, the minimum principal stress component, was orientated horizontally, perpendicular to the strike of the dykes. σ_1, the maximum principal stress component, was therefore orientated either vertically (equivalent to extension) or horizontally (equivalent to strike-slip motion). Using principally the orientation of dykes, Zoback *et al.* (1981), Eaton (1982, 1984) and Laughlin *et al.* (1983) have been able to indicate the lithospheric extension directions in the Basin and Range Province and Mexico, from the mid-Tertiary to the present. From regional considerations, especially associated normal faults in the Basin and Range Province, extensional tectonics with σ_1 being vertical can be inferred.

In another application of this principle, Naka-mura (1977) considered the alignment of flank eruptions, elongation of volcanoes and preferred orientation development of radial dykes about central volcanoes in modern convergent margin volcanic arc systems. In these instances, alignments of dykes and flank eruptions and elongation direction are largely perpendicular to the associated trenches, and are therefore generally parallel to the direction of plate convergence. In these circumstances σ_1 is horizontal and orientated parallel to the direction of convergence, and σ_3 is also horizon-

tal, indicating the direction of dyke dilatation. Such a situation implies a local, mildly compressive, strike-slip stress field configuration, which is not, however, strong enough to prevent magma rise through the lithosphere, as, for example, at the peak of major compressional orogenic deformational events.

Finally, locations of volcanism can frequently be related to major fractures in the lithosphere or intersections of volcano-tectonic lineaments (e.g. Wolfe & Self 1982). In the simplest case, seamounts may occur along lithospheric fracture zones that are or were active transform faults (e.g. Searle 1984). In the more complex situation, volcanoes may occur along on-land surface projections of fractures or above hinge-type transform faults in the subducted plate which divide the downgoing plate into segments of differing dip or age, or both (Nixon 1982). For example, in the Central Mexican convergent arc system, all of these factors contribute to a complex spatial pattern of volcanism as well as an anomalously(?) diverse array of compositional characteristics in the arc volcanics (Nixon 1982).

15.13 An approach to evaluating the tectonic context of ancient successions

Evaluation of the tectonic setting of ancient terrains has to be based on an appraisal of the total geologic framework, not simply the geochemistry or any other single factor. The following are suggested guidelines, based on the approach adopted by Cas (1983b) in reviewing the palaeogeographic and tectonic development of the Palaeozoic Lachlan Fold Belt of southeastern Australia.

(a) Divide the time interval of interest into the smallest possible time-slices for which sufficiently detailed control of the ages of stratigraphic formations is available. For the Palaeozoic, time slices of 5 million years or less duration will be rare; 10–20-million-year time-slices will be more common, simply because of lack of adequate fossil coverage, or detailed enough biozonation.

(b) Construct a base map for the *whole terrain* of

interest, even if the specific area of interest is only a minor part.

(c) On multiple copies of the base map, plot the outcrops of formations relevant to geology for each time-slice. For each formation depict environmentally significant sedimentary facies, volcanics (differentiating their compositions) and plutonics, also differentiating their compositions. Only plutonics with ages falling within a particular time-slice should be plotted.

(d) Find the best available modern palaeogeographic analogue for the facies configuration plotted. The analogue must be of the same scale, and should have an equivalent facies distribution pattern.

(e) Assuming that the tectonic development of the modern analogue is known, identify the important dynamic elements in the geological framework of the *modern* analogue that support the accepted tectonic history. This should include structural, metamorphic, petrological (particular suites), facies, and geometrical aspects.

(f) Test whether evidence for equivalent dynamic elements can be found in the *ancient* terrain of interest. Evaluate the degree of correspondence and the degree to which the two can be considered to be tectonic analogues. Equally importantly, evaluate the degree to which they differ and the significance of these differences. The present is therefore used to test the past critically, as well as to model it.

15.14 Further reading

H. Williams and McBirney (1979) cover some of the aspects discussed above, while Cas (1983b) attempts to evaluate the tectonic setting of magmatic activity and sedimentation critically in an ancient Palaeozoic orogenic terrain. Gill (1981) and Thorpe (1982) both consider the distribution of andesitic volcanoes and their relationship to plate tectonic settings. Shaw (1980) is essential reading on the stress-field conditions associated with the migration of magmas in the subsurface.

Methods used in studying modern pyroclastic deposits

I.1 Physical analysis

Geologists who work solely on ancient volcanic rocks often have only a limited conception of the techniques employed to study Recent unconsolidated pyroclastic deposits, and so may not fully understand how the data are obtained or expressed. Many of the problems encountered in the study of modern pyroclastic deposits are similar to those found in sedimentary rocks, where grainsize, grain shape, geometry of the deposit and internal fabric must be the tools used to determine the physical processes controlling their formation and deposition. The pioneers in this type of approach were undoubtedly Japanese volcanologists (e.g. Kuno 1941, Aramaki 1956, Katsui 1959, Murai 1961, Kuno *et al.* 1964), whereas G. P. L. Walker (e.g. 1971, 1973b) can be credited with extending and developing the approach.

The following are properties that are now routinely measured in the physical analysis of modern pyroclastic deposits:

thickness
maximum grainsize
grainsize distribution
proportions of components
crystal content of pumice clasts
density and porosity

The methods used to measure these, and the major uses of these measurements are set out below.

I.1.1 THICKNESS

Maximum thickness of a pyroclastic fall deposit is measured in centimetres or metres, and the measurements are used to construct an isopach map. In the figures accompanying Chapter 6 there are a number of examples of such maps, which in many cases, are a meaningful indication of:

(a) the vent position,
(b) the dispersal, which can be related to the type of eruption, and
(c) the volume of the deposit.

Construction of an isopach map entails mapping out the deposit, sometimes over large areas. However, by mapping we do not mean tracing lithological boundaries between deposits, as these are usually so complex that no attempt is made to draw them. Thus, a map of a Recent pyroclastic fall deposit generally shows its inferred original distribution, and not its present outcrop pattern, as between datum points the deposit could be partially or completely eroded. Indeed, non-welded pyroclastic deposits may be ephemeral (Ch. 10). In addition, outcrops may be so rapidly overgrown and badly weathered that they cannot be used. For example, new roadcuts in the tropical Caribbean are sometimes completely overgrown within four or five years. Because pyroclastic fall deposits can change markedly laterally, it is preferable to use reasonably closely spaced datum points. Although spacing ultimately depends on the type of eruption, for the large plinian fall deposits localities within 1 km of each other are favoured. Within these distances correlation is more certain, and internal changes can be carefully documented. This is very important in correlating deposits between localities and in understanding the eruption and its stratigraphy.

The volumes of air-fall deposits have been calculated from isopach maps in various ways. A common method involves measuring the area enclosed by each isopach and then to plot area against thickness on a log-log 'area plot' (Fig. 6.18). A curve, or two straight lines (Rose et al. 1973), are fitted to the data, and integration of this curve gives the volume. Other methods include plots of volume against thickness and plots based on theoretical isopach shape (Froggatt 1982). All of these methods involve extrapolation of isopachs to the low-thickness distal limits of the deposit where outcrop may be poorly eroded. This introduces major uncertainty for larger, more widely dispersed types of deposits, especially where secondary thickening might have been important (Ch. 6). To try to resolve this, G. P. L. Walker (1980, 1981b, c) developed an independent method for estimating the total volume of plinian deposits, based on crystal concentration studies (see below) of the Taupo ultraplinian, and Waimihia and Hatepe plinian deposits. Once the total volume erupted had been estimated from the proportion of free crystals relative to the magmatic ratio as represented in pumice clasts, a straight-line extrapolation could be made at the low thickness end on an area plot to a selected limiting thickness value, so giving the same volume as would be calculated by integration of the area curve. It was found that for all three deposits extrapolations to the same lower limiting thickness of 1 µm had nearly identical slopes. Total erupted volumes of other plinian deposits can be conveniently estimated on an area plot by extrapolation parallel to this slope, using the same limiting thickness of 1 µm (G. P. L. Walker 1981b; Table 6.2).

Measurements of the variation in the thickness of pyroclastic flow and surge deposits are less meaningful in terms of an indicator of vent location. This is because both are gravity-controlled mass flows, which therefore tend to pond in depressions. However, thickness is important in calculation of the volumes of such deposits.

I.1.2 MAXIMUM GRAINSIZE

Measuring the average maximum juvenile and lithic clast size is an important field technique, which involves measuring, at numerous localities, the long axis of several of the largest clasts in a deposit. In some detailed sections the variation in grainsize between different layers of one deposit is measured. Usually the sizes of the three or five largest clasts are then averaged, and this would closely approximate the coarsest one-percentile often quoted by sedimentologists. Average maximum pumice (or scoria) and lithic sizes can be plotted up as isopleth maps (Chs 6–8). As with isopach maps, maximum-size isopleth maps are important in locating the vent from which pyroclastic fall deposits were erupted, and for comparing their dispersal in order to characterise the type of eruption. However, for pyroclastic fall deposits, such isopleth maps have certain advantages over an isopach map, because at some localities it may not be the original depositional thickness that is being measured. The top of a deposit may have been eroded by either a later surge or flow, which is sometimes common with near-vent plinian deposits, or by later local erosion or soil-forming processes. It may also have been overthickened by secondary slumping, especially if the fall was deposited on a steep slope. Also, some extremely widely dispersed (ultraplinian) deposits may be thickest just down-wind of the vent (Chs 6 & 8), and secondary thickening of distal ash may occur (Ch. 6).

Measurements of maximum clast size are also used to analyse the energetics of pyroclastic fall eruptions. This is especially so for large ballistic clasts; that is, those clasts which are so heavy that they follow ballistic trajectories and are unaffected by wind drift. The distance at which ballistics fall from the vent (that is, their range) can be used to estimate initial gas thrust velocities from the vent, or the muzzle velocities of the ballistics from the vent (Ch. 6). L. Wilson (1972) presented tables of calculated ranges for particles of varying radii and density, launched at speeds between 10 and 1000 m s^{-1} and various angular elevations. These are reproduced in Table I.1, and cover most ballistic clast sizes (Ch. 6).

For plinian deposits the muzzle velocity can be conveniently calculated from measurements of maximum clast size using the equation of L. Wilson (1976, 1978):

$$u_0^2 = \frac{(8gr_0\sigma_0)}{3C\varrho_0} \tag{I.1}$$

where u_0 is the velocity of the gas (or muzzle velocity), C is the drag coefficient (~1 for plinian velocities), ϱ_0 is the

Table I.1 Logarithms (base 10) of the ranges of larger pyroclastic particles (after L. Wilson 1972).

Velocity (m s^{-1})	Density (g cm^{-3})	Launched at 45° radius (cm)					Launched at 66° radius (cm)					Launched at 97° radius (cm)				
		1.0	3.0	10.0	30.0	100.0	1.0	3.0	10.0	30.0	100.0	1.0	3.0	10.0	30.0	100.0
10	3.5	2.9556		3.0032		3.0075	2.8261		2.8736		2.8786	1.9790		2.0211		2.0268
	2.5	2.9369		3.0017		3.0070	2.8073		2.8719		2.8785	1.9614		2.0207		2.0267
	1.0	2.8559		2.9914		3.0060	2.7242		2.8601		2.8770	1.8830		2.0099		2.0254
	0.5	2.7579		2.9760		3.0043	2.6222		2.8427		2.8753	1.7846		1.9937		2.0236
30	3.5	3.6440		3.9315		3.9583	3.4896		3.8010		3.8290	2.6641		2.9505		2.9775
	2.5	3.5719		3.9208		3.9571	3.4235		3.7897		3.8276	2.5897		2.9395		2.9762
	1.0	3.3383		3.8680		3.9483	3.1834		3.7329		3.8180	2.3497		2.8843		2.9673
	0.5	3.1364		3.7965		3.9356	2.9752		3.6544		3.8036	2.1373		2.8070		2.9540
100	3.5	4.0418		4.6867		4.9660	3.8644		4.5529		4.8358	3.0228		3.7156		3.9873
	2.5	3.7372		4.6173		4.9514	3.7468		4.4821		4.8195	2.9034		3.6437		3.9721
	1.0	3.6042		4.3999		4.8817	3.4194		4.2548		4.7461	2.5717		3.4116		3.9026
	0.5	3.3549		4.2139		4.7912	3.1673		4.0556		4.6504	2.4068		3.2078		3.8097
300	3.5	4.2389	4.7075	5.0257	5.3861	5.6581	4.0410	4.5111	4.8673	5.2357	5.5274	3.1877	3.6669	4.0293	4.3988	4.6906
	2.5	4.1110	4.5849	4.9177	5.2957	5.5914	3.9111	4.3852	4.7573	5.1368	5.4571	3.0568	3.5394	3.9157	4.3000	4.6212
	1.0	3.7592	4.2319	4.6183	5.0127	5.3732	3.5575	4.0304	4.4499	4.8381	5.2261	2.6997	3.1770	3.5993	4.0006	4.3900
	0.5	3.4931	3.9647	4.3891	4.7744	5.1779	3.2906	3.7590	4.2103	4.5920	5.0199	2.4302	2.9015	3.3549	3.7521	4.1815
600	3.5	4.2959	4.7744	5.1323	5.5404	5.9162	4.0926	4.5719	4.9631	5.3820	5.7887	3.2356	3.7238	4.1204	4.5432	4.9575
	2.5	4.1652	4.6458	5.0139	5.4302	5.8157	3.9602	4.4406	4.8427	5.2624	5.6795	3.1024	3.5912	3.9964	4.4227	4.8464
	1.0	3.8073	4.2817	4.6933	5.1087	5.5245	3.6013	4.0758	4.5155	4.9256	5.3690	2.7404	3.2191	3.6607	4.0836	4.5308
	0.5	3.5373	4.0098	4.4526	4.8524	5.2928	3.3309	3.8005	4.2660	4.6621	5.1253	2.4679	2.9400	3.4069	3.8174	4.2831
1000	3.5	4.3426	4.8275	5.2112	5.6510	6.1029	4.1354	4.6207	5.0348	5.4885	5.9963	3.2755	3.7693	4.1881	4.6482	5.1767
	2.5	4.2100	4.6948	5.0861	5.5273	5.9760	4.0014	4.4857	4.9075	5.3542	5.8474	3.1407	3.6332	4.0573	4.5119	5.0191
	1.0	3.8477	4.3231	4.7517	5.1810	5.6329	3.6384	4.1139	4.5670	4.9921	5.4732	2.7752	3.2547	3.7089	4.1463	4.6333
	0.5	3.5749	4.0479	4.5032	4.9130	5.3772	3.3654	3.8358	4.3108	4.7173	5.2033	2.5003	2.9732	3.4489	3.8687	4.3578

effective density of the volcanic gas in the vent (0.25 kg m^{-3} for dusty gas), g is the acceleration due to gravity, r_0 is the radius of the average maximum clast at vent and σ_0 is its density (generally taken as 2.5 g cm^{-3} for lithics). Because of inaccessibility, it is usually impossible to measure r_0, but it can be estimated by plotting a graph of the product $r_0\sigma_0$ of the largest clasts against distance from the vent, and extrapolation to the zero range (L. Wilson 1978). Applying Equation I.1 gives the maximum muzzle velocity during the eruption, and the average velocity is taken as half this value.

Measurements of the average maximum lithic size are also important in locating the vent position for some pyroclastic flows and surges, and their distribution can be important in understanding and quantifying some of the transport and eruption processes (Chs 7 & 8). Maximum pumice size seems to be of less value in locating vents in this type of deposit.

I.1.3 GRAINSIZE DISTRIBUTION

Mechanical or granulometric analyses are used as the main source of data when examining the grainsize variations in non-welded and unconsolidated pyroclastic deposits. Generally, the methods described by G. P. L. Walker (1971) are followed. Analyses are made with a set of sieves with mesh sizes spaced at one-phi (ϕ) intervals (where $\phi = -\log_2 d$, d being the grainsize in millimetres) and ranging in size from -5 to 4 phi (32 to $\frac{1}{16}$ mm). Sieving is usually carried out by hand to avoid excessive breakage of juvenile vesiculated fragments, which can occur during mechanical sieving, as indicated by G. P. L. Walker (1971). The material retained in each sieve (each size class) is then weighed to 0.01 g on a laboratory balance and the weight percentage calculated. Sometimes the 16 mm and 32 mm size classes are sieved in the field, and weighed on a portable balance to 0.1 g. In this case the <16 mm fraction can be split, so reducing the sample size that needs to be transported. For measuring clasts coarser than 64 mm a number of

Table I.2 Details of sieve analyses of a sample of a pyroclastic fall, surge and flow deposit. The samples are all from the Upper Bandelier Tuff collected from the locality shown in Plate 8. These data are used as a basis for the graphical analysis of the size distributions shown in Figure I.1.

Grainsize		Fall		Surge		Flow	
(mm)	(ϕ)	wt%	Cumulative wt%	wt%	Cumulative wt%	wt%	Cumulative wt%
>16	>−4	2.49	2.49			7.96	7.96
>8	>−3	7.01	9.50			2.88	10.84
>4	>−2	11.27	20.77			5.69	16.53
>2	>−1	12.25	33.02	2.48	2.48	5.87	22.40
>1	>0	28.17	61.19	6.10	8.58	13.57	35.97
>0.5	>1	23.61	84.80	11.30	19.88	17.56	53.53
>0.25	>2	9.10	93.90	15.23	35.11	10.80	64.33
>0.125	>3	2.92	96.82	19.34	54.45	10.58	74.91
>0.0625	>4	1.38	98.20	23.50	77.95	13.97	88.88
<0.0625	<4	1.80	100.00	22.05	100.00	11.13	100.01

techniques are used: field sieves can be used if available, linear point traverses can be made in the field, or photographs of the deposit can be taken and used to determine the proportion of larger clasts. With these last two methods clast abundances are often expressed as a volume percentage rather than weight percentage, but if the average density of clasts is measured in the laboratory, then conversions to weight percentages can be made. Whenever different techniques are used, or splits taken, the results must be integrated.

There are no set rules governing the size of sample that should be collected for a routine sieve analysis of pyroclastic and volcaniclastic deposits. In many cases this is determined by the practicalities of the amount of material that can be transported back to a field camp or laboratory. Obviously, the sample size needed to give a representative sieve analysis of a deposit becomes larger with increasing maximum grainsize, and is also larger if the sorting in a deposit is apparently poor. For very coarse deposits, samples weighing several kilogrammes may be required, and these would have to be initially sieved in the field and a split of the finer sizes taken. For coarse plinian fall deposits (Ch. 6) within a few kilometres of the vent, samples between 0.5 and 2.0 kg would be collected. In some very poorly sorted deposits larger clasts can be measured (for instance, by linear traverses) and then much smaller samples of just the matrix taken. For fine-grained deposits only containing ash-sized particles (<2 mm) samples weighing a few grammes to a few tens of grammes may be adequate.

From the raw grainsize data (Table I.2) the usual procedure is to construct cumulative curves of the grainsize distribution on arithmetic probability paper (Fig. I.1), and to determine the Inman (1952) parameters of median diameter (Md_ϕ), graphical standard deviation (σ_ϕ), which is a measure of sorting and occasionally first-order skewness (α_ϕ), which is a measure of the asymmetry of the distribution (Table I.3). The relevant formulae are:

$$Md_\phi = \phi_{50} \tag{I.2}$$

$$\sigma_\phi = (\phi_{84} - \phi_{16})/2 \tag{I.3}$$

$$\alpha_\phi = [(\phi_{84} + \phi_{16}) - Md_\phi]/\sigma_\phi \tag{I.4}$$

A standard practice is then to plot Md_ϕ against σ_ϕ (Fig. 5.3), as was done by G. P. L. Walker (1971) in an important study which set out the major grainsize differences between pyroclastic fall and pyroclastic flow deposits. Because of their simplicity, these two parameters are still mainly chosen and most of the major grainsize studies of pyroclastic deposits have used them (Murai 1961, G. P. L. Walker 1971, Sparks 1976). The validity of these statistics relies on the assumption that the grainsize distribution is approxi-

Table I.3 Grainsize parameters for our three pyroclastic samples, derived graphically from the cumulative curves in Figure I.1(a).

Inman parameter	Fall	Surge	Flow
Md_ϕ	−0.35	2.8	0.8
σ_ϕ	1.65	1.85	2.8

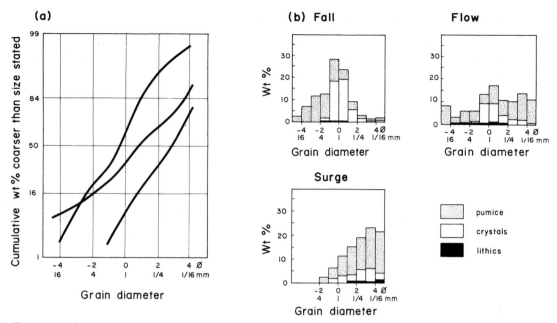

Figure I.1 Graphical representation of the three grainsize analyses from samples of the Upper Bandelier Tuff in Table I.2. (a) Cumulative plot on probability ordinate. The 16th, 50th and 84th percentiles are drawn, and their intersection with the grainsize distribution curves gives the grain diameters used to derive the Inman parameters in Table I.3. (b) Histograms for components separated by methods described in the text. The components have their own grainsize distributions, while each deposit has an overall distribution which is the combined distributions of the three components. The fall deposit is well sorted for a pyroclastic deposit, and is distinctively unimodal, having a high proportion of crystals within a limited size range in the 0.5 and 1 mm size classes. The flow deposit is poorly sorted and polymodal. The sub-populations reflect more than one transport process affecting the various grainsizes and components differently in the moving pyroclastic flow (Ch. 7); note that again there is a peak in the proportion of crystals in the 0.5 and 1 mm size classes. The surge deposit is unimodal, but the distribution has an extended coarse tail, or is negatively skewed. It is quite well sorted, but not as good as the fall deposit. This sample is also a core sample through several laminae which make up this depositional unit.

mately log normal. In many analyses, the central 68% of the distribution approximates a straight line, and it is argued that the statistics are useful for comparison between different samples. If used with care, such statistical information can also be used to aid genetic interpretation of pyroclastic deposits (Sparks 1976, Bond & Sparks 1976).

Most pyroclastic deposits, when compared with normal sedimentary grain aggregates, are poorly sorted (Ch. 1). This has led to unfortunate differences in the descriptive assessment of sorting given by sedimentologists and volcanologists to sedimentary and pyroclastic deposits, respectively. To most sedimentologists, any deposit with a value of $\sigma_\phi > 1.0$ would be described as poorly sorted. To a volcanologist, the division between good and poor sorting in pyroclastic deposits is $\sigma_\phi = 2.0$, and this value partly stems from the original Md_ϕ/σ_ϕ plot of G. P. L.

Walker (1971), which showed that better sorted pyroclastic fall deposits generally had values of $\sigma_\phi < 2.0$, while less well sorted pyroclastic flow deposits had values greater than 2.0 (Fig. 5.3). Table I.4 shows the essential

Table I.4 Differences in descriptive summaries of sorting used by sedimentologists and volcanologists.

Sorting (σ_ϕ)	Sedimentary deposits	Pyroclastic deposits
0–1	very well sorted to moderately sorted	very well sorted
1–2	poorly sorted	well sorted
2–4	very poorly sorted	poorly sorted
>4	extremely poorly sorted	very poorly sorted

differences in descriptive summaries of sorting between volcanologists and sedimentologists. Also, see Chapter 1 for a discussion on the differences between size sorting and hydraulic particle sorting.

Fluidisation experiments on ignimbrite materials have recently suggested that the most useful statistical parameters to be used on ignimbrites are those of Folk and Ward (1957), where:

$$M_Z = \frac{\phi_{16} + \phi_{50} + \phi_{84}}{3} = \text{graphic mean} \qquad (I.5)$$

$$\phi_I = \frac{\phi_{84} - \phi_{16}}{4} + \frac{\phi_{95} - \phi_5}{6.6} = \text{inclusive graphic} \qquad (I.6)$$
$$\text{standard deviation}$$
$$\text{(a measure of sorting)}$$

During experiments, C. J. N. Wilson (1981, Ch. 7) found that the addition of fines to a coarse sample could have changed its fluidisation behaviour. Although the dynamic behaviour of the sample had changed, as had its grainsize make-up, these grainsize changes were not detected in the Inman parameters, making their use in interpreting depositional processes and conditions doubtful. At the other extreme, more accurate sedimentological parameters, such as the method of moments, proved oversensitive. Adding a small amount of fines to a coarse sample may make no appreciable change to its fluidisation behaviour, yet have an inordinate effect on method of moments grainsize parameters. C. J. N. Wilson (1981) concluded that the Folk and Ward parameters were the best compromise, and the use of these are a necessary step in studies attempting to understand the dynamics of ignimbrites.

For samples containing a large amount of fine ash (e.g. >50% finer than $\frac{1}{16}$ mm, which is, conventionally, the finest grainsize sieved) analyses can be completed by pipetting or with a Coulter counter. For samples with lesser amounts of ash the distribution curve is usually simply extrapolated to ϕ_{84} as a straight line. However, for the more-refined studies that are now being carried out on pyroclastic deposits, and with the increased availability of Coulter counters, accurate analysis of the fine end of the grainsize distribution is desirable and easier than it used to be. With more-detailed grainsize studies it will be more appropriate to sieve at half-phi mesh intervals, and perhaps, in some cases, even quarter-phi intervals.

For further information and discussion of the size properties of grain aggregates in general, the reader is referred to the relevant parts in standard sedimentology textbooks. Particularly useful is the unique manual of Folk (1980), and also Pettijohn et al. (1972) and Leeder (1982), and the references therein.

I.1.4 PROPORTIONS OF COMPONENTS

The relative proportions of the different components in a pyroclastic deposit reflect its mode of formation, and details of the transport process. Different techniques are used to separate pumice, crystals and lithics in the different size classes (Fig. I.1b). The larger size classes (>4 mm) are hand picked, but forceps are used for the fine end of this size range. Quite often with the >16 mm size classes this is done in the field, if field sieving and weighing can be carried out. For the size classes 2 mm to 0.5 mm, hand picking is carried out under a binocular microscope using fine forceps or a camel-hair brush. The main problem is to separate out as much material as is needed to produce a satisfactory result, while keeping to a minimum the time involved, so as to be able to treat a large enough number of samples. This usually involves making tests to determine the minimum weight of split sample that will give consistent results, or the minimum weight for routine analysis (for the 2 mm class this is between 5 and 10 g). Weighings are usually carried out accurately on an analytical balance. In the finest size classes, grains are usually counted either under a binocular microscope (0.25 and 0.125 mm) or under a petrological microscope (0.063 and <0.063 mm). This entails, first determining the minimum number of counts needed for routine analysis (for the 0.25 mm size class this is usually up to about 500 grains) and, secondly, determining a conversion factor for pumice to convert the counted percentage of pumice grains present into equivalent weight percentages. This is usually arrived at by comparing the weights of equal numbers of pumice and lithic fragments counted from the 2 mm size class. Conversion factors are usually between about 0.5 and 0.75. However, there are no standard techniques used in component analysis, and workers often substitute their own variations depending on the needs of the study (for instance, water panning to separate pumice from crystals and lithics first).

Measurement of the proportions of components in an air-fall deposit enables particles to be grouped according to their terminal fall velocities (Ch. 6). Terminal fall velocities have been determined experimentally for various sizes of pumice, lithic clasts and crystals, and theoretically computed for a range of sizes and densities by G. P. L. Walker et al. (1971). Theoretical curves for cylindrical particles, which were found to approximate most closely the behaviour of pyroclastic particles, are shown in Figure I.2. L. Wilson (1972) also computed the fall times of particles of various sizes and density corresponding to five release heights (Fig. I.3).

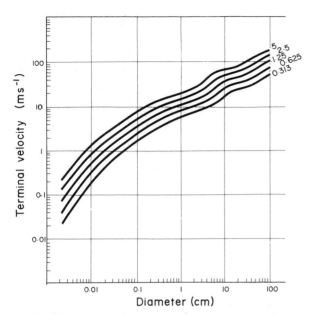

Figure I.2 Computed terminal fall velocities for cylindrical particles, which show good agreement with the behaviour of pyroclastic particles. The curves are for grains ranging in density from 0.313 to 5 g cm^{-3}. (After G. P. L. Walker *et al.* 1971.)

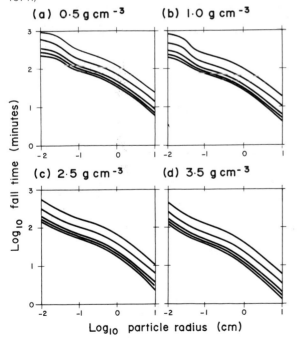

Figure I.3 Computed fall times of particles ranging in density from 0.5 to 3.5 g cm^{-3}. The curves (from bottom to top) are for particles released from heights of 5, 10, 20, 30 and 50 km. (After L. Wilson 1972.)

I.1.5 CRYSTAL CONTENT OF PUMICE

The weight percentage of crystals separated from artificially crushed large pumice clasts is assumed to represent the original magmatic crystal content (G. P. L. Walker 1972). Together with component analysis, these data are important in determining the amount of crystal enrichment or depletion, or glass (vitric) enrichment or depletion in particular types of porphyritic pumice deposit. Enrichment or depletion in either of these components is controlled by, and therefore can be used to assess, aeolian fractionation processes in falls, and transport processes in flows and surges. Also, from these data, total erupted volumes can be better calculated. Most pumice deposits (falls, flows and surges) have lost some vitric component, usually transported far beyond where the thickness of the deposit can be measured or isopachs drawn, and therefore the amount of crystal enrichment is a means of estimating this loss.

Usually, large weighed pumice clasts are crushed to free all the crystals. Tests can be made to determine the minimum clast size that gives the true magmatic crystal : glass ratio. Ratios may be inconsistent in small clasts. For most routine studies, a number of clasts from the >16 mm size classes are suitable; with very coarsely porphyritic pumices larger clasts may be needed. Crystals can then be separated from the vitric fraction by panning under water, or if the vitric fragments prove too dense to be hydraulically separated this can be supplemented by hand picking or counting with a binocular microscope. The weight of the loose crystals can then be expressed as a percentage weight of the original clasts or, together with results from component analysis, used to define the enrichment of crystals in a deposit, expressed as an enrichment factor, EF (G. P. L. Walker 1972), given by

$$\text{EF} = \frac{C_2}{P_2} \times \frac{P_1}{C_1} \qquad (\text{I.7})$$

where C_1/P_1 is the weight ratio of free crystals to glass in artificially crushed pumice and C_2/P_2 is the same ratio in the deposit. For pyroclastic deposits that are depleted in crystals relative to the magmatic proportion, it is more appropriate to define a depletion factor, DF, which is the reciprocal of EF, and quantifies the enrichment in the glass component.

The weight percentage of vitric material lost (VL) during the eruption and emplacement of an ignimbrite is given by

$$\text{VL} = \frac{K}{100}\left(P_1 - \frac{C_1}{C_2}P_2\right) \qquad (\text{I.8})$$

where K is the weight percentage of grainsizes in the ignimbrite finer than 2 mm (crystal concentration data only applying to the matrix). This material is lost into a co-ignimbrite ash-fall (Chs 6 & 8), the volume of which needs to be added to that of the ignimbrite to estimate the total volume erupted during the ignimbrite-forming event.

For widely dispersed pumice fall deposits, crystal concentration studies can be used to estimate the total mass and volume erupted, without the need for extrapolation of isopachs at the distal limits (as discussed earlier). In practice, using the isopach map of the deposit and measured bulk densities of samples of the deposit (see below), an isopleth map is constructed to show the mass of deposit per square centimetre (Fig. I.4a). From sieve analyses a second map can be derived showing the mass per unit area of pumice that is <2 mm in size (Fig. I.4b). Integration of this map, by estimating the value at the intersection points of grid lines, yields the total mass of <2 mm pumice. From the crystal content of sieved samples, another map showing the mass per unit area of free crystals is derived (Fig. I.4c), and integration of this

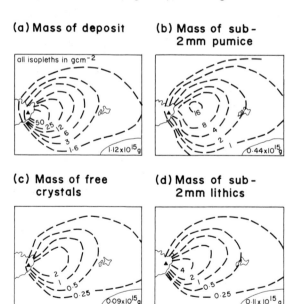

(a) Mass of deposit

all isopleths in gcm⁻²

1.12×10^{15} g

(b) Mass of sub-2 mm pumice

0.44×10^{15} g

(c) Mass of free crystals

0.09×10^{15} g

(d) Mass of sub-2 mm lithics

0.11×10^{15} g

Figure I.4 Maps showing the basic data for the determination of the total mass and volume of the Hatepe plinian deposit erupted from Lake Taupo, New Zealand (Chs 6 & 8). Map (a) is derived from isopachs of the deposit (Fig. 8.51a) and bulk density data. The other maps give isopleths (g cm⁻²) for the different components in the fractions of the deposit finer than 2 mm. The figures in the bottom right-hand corner of each map give the total mass for the on-land part of the deposit. (After G. P. L. Walker 1981c.)

map yields the total mass of free crystals (C') in the accessible parts of the deposit. A fourth map is constructed from the lithic content of sieved samples, and the total mass of <2 mm lithics ($L'_{<2}$) is derived. The method used to calculate total erupted mass and volume is summarised in Table I.5 using the Hatepe pumice as our example (Fig. I.4; G. P. L. Walker 1981c). The method depends on the fact that liberated crystals fall closer to source than similar sized pumice or glass shards (Fig. I.3), and because of their rather restricted size range crystals are not a large component in the most widely dispersed size classes. Assuming that C' is equal to the total quantity of crystals liberated (C), the total erupted quantity of vitric particles in the <2 mm size classes ($P_{<2}$) can be determined. A second calculation assumes 20% of the crystals erupted fell outside the mapped area.

I.1.6 DENSITY AND POROSITY

The standard procedure for determining the density and porosity of a welded tuff (or lava) sample is, first, to oven-dry the sample at about 100°C for 24 h and then to allow it to cool in a desiccator, after which it is weighed to determine the dry weight in air (M_1). The sample is then placed in a container from which the air is evacuated, to extract air from the pore spaces. This container is flooded with deaerated water and the sample is left immersed under pressure for two days to allow water to be absorbed. The sample is removed from the water and quickly weighed in air, after removing the excess water from the surface, to give the wet weight in air (M_2). The sample is then weighed while immersed in water to obtain the wet weight in water (M_3). Then:

$$\text{density} \quad = M_1/(M_2 - M_3) \tag{I.9}$$

$$\text{porosity} \quad = \frac{M_2 - M_1}{M_2 - M_3} \times 100\% \tag{I.10}$$

However, the porosity measured by this method only measures open, connected pore space, and unconnected vesicles formed before or after emplacement are not included.

To determine the bulk density of an unconsolidated pyroclastic deposit, a dried sample is placed in a suitably-sized beaker with a graduated volumetric scale. The beaker is tapped gently to ensure that all the void space is filled, and when no further compaction occurs the volume is measured. For coarse plinian deposits a voidage correction will be required. The sample is weighed, and the weight divided by the volume gives the density. A useful technique actually to collect samples of

Table I.5 Mass and volume calculations for the Hatepe plinian deposit based on crystal concentration studies (after G. P. L. Walker 1981c).

V' volume within mapped area* (km^3)	2.33
M' mass within mapped area*	1.13
$P'_{<2}$ <2 mm pumice*†	0.44
C' free crystals	0.09
$L'_{<2}$ <2 mm lithics*	0.11
Calculation assuming $C' = C$	
$P_{<2}$ total <2 mm pumice†‡	2.64
mass outside mapped area	
P'' pumice, all <2 mm†§	2.20
L'' lithics, all <2 mm¶	0.37
M'' total outside mapped area	2.57
V'' volume outside mapped area‖ (km^3)	3.67
M total mass of deposit (=$M' + M''$)	3.70
V total volume (= $V' + V''$) (km^3)	6.00
DRE volume (assuming $\varrho = 2.5 \, \text{g cm}^{-3}$) (km^3)	1.48
Calculation assuming $C' = 80\%$ of C	
$P_{<2}$ total <2 mm pumice†‡	3.22
mass outside mapped area	
P'' pumice, all <2 mm†§	2.78
C'' free crystals	0.02
L'' lithics, all <2 mm¶	0.47
M'' total outside mapped area	3.27
V'' volume outside mapped area‖ (km^3)	4.67
M total mass of deposit (= $M' + M''$)	4.40
V total volume (= $V' + V''$) (km^3)	7.00
DRE volume (assuming $\varrho = 2.5 \, \text{g cm}^{-3}$) (km^3)	1.76

Values of mass are all in units of 10^{15} g. *By integration of the appropriate map (Figs 8.51a & I.4). †Including glass shards. ‡Equals $C' \times$ magmatic glass : crystal ratio (96.7/3.3). §By difference, equals $P_{<2} - P'_{<2}$.

¶Some 8% lithics have also been assumed lost outside the mapped area. ‖Assuming the same bulk density (0.7 g cm^{-3}) as in the eastern part of the mapped area.

known volume in the field is by forcing a tub of known volume into the outcrop.

In certain studies, knowledge of the density of pumice clasts is required, sometimes in the different size classes. In the 16 mm to 0.5 mm classes, bulk pumice density is determined by placing samples of picked pumice from one size class into a small graduated beaker or measuring cylinder and gently tapping until no further compaction occurs. The sample weight is then divided by volume. Usually some tests are made initially to determine a voidage correction for individual size classes, this correction obviously increasing with increasing grainsize.

For the >32 mm size classes pumice clast density can be measured by the method described for welded tuffs

and lavas above or, if the apparatus is not available, by simple displacement in water. Each clast from a sample is weighed individually (oven-dry), and then soaked in water for at least half an hour to ensure that all connected vesicle space has been flooded, otherwise intake during measurement would increase the apparent volume of the clast. The clast is then immersed in a container or measuring cylinder and the volume of water displaced equals the volume of the clast. Other methods used involve coating the clasts in *waterglass*, or cutting cubes out of the clasts (but both of these methods destroy the pumices for further use), or approximating their volumes to that of equivalent ellipsoids.

Sometimes it is necessary to know the bulk density of the matrix (<2 mm) of samples of pyroclastic flow deposits. This can again be determined using a measuring cylinder; no voidage correction is necessary.

I.2 Stratigraphic analysis

This type of analysis comes under the broad heading of **tephrochronology**. However, tephrochronology has a wide variety of applications, and is an important tool in a number of disciplines. For example, ash layers have been used in dating archaeological sites, measuring rates of sedimentation in oceanic and other sedimentary basins, and in palaeoecological studies. The pioneer in this field was Sigvaldur Thorarinsson, who introduced the term 'tephra' and the study of tephrochronology in 1944 in his doctoral thesis. He promoted the use of tephra as an important tool in volcanological, pollen-analytical, glaciological, geomorphological and archaeological research. For a historical perspective, the reader is referred to Thorarinsson (1981).

The aim of stratigraphic analysis of modern pyroclastic successions is to divide them into eruptive units and to facilitate their correlation. This can involve much more than just determining the '*ash stratigraphy*', because the measurements of thickness, grainsize and constitution of a deposit, which are necessary for correlation, are of great volcanological value in assessing the style and scale of explosive volcanic activity, and in understanding processes (Section I.1). Two papers which set out this approach are G. P. L. Walker and Croasdale (1971) and Booth *et al.* (1978). However, we should acknowledge some of the pioneering Japanese work in this field, for example, Nakamura (1962) and Aramaki (1963).

The deposits of different eruptions can be separated from one another by recognition of intervening:

(a)   soils,
(b)   erosion surfaces,
(c)   (epiclastic) sediments and
(d)   lavas.

Different deposits can be identified from one another by differences in:

(a)   composition and mineralogy,
(b)   grainsize,
(c)   thickness,
(d)   colour,
(e)   degree and style of welding and
(f)   relative stratigraphic position.

It is often found that no single characteristic is indicative of a particular fall, flow or surge deposit, and some are so similar that they can only be distinguished once their relative stratigraphic position to a distinctive or *key deposit* have been determined.

Stratigraphic relations can be very complex (Chs 1, 13 & 14). The usual technique is to construct logged sections of all available outcrops (including digging pits, Plate 5) and piece together the pyroclastic stratigraphy from a number of key locations (e.g. Figs 8.7 & 13.40). In this way studies can be made on the whole succession, or just on the separate layers accumulated during different phases of the same eruption. The deposits of different eruptions are often dispersed differently around a volcano, and transport and depositional mechanisms (fall, flow or surge) will also control distribution. The deposits accumulated during the *same* eruption can show just as many spatial complexities in their distribution. It is therefore very unlikely that any one section will show the whole stratigraphy of a volcano, or even a large part of it. Where volcanoes are in close proximity, deposits erupted from different centres will also overlap and interfinger. The whole stratigraphy is further complicated by epiclastic processes of reworking and mass-wastage.

It is therefore no surprise that there are very few outcrop or contact maps drawn of individual pyroclastic deposits. Most maps group together large parts of the stratigraphy. For example, the terms '*newer*' or '*older pyroclastics*' may be used to subdivide the stratigraphy, but each probably represents the deposits of many eruptions, and their epiclastic derivatives.

Where possible, stratigraphic sections should be dated. This can be most important for correlation, and for such things as volcanic hazard assessment and determination of magma production rates. The three most important dating methods used are ^{14}C dating, which has been most important for young deposits (<50 000 years BP); and **fission-track** and **K–Ar** dating, which have been used for Quaternary pyroclastic deposits. The fission-track method can routinely date glass shards and zircon older than 100 000 years. The usefulness of K–Ar dating depends on the material to be dated. Sanidine can be routinely dated where ages are older than 70 000 years; the practical younger limit of plagioclase is 200 000 years; in rare cases some minerals with high potassium contents can be reliably dated if they are as young as 30 000 years. For a full discussion of these dating techniques applied to Quaternary tephra, and their limitations, see the excellent review by Naeser *et al.* (1981).

In some studies of pyroclastic deposits **geochemical fingerprinting** can be essential for correlation and tephrochronology, and has been very successively used, for example, in correlation of deep-sea ash layers (Ch. 9). Rapid and routine electron microprobe analysis of individual glass shards now provides a particularly powerful correlative tool. Again, as with dating, a discussion of these methods is beyond the scope of this book. For the reader interested in geochemical correlation there is an excellent review by Westgate and Gorton (1981). Naeser *et al.* (1981), Westgate and Gorton (1981) and Thorarinsson (1981) all feature in Self and Sparks (1981), to which we refer the reader for a number of other papers detailing methods that can be used in correlation, and for an up-to-date picture of the applications of tephra studies and tephrochronology.

Grainsize textural classes of volcaniclastic rocks, some possible origins, and suggested diagnostic characteristics

Grainsize–textural class		Origin	Essential characteristics	Preservation potential	Recognition potential
A Conglomerate – closed framework (rounded clasts essential)	1	Epiclastic reworking (fluvial, shoreline) (Ch. 10)	heterogeneous clast composition; tractional structures; well-rounded clasts; context with and within sedimentary succession	very good	very good
	2	Epiclastic mass-flow redeposition (subaqueous) (Ch. 10)	heterogeneous clast composition; disorganised to graded– stratified facies (Ch. 10); association with other mass-flow facies	very good	very good
	3	Pumice and scoria concentration zones in ignimbrites (upper part of *layer 2b*) and scoria-flow deposits (Ch. 7)	homogeneous composition; clast support of pumice or scoria; sheet to lensoidal geometry; fines depleted; crystal-enriched if magma porphyritic; intercalated with other recognisable ignimbrite facies; usually at tops of flow units; thickness <2 m	good in welded ignimbrites	moderate
	4	Fines-depleted ignimbrite (Ch. 7)	homogeneous composition; crystal-rich matrix if magma porphyritic; massive – occasional bedding; thickness – several to >10 m; succeeded by volcanic breccia (lithic-rich ground layer of ignimbrite)	low	poor

Grainsize textural class	Origin	Essential characteristics	Preservation potential	Recognition potential
B Conglomerate – open framework (rounded clasts essential)	5 Epiclastic reworking and mass-flow redeposition (deposits with granular matrix) (Ch. 7)	similar to 1 and 2	very good	very good
	6 Cohesive pebbly mud flows and lahars (Ch. 7)	pebbly mudstones texturally; composition of clasts heterogeneous to homogeneous; internally massive; up to a few tens of metres thick; lack evidence of hot state emplacement (hot blocks, thermal colour alteration thermal remanent magnetisation); no gas segregation structures	good if formed	poor
	7 Non-welded (uncollapsed pumice) ignimbrite and scoria-flow deposits (Ch. 5) (Fig. 5.24a)	compositionally homogeneous (subject to variation in content and composition of lithics which may form breccia horizons); internally massive (with exception of ignimbrite veneer deposits in violent ignimbrites (Ch. 7) which may be crudely layered); up to a few tens of metres thick; may contain gas segregation pipes & pods with clast-supported fabric; accretionary lapilli may be present; gradational downwards into 4 or 17	often very poor	poor
C Breccia – closed framework (angular clasts)	8 Epiclastic redeposition and mass-wastage (includes gravitational collapse, including caldera margin collapse breccias (Chs 8 & 13) (Ch. 10) (Fig. 10.10)	compositionally homogeneous to heterogeneous; disorganised to graded–stratified facies for redeposited units; massive to diffusely layered for mass-wastage (e.g. scree slopes, avalanches, surface mounds on debris flows); local lobate geometry to more extensive for redeposited facies and where large-scale sector collapse has occurred; thickness up to hundreds of metres; associated epiclastic facies may contain tractional structures	moderate	moderate
	9 Aa lavas (Ch. 4) (Fig. 4.6)	compositionally homogeneous (basaltic); very irregular spinose clast morphology; variation in vesicularity; accidental clasts incorporated from substrate; margins brecciated and interior massive; usually less than 10 m thick	poor for spinose top surface	moderate

Grainsize textural class	Origin	Essential characteristics	Preservation potential	Recognition potential
	10 Block lavas and autobrecciated lavas (Ch. 4) (Figs 3.26, 4.18a & b)	as for 9 except that clasts are angular blocks; intermediate or silicic composition; thickness up to 100 m or more	good	good
	11 Lava dome/flow-front talus deposits (Chs 4 & 10) (Fig. 10.12)	as for 10; diffuse layering in scree slope talus deposits; association with dome lava	good	good
	12 Agglutinates (Chs 3 & 5) (Figs 3.13 & 6.8)	homogeneous composition (basaltic, rarely peralkaline); moulded fluidal clast shapes and accommodation (Chs 3 & 5); sector to annular geometry around vent; variable thickness up to tens of metres; interbedded massive lavas (clastogenic lavas, Ch. 4)	moderate	very good
	13 Agglomerates (Chs 5 & 12) (Fig. 6.6)	only diagnostic criterion is shaped bombs or 'hot' breadcrusted or jointed blocks that have not been redeposited	poor	very difficult
	14 Quench-fragmented lavas, cryptodomes, shallow intrusives (hyaloclastites) (Chs 3 & 4) (Figs 3.12 & 25)	compositionally homogeneous; very angular to splintery clasts; coarse blocks to finely granulated glassy aggregates; may be crystal-rich if porphyritic	good	very good
		'jigsaw puzzle' fit of clasts where there has been no redistribution from site of fragmentation by turbulent mixing; gradational to intercalated with unfragmented lava (massive, pillowed, jointed); may be pervasively altered in ancient rocks	good	very good
	15 Hydrothermal explosion breccias (Chs 3 & 13)	diverse clast types and morphology; clasts variably altered; matrix of hydrothermally altered clays; may be associated with surge deposits; accretionary lapilli may occur	poor	poor
	16 Hydraulic fracture breccias (Ch. 14) (Fig. 14.5)	compositionally homogeneous to partially heterogeneous; clasts variably altered; angular to splintery clasts; 'jigsaw puzzle' fit of clasts where little transport of clasts has occurred; confined to cross-cutting zones centimetres to metres wide	very good	very good

Grainsize textural class	Origin	Essential characteristics	Preservation potential	Recognition potential
	17 Pumice-fall deposits (subplinian, plinian, ultraplinian) (Chs 5 & 6) (Figs 6.14, 41c & 8.49, Plate 8)	homogeneous clast composition (but variable accessory lithics); identical crystal types in both pumice clasts and matrix; massive to diffusely layered; no cross-stratification; thickness up to 25 m, but usually< <10 m; susceptible to weathering and alteration with breakdown of glass to clays, etc.; susceptible to tectonic deformation and layer shortening; where welded, eutaxitic texture also present, and local distribution around the vent	poor except where covered by co-eruptive welded ignimbrites; excellent where welded	poor, recognised by context; good for welded deposits
	18 Scoria-fall deposits (hawaiian, strombolian) (Chs 5 & 6) (Figs 3.16, 5.4, 6.6 & 6.10, Plate 5)	as for 16, but even more susceptible to weathering and alteration	very poor	very poor
	19 Lithic concentration zones (base of *layer 2b*) and ground layers of violent ignimbrites (Chs 7 & 8) (Figs 7.10 & 26)	homogeneous to heterogeneous lithic clast composition; gradational upwards into matrix-supported and lithic-poor breccia (upper part of *layer 2b*); interbedded with other ignimbrite facies – underlain by basal layer (sand to microbreccia grainsize); thickness generally <1 m; ground layer of violent ignimbrites may overlie 4 or 21, and is sharply overlain by *layer 2b* ignimbrite facies	good for lithic concentration zones in welded ignimbrites; otherwise poor	good if preserved
	20 Co-ignimbrite breccias (lag breccias and ground breccias) (Ch. 8) (Fig. 8.20)	as for 19, but deposits thicker and clasts coarser; thickness up to 20+ m(?); upper contact sharp to gradational into open framework co-ignimbrite breccias and other ignimbrite facies	good if capped by welded ignimbrite	good if preserved
	21 Fines-depleted ignimbrite (Chs 7 & 8) (Figs 7.28 & 30b)	as for 4, but pumice clasts angular	poor	poor

Grainsize textural class	Origin	Essential characteristics	Preservation potential	Recognition potential
D Breccia – open framework (angular clasts essential)	22 Glacial till and moraines (diamictites) (Ch. 10) (Figs 10.2, 13 & 15)	heterogeneous clast composition; clast shape variable from angular to rounded; matrix includes large proportion of fine rock powder; unlikely to contain significant pumice or shards; massive to crudely bedded; associated striated pavements, pebbles and fluvioglacial facies; variable thickness	moderate	moderate to good
	23 Glacial dropstone deposits (Ch. 10) (Figs 10.2 & 13)	as for 22, but thinner and matrix may be coarser, and contained within lacustrine and marine facies; dropstones may show impact sags; may be reworked	good	very good (structure distinguished from pyroclastic bomb sags by context)
	24 Epiclastic reworking and/or mass-flow redeposition with granular matrix (Ch. 10) (Figs 10.28a & 31b)	as for 5 (also see 1), but clasts angular to sub-rounded	very good	moderate
	25 Cohesive debris flows and lahars (Ch. 10) (Figs 2.13, 10.30 & 31)	as for 6, but clasts angular to sub-rounded	very good	moderate
	26 Ignimbrite (*layer 2b*), and other (denser clast) pyroclastic flow deposits (block and ash flows, scoria flows) (Chs 5, 7 & 8) (Figs 5.14, 15, 16, 7.31, 8.38 & 10.32, Plate 8)	homogeneous clast composition (but variable accessory and accidental lithics); crystal types same in pumice clasts and matrix; massive depositional units (with exception of veneer deposits in violent ignimbrites which show crude stratification); thickness variable – ignimbrites <5 m to hundreds of metres; denser clast flow deposits up to several tens of metres; evidence of hot state emplacement (see 6), and in the case of welded ignimbrites, development of eutaxitic texture and columnar jointing; gas segregation pipes and pods (with clast support); association with other ignimbrite facies (*layer 2a*) and co-eruptive fall and surge deposits	excellent for *welded* ignimbrites, otherwise poor	excellent for *welded* ignimbrites, otherwise poor
	27 Co-ignimbrite breccias and proximal ignimbrites (Ch. 8) (Fig. 8.20)	as for 18 and 20 but matrix-supported; presence of large segregation pipes and pods (metre-sized)	good if capped by welded ignimbrite	good if preserved

Grainsize textural class	Origin	Essential characteristics	Preservation potential	Recognition potential
	28 Near-vent base surge deposits (Chs 5 & 7) (Figs 5.21, 22, 7.40, 43)	compositionally homogeneous to heterogeneous; variable vesicularity of juvenile clasts (Ch. 3); presence of ballistics and impact sags; cored lapilli; massive, bedded and cross-bedded internal structures; thickness of multiple base-surge piles (tuff rings) up to tens of metres	poor	good if preserved
	29 Ground or ash-cloud surge deposits (Chs 5 & 7) (Fig. 5.23, Plate 8)	compositionally homogeneous to heterogeneous microbreccias (dependent on composition of parent pyroclastic flows and lithic content); stratified and cross-stratified; position below and above, respectively, pyroclastic flow facies; thickness generally <2 m	good when capped by or within welded ignimbrite succession	good, but not in tectonically deformed units
	30 Giant pumice beds (Ch. 13) (Fig. 13.46)	uniform composition of pumice clasts; enclosing matrix sediments are stratified; lacustrine (or marine) setting; radial jointing in some individual clasts; chilled glassy sheath on margins of some clasts; clasts up to several metres	moderate within thick caldera lake successions	very good in undeformed terrains
E Sandstones (sand-sized framework grains predominant)	31 Epiclastic reworking (Ch. 10) (Figs 10.19 & 24)	abundant tractional structures; cross-stratification is either high angle of repose or hummocky cross-stratification (cf. surge cross-stratification); body and trace fossils	very good	very good
	32 Epiclastic mass-flow redeposition (Ch. 10) (Fig. 10.28, Plate 11)	mass-flow facies characteristics; body and trace fossils	very good	very good
	33 Weathered and/or devitrified lava/dykes (Ch. 14) (Fig. 2.10)	generally granular texture; even distribution of phenocrysts if crystallised; thick massive character; (?)relic flow banding; lithophysae, spherulites (Chs 4 & 14); radiate fibrous to granophyric groundmass of quartz and feldspar	very good	recognition as lavas difficult in instances
	34 Fine-grained ignimbrite (Chs 5 & 8) (Fig. 5.16c)	gross granular texture; thick massive character; rare shard textures in thin section and may be eutaxitic shard texture if originally welded; gradational into other ignimbrite facies (lithic concentration zones, gas segregation structures)	poor unless welded	recognition of origin may be difficult

Grainsize textural class	Origin	Essential characteristics	Preservation potential	Recognition potential
	35 Air-fall ashes and tuffs (Chs 5 & 6) (Figs 6.32 & 35)	homogeneous composition; rare shards preserved; possible internal diffuse lamination; thickness generally <1 m; accretionary lapilli	good if in welded ignimbrite successions (co-ignimbrite ashes) and in lacustrine and *deep* marine successions	moderate; difficulty in distinguishing from re-deposited origin for subaqueous ashes
	36 Base-surge deposits (Chs 5 & 7) (Figs 7.40, 43)	as for 28, but finer-grained; presence of cogenetic air-fall ash layers with accretionary lapilli	poor	good if preserved
	37 Ground and ash-cloud surge deposits (Chs 5 & 7) (Plate 8)	as for 29 but finer	see 29	see 29
F Mudstones (mud-sized grade predominant	38 Epiclastic (Ch. 10) (Fig. 10.20)	as for 31 and 32	very good	very good
	39 Fine-grained ignimbrite (Chs 5 & 8) (Fig. 5.16c)	as for 34	poor unless welded	recognition may be difficult
	40 Air-fall ashes and tuffs (Chs 5 & 6) (Figs 6.35 & 8.52)	as for 35	see 35	see 35
	41 Surge deposits (Chs 5 & 7)	as for 36 and 37	see 36 and 37	

REFERENCES

Åberg, G., L. Aguirre, B. Levi and J. O. Nÿstrom 1984. Spreading-subsidence and generation of ensialic marginal basins: an example from the Early Cretaceous of central Chile. In Kokelaar and Howells (1984), 185–93.

Allen, J. R. L. 1968. *Current ripples. Their relations to patterns and sediment motion.* Amsterdam: North-Holland.

Allen, J. R. L. 1970a. *Physical processes of sedimentation — an introduction.* London: Allen & Unwin.

Allen, J. R. L. 1970b. The sequence of sedimentary structures in turbidites, with special reference to dunes. *Scot. J. Geol.* 6, 146–61.

Allen, J. R. L. 1971. Mixing at turbidity current heads, and its geological implications. *J. Sed. Petrol.* 41, 97–113.

Allen, J. R. L. 1982. *Developments in sedimentology.* Vols 30A & B: *Sedimentary structures. Their character and physical basis.* Amsterdam: Elsevier.

Almond, D. C. 1971. Ignimbrite vents in Sabaloka Cauldron, Sudan. *Geol. Mag.* 108, 159–76.

Amos, R. C., S. Self and B. Crowe 1981. Pyroclastic activity of Sunset Crater: evidence for a large magnitude, high dispersal strombolian eruption. *EOS* 62, 1085.

Anderson, D. J. and D. H. Lindsley 1981. A valid Margules formulation for an asymmetric ternary solution: revision of the olivine-ilmenite thermometer, with application. *Geochim. Cosmochim. Acta* 45, 847–53.

Anderson, R. N., J. Honnorez, K. Becker, A. C. Adamson, J. C. Alt, R. Emmermann, P. D. Kempton, H. Kinoshita, C. Laverne, M. J. Mottl and R. L. Newmark 1982. DSDP Hole 504B, the first reference section over 1 km through layer 2 of the oceanic crust. *Nature* 300, 589–94.

Andrews, J. E., G. H. Packham *et al.* 1975. *Initial reports of the Deep-Sea Drilling Project.* Vol. 30. Washington, DC: US Government Printing Office.

Aramaki, S. 1956. The activity of Asama volcano, Pt. 1. *Jap. J. Geol. Geogr.* 27, 189–229.

Aramaki, S. 1963. Geology of Asama volcano. *Fac. Sci. Univ. Tokyo J.* 2, 229–443.

Aramaki, S. and M. Yamasaki 1963. Pyroclastic flows in Japan. *Bull. Volcanol.* 26, 89–99.

Archambault, C. and J. C. Tanguy 1976. Comparative temperature measurements on Mount Etna lavas: problems and techniques. *J. Volcanol. Geotherm. Res.* 1, 113–25.

Arndt, N. T. and E. G. Nisbet (eds) 1982. *Komatiites.* London: Allen & Unwin.

Arndt, N. T., A. J. Naldrett and D. R. Pyke 1977. Komatiitic and iron-rich tholeiitic lavas of Munro Township, northeast Ontario. *J. Petrol.* 18, 319–69.

Atkinson, A., T. J. Griffin and P. J. Stephenson 1975. A major lava tube system from Undara volcano, north Queensland. *Bull. Volcanol.* 39, 1–28.

Aziz-Ur-Rahman and I. McDougall 1972. Potassium argon ages on the Newer Volcanics of Victoria. *Proc. R. Soc. Victoria* 85, 61–70.

Bacon, C. R. 1983. Eruptive history of Mount Mazama and Crater Lake caldera, Cascade Range, U.S.A. In *Arc volcanism,* S. Aramaki and I. Kushiro (eds), *J. Volcanol. Geotherm. Res.* 18, 57–115.

Bailey, R. A., G. B. Dalrymple and M. A. Lanphere 1976. Volcanism, structure and geochronology of Long Valley caldera, Mono County, California. *J. Geophys. Res.* 81, 725–44.

Baker, B. H., P. A. Mohr and L. A. J. Williams 1972. *Geology of the eastern rift system of Africa.* Geol. Soc. Am. Spec. Pap., no. 136.

Baker, M. C. W. 1981. The nature and distribution of Upper Cenozoic ignimbrite centres in the central Andes. *J. Volcanol. Geotherm. Res.* 11, 293–315.

Baker, P. E. 1969. The geological history of Mt. Misery volcano, St. Kitts, West Indies. *Overseas Geol. Miner. Resources* 10, 207–17.

Ball, E. E. and R. W. Johnson 1976. Volcanic history of Long Island, Papua New Guinea. In R. W. Johnson (1976), 133–47.

Ballance, P. F. and H. G. Reading (eds) 1980. *Sedimentation in oblique-slip mobile zones*. Int. Assoc. Sed. Spec. Publn 4. London: Blackwell.

Ballard, R. D. and T. H. van Andel 1977. Morphology and tectonics of the inner rift valley at lat. 36°50′ N on the Mid-Atlantic Ridge. *Geol. Soc. Am. Bull.* 88, 507–30.

Ballard, R. D., R. D. Holcomb and T. H. van Andel 1979. The Galapagos Rift at 86°W: 3. Sheet flows, collapse pits and lava lakes of the rift valley. *J. Geophys. Res.* 84, 5407–22.

Ballard, R. D., R. Hekinian and J. Francheteau 1984. Geological setting of hydrothermal activity at 12 degrees 50′ N on the East Pacific Rise — a submersible study. *Earth & Planet. Sci. Letts.* 69, 176–86.

Banks, N. G. and R. P. Hoblitt 1981. Summary of temperature studies of 1980 deposits. In Lipman & Mullineaux (1981), 295–313.

Barberi, F., F. Innocenti, L. Lirer, R. Munro, T. Pescatore and R. Santacroce 1978. The Campanian ignimbrite: a major prehistoric eruption in the Neapolitan area (Italy). *Bull. Volcanol.* 41, 10–32.

Bartholomew, D. S. and J. Tarney 1984. Crustal extension in the southern Andes (45°–46°S). In Kokelaar & Howells (1984).

Basaltic Volcanism Study Project 1981. *Basaltic volcanism on the terrestrial planets*. New York: Pergamon.

Batiza, R. and D. Vanko 1983. Volcanic development of small oceanic central volcanoes on the flanks of the East Pacific Rise inferred from narrow-beam echo-sounder surveys. *Marine Geol.* 54, 53–90.

Bauer, G. R. 1970. The geology of Tofua Island, Tonga. *Pacific Sci.* 24, 333–50.

Beane, R. E. 1982. Hydrothermal alteration in silicate rocks: southwestern North America. In *Advances in geology of the porphyry copper deposits: southwestern North America*, S. R. Titley (ed.), 117–37. Tucson: Univ. Arizona Press.

Bennett, F. D. 1974. On volcanic ash formation. *Am. J. Sci.* 274, 1–120.

Benson, G. T. and L. R. Kittleman 1968. Geometry of flow layering in silicic lavas. *Am. J. Sci.* 266, 265–76.

Bibee, L. D., G. G. Shor and R. S. Lu 1980. Inter-arc spreading in the Mariana Trough. *Marine Geol.* 35, 183–97.

Bierwirth, P. N. 1982. *Experimental welding of volcanic ash*. Unpubl. B.Sc. Hons. thesis, Monash University.

Birch, W. D. 1978. Petrogenesis of some Palaeozoic rhyolites in Victoria. *J. Geol. Soc. Aust.* 25, 75–87.

Black, P.M. 1970. Observations on White Island. *Bull. Volcanol.* 34, 158–67.

Blackburn, E., L. Wilson and R. S. J. Sparks 1976. Mechanisms and dynamics of strombolian activity. *J. Geol. Soc. Lond.* 132, 429–40.

Blackburn, G., G. B. Alison and F. W. J. Learey 1982. Further evidence on the age of tuff at Mt. Gambier. *Trans R. Soc. S. Aust.* 106, 163–8.

Blake, S. 1981a. Volcanism and the dynamics of open magma chambers. *Nature* 289, 783–5.

Blake, S. 1981b. Eruption from zoned magma chambers. *J. Geol. Soc. Lond.* 138, 281–7.

Blatt, H., G. V. Middleton and R. Murray 1980. *Origin of sedimentary rocks*. Englewood Cliffs, New Jersey: Prentice-Hall.

Bloomfield, K., G. Sanchez Rubio and L. Wilson 1977. Plinian eruptions of Nevado de Toluca volcano, Mexico. *Geol. Rundsch.* 66, 120–46.

Board, S. J., C. L. Farmer and D. H. Poole 1974. Fragmentation in thermal explosions. *Int. J. Heat Mass Transfer* 17, 331–9.

Bonatti, E. 1967. Mechanisms of deep-sea volcanism in the South Pacific. In *Researches in geochemistry*, Vol. 2, P. H. Abelson (ed.), 453–91. New York: Wiley.

Bond, A. and R. S. J. Sparks 1976. The Minoan eruption of Santorini, Greece. *J. Geol Soc. Lond.* 132, 1–16.

Booth, B. 1973. The Granadilla pumice deposit of southern Tenerife, Canary Islands. *Proc. Geol Assoc. Lond.* 84, 353–70.

Booth, B., R. Croasdale and G. P. L. Walker 1978. A quantitative study of five thousand years of volcanism on São Miguel, Azores. *Phil Trans R. Soc. Lond. (A)* 288, 271–319.

Bottinga, Y. and D. F. Weill 1970. Densities of liquid silicate systems calculated from partial molar volumes of oxide components. *Am. J. Sci.* 269, 169–82.

Bottinga, Y. and D. F. Weill 1972. The viscosity of magmatic silicate liquids: A model for calculation. *Am. J. Sci.* 272, 438–75.

Bouma, A. H. 1962. *Sedimentology of some flysch deposits*. Amsterdam: Elsevier.

Bracey, D. R. and T. A. Ogden 1972. Southern Mariana arc: geophysical observations and hypothesis of evolution. *Geol Soc. Am. Bull.* 83, 1509–22.

Branch, C. D. 1976. Development of porphyry copper and stratiform volcanogenic ore bodies during the life cycle of andesitic stratovolcanoes. In R. W. Johnson (1976), 337–42.

Brazier, S. A., A. N. Davis, H. Sigurdsson and

R. S. J. Sparks 1982. Fall-out and deposition of volcanic ash during the 1979 explosive eruption of the Soufrière of St. Vincent. *J. Volcanol. Geotherm. Res.* 14, 335–59.

Brazier, S., R. S. J. Sparks, S. N. Carey, H. Sigurdsson and J. A. Westgate 1983. Bimodal grain size distribution and secondary thickening in air-fall ash layers. *Nature* 301, 115–9.

Briden, J. C., D. C. Rex, A. M. Fallar and J. F. Tomblin 1979. K-Ar geochronology and paleomagnetism of volcanic rocks in the Lesser Antilles island arc. *Phil Trans R. Soc. Lond. (A)* 291, 485–528.

Briggs, N. D. 1976. Welding and crystallisation zonation in Whakamaru ignimbrite, central North Island, New Zealand. *N.Z. J. Geol. Geophys.* 19, 189–212.

Bristow, J. W. and E. P. Saggerson 1983. A review of Karoo vulcanicity in southern Africa. *Bull. Volcanol.* 46, 135–59.

Brookfield, M. E. 1984. Eolian facies. In R. G. Walker (1984), 91–103.

Brooks, D. A., R. L. Carlson, D. L. Harry, P. J. Melia, R. P. Moore, J. E. Rayhorn and S. G. Tull 1984. Characteristics of back-arc regions. *Tectonophysics* 102, 1–16.

Brugman, M. M. and M. F. Meier 1981. The 1980 eruptions of Mount St. Helens. Response of glaciers to the eruptions of Mount St. Helens. In Lipman & Mullineaux (1981), 743–56.

Bryan, W. B., G. D. Stice and A. Ewart 1972. Geology, petrography and geochemistry of the volcanic islands of Tonga. *J. Geophys. Res.* 77, 1566–85.

Buchanan, D. J. 1974. A model for fuel-coolant interactions. *J. Phys. D: Appl. Phys.* 7, 1441–57.

Bullard, F. M. 1947. Studies on Parícutin volcano, Michoacan, Mexico. *Geol Soc. Am. Bull.* 58, 433–50.

Bullard, F. M. 1976. *Volcanoes of the earth.* St Lucia: Univ. Queensland Press.

Bultitude, R. J. 1976a. Eruptive history of Bagana volcano, Papua New Guinea, between 1882 and 1975. In R. W. Johnson (1976), 317–36.

Bultitude, R. J. 1976b. Flood basalts of probable early Cambrian age in northern Australia. In R. W. Johnson (1976), 1–20.

Burke, K. 1978. Evolution of continental rift systems in the light of plate tectonics. In Ramberg & Neumann (1978), 1–9.

Burnham, C. W. 1972. The energy of explosive eruptions. *Earth, Mineral Sci. (Pennsylvania State Univ.)* 41, 69–70.

Burnham, C. W. 1979. The importance of volatile constituents. In *The evolution of the igneous rocks,* H. S. Yoder, Jr (ed.), 439–82. Princeton, New Jersey: Princeton University Press.

Burnham, C. W. 1983. Deep submarine pyroclastic eruptions. In Ohmoto and Skinner (1983), 142–8.

Burns, R. E., J. E. Andrews *et al.* 1973. *Initial reports of the Deep-Sea Drilling Project.* Vol. 21. Washington, DC: US Government Printing Office.

Busby-Spera, C. 1984. Large-volume rhyolite ash-flow eruptions and submarine caldera collapse in the lower Mesozoic Sierra Nevada, California. *J. Geophys. Res.* 89, 8417–27.

Bussell, M. A. 1983. Timing of Tectonic and magmatic events in the central Andes of Peru. *J. Geol Soc. Lond.* 140, 279–86.

Byers, F. M. Jr, W. J. Carr, P. P. Orkitt, W. D. Quinlivan and K. A. Sargent 1976. *Volcanic suites and related cauldrons of Timber Mountain — Oasis Valley Caldera Complex, southern Nevada.* US Geol Surv. Prof. Pap., no. 919.

Carey, S. N. and H. Sigurdsson 1978. Deep-sea evidence for distribution of tephra from the mixed magma eruption of the Soufrière on St. Vincent 1902: ash turbidites and air fall. *Geology* 6, 271–4.

Carey, S. N. and H. Sigurdsson 1980. The Roseau Ash: deep-sea tephra deposits from a major eruption on Dominica, Lesser Antilles arc. *J. Volcanol. Geotherm. Res.* 7, 67–86.

Carey, S. N. and H. Sigurdsson 1982. Influence of particle aggregation on deposition of distal tephra from the May 18, 1980, eruption of Mount St. Helens volcano. *J. Geophys. Res.* 87, 7061–72.

Carey, S. N. and H. Sigurdsson 1984. A model of volcanogenic sedimentation in marginal basins. In Kokelaar & Howells (1984), 37–58.

Carmichael, I. S. E., F. J. Turner and J. Verhoogen 1974. *Igneous petrology.* New York: McGraw-Hill.

Carozzi, A. V. 1960. *Microscopic sedimentary petrography.* New York: Wiley.

Carr, R. G. 1981. A scanning electron microscope study of post-depositional changes in the Matahina ignimbrite, North Island, New Zealand. *N.Z. J. Geol. Geophys.* 24, 429–34.

Carter, R. M. 1975. A discussion and classification of subaqueous mass transport with particular application to grain-flow, slurry-flow, and fluxoturbidites. *Earth Sci. Rev.* 11, 145–77.

Cas, R. A. F. 1977. *The Merrions Tuff — its genesis and palaeogeographic setting.* Unpubl. Ph.D. thesis, Macquarie University.

Cas, R. A. F. 1978a. Silicic lavas in Palaeozoic flysch-like deposits in New South Wales, Australia: behaviour of deep subsqueous silicic flows. *Geol Soc. Am. Bull.* 89, 1708–14.

Cas, R. A. F. 1978b. Basin characteristics of the Early Devonian part of the Hill End Trough based on a stratigraphic analysis of the Merrions Tuff. *J. Geol Soc. Aust.* 24, 381–401.

Cas, R. A. F. 1979. Mass-flow arenites from a Palaeozoic interarc basin, New South Wales, Australia: mode and environment of emplacement. *J. Sed. Petrol.* **49**, 29–44.

Cas, R. A. F. 1983a. Submarine 'crystal-tuffs': their origin using a Lower Devonian example from southeastern Australia. *Geol Mag.* **120**, 471–86.

Cas, R. A. F. 1983b. *A review of the palaeogeographic and tectonic development of the Palaeozoic Lachlan Fold Belt of southeastern Australia.* Geol Soc. Aust. Spec. Publn, no. 10.

Cas, R. A. F. and J. G. Jones 1979. Palaeozoic interarc basin in eastern Australia and a modern New Zealand analogue. *N.Z. J. Geol. Geophys.* **22**, 71–81.

Cas, R. A. F., R. H. Flood and S. E. Shaw 1976. Hill End Trough: new radiometric ages. *Search (Aust.)* 7, 205–7.

Cas, R. A. F., C. McA. Powell and K. A. W. Crook 1980. Ordovician palaeogeography of the Lachlan Fold Belt: a modern analogue and tectonic constraints. *J. Geol Soc. Aust.* **27**, 19–31.

Cas, R. A. F., C. McA. Powell, C. L. Fergusson, J. G. Jones, W. D. Roots and J. Fergusson 1981. The Kowmung Volcaniclastics: a deep-water sequence of mass-flow origin. *J. Geol Soc. Aust.* **28**, 271–88.

Cathles, L. M., A. L. Guber, T. C. Lenagh and F. Ö. Dudas 1983. Kuroko-type massive sulfide deposits of Japan: products of an aborted island-arc rift. In H. Ohmoto and B. J. Skinner (1983), 96–114.

Chapin, C. E. and W. E. Elston (eds) 1979. *Ash flow tuffs.* Geol Soc. Am. Spec. Pap., no. 180.

Chapin, C. E. and G. R. Lowell 1979. Primary and secondary flow structures in ash-flow tuffs of the Gribbles run paleovalley, Central Colorado. In Chapin & Elston (1979), 137–54.

Chappell, B. W. and A. J. R. White 1974. Two contrasting granite types. *Pacific Geol.* **8**, 173–4.

Cheshire, S. G. and J. D. Bell 1977. The Speedwell vent, Castleton, Derbyshire: a Carboniferous littoral cone. *Proc. Yorks. Geol Soc.* **41**, 173–84.

Choubey, V. D. 1973. Long-distance correlation of Deccan basalt flows, central India. *Geol Soc. Am. Bull.* **84**, 2785–90.

Chouet, B., N. Hamisevicz and T. R. McGetchin 1974. Photoballistics of volcanic jet activity at Stromboli, Italy. *J. Geophys. Res.* **79**, 4961–76.

Christiansen, R. L. 1979. Cooling units and composite sheets in relation to caldera structure. In Chapin & Elston (1979), 29–42.

Christiansen, R. L. and P. W. Lipman 1966. Emplacement and thermal history of a rhyolite lava flow near Fortymile Canyon, southern Nevada. *Geol Soc. Am. Bull.* **77**, 671–84.

Christiansen, R. L. and P. W. Lipman 1972. Cenozoic volcanism and plate-tectonic evolution of the western United States. II. Late Cenozoic. *Phil Trans R. Soc. Lond. A* **271**, 249–84.

Christiansen, R. L. and D. W. Peterson 1981. The

1980 eruptions of Mount St. Helens. Chronology of the 1980 eruptive activity. In Lipman & Mullineaux (1981), 17–67.

Clemens, J. D. and V. J. Wall 1981. Origin and crystallisation of some peraluminous (S-type) granite magmas. *Can. Mineral.* **19**, 111–31.

Clemens, J. D. and V. J. Wall 1984. Origin and evolution of a peraluminous silicic ignimbrite suite: the Violet Town Volcanics. *Contr. Miner. Petrol.* **88**, 354–71.

Clifton, H. E., R. E. Hunter and R. L. Phillips 1971. Depositional structures and processes in the non-barred high energy near shore. *J. Sed. Petrol.* **41**, 651–70.

Clough, B. J. 1981. *The geology of La Primavera volcano, Mexico.* Unpubl. Ph.D. thesis, Imperial College, University of London.

Clough, B. J., J. V. Wright and G. P. L. Walker 1981. An unusual bed of giant pumice in Mexico. *Nature* **289**, 49–50.

Clough, B. J., J. V. Wright and G. P. L. Walker 1982. Morphology and dimensions of the young comendite lavas of La Primavera Volcano, Mexico. *Geol Mag.* **119**, 477–85.

Coates, R. E. 1968. The Circle Creek rhyolite, a volcanic dome complex in Northern Elko county, Nevada. In Coats *et al.* (1968), 69–107.

Coats, R. R., R. L. Hay and C. A. Anderson (eds) 1968. *Studies in volcanology (Howell Williams volume).* Geol Soc. Am. Mem., no. 116.

Cole, J. W. 1970. Structure and eruptive history of the Tarawera volcanic complex. *N.Z. J. Geol. Geophys.* **13**, 879–902.

Cole, J. W. 1979. Structure, petrology and genesis of Cenozoic volcanism, Taupo volcanic zone, New Zealand — a review. *N.Z. J. Geol. Geophys.* **22**, 631–57.

Cole, J. W. 1981. Genesis of lava of the Taupo Volcanic Zone, North Island, New Zealand. *J. Volcanol. Geotherm. Res.* **1**, 317–37.

Cole, J. W. 1984. Taupo–Rotorua depression — an ensialic marginal basin of North Island, New Zealand. In Kokelaar & Howells (1984), 109–20.

Coleman, R. G. 1971. Plate tectonic emplacement of upper mantle peridotites along continental edges. *J. Geophys. Res.* **76**, 1212–22.

Coleman, R. G. 1977. *Ophiolites.* Berlin: Springer.

Colgate, S. A. and T. Sigurgeirsson 1973. Dynamic mixing of water and lava. *Nature* **244**, 552–4.

Colley, H. 1976. Classification and exploration guide for Kuroko-type deposits based on occurrences in Fiji. *Trans Inst. Min. Metal.* **85**, B190–9.

Collins, W. J., S. D. Beams, A. J. R. White and B. W. Chappell 1982. Nature and origin of A-type granites with particular reference to southeastern Australia. *Contr. Miner. Petrol.* **80**, 189–200.

Collinson, J. D. and D. B. Thompson 1982. *Sedimentary structures*. London: Allen & Unwin.

Compston, W. and B. W. Chappell 1979. Sr-isotope evolution of granitoid source rocks in: *The Earth, its origin, structure and evolution*, M. W. McElhinny (ed.), 377–426. New York: Academic Press.

Conybeare, C. E. B. and K. A. W. Crook 1968. *Manual of sedimentary structures*. Bur. Miner. Resour. (Australia) Geol. Geophys. Bull., no. 102.

Cooper, A. K., M. S. Marlow and D. W. Scholl 1977. The Bering sea — a multiforious marginal basin. In *Island arcs, deep sea trenches and back-arc basins*, M. Talwani and W. C. Pitman, III (eds), 437–50. Washington, DC: Am. Geophys. Union.

Corradini, M. L. 1981. Phenomological modelling of the triggering phase of small-scale steam explosion experiments. *Nucl. Sci. Engng.* 78, 154–70.

Cousineau, P. and E. Dimroth 1982. Interpretation of the relations between massive, pillowed and brecciated facies in an Archean submarine andesite volcano-Amulet andesite, Rouyn–Noranda, Canada. *J. Volcanol. Geotherm. Res.* 13, 83–102.

Cox, K. G., J. D. Bell and R. J. Pankhurst 1979. *The interpretation of igneous rocks*. London: Allen & Unwin.

Cox, S. F. 1981. The stratigraphic and structural setting of the Mt. Lyell volcanic-hosted sulfide deposits. *Econ. Geol.* 76, 231–45.

Crandell, D. R. 1971. *Post-glacial lahars from Mount Rainier volcano, Washington*. US Geol Surv. Prof. Pap., no. 677.

Crandell, D. R. and R. D. Miller 1964. *Post hysithermal glacier advances at Mount Rainier, Washington*, 110–4. US Geol Surv. Prof. Pap., no. 501–D.

Crandell, D. R. and R. D. Miller 1974. *Quaternary stratigraphy and extent of glaciation in the Mount Rainier region, Washington*. US Geol Surv. Prof. Pap., no. 847.

Cross, T. A. and R. H. Pilger 1982. Crontrols of subduction geometry, location of magmatic arcs, and tectonics of arc and back-arc regions. *Geol Soc. Am. Bull.* 93, 545–62.

Crowe, B. M. and R. V. Fisher 1973. Sedimentary structures in base-surge deposits with special reference to cross-bedding, Ubehebe Craters, Death valley, California. *Geol. Soc. Am. Bull.* 84, 663–82.

Cummings, D. 1964. *Eddies as indicators of local flow direction in rhyolite, D70–2*. US Geol Surv. Prof. Pap., no. 475–D.

Curtis, G. H. 1968. The stratigraphy of the ejecta from the 1912 eruption of Mount Katmai and Novarupta, Alaska. In Coats *et al.* (1968), 153–210.

Dalziel, I. W. D. 1981. Back-arc extension in the southern Andes: a review and critical reappraisal. *Phil Trans R. Soc. Lond. (A)* 300, 319–35.

Davies, D. K., M. W. Quearry and S. B. Bonis 1978a. Glowing avalanches from the 1974 eruption of volcano Fuego, Guatemala. *Geol Soc. Am. Bull.* 89, 369–84.

Davies, D. K., W. R. Almon, S. B. Bonis and B. E. Hunter 1978b. Deposition and diagenesis of Tertiary-Holocene volcaniclastics, Guatemala. In *Aspects of diagenesis*, P. A. Scholle and P. R. Schluger (eds), 281–306. SEPM Spec. Publn, no. 26.

Davies, D. K., R. K. Vessell, R. C. Miles, M. G. Foley and S. D. Bonis 1978c. Fluvial transport and downstream sediment modifications in an active volcanic region. In Miall (1978), 61–84.

Davies, H. L. and L. E. Smith 1971. Geology of eastern Papua. *Geol Soc. Am. Bull.* 83, 3299–312.

Dawson, J. B. and D. G. Powell 1969. The Natron-Engaruka explosion crater area, Northern Tanzania. *Bull. Volcanol.* 33, 791–817.

Day, R. A. 1983. *Petrology and geochemistry of the Older Volcanics, Victoria (distribution, characterisation and petrogenesis)*. Unpubl. Ph.D. thesis, Monash University.

Delaney, P. T. and D. D. Pollard 1982. Solidification of basaltic magma during flow in a dyke. *Am. J. Sci.* 282, 856–85.

de Rosen-Spence, A. F., G. Provost, E. Dimroth, K. Gochnaucr and V. Owen 1980. Archean subaqueous felsic flows, Rouyn–Noranda, Quebec, Canada, their Quaternary equivalents. *Precamb. Res.* 12, 43–77.

de Wit, M. J. and C. Stern 1978. Pillow talk. *J. Volcanol. Geotherm. Res.* 4, 55–80.

Dewey, J. F. 1963. The Lower Palaeozoic stratigraphy of Central Murrisk, County Mayo, Ireland, and the evolution of the South Mayo Trough. *Q. J. Geol Soc. London.* 119, 313–43.

Dickinson, W. R. (ed.) 1974. *Tectonics and sedimentation*. SEPM Spec. Publn 22, 1–27.

Dickinson, W. R. 1979. Cenozoic plate tectonic setting of the Cordilleran region in the United States. In *Cenozoic palaeogeography of the western United States*, J. M. Armentrout, M. R. Cole and H. Terbest (eds), 1–13. Pacific Coast Palaeogeography Symp. 3.

Dimroth, E., P. Cousineau, M. Leduc and Y. Sanschagrin 1978. Structure and organisation of Archean subaqueous basalt flows, Rouyn–Noranda area, Quebec, Canada. *Can. J. Earth Sci.* 15, 902–18.

Di Paola, G. M. 1972. The Ethiopian Rift Valley (between 7°00′ and 8°40′ lat. North). *Bull. Volcanol.* 36, 517–60.

Donaldson, C. H. 1982. Spinifex-textured komatiites: a review of textures, compositions and layering. In Arndt & Nisbet (1982), 213–44.

Douglas, J. G. and J. A. Ferguson (eds) 1976. *Geology of Victoria*. Geol. Soc. Aust. Spec. Publn, no. 5.

Doumas, C. (ed.) 1978. *Thera and the Aegean World*. Vol. I. London: Thera and the Aegean World.

Doumas, C. (ed.) 1980. *Thera and the Aegean World*. Vol. II. London: Thera and the Aegean World.

Doyle, L. J. and O. H. Pilkey (eds) 1979. *Geology of continental slopes*. SEPM Spec. Publn, no. 27.

Drake, R. E. 1976. Chronology of Cenozoic igneous and tectonic events in the central Chilean Andes — Latitudes 35°30' to 36°S. *J. Volcanol. Geotherm. Res.* 1, 265–84.

Drexler, J. W., W. I. Rose Jr, R. S. J. Sparks and M. T. Ledbetter 1980. The Los Chocoyos Ash, Guatemala: a major stratigraphic marker in middle America and three ocean basins. *Quatern. Res.* 13, 327–45.

Druitt, T. H. 1985. Vent evolution and lag breccia formation during the Cape Riva eruption of Santorini, Greece. *J. Geol.* 93, 439–54.

Druitt, T. H. and R. S. J. Sparks 1982. A proximal ignimbrite breccia facies on Santorini volcano, Greece. *J. Volcanol. Geotherm. Res.* 13, 147–71.

Druitt, T. H. and R. S. J. Sparks 1985. On the formation of calderas during ignimbrite eruptions. *Nature* 310, 679–81.

Duffield, W. A., E. K. Gibson and G. H. Heiken 1977. Some characteristics of Pele's hair. *J. Res. U.S. Geol. Surv.* 5, 93–101.

Duffield, W. A., C. R. Bacon and G. R. Roquemore 1979. Origin of reverse-graded bedding in air-fall pumice, Coso Range, California. *J. Volcanol. Geotherm. Res.* 5, 35–48.

Duffield, W. A., L. Stieltjes and J. Varet 1982. Huge landslide blocks in the growth of Piton de la Fournaise, La Réunion and Kilauea volcano, Hawaii. *J. Volcanol. Geotherm. Res.* 12, 147–60.

Duncan, A. R. and H. M. Pantin 1969. Evidence for submarine geothermal activity in the Bay of Plenty. *N.Z. J. Marine Freshwat. Res.* 3, 602–6.

Dunett, D. 1969. A technique of finite strain analysis using elliptical particles. *Tectonophysics* 7, 114–36.

Duyverman, H. J. and M. J. Roobol 1981. Gas pipes in Eocambrian volcanic breccias. *Geol Mag.* 118, 265–70.

Eaton, G. P. 1982. The Basin and Range Province: origin and tectonic significance. *A. Rev. Earth Planet. Sci.* 10, 409–40.

Eaton, G. P. 1984. The Miocene Great Basin of western North America as an extending back-arc region. *Tectonophysics* 102, 275–95.

Edney, W. 1984. *The geology of the Tower Hill volcanic centre, western Victoria*. Unpubl. M.Sc. thesis, Monash University.

Eggler, D. H. and M. Rosenhauer 1978. Carbon dioxide in silicate melts: II. Solubilities of CO_2 and H_2O in $CaMgSi_2O_6$ (diopside) liquids and vapours at pressures to 40 kb. *Am. J. Sci.* 278, 64–94.

Eguchi, T. 1984. Seismotectonics around the Mariana Trough. *Tectonophysics* 102, 33–52.

Eichelberger, J. C. 1980. Vesiculation of mafic magma during replenishment of silicic magma reservoirs. *Nature* 288, 446–50.

Eichelberger, J. C. and F. G. Koch 1979. Lithic fragments in the Bandelier Tuff, Jemez Mountains, New Mexico. *J. Volcanol. Geotherm. Res.* 5, 115–34.

Elliot, D. 1970. Determination of finite strain and initial shape from deformed elliptical objects. *Geol Soc. Am. Bull.* 81, 2221–36.

Ellwood, B. B. 1982. Estimates of flow direction for calc-alkaline welded tuffs and palaeomagnetic data reliability from anisotropy of magnetic susceptibility measurements: central San Juan Mountains, southwest Colorado. *Earth Planet. Sci. Lett.* 59, 303–14.

Elston, W. E. and E. I. Smith 1970. Determination of flow direction of rhyolite ash-flow tuffs from fluidal textures. *Geol Soc. Am. Bull.* 81, 3393–406.

Ewart, A. 1979. A review of the mineralogy and chemistry of Tertiary–Recent dacitic, latitic, rhyolitic and related salic volcanic rocks. In *Trondhjemites, dacites and related rocks*, F. Baker (ed.), 113–21. The Hague: Elsevier.

Ewart, A. and J. J. Stipp 1968. Petrogenesis of the volcanic rocks of the central North Island, New Zealand, as indicated by a study of Sr^{87}/Sr^{86} ratios and Sr, Rb, K, U, and Th abundances. *Geochim. et Cosmochim. Acta* 32, 699–736.

Ewart, A., W. B. Bryan and J. B. Gill 1973. Mineralogy and geochemistry of the younger volcanic islands of Tonga, S.W. Pacific. *J. Petrol.* 14, 429–65.

Ewart, A., R. N. Brothers and A. Mateen 1977. An outline of the geology and geochemistry and the possible petrogenetic evolution of the volcanic rocks of the Tonga–Kermadec–New Zealand island arc. *J. Volcanol. Geotherm. Res.* 2, 205–50.

Eyles, N., C. H. Eyles and A. D. Miall 1983. Lithofacies types and vertical profile models: an alternative approach to the description and environmental interpretation of glacial diamict and diamicitite sequences. *Sedimentology* 30, 393–410.

Eyles, N. and A. D. Miall 1984. Glacial facies. In R. G. Walker (1984), 15–34.

Fahnestock, R. K. 1963. *Morphology and hydrology of a glacier stream — White River, Mount Rainier, Washington*. US Geol Surv. Prof. Pap., no. 422–A.

Fahnestock, R. K. 1978. Little Tahoma Peak rockfalls and avalanches, Mount Rainier, Washington, U.S.A. In Voight (1978), 181–96.

Feigenson, M. D. and F. J. Spera 1981. Dynamical model for temporal variation in magma type and eruption interval at Kohala volcano, Hawaii. *Geology* 9, 531–3.

Fenner, C. N. 1920. The Katmai region, Alaska, and the great eruption of 1912. *J. Geol.* 28, 569–606.

Fenner, C. N. 1948. Incandescent tuff flows in southern Peru. *Geol Soc. Am. Bull.* 59, 879–93.

Fenner, R. E. 1932. Concerning basaltic glass. *Am. Mineral.* 17, 104–7.

Fergusson, C. L., R. A. F. Cas, W. J. Collins, G. Y. Craig, K. A. W. Crook, C. McA. Powell, P. A. Scott and G. C. Young 1979. The Late Devonian Boyd Volcanic Complex, Eden, N.S.W.: a reinterpretation. *J. Geol Soc. Aust.* 26, 87–105.

Fernandez, H. E. 1969. Notes on the submarine ash-flow tuff in Siargao Island, Surigao del Norte (Philippines). *Philippine Geol.* 23, 29–36.

Fink, J. H. 1980a. Surface folding and viscosity of rhyolite flows. *Geology* 8, 250–4.

Fink, J. H. 1980b. Gravity instability in the Holocene Big and Little Glass Mountain rhyolitic obsidian flows, northern California. *Tectonophysics* 66, 147–66.

Fink, J. H. 1983. Structure and emplacement of a rhyolitic obsidian flow — Little Glass Mountain, Medicine Lake Highland, Northern California. *Geol Soc. Am. Bull.* 94, 362–80.

Fink, J. H. and R. C. Fletcher 1978. Ropy pahoehoe: surface folding of a viscous fluid. *J. Volcanol. Geotherm. Res.* 4, 151–70.

Fisher, R. V. 1960. Classification of volcanic breccias. *Geol Soc. Am. Bull.* 71, 973–82.

Fisher, R. V. 1961. Proposed classification of volcano-clastic sediments and rocks. *Geol Soc. Am. Bull.* 72, 1409–14.

Fisher, R. V. 1966a. Mechanism of deposition from pyroclastic flows. *Am. J. Sci.* 264, 350–63.

Fisher, R. V. 1966b. Rocks composed of volcanic fragments and their classification. *Earth-Sci. Rev.* 1, 287–98.

Fisher, R. V. 1968. Puu Hou littoral cones, Hawaii. *Geol. Rundsch.* 57, 837–64.

Fisher, R. V. 1977. Erosion by volcanic base-surge density currents: U-shaped channels. *Geol Soc. Am. Bull.* 88, 1287–97.

Fisher, R. V. 1979. Models for pyroclastic surges and pyroclastic flows. *J. Volcanol. Geotherm. Res.* 6, 305–18.

Fisher, R. V. 1984. Submarine volcaniclastic rocks. In Kokelaar & Howells (1984), 5–27.

Fisher, R. V. and D. W. Charleton 1976. Mid-Miocene Blance Formation, Santa Cruz Island, California. In *Aspects of the geological history of the California continental borderland*, D. G. Howell (ed.), 228–40. Pacific Section Am. Assoc. Petroleum Geologists, Misc. Publns, no. 24.

Fisher, R. V. and G. Heiken 1982. Mt Pelée, Martinique: May 8 and 20, 1902, pyroclastic flows and surges. *J. Volcanol. Geotherm. Res.* 13, 339–71.

Fisher, R. V. and H.-U. Schmincke 1984. *Pyroclastic rocks.* Berlin: Springer-Verlag.

Fisher, R. V. and A. C. Waters 1970. Base surge bedforms in maar volcanoes. *Am. J. Sci.* 268, 157–80.

Fisher, R. V., H.-U. Schmincke and P. V. Bogard 1983. Origin and emplacement of a pyroclastic flow and surge unit at Laacher See, Germany. *J. Volcanol. Geotherm. Res.* 17, 375–92.

Fisher, R. V., A. L. Smith and M. J. Roobol 1980. Destruction of St. Pierre, Martinique by ash-cloud surges, May 8 and 20, 1902. *Geology* 8, 472–6.

Fiske, R. S. 1963. Subaqueous pyroclastic flows in the Ohanapecosh Formation, Washington. *Geol Soc. Am. Bull.* 74, 391–406.

Fiske, R. S. 1969. Recognition and significance of pumice in marine pyroclastic rocks. *Geol Soc. Am. Bull.* 80, 1–8.

Fiske, R. S. and T. Matsuda 1964. Submarine equivalents of ash flows in the Tokiwa Formation, Japan. *Am. J. Sci.* 262, 76–106.

Fiske, R. S., C. A. Hopson and A. C. Waters 1963. *Geology of Mount Rainier National Park, Washington.* US Geol Surv. Prof. Pap., no. 444.

Folk, R. L. 1980. *Petrology of sedimentary rocks.* Austin: Hemphills.

Folk, R. L. and W. C. Ward 1957. Brazos river bar: a study of the significance of grainsize parameters. *J. Sed. Petrol.* 27, 3–26.

Fornari, D. J., D. W. Peterson, J. P. Lockwood, A. Malahoff and B. C. Heezen 1979. Submarine extension of the southwest rift zone of Mauna Loa volcano, Hawaii: visual observations from U.S. Navy deep submergence vehicle *DSV Sea Cliff. Geol Soc. Am. Bull.* 90, 435–43.

Francheteau, J., H. D. Needham, P. Choukroune, T. Juteau, M. Seguret, R. D. Ballard, P. J. Fox, W. Normark, A. Carranza, D. Cordoba, J. Guerrero, C. Rangin, H. Bougault, P. Cambon and R. Hekinian 1979. Massive deep-sea sulphide ore deposits discovered on the East Pacific Rise. *Nature* 277, 523–8.

Francis, E. H. 1983. Magma and sediment — II. Problems of interpreting palaeovolcanics buried in the stratigraphic column. *J. Geol Soc. Lond.* 140, 165–83.

Francis, E. H. and M. F. Howells 1973. Transgressive welded ash-flows tuffs among the Ordovician sediments of N.E. Snowdonia. *J. Geol Soc. Lond.* 129, 621–41.

Francis, P. W. and M. C. W. Baker 1977. Mobility of pyroclastic flows. *Nature* 270, 164–5.

Francis, P. W., M. Gardeweg, C. F. Ramirez and D. H. Rothery 1985. Catastrophic debris avalanche deposit of Socompa volcano, northern Chile. *Geology* 13, 600–3.

Francis, P. W., L. O'Callaghan, G. A. Kretzchmar, R. S. Thorpe, R. S. J. Sparks, R. N. Page, R. E. de Barrio, G. Gillou and O. E. Gonzalez 1983. The Cerro Galan ignimbrite. *Nature* 301, 51–3.

Francis, P. W., M. J. Roobol, G. P. L. Walker, P. R. Cobold and M. P. Coward 1974. The San Pedro and San Pablo volcanoes of North Chile and their hot avalanche deposits. *Geol. Rundsch.* 63, 357–88.

Franklin, J. M., J. W. Lydon and D. F. Sangster 1981. Volcanic-associated sulphide deposits. In *Economic Geology seventy-fifth anniversary volume*, B. J. Skinner (ed.), 485–627. New Haven, Connecticut: The Economic Geology Publ. Co.

Freundt, A. and H.-U. Schmincke 1985. Hierarchy of facies of pyroclastic flow deposits generated by Laacher See-type eruptions. *Geology* 13, 278–81.

Frey, F. A. and D. A. Clague 1984. Geochemistry of diverse basalt types from Loihi Seamount, Hawaii — petrogenetic implications. *Earth Planet. Sci. Lett.* 66, 337–55.

Friedman, G. M. and J. E. Sanders 1978. *Principles of sedimentology*. New York: Wiley.

Friedman, I. and W. Long 1976. Hydration rate of obsidian. *Science* 191, 347–52.

Friedman, I., W. Long and R. L. Smith 1963. Viscosity and water contents of rhyolite glass. *J. Geophys. Res.* 68, 6523–35.

Friedman, J. D., G. R. Olhoeft, G. R. Johnson and D. Frank 1981. Heat content and thermal energy of the June dacitic dome in relation to total energy yield May-October 1980. In Lipman & Mullineaux (1981), 557–67.

Froggatt, P. C. 1982. Review of methods of estimating rhyolitic tephra volumes: applications to the Taupo volcanic zone, New Zealand. *J. Volcanol. Geotherm. Res.* 14, 301–18.

Froggatt, P. C., C. J. N. Wilson and G. P. L. Walker 1981. Orientation of logs in the Taupo Ignimbrite as indicator of flow direction and vent position. *Geology* 9, 109–11.

Fudali, R. F. and W. G. Melson 1972. Ejecta velocities, magma chamber pressure and kinetic energy associated with the 1968 eruption of Arenal volcano. *Bull. Volcanol.* 35, 383–401.

Fuller, R. E. 1931. The aqueous chilling of basaltic lava on the Columbia River Plateau. *Am. J. Sci.* 21, 281–300.

Furnes, H. and B. A. Sturt 1976. Beach/shallow marine hyaloclastite deposits and their geological significance — an example from Gran Canaria. *J. Geol.* 84, 439–53.

Furnes, H., I. B. Friedliefsson and F. B. Atkins 1980. Subglacial volcanics — on the formation of acid hyaloclastites. *J. Volcanol. Geotherm. Res.* 8, 95–110.

Gibson, I. L. and H. Tazieff 1967. Additional theory on the origin of fiamme in ignimbrites. *Nature* 215, 1473–4.

Gibson, I. L. and G. P. L. Walker 1963. Some composite rhyolite/basalt lavas and related composite dykes in eastern Iceland. *Proc. Geol. Assoc.* 74, 301–18.

Gilbert, C. M. 1938. Welded tuff in eastern California. *Geol Soc. Am. Bull.* 49, 1829–62.

Gill, J. B. 1976. Composition and age of Lau Basin and Ridge volcanic rocks: implications for evolution of an inter-arc basin and remnant arc. *Geol Soc. Am. Bull.* 87, 1384–95.

Gill, J. B. 1981. *Orogenic andesites and plate tectonics*. Berlin: Springer.

Gill, J. B., A. L. Stork and P. M. Whelan 1984. Volcanism accompanying back-arc development in the southwest Pacific. *Tectonophysics* 102, 207–24.

Gorshkov, G. S. 1959. Gigantic eruption of the volcano Bezymianny. *Bull. Volcanol.* 20, 77–112.

Gorshkov, G. S. and Y. M. Dubik 1970. Gigantic directed blast at Sheveluch volcano (Kamchatka). *Bull. Volcanol.* 34, 261–8.

Greeley, R. 1977a. Volcanic morphology. In Greeley & King (1977), 5–22.

Greeley, R. 1977b. Basaltic 'plains' volcanism. In Greeley & King (1977), 23–44.

Greeley, R. 1982. The Snake River Plain, Idaho: representative of a new category of volcanism. *J. Geophys. Res.* 87, 2705–12.

Greeley, R. and J. S. King (eds) 1977. *Volcanism of the eastern Snake River Plain, Idaho: a comparative planetary geology guidebook*. Washington, DC: NASA.

Green, D. H. and A. E. Ringwood 1967. The genesis of basaltic magmas. *Contr. Miner. Petrol.* 15, 103–90.

Greene, R. C. and D. Plouff 1981. Location of a caldera source for the Soldier Meadow Tuff, northwestern Nevada, indicated by gravity and aeromagnetic data: Summary. *Geol Soc. Am. Bull.* 92, 4–6.

Gregg, D. R. 1960. The geology of the Tongariro subdivision. *N.Z. Geol Surv. Bull.* 40.

Grindley, G. W. 1960. *Sheet 8, Taupo. Geological map of New Zealand 1 : 250,000*. Wellington: Dept Sci. Ind. Res.

Guber, A. L. and S. Merill 1983. Palaeobathymetric significance of the foraminifera from the Hokuroku district. In *The Kuroko and related volcanogenic massive sulphide deposits*, H. Ohmoto and B. J. Skinner (eds), 55–70. Econ. Geol. Mon. 5.

Guest, J. E. and J. Sanchez 1969. A large dacitic lava flow in northern Chile. *Bull. Volcanol.* 33, 778–90.

Gunnarsson, A. 1973. *Volcano. Ordeal by fire in Iceland's Westmann Islands*. Reykjavik: Iceland Review.

Hall, S. H. 1978. *The stratigraphy of northern Lipari and the structure of the Rocche Rosse rhyolite flow and its implications*. Unpubl. B.Sc. thesis, University of Leeds.

Hamilton, W. 1979. *Tectonics of the Indonesian region.* Geol Surv. Prof. Pap., no. 1078.

Hampton, M. A. 1972. The role of subaqueous debris flow in generating turbidity currents. *J. Sed. Petrol.* **42**, 775–93.

Hampton, M. A. 1975. Competence of fine-grained debris flow. *J. Sed. Petrol.* **45**, 834–44.

Hampton, M. A. 1979. Buoyancy in debris flows. *J. Sed. Petrol.* **49**, 753–8.

Hargraves, R. B. (ed.) 1980. *Physics of magmatic processes.* Princeton, New Jersey: Princeton University Press.

Hargreaves, R. and L. D. Ayres 1979. Morphology of Archean metabasalt flows, Utik Lake, Manitoba. *Can. J. Earth Sci.* **16**, 1452–66.

Harms, J. C. and R. K. Fahnestock 1965. Stratification, bed forms, and flow phenomena (with an example from the Rio Grande). In *Primary sedimentary structures and their hydrodynamic interpretation*, G. V. Middleton (ed.), 84–15. SEPM Spec. Publn, no. 12.

Harms, J. C., J. B. Southard and R. G. Walker 1982. *Structures and sequences in clastic rocks.* SEPM Short Course no. 9.

Harris, D. M., W. I. Rose Jr, R. Rose and M. R. Thompson 1981. Radar observations of ash eruptions. In Lipman & Mullineaux (1981), 323–33.

Hatch, F. H., A. K. Wells and M. K. Wells 1972. *Petrology of the igneous rocks.* London: George Allen & Unwin.

Hawkins, J. W. and J. Melchior 1984. Petrology of basalts from Loihi Seamount, Hawaii. *Earth Planet. Sci. Lett.* **66**, 356–68.

Hawkins, J. W., S. H. Bloomer, C. A. Evans and J. T. Melchior 1984. Evolution of intra-oceanic arc-trench systems. *Tectonophysics* **102**, 175–205.

Hay, R. L. 1959. Formation of the crystal-rich glowing avalanche deposits of St. Vincent, B.W.I. *J. Geol.* **67**, 540–62.

Hay, R. L. and A. Iijima 1968. Nature and origin of palagonite tuffs of the Honolulu group on Oahu, Hawaii. In Coats *et al.* (1968), 331–76.

Hay, R. L., W. Hildreth and R. N. Lambe 1979. Globule ignimbrite of Mount Suswa, Kenya. In Chapin & Elston (1979), 167–75.

Hays, J. D. and D. Ninkovich 1970. *North Pacific deep-sea ash chronology and age of present Aleutian under-thrusting*, 263–90. Geol Soc. Am. Mem., no. 126.

Healy, J. 1962. Structure and volcanism in the Taupo volcanic zone, New Zealand. In *Crust of the Pacific Basin*, G. A. Macdonald and H. Kuno (eds), 151–157. Am. Geophys. Un., Mon. 6.

Healy, J. 1963. Welded pyroclastic rock at Tongariro. *N.Z. J. Geol. Geophys.* **6**, 712–4.

Healy, J., J. C. Schofield and B. N. Thompson 1964. *Geological map of New Zealand 1 : 250,000 Sheet 5 Rotorua.* Wellington: N.Z. Dept Sci. Ind. Res.

Healy, J., E. F. Lloyd, D. E. H. Rishworth, C. P. Wood, R. B. Glover and R. R. Dibble 1978. *The eruption of Ruapehu, New Zealand, on 22 June 1969.* N.Z. D.S.I.R. Bull., no. 224.

Heiken, G. H. 1971. Tuff rings: examples from Fort Rock–Christmas Lake Valley basin, south central Oregon. *J. Geophys. Res.* **76**, 5615–26.

Heiken, G. H. 1972. Morphology and petrology of volcanic ashes. *Geol Soc. Am. Bull.* **83**, 1961–88.

Heiken, G. H. 1974. *An atlas of volcanic ash.* Smithsonian Contr. Earth Sci., no. 12.

Heiken, G. H. 1978. Characteristics of tephra from Cinder Cone, Lassen Volcanic National Park, California. *Bull. Volcanol.* **41**, 119–30.

Heim, A. 1932. *Bergsturz und Menschenleben.* Zurich: Fretz und Wasmuth.

Hekinian, R., M. Fevrier, J. L. Bischoll, P. Picot and W. C. Shanks 1980. Sulphide deposits from the East Pacific Rise near 21°N. *Science* **207**, 1433–44.

Henley, R. W. and A. J. Ellis 1983. Geothermal systems ancient and modern: a geochemical review. *Earth Sci. Rev.* **19**, 1–50.

Hermance, J. F. 1983. The Long Valley/Mono Basin complex in eastern California: status of present knowledge and future research needs. *Rev. Geophys. Space Phys.* **21**, 1545–65.

Hess, P. C. 1980. Polymerisation model for silicate melts. In Hargraves (1980), 1–48.

Hickson, C. J., P. Hickson and W. C. Barnes 1982. Weighted vector analysis applied to surge deposits from the May 18, 1980 eruption of Mt. St. Helens, Washington. *Can. J. Earth Sci.* **19**, 829–36.

Hildreth, W. 1981. Gradients in silicic magma chambers: implications for lithospheric magmatism. *J. Geophys. Res.* **86**, 10153–92.

Hiscott, R. N. and G. V. Middleton 1979. Depositional mechanics of thick-bedded sandstones at the base of a submarine slope, Tourelle Formation (Lower Ordovician), Quebec, Canada. In Doyle & Pilkey (1979), 307–26.

Hobbs, B. E., W. D. Means and P. F. Williams 1976. *An outline of structural geology.* New York: Wiley.

Hoblitt, R. P. and K. S. Kellogg 1979. Emplacement temperatures of unsorted and unstratified deposits of volcanic rock debris as determined by palaeo-magnetic techniques. *Geol Soc. Am. Bull.* **90**, 633–42.

Hoblitt, R. P. and C. D. Miller 1984. Comment on Walker & McBroome 1983. *Geology* **12**, 692–3.

Hoblitt, R. P., C. D. Miller and J. W. Vallance 1981. Origin and stratigraphy of the deposit produced by the May 18 directed blast. In Lipman & Mullineaux (1981), 401–19.

Hogg, S. E. 1982. Sheetfloods, sheetwash, sheetflow, or . . .? *Earth Sci. Rev.* **18**, 59–76.

Hollister, C. D., M. F. Glenn and P. F. Lonsdale 1978.

Morphology of seamounts in the western Pacific and Philippine Basin from multi-beam sonar data. *Earth Planet. Sci. Lett.* 41, 405–18.

Holmes, A. 1920. *The nomenclature of petrology.* London: T. Murby.

Honnorez, J. and P. Kirst 1975. Submarine basaltic volcanism: morphometric parameters for discriminating hyaloclastites from hyalotuffs. *Bull. Volcanol.* 39, 1–25.

Houghton, B. F. and W. R. Hackett 1984. Strombolian and phreatomagmatic deposits of Ohakune craters, Ruapehu, New Zealand: a complex interaction between external water and rising basaltic magma. *J. Volcanol. Geotherm. Res.* 21, 207–31.

Howard, K. 1973. Avalanche mode of motion: implications from lunar examples. *Science* 180, 1052–5.

Howells, M. F., B. E. Leveridge and C. D. R. Evans 1973. *Ordovician ash-flow tuffs in eastern Snowdonia.* Inst. Geol. Sci. Rep. No. 73/3.

Howells, M. F., B. E. Leveridge, R. Addison, C. D. R. Evans and M. J. C. Nutt 1979. The Capel Curig Volcanic Formation Snowdonia, North Wales; variations in ash-flow tuffs related to emplacement environment. In *The Caledonides of the British Isles — reviewed,* A. L. Harris, C. H. Holland and B. E. Leake (eds), 611–8. Geol Soc. Lond. Spec. Publn, no. 8.

Howells, M. F., S. D. G. Campbell and A. J. Reedman 1985. Isolated pods of subaqueous welded ash-flow tuff: a distal facies of the Capel Curig Volcanic Formation (Ordovician), North Wales. *Geol Mag.* 122, 175–80.

Hsü, K. J. 1975. Catastrophic debris streams (sturzstroms) generated by rock falls. *Geol Soc. Am. Bull.* 86, 129–40.

Hsü, K. J. 1978. Albert Heim: observations on landslides and relevance to modern interpretations. In Voight (1978), 70–93.

Huang, T. C., N. D. Watkins and D. M. Shaw 1975. Atmospherically transported glass in deep-sea sediments: volcanism in sub-Antarctic latitudes of the South Pacific during late Pliocene and Pleistocene time. *Geol Soc. Am. Bull.* 86, 1305–15.

Huang, T. C., N. D. Watkins and L. Wilson 1979. Deep-sea tephra from the Azores during the past 300,000 years: eruptive cloud height and ash volume estimates. *Geol Soc. Am. Bull.* 90, part II, 235–88.

Hughes, C. J. 1983. *Igneous Petrology.* Amsterdam: Elsevier.

Hulme, G. 1974. The interpretation of lava flow morphology. *Geophys. J. R. Astr. Soc.* 39, 361–83.

Hunter, R. E. 1985. A kinematic model for the structure of lee side deposits. *Sedimentology* 32, 409–22.

Huppert, H. E. and R. S. J. Sparks 1980a. The fluid dynamics of a basaltic magma chamber replenished by influx of hot, dense ultrabasic magma. *Contrib. Miner. Petrol.* 75, 279–89.

Huppert, H. E. and R. S. J. Sparks 1980b. Restrictions on the composition of mid-ocean ridge basalts: a fluid dynamical investigation. *Nature* 286, 46–8.

Huppert, H. E. and R. S. J. Sparks 1984. Double-diffusive convection due to crystallisation in magmas. *A. Rev. Earth Planet. Sci.* 12, 11–37.

Huppert, H. E., R. S. J. Sparks and J. S. Turner 1982a. Effects of volatiles on mixing in calc-alkaline magma systems. *Nature* 297, 554–7.

Huppert, H. E., J. B. Shepherd, H. Sigurdsson and R. S. J. Sparks 1982b. On lava dome growth, with application to the 1979 lava extrusion of the Soufriere of St. Vincent. *J. Volcanol. Geotherm. Res.* 14, 199–222.

Huppert, H. E., R. S. J. Sparks, J. S. Turner and N. T. Arndt 1984. Emplacement and cooling of komatiite lavas. *Nature* 309, 19–22.

Hussong, D. M., S. Uyeda *et al.* 1981. *Initial reports of the Deep-Sea Drilling Project.* Vol. 60. Washington, DC: US Government Printing Office.

Hutchinson, R. W. 1973. Volcanogenic sulphide deposits and their metallogenic significance. *Econ. Geol.* 68, 1223–45.

Hutchinson, R. W. 1980. Massive base metal sulphide deposits as guides to tectonic evolution. In D. W. Strangway (1980), 659–81.

Hutchinson, R. W. 1984. Hydrothermal concepts — the old and the new. *Econ. Geol.* 78, 1734–41.

Inman, D. L. 1952. Measures for describing the size distribution of sediments. *J. Sed. Petrol.* 22, 125–45.

International Association of Volcanology and Chemistry of the Earth's Interior (IAVCEI) 1951–present. *Catalogue of active volcanoes of the world, including solfatara fields.* 22 volumes to date. Rome: IAVCEI.

Irvine, T. N. and W. R. A. Baragar 1971. A guide to the chemical classification of the common volcanic rocks. *Can. J. Earth Sci.* 8, 523–48.

Izett, G. A. 1981. Volcanic ash beds: recorders of Upper Cenozoic silicic volcanism in the western United States. *J. Geophys. Res.* 86, 10200–23.

Jakobsson, S. 1978. Environmental factors controlling the palagonitisation of the Surtsey tephra, Iceland. *Bull. Geol. Soc. Denmark* 27, 91–105.

Jakobsson, S. P. and J. G. Moore 1980. Unique hole shows how volcano grew. *Geotimes* April 1980, 14–16.

James, D. E. 1971. Andean crustal and upper mantle structure. *J. Geophys. Res.* 76, 3246–71.

Janda, R. J., K. M. Scott, K. M. Nolan and H. A. Martinson 1981. The 1980 eruptions of Mount St. Helens lahar movement, effects and deposits. In Lipman & Mullineaux (1981), 461–78.

Johnson, A. M. 1970. *Physical processes in geology*. San Francisco: Freeman, Cooper.

Johnson, R. W. (ed.) 1976. *Volcanism in Australasia*. Amsterdam: Elsevier.

Johnson, R. W. (ed.) 1981. *Cooke-Ravian volume of volcanological papers*. Geol. Surv. Papua New Guinea Mem. 10.

Jones, B. L., S. S. W. Chinn and J. C. Brice 1984. Olekele rock avalanche, island of Kauai, Hawaii. *Geology* 12, 209–11.

Jones, J. G. 1966. Intraglacial volcanoes of south-west Iceland and their significance in the interpretation of the form of marine basaltic volcanoes. *Nature* 212, 586–8.

Jones, J. G. 1967a. A lacustrine volcano of central France, and the nature of peperites. *Proc. Geol Assoc.* 80, 177–88.

Jones, J. G. 1967b. Clastic rocks of Espiritu Santo Island, New Hebrides. *Geol Soc. Am. Bull.* 78, 1281–8.

Jones, J. G. 1968. Pillow lava and pahoehoe. *J. Geol.* 76, 485–8.

Jones, J. G. 1969. Intraglacial volcanoes of the Laugarvatn region, southwest Iceland, I. *Q. J. Geol Soc. Lond.* 124, 197–211.

Jones, J. G. 1970. Intraglacial volcanoes of the Laugarvatn region, southwest Iceland, II. *J. Geol.* 78, 127–40.

Jones, J. G. and P. H. H. Nelson 1970. The flow of basalt lava from air into water — its structural expression and stratigraphic significance. *Geol Mag.* 107, 13–21.

Joyce, E. B. 1975. Quaternary volcanism and tectonics in southeastern Australia. In *Quaternary studies*, R. P. Suggate and M. M. Cresswell (eds), 169–76. Wellington: R. Soc. New Zealand.

Kamata, H. and K. Mimura 1983. Flow directions inferred from imbrication in the Handa pyroclastic flow deposit in Japan. *Bull. Volcanol.* 46, 277–2.

Kantha, L. H. 1980. A note on the effect of viscosity on double-diffusive processes. *J. Geophys. Res.* 85, 4398–404.

Kantha, L. H. 1981. 'Basalt fingers' — origin of columnar joints? *Geol Mag.* 118, 251–64.

Karig, D. E. 1970. Ridges and basins of the Tonga-Kermadec island arc system. *J. Geophys. Res.* 75, 239–54.

Karig, D. E. 1971. Structural history of the Mariana island arc system. *Geol Soc. Am. Bull.* 82, 323–44.

Karig, D. E. 1972. Remnant arcs. *Geol Soc. Am. Bull.* 83, 1057–68.

Karig, D. E. 1974. Evolution of arc systems in the western Pacific. *A. Rev. Earth Planet. Sci.* 2, 51–75.

Karig, D. E. 1982. Initiation of subduction zones: implications for arc evolution and ophiolite development. In Leggett (1982), 563–76.

Karig, D. E. and G. F. Sharman 1975. Subduction and accretion in trenches. *Geol Soc. Am. Bull.* 86, 377–89.

Karig, D. E., J. C. Ingle *et al.* 1975. *Initial reports of the Deep-Sea Drilling Project*. Vol. 31. Washington, DC: US Government Printing Office.

Katili, J. A. 1973. On fitting certain geological and geophysical features of the Indonesian island arc to the new global tectonics. In *The Western Pacific: island arcs, marginal seas, geochemistry*, P. J. Coleman (ed.), 287–305. University of Western Australia Press.

Kato, I., I. Muroi, T. Yamazaki and M. Abe 1971. Subaqueous pyroclastic flow deposits in the Upper Donzurubo Formation, Nijosan district, Osaka, Japan. *J. Geol Soc. Jpn* 77, 193–206.

Katsui, Y. 1959. On the Shikotsu pumice-fall deposit. *Bull. Volcanol. Soc. Jap.* 4, 33–48.

Kawachi, Y., I. J. Pringle and D. S. Coombs 1983. Pillow lavas of the Eocene Oamaru volcano, North Otago. Tour Bh 3. *Pacific Sci. Congr., Dunedin, New Zealand*.

Kent, D. V., D. Ninkovich, T. Pescatore and R. S. J. Sparks 1981. Palaeomagnetic determination of emplacement temperature of Vesuvius A.D. 79 pyroclastic deposits. *Nature* 290, 393–6.

Kent, P. E. 1966. The transport mechanism of catastrophic rock falls. *J. Geol.* 74, 79–83.

Kieffer, G. 1971. Aperçu sur la morphologie des régions volcaniques du Massif Central. In: *Symposium Jean Jung: Géologie, géomorphologie et structure profonde du Massif Central français*, Clermont-Ferrand, 479–510.

Kienle, J. and G. E. Shaw 1979. Plume dynamics, thermal energy and long-distance transport of vulcanian eruption clouds from Augustine volcano, Alaska. *J. Volcanol. Geotherm. Res.* 6, 139–64.

Kienle, J., P. R. Kyle, S. Self, R. J. Motyka and V. Lorenz 1980. Ukinrek maars, Alaska, I. April 1977 eruption sequence, petrology and tectonic setting. *J. Volcanol. Geotherm. Res.* 7, 11–37.

King, B. C. 1970. Vulcanicity and rift tectonics in East Africa. In *African magmatism and tectonics*, T. W. Clifford and I. G. Gass (eds), 263–83. Edinburgh: Oliver & Boyd.

Klein, G. de V. 1975. Sedimentary tectonics in southwest Pacific marginal basins based on leg 30 Deep Sea Drilling Project cores from the South Fiji, Hebrides, and Coral Sea Basins. *Geol Soc. Am. Bull.* 86, 1012–8.

Klein, G. de V. and Y. Lee 1984. A preliminary assessment of geodynamic controls on depositional systems and sandstone diagenesis in back-arc basins, western Pacific Ocean. *Tectonophysics* 102, 119–52.

Kokelaar, B. P. 1982. Fluidisation of wet sediments

during the emplacement and cooling of various igneous bodies. *J. Geol Soc. Lond.* **139**, 21–33.

Kokelaar, B. P. 1983. The mechanism of surtseyan volcanism. *J. Geol Soc. Lond.* **140**, 939–44.

Kokelaar, B. P. and M. F. Howells (eds) 1984. *Marginal basin geology: volcanic and associated sedimentary and tectonic processes in modern and ancient marginal basins.* Geol Soc. Lond. Spec. Publn, no. 16.

Kokelaar, B. P., M. F. Howells, R. E. Bevins and R. A. Roach 1984. Volcanic and associated sedimentary and tectonic processes in the Ordovician marginal basin of Wales: a field guide. In Kokelaar & Howells (1984), 291–322.

Kokelaar, B. P., R. E. Bevins and R. A. Roach 1985. Submarine silicic volcanism and associated sedimentary and tectonic processes, Ramsey Island, SW Wales. *J. Geol Soc. Lond.* **142**, 591–613.

Komar, P. D. 1970. The competence of turbidity currents. *Geol Soc. Am. Bull.* **81**, 1555–62.

Kono, Y. and Y. Osima 1971. Numerical experiments on the welding processes in pyroclastic flow deposits. *Bull. Volc. Soc. Jpn* **16**, 1–14.

Korgen, B. J. 1972. *Geological oceanography — student notes to 35 mm colour slide set.* New York: Harper & Row.

Korringa, M. K. 1973. Linear vent area of the Soldier Meadow Tuff, an ash-flow sheet in northwestern Nevada. *Geol Soc. Am. Bull.* **84**, 3849–66.

Kouda, R. and H. Koide 1978. Ring structures, resurgent cauldron, and ore deposits in the Hokuroku volcanic field, northern Akita, Japan. *Min. Geol.* **28**, 233–44.

Kozu, S. 1934. The great activity of Komagatake 1929. *Mineralog. Petrog. Mitteil.* **45**, 133–74.

Krishnamurthy, P. and G. R. Udas 1981. Regional geochemical characters of the Deccan Trap lavas and their genetic implications. In K. V. Subbarao and R. N. Sukheswala (1981), 394–418.

Kuenzi, W. D., O. H. Horst and R. V. McGehee 1979. Effect of volcanic activity in fluvial-deltaic sedimentation in a modern arc-trench gap, southwestern Guatemala. *Geol Soc. Am. Bull.* **90**, 827–38.

Kulm, L. D., J. M. Resig, T. M. Thornburg and H.-J. Schrader 1982. Cenozoic structure, stratigraphy and tectonics of the central Peru forearc. In Leggett (1982), 151–69.

Kunii, D. and O. Levenspiel 1969. *Fluidisation engineering.* New York: Wiley.

Kuno, H. 1941. Characteristics of deposits formed by pumice flows and those by ejected pumice. *Bull. Earthq. Res. Inst. Univ. Tokyo* **19**, 144–8.

Kuno, H., T. Ishikawa, Y. Katsui, Y. Yago, M. Yamasaki and S. Taneda 1964. Sorting of pumice and lithic fragments as a key eruptive and emplacement mechanisms. *Jap. J. Geol. Geogr.* **35**, 223–38.

Kushiro, I. 1976. Changes in viscosity and structure of melt of $NaAlSi_2O_6$ composition at high pressures. *J. Geophys. Res.* **81**, 6347–50.

Kushiro, I. 1978. Density and viscosity of hydrous calc-alkalic andesite magma at high pressures. *Carnegie Inst. Washington Ybk* **77**, 675–77.

Kushiro, I. 1980. Viscosity, density and structure of silicate melts at high pressures, and their petrological applications. In Hargraves (1980), 93–120.

Kushiro, I., H. S. Yoder Jr and B. O. Mysen 1976. Viscosities of basalt and andesite melts at high pressures. *J. Geophys. Res.* **81**, 6351–6.

La Croix, A. 1903. Sur les principaux résultats de la mission de la Martinique. *C.R. Acad. Sci. Paris* **135**, 871–76.

La Croix, A. 1904. *La Montagne Pelée et ses éruptions.* Paris: Masson.

La Croix, A. 1930. Remarques sur les matériaux de projection des volcans et sur la génèse des roches pyroclastiques qu'ils constituent. *Soc. Geol. Fr., Livre Jubliaire Centenaire* **2**, 431–72.

Lajoie, J. 1984. Volcaniclastic rocks. In R. G. Walker (1984), 39–52.

Larsson, W. 1936. Vulkanische äsche vom Ausbruch des ChilenischenVolkans Quizapú. *Bull. Geol Inst. Univ. Upsala* **26**, 27–52.

Laughlin, A. W., M. J. Aldrich and D. T. Vaniman 1983. Tectonic implications of mid-Tertiary dykes in west-central New Mexico. *Geology* **11**, 45–8.

Laughton, A. S., W. A. Berggren *et al.* 1972. *Initial reports of the Deep Sea Drilling Project Vol. XII.* Washington, DC: US Government Printing Office.

Ledbetter, M. T. 1985. Tephrochronology of marine tephra adjacent Central America. *Geol Soc. Am. Bull.* **96**, 77–82.

Ledbetter, M. T. and R. S. J. Sparks 1979. Duration of large-magnitude explosive eruptions deduced from graded bedding in deep-sea ash layers. *Geology* **7**, 240–4.

Leeder, M. R. 1982. *Sedimentology. Process and product.* London: George Allen & Unwin.

Leggett, J. K. (ed.) 1982. *Trench-forearc geology: sedimentation and tectonics on modern and ancient active plate margins.* Geol Soc. Lond. Publn, no. 10.

Lewis, J. V. 1914. Origin of pillow lavas. *Geol Soc. Am. Bull.* **25**, 591–654.

Leys, C. A. 1982. *Volcanic and sedimentary processes in phreatomagmatic volcanoes.* Unpubl. Ph.D. thesis, University of Leeds.

Lipman, P. W. 1965. *Chemical comparison of glassy and crystalline volcanic rocks.* US Geol Surv. Bull., no. 1201–D.

Lipman, P. W. 1967. Mineral and chemical variations within an ash-flow sheet from Aso Caldera, South Western Japan. *Contrib. Miner. Petrol.* **16**, 300–27.

Lipman, P. W. 1975. *Evolution of the Platoro caldera complex and related volcanic rocks, southeastern San Juan Mountains, Colorado.* US Geol Surv. Prof. Pap., no. 852.

Lipman, P. W. 1976. Caldera-collapse breccias in the western San Juan Mountains, Colorado. *Geol Soc. Am. Bull.* 87, 1397–1410.

Lipman, P. W. 1980. Rates of volcanic activity along the southwest rift zone of Mauna Loa volcano, Hawaii. *Bull. Volcanol.* 43, 703–25.

Lipman, P. W. 1984. The roots of ash flow calderas in western North America: windows into the tops of granitic batholiths. *J. Geophys. Res.* 89, 8801–41.

Lipman, P. W. and R. L. Christiansen 1964. *Zonal features of an ash-flow sheet in the Piapi Canyon Formation, southern Nevada,* B74–8. US Geol Surv. Prof. Pap., no. 501B.

Lipman, P. W. and D. R. Mullineaux (eds) 1981. *The 1980 eruptions of Mount St. Helens, Washington.* US Geol Surv. Prof. Pap., no. 1250.

Lipman, P. W. and R. L. Christiansen and J. T. O'Connor 1966. *A compositionally zoned ash-flow sheet in southern Nevada.* US Geol Surv. Prof. Pap., no. 524–F.

Lipman, P. W., S. Self and G. Heiken (eds) 1984. *Calderas and associated igneous rocks.* J. Geophys. Res. 89 B10, 8219–342.

Lirer, L., T. Pescatore, B. Booth and G. P. L. Walker 1973. Two plinian pumice-fall deposits from Somma Vesuvius, Italy. *Geol Soc. Am. Bull.* 84, 759–72.

Lisitzin, A. P. 1962. Bottom sediments of the Antarctic. *Am. Geophys. Un. Geophys. Mon.* 7, 81–8.

Lisle, R. J. 1977. Estimation of the tectonic strain ratio from the mean shape of deformed elliptical markers. *Geol. Mijnbouw* 56, 140–4.

Lock, B. E. 1972. A Lower Palaeozoic rheo-ignimbrite from White Bay, Newfoundland. *Can. J. Earth Sci.* 9, 1495–503.

Lockwood, J. P. and P. W. Lipman 1980. Recovery of datable charcoal beneath young lavas: lessons from Hawaii. *Bull. Volcanol.* 43, 609–15.

Lofgren, G. 1970. Experimental devitrification rates of rhyolitic glass. *Geol Soc. Am. Bull.* 81, 553–60.

Lofgren, G. 1971a. Spherulitic textures in glassy and crystalline rocks. *J. Geophys. Res.* 76, 5635–48.

Lofgren, G. 1971b. Experimentally produced devitrification textures in natural rhyolitic glass. *Geol Soc. Am. Bull.* 82, 111–24.

Loney, R. A. 1968. Flow structure and composition of the Southern Coulée, Mono craters, California — a pumiceous rhyolite flow. In Coats *et al.* (1968), 153–210.

Lonsdale, P. and R. Batiza 1980. Hyaloclastite and lava flows on young seamounts examined with a submersible. *Geol Soc. Am. Bull.* 91, 545–54.

Loomis, B. F. 1948. *Pictorial history of the Lassen volcano.* National Park Service, California: Department of the Interior.

Lorenz, V. 1973. On the formation of maars. *Bull. Volcanol.* 37, 183–204.

Lorenz, V. 1974. Vesiculated tuffs and associated features. *Sedimentology* 21, 273–91.

Lorenz, V., A. R. McBirney and H. Williams 1970. *An investigation of volcanic depressions. Part III: Maars, tuff-rings, tuff-cones and diatremes.* NASA Progress Rep. NGR–38–003–012.

Lowder, G. G. and I. S. E. Carmichael 1970. The volcanoes and caldera of Talasea, New Britain: geology and petrology. *Geol Soc. Am. Bull.* 81, 17–38.

Lowe, D. R. 1975. Water escape structures in coarse-grained sediments. *Sedimentology* 22, 157–204.

Lowe, D. R. 1976. Grain flow and grain flow deposits. *J. Sed. Petrol.* 46, 188–99.

Lowe, D. R. 1979. Sediment gravity flows: their classification and some problems of application to natural flows and deposits. In Doyle & Pilkey (1979), 75–82.

Lowe, D. R. 1982. Sediment gravity flows: II, depositional models with special reference to the deposits of high density turbidity currents. *J. Sed. Petrol.* 52, 279–97.

Lowman, R. D. W. and T. W. Bloxam 1981. The petrology of the Lower Palaeozoic Fishguard Volcanic Group and associated rocks E. of Fishguard, N. Pembrokeshire (Dyfed), South Wales. *J. Geol Soc. Lond.* 138, 47–68.

Luyendyk, B. P. and K. C. Macdonald 1977. Physiography and structure of the inner floor of the Famous rift valley: observations with a deep-towed instrument package. *Geol Soc. Am. Bull.* 88, 648–63.

McBirney, A. R. 1963. Factors governing the nature of submarine volcanism. *Bull. Volcanol.* 26, 455–69.

McBirney, A. R. 1971. Oceanic volcanism: a review. *Geophys. Space Phys. Rev.* 9, 523–56.

McBirney, A. R. 1973. Factors governing the intensity of explosive andesitic eruptions. *Bull. Volcanol.* 37, 443–53.

McBirney, A. R. 1980. Mixing and unmixing of magmas. *J. Volcanol. Geotherm. Res.* 7, 357–71.

McBirney, A. R. and T. Murase 1970. Factors governing the formation of pyroclastic rocks. *Bull. Volcanol.* 34, 372–84.

McBirney, A. R. and T. Murase 1984. Rheological properties of magmas. *A. Rev. Earth Planet. Sci.* 12, 337–57.

McBirney, A. R. and R. M. Noyes 1979. Crystallisation and layering of the Skaergaard intrusion. *J. Petrol.* 20, 487–554.

Macdonald, G. A. 1967. Forms and structures of extrusive basalt rocks. In *The Poldervaart Treatise on rocks of basaltic composition*, H. H. Hess and A. Poldervaart (eds), Vol. 1, 1–61. New York: Interscience.

Macdonald, G. A. 1968. Composition and origin of Hawaiian lavas. *Geol Soc. Am. Bull.* 116, 477–522.

Macdonald, G. A. 1972. *Volcanoes*. Englewood Cliffs, New Jersey: Prentice-Hall.

Macdonald, G. A. and A. T. Abbott 1979. *Volcanoes in the sea — the geology of Hawaii*. Honolulu: University Press of Hawaii.

Macdonald, G. A. and A. Alcaraz 1956. Nuées ardentes of the 1948–1953 eruption of Hibok-Hibok. *Bull. Volcanol.* 18, 169–78.

Macdonald, K. C. 1982. Mid-ocean ridges: fine-scale tectonic, volcanic and hydrothermal processes within the plate boundary zone. *A. Rev. Earth Planet. Sci.* 10, 155–90.

Macdonald, K. C. and B. P. Luyendyk 1977. Deep-tow studies of the structure of the mid-Atlantic Ridge crest near Lat. 37°N. *Geol Soc. Am. Bull.* 88, 621–36.

McDougall, I. 1974. Potassium argon ages from lavas of the Hawaiian Islands. *Geol Soc. Am. Bull.* 75, 107–28.

McGeary, S., A. Nur and Z. Ben-Avraham 1985. Spatial gaps in arc volcanism: the effect of collision or subduction of oceanic plateaus. *Tectonophysics* 119, 195–221.

McGetchin, T. R. and M. Settle 1975. Cinder cone separation distances: implications for the depth of formation of gabbroic xenoliths. *EOS* 56, 1070.

McGetchin, T. R. and W. G. Ullrich 1973. Xenoliths in maars and diatremes with inferences for the Moon, Mars and Venus. *J. Geophys. Res.* 78, 1833–53.

McGetchin, T. R., M. Settle and B. H. Chouet 1974. Cinder cone growth modeled after Northeast crater, Mt. Etna, Sicily. *J. Geophys. Res.* 79, 3257–72.

MacKenzie, W. S., C. H. Donaldson and C. Guilford 1982. *Atlas of igneous rocks and their textures*. London: Longman.

Macpherson, D. W. 1984. A model for predicting the volumes of vesicles in submarine basalts. *J. Geol.* 92, 73–82.

Macpherson, G. J. 1983. The Snow Mountain Volcanic Complex: an on-land seamount in the Franciscan terrain, California. *J. Geol.* 91, 73–92.

McPhie, J. 1983. Outflow ignimbrite sheets from Late Carboniferous calderas, Currabubula Formation, New South Wales, Australia. *Geol Mag.* 120, 487–503.

McTaggart, K. C. 1960. The mobility of nuées ardentes. *Am. J. Sci.* 258, 369–82.

Mahood, G. A. 1980. Geological evolution of a Pleistocene Rhyolite Center — Sierra La Primavera, Jalisco, Mexico. *J. Volcanol. Geotherm. Res.* 8, 199–230.

Mahood, G. A. and W. Hildreth 1983. Nested calderas and trapdoor uplift at Pantelleria, Strait of Sicily. *Geology II*, 722–726.

Marsh, B. D. 1981. On the crystallinity, probability of occurrence, and rheology of lava and magma. *Contrib. Mineral. Petrol.* 78, 85–98.

Marshall, P. 1935. Acid rocks of the Taupo-Rotorua volcanic district. *Trans. R. Soc. N.Z.* 64, 323–66.

Martin, D. P. and W. I. Rose Jr 1981. Behavioural patterns of Fuego volcano, Guatemala. *J. Volcanol. Geotherm. Res.* 10, 67–81.

Mathews, W. H. 1947. 'Tuyas.' Flat topped volcanoes in northern British Columbia. *Am. J. Sci.* 245, 560–70.

Mathisen, M. E. and C. F. Vondra 1983. The fluvial and pyroclastic deposits of the Cagayan Basin, northern Luzon, Philippines — an example of non-marine volcaniclastic sedimentation in an interarc basin. *Sedimentology* 30, 369–92.

Mattson, P. H. and W. Alvarez 1973. Base surge deposits in Pleistocene volcanic ash near Rome. *Bull. Volcanol.* 37, 553–72.

Menard, H. W. 1964. *Marine geology of the Pacific*. New York: McGraw-Hill.

Menard, H. W. 1969. Growth of drifting volcanoes. *J. Geophys. Res.* 74, 4827–37.

Miall, A. D. (ed.) 1978. *Fluvial sedimentology*. Can. Soc. Petrol. Geol. Mem., no. 5.

Michel, R. 1948. Etude géologique du plateau de Gergoria. *Rev. Sci. Nat. Auvergne* 14, 1–68.

Michel-Levy, A. 1890. Compte rendu de l'excursion du 16 Septembre à Gergorie et Veyre-Monton. *Bull. Soc. Geol. Fr.* 18, 891–7.

Middleton, G. V. 1966. Experiments on density and turbidity currents, I. *Can. J. Earth Sci.* 3, 523–46.

Middleton, G. V. and M. A. Hampton 1976. Subaqueous sediment transport and deposition by sediment gravity flows. In *Marine sediment transport and environmental management*, D. J. Stanley and D. J. P. Swift (eds), 197–218. New York: Wiley.

Middleton, G. V. and J. B. Southard 1978. *Mechanics of sediment movement*. SEPM (Eastern Section) short course, no. 3.

Miller, T. P. and R. L. Smith 1977. Spectacular mobility of ash flows around Aniakchak and Fisher Calderas, Alaska. *Geology* 5, 173–6.

Mills, H. H. 1976. Estimated erosion rates on Mount Rainier, Washington. *Geology* 4, 401–6.

Minakami, T. 1950. On explosive activities of andesitic volcanoes and their forerunning phenomena. *Bull. Volcanol.* 10, 59–87.

Minakami, T., T. Ishikawa and K. Yagi 1951. The 1944 eruption of Volcano Usu in Hokkaido, Japan. *Bull. Volcanol.* 11, 45–160.

Mitchell, A. H. G. 1970. Facies of an early Miocene volcanic arc, Malekula Island, New Hebrides. *Sedimentology* 14, 201–43.

Miyashiro, A. 1972. Metamorphism and related magmatism in plate tectonics. *Am. J. Sci.* 272, 629–56.

Miyashiro, A. 1974. Volcanic rock series in island arcs and active continental margins. *Am. J. Sci.* 274, 321–55.

Moberly, R., G. L. Shepherd and W. T. Coulbourn 1982. Forearc and other basins, continental margin of northern and southern Peru and adjacent Ecuador and Chile. In Leggett (1982), 171–89.

Mohr, P. 1983. Ethiopian flood basalt province. *Nature* 303, 577–84.

Mohr, P. A. and C. A. Wood 1976. Volcano spacings and lithospheric attenuation in the Eastern Rift of Africa. *Earth Planet. Sci. Letts.* 33, 126–44.

Moore, D. G. 1978. Submarine slides. In Voight (1978), 563–604.

Moore, G. F. and D. E. Karig, 1980. Structural geology of Nias Island, Indonesia: implications for subduction zone tectonics. *Am. J. Sci.* 280, 193–223.

Moore, J. C., J. S. Watkins, T. H. Shipley, S. B. Bachman, F. W. Beghtel, A. Butt, B. M. Didyk, J. K. Leggett, N. Lundberg, K. J. McMillen, N. Niitsuma, L. E. Shephard, J.-F. Stephan and H. Stradner 1979. Progressive accretion in the Middle America Trench, southern Mexico. *Nature* 281, 638–42.

Moore, J. G. 1964. *Giant submarine landslides on the Hawaiian ridge*, 95–8. US Geol Surv. Res. Pap., 501–D.

Moore, J. G. 1966. *Rate of palagonitisation of submarine basalt adjacent to Hawaii*, D163–71. US Geol Surv. Prof. Pap., no. 550–D.

Moore, J. G. 1967. Base surge in recent volcanic eruptions. *Bull. Volcanol.* 30, 337–63.

Moore, J. G. 1975. Mechanism of formation of pillow lava. *Am. Sci.* 63, 269–77.

Moore, J. G. and R. S. Fiske 1969. Volcanic substructure inferred from dredge samples and ocean bottom photographs, Hawaii. *Geol Soc. Am. Bull.* 80, 1191–202.

Moore, J. G. and W. G. Melson 1969. Nuées ardentes of the 1968 eruption of Mayon Volcano, Philippines. *Bull. Volcanol.* 33, 600–20.

Moore, J. G. and D. L. Peck 1962. Accretionary lapilli in volcanic rocks of the western United States. *J. Geol.* 70, 182–93.

Moore, J. G. and T. W. Sisson 1981. Deposits and effects of the May 18 pyroclastic surge. In Lipman & Mullineaux (1981), 421–38.

Moore, J. G., K. Nakamura and A. Alcaraz 1966. The 1965 eruption of Taal volcano. *Science* 155, 955–60.

Moore, J. G., R. L. Phillips, R. W. Grigg, D. W. Peterson and D. A. Swanson 1973. Flow of lava into the sea, 1969–71, Kilauea volcano, Hawaii. *Geol Soc. Am. Bull.* 84, 537–46.

Moorhouse, W. W. 1970. *A comparative atlas of textures of Archean and younger volcanic rocks.* Geol. Assoc. Can. Spec. Pap., no. 8.

Moores, E. M. 1982. Origin and emplacement of ophiolites. *Rev. Geophys. Space Phys.* 20, 735–60.

Morton, B. R., G. Taylor and J. S. Turner 1956. Turbulent gravitational convection from maintained and instantaneous sources. *Proc. R. Soc. Lond. (A)* 234, 1–23.

Mullineaux, D. R. and D. R. Crandell 1962. Recent lahars from Mount St. Helens, Washington. *Geol Soc. Am. Bull.* 73, 855–70.

Murai, I. 1961. A study of the textural characteristics of pyroclastic flow deposits in Japan. *Bull. Earthq. Res. Inst.* 39, 133–254.

Murase, T. 1962. Viscosity and related properties of volcanic rocks at 800° to 1400°C. *Hokkaido Univ. Fac. Sci. J., Ser. 7* 1, 487–584.

Murase, T. and A. R. McBirney 1973. Properties of some common igneous rocks and their melts at high temperatures. *Geol Soc. Am. Bull.* 84, 3563–92.

Murata, K. J., C. Dondoli and R. Saenz 1966. The 1963–65 eruption of Irazu volcano Costa Rica. (The period from March 1963–October 1964.) *Bull. Volcanol.* 29, 765–96.

Mutti, E. 1965. Submarine flood tuffs (ignimbrites) associated with turbidites in Oligocene deposits of Rhodes Island (Greece). *Sedimentology* 5, 265–88.

Mutti, E. and F. Ricci Lucchi 1978. Turbidites of the Northern Apennines. *Int. Geol Rev.* 20, 125–66.

Mysen, B. O. 1977. The solubility of H_2O and CO_2 under predicted magma genesis conditions and some petrological and geophysical implications. *Rev. Geophys. Space Phys.* 15, 351–61.

Mysen, B. O., D. Virgo and F. A. Seifert 1982. The structure of silicate melts: implications for chemical and physical properties of natural magma. *Rev. Geophys. Space Phys.* 20, 353–83.

Naeser, C. W., N. D. Briggs, J. D. Obradovich and G. A. Izett 1981. Geochronology of Quaternary tephra deposits. In Self & Sparks (1981), 13–47.

Nairn, I. A. 1972. Rotoehu Ash and the Rotoiti Breccia formation, Taupo volcanic zone, New Zealand. *N.Z. J. Geol. Geophys.* 15, 251–61.

Nairn, I. A. 1976. Atmospheric shock waves and condensation clouds from Ngauruhoe explosive eruptions. *Nature* 259, 190–2.

Nairn, I. A. 1979. Rotomahana– Waimangu eruption 1886: base surge and basalt magma. *N.Z. J. Geol. Geophys.* 22, 363–78.

Nairn, I. A. 1981. *Some studies of the geology, volcanic history and geothermal resources of the Okataina volcanic centre, Taupo volcanic zone, New Zealand.* Unpubl. Ph.D. thesis, Victoria University.

Nairn, I. A. and S. Self 1978. Explosive eruptions and pyroclastic avalanches from Ngauruhoe in February 1975. *J. Volcanol. Geotherm. Res.* 3, 39–60.

Nairn, I. A. and S. Wiradiradja 1980. Late Quaternary hydrothermal explosion breccias at Kawerau geothermal field, New Zealand. *Bull. Volcanol.* 43, 1–14.

Nairn, I. A., C. P. Wood and C. A. Y. Hewson 1979. Phreatic eruptions of Ruapehu: April 1975. *N.Z. J. Geol. Geophys.* 22, 155–73.

Nakamura, K. 1962. Volcano-stratigraphic study of Oshima volcano, Izu. *Bull. Earthq. Res. Inst. Univ. Tokyo.* 42, 649–728.

Nakamura, K. 1977. Volcanoes as possible indicators of tectonics stress orientation — principle and proposal. *J. Volcanol. Geotherm. Res.* 2, 1–16.

Nakamura, Y. 1978. Geology and petrology of Bandai and Nekoma volcanoes. *Tohuku Univ. Sci. Rep., Ser. 3* 14, 67–119.

Naldrett, A. K. and A. J. Macdonald 1980. Tectonic settings of Ni-Cu sulphide ores: their importance in genesis and exploration. In D. W. Strangway (ed.), 631–57.

Nardin, T. R., F. J. Hein, D. S. Gorsline and B. D. Edwards 1979. A review of mass movement processes, sediment and acoustic characteristics, and contrasts in slope and base-of-slope systems versus canyon-fan-basin floor systems. In Doyle & Pilkey (1979), 61–73.

Nathan, S. (ed.) 1976. Volcanic and geothermal geology of the central North Island, New Zealand. *25th Int. Geol. Congr. Excurs. Guide* 55A & 56C.

Neall, V. E. 1979. *Sheets P19, P20 and P21 New Plymouth, Egmont and Manaia (1st ed). Geological Map of New Zealand 1 : 500,000*. 3 maps and notes (36 pp.). Wellington: N.Z. Dept Sci. Ind. Res.

Nelson, S. A. 1981. The possible role of thermal feedback in the eruption of siliceous magmas. *J. Volcanol. Geotherm. Res.* 11, 127–37.

Nelson, S. A. and I. S. E. Carmichael 1979. Partial molar volumes of oxide components in silicate liquids. *Contrib. Miner. Petrol.* 71, 117–24.

Nesbitt, R. W., J. Bor-ming and A. C. Purvis 1982. Komatiites: an early Precambrian phenomenon. *J. Volcanol. Geotherm. Res.* 14, 31–45.

Neumann, E. R. and I. B. Ramberg (eds) 1977. *Petrology and geochemistry of continental rifts*. Holland: D. Reidel.

Neumann van Padang, M. 1933. De uitbarsting van den Merapi (Midden Java) in de jaren 1930–31. *Ned. Indes. Dienst Mijnbouwk. Vulkan. Seism. Mededel.*, no. 12.

Newhall, C. G. and W. G. Melson 1983. Explosive activity associated with the growth of volcanic domes. *J. Volcanol. Geotherm. Res.* 17, 111–31.

Newmark, R. L., R. N. Anderson, D. Moos and M. D. Zoback 1985. Structure, porosity and stress regime of the upper oceanic crust: sonic and ultrasonic logging of DSDP Hole 504B. *Tectonophysics* 118, 1–42.

Niem, A. R. 1977. Mississippian pyroclastic flow and ash-fall deposits in the deep-marine Ouachita flysch basin, Oklahoma and Arkansas. *Geol Soc. Am. Bull.* 88, 49–61.

Ninkovich, D. and B. C. Heezen 1965. Santorini tephra. In: *Submarine geology and geophysics*, W. F. Whittand and R. Bradshaw (eds), 413–53. Proc. 17th Symp. Colston Res. Soc. London: Butterworths.

Ninkovich, D. and N. J. Shackleton 1975. Distribution, stratigraphic position and age of ash layer 'L', in the Panama Basin region. *Earth Planet. Sci. Lett.* 27, 20–34.

Ninkovich, D., R. S. J. Sparks and M. J. Ledbetter 1978. The exceptional magnitude and intensity of the Toba eruption, Sumatra: an example of the use of deep-sea tephra layers as a geological tool. *Bull. Volcanol.* 41, 286–98.

Ninkovich, D., N. D. Opdyke, B. C. Heezen and J. H. Foster 1966. Paleomagnetic stratigraphy, rates of deposition and tephro-chronology in the North Pacific deep-sea sediments. *Earth Planet. Sci. Lett.* 1, 476–92.

Nixon, G. T. 1982. The relationship between Quaternary volcanism in central Mexico and the seismicity and structure of subducted ocean lithosphere. *Geol Soc. Am. Bull.* 93, 514–23.

Noble, D. C. 1967. Sodium, potassium and ferrous iron contents of some secondarily hydrated natural silicic glasses. *Am. Mineral.* 52, 230–85.

Noble, D. C., E. H. Mckee, E. Farrer and U. Petersen 1974. Episodic Cenozoic volcanism and tectonism in Andes of Peru. *Earth & Planet. Sci. Letts.* 21, 213–20.

Nur, A. and Z. Ben-Avraham 1983. Volcanic gaps due to oblique consumption of aseimic ridges. *Tectonophysics* 99, 355–62.

O'Hara, M. J. and R. E. Matthews 1981. Geochemical evolution in an advancing, periodically replenished, periodically tapped, continuously fractionated magma chamber. *J. Geol Soc. Lond.* 138, 237–77.

Ohmoto, H. 1978. Submarine calderas: a key to the formation of volcanogenic massive sulphide deposits? *Mining Geol. (Jap)* 28, 219–31.

Ohmoto, H. and B. J. Skinner (eds) 1983. *The Kuroko and related volcanogenic massive sulphide deposits*. Econ. Geol. Mon. 5.

Ohmoto, H. and T. Takahashi 1983. Geologic setting of the Kuroko deposits, Japan. III Submarine calderas and Kuroko genesis. In Ohmoto & Skinner (1983), 39–54.

Ollier, C. D. 1967a. Maars, their characteristics, varieties and definition. *Bull. Volcanol.* 31, 45–73.

Ollier, C. D. 1967b. Landforms of the Newer Volcanic Province of Victoria. In *Landform studies from Australia and New Guinea*. J. N. Jennings and J. A. Mabbutt (eds), 315–39. Canberra: Aust. Natn. Univ. Press.

Ollier, C. D. 1969. *Volcanoes*. Canberra: Australian National University Press.

Ollier, C. D. and M. C. Brown 1965. Lava caves of Victoria. *Bull. Volcanol.* 28, 215–29.

Ollier, C. D. and E. B. Joyce 1964. Volcanic physiography of the Western Plains of Victoria. *Proc. R. Soc. Victoria* 77, 357–766.

Orsi, G. and M. F. Sheridan 1986, The Green Tuff of Pantelleria; an example of rheoignimbrite. (abstr.). *Abstr. IAVCEI Int. Volcanol. Cong, New Zealand.*

Packham, G. H. 1968. The Lower and Middle Palaeozoic stratigraphy and sedimentary tectonics of the Sofala–Hill End–Euchareena region N.S.W. *Proc. Linn. Soc. N.S.W.* 93, 111–63.

Pantin, H. M. and A. R. Duncan 1969. Evidence for submarine geothermal activity in the Bay of Plenty. *N.Z. J. Marine Freshwater Res.* 3, 473–505.

Parsons, W. H. 1969. Criteria for the recognition of volcanic breccias: Review. *Geol Soc. Am. Mem.* 115, 263–304.

Paton, T. R. 1978. *The formation of soil material.* London: Allen & Unwin.

Peacock, M. A. and R. R. Fuller 1928. Chlorophaeite, sideromelane and palagonite from the Columbia River Plateau. *Am. Mineral.* 13, 360–83.

Pearce, J. A. and J. R. Cann 1973. Tectonic setting of basic volcanic rocks determined using trace element analyses. *Earth Planet. Sci. Lett.* 19, 290–300.

Peckover, R. S., D. J. Buchanan and D. Ashby 1973. Fuel-coolant interactions in submarine vulcanism. *Nature* 245, 307–8.

Penrose Conference 1972. Ophiolites. *Geotimes* 12, 24–5.

Perret, F. A. 1937. *The eruption of Mt. Pelée 1929–1932.* Carnegie Inst. Washington Publn, no. 458.

Peterson, D. W. and D. A. Swanson 1974. Observed formation of lava tubes during 1970–71 at Kilauea volcano, Hawaii. *Stud. Speleol.* 2, 209–23.

Peterson, D. W. and R. I. Tilling 1980. Transition of basaltic lava from pahoehoe to aa, Kilauea volcano, Hawaii: field observations and key factors. *J. Volcanol. Geotherm. Res.* 7, 271–93.

Pettijohn, F. J. 1975. *Sedimentary rocks.* New York: Harper & Row.

Pettijohn, F. J., P. E. Potter and R. Siever 1972. *Sand and sandstones.* New York: Springer.

Phillips, W. J. 1972. Hydraulic fracturing and mineralisation. *J. Geol Soc. Lond.* 128, 337–59.

Pichler, H. 1965. Acid hyaloclastites. *Bull. Volcanol.* 28, 293–310.

Pichler, H. and W. L. Friedrich 1980. Mechanism of the Minoan eruption of Santorini. In Doumas (1980), 15–30.

Pichler, H. and S. Kussmaul 1980. Comments on the geological map of the Santorini Islands. In Doumas (1980), Vol. II, 413–27.

Pike, R. J. and G. D. Clow 1981. *Revised classification of terrestrial volcanoes and a catalog of topographic dimensions with new results on edifice volume.* US Geol Surv. Open File Rep., OF 81–1038.

Pilger, R. H. 1982. The origin of hotspot traces: evidence from eastern Australia. *J. Geophys. Res.* 37(B3), 1825–34.

Pinkerton, H. and R. S. J. Sparks 1976. The 1975 subterminal lavas, Mount Etna: a case history of the formation of a compound lava field. *J. Volcanol. Geotherm. Res.* I, 167–82.

Pinkerton, H. and R. S. J. Sparks 1978. Field measurements of the rheology of lava. *Nature* 276, 383–6.

Pirrson, L. V. 1915. The microscopical characters of volcanic tuffs — a study for students. *Am. J. Sci. (4th Ser.)* 40, 191–211.

Plafker, G. and G. E. Ericksen 1978. Nevados Huascarán avalanches, Peru. In Voight (1978), 277–314.

Porter, S. C. 1972. Distribution, morphology and size frequency of cinder cones on Mauna Kea volcano, Hawaii. *Geol Soc. Am. Bull.* 83, 3607–12.

Porter, S. C. 1973. Stratigraphy and chronology of Late Quaternary tephra along the south rift zone of Mauna Kea volcano, Hawaii. *Geol Soc. Am. Bull.* 84, 1923–40.

Porter, S. C. 1979. Quaternary stratigraphy and chronology of Mauna Kea volcano, Hawaii: 38,000 yr. record of mid-Pacific volcanism and ice-cap glaciation. *Geol Soc. Am. Bull.* 90, 609–11.

Postma, G. 1983. Water escape structures in the context of a depositional model of a mass flow dominated conglomeratic fan-delta (Abrioja Formation Pliocene, Almeria Basin, SE Spain). *Sedimentology* 30, 91–103.

Potter, P. E. and F. J. Pettijohn 1963. *Atlas and glossary of sedimentary structures.* Berlin: Springer.

Press, F. and R. Siever 1978. *Earth.* New York: W. H. Freeman.

Pyke, D. R., A. J. Naldrett and O. R. Ecktrand 1973. Archaean ultramafic flows in Munro Township, Ontario. *Geol Soc. Am. Bull.* 84, 955–78.

Raam, A. 1968. Petrology and diagenesis of Broughton Sandstone (Permian), Kiama district, New South Wales. *J. Sed. Petrol.* 38, 319–31.

Ragan, D. H. and M. F. Sheridan 1972. Compaction of the Bishop Tuff, California. *Geol Soc. Am. Bull.* 83, 95–106.

Ramberg, I. B. and E. R. Neumann (eds) 1978. *Tectonics and geophysics of continental rifts*, Part 2. Dortrecht: Reidel.

Rampino, M. R., S. Self and R. W. Fairbridge 1979. Can rapid climatic change cause volcanic eruptions? *Science* 206, 826–9.

Ramsay, J. G. 1967. *Folding and fracturing of rocks.* New York: McGraw-Hill.

Raymond, C. F. 1978. Mechanics of glacier movement. In Voight (1978), 793–833.

Rea, W. J. 1974. The volcanic geology and petrology of Montserrat, West Indies. *J. Geol Soc. Lond.* **130**, 341–66.

Reading, H. G. (ed.) 1978. *Sedimentary environments and facies*. Oxford: Blackwell Scientific.

Reynolds, D. L. 1954. Fluidisation as a geological process and its bearing on the problems of intrusive granites. *Am. J. Sci.* **252**, 577–613.

Reynolds, M. A. and J. G. Best 1976. Summary of 1953–57 eruption of Tuluman volcano, Papua New Guinea. In *Volcanism in Australasia*, R. W. Johnson (ed.), 287–96. Amsterdam: Elsevier.

Reynolds, M. A., J. G. Best and R. W. Johnson 1980. 1953–57 eruption of Tuluman volcano: rhyolitic volcanic activity in the northern Bismarck Sea. *Geol Surv. Papua New Guinea Mem.* **7**.

Rhodes, R. C. and E. I. Smith 1972. Distribution and directional fabric of ash-flow sheets in the northwestern Mogollon Plateau, New Mexico. *Geol Soc. Am. Bull.* **83**, 1863–8.

Richards, A. F. 1958. Trans-Pacific distribution of floating pumice from Isla San Benedicto, Mexico. *Deep-Sea Res.* **5**, 29–35.

Richards, A. F. 1959. Geology of the Islas Revillagidedo, Mexico, 1. Birth and development of Volcán Bárcena, Isla San Benedicto (1). *Bull. Volcanol.* **22**, 73–123.

Richards, A. F. 1965. Geology of the Islas Revillagidedo, 3. Effects of erosion on Isla San Benedicto 1952–61 following the birth of Volcán Bárcena. *Bull. Volcanol.* **28**, 381–403.

Richardson, S. 1978. *The geology of southern Lipari, with particular reference to the rhyolite tholoids in the extreme south of the island*. Unpubl. B.Sc. thesis, University of Leeds.

Richet, P., Y. Bottinga, L. Denielou, J. P. Petitet and C. Tequi 1982. Thermodynamic properties of quartz, cristobalite and amorphous SiO_2: drop calorimetry measurements between 1000 and 1800 K and a review from 0 to 2000 K. *Geochim. Cosmochim. Acta* **46**, 2639–58.

Richter, D. H., J. P. Eaton, K. J. Murata, W. U. Ault and H. J. Krivoy 1970. *Chronological narrative of the 1959–60 eruption of Kilauea volcano, Hawaii*. US Geol Surv. Prof. Pap., No. 537.

Rickard, M. J. 1984. Pluton spacing and the thickness of crustal layers in Baja, California. *Tectonophysics* **101**, 167–72.

Rickard, M. J. and P. Ward 1981. Palaeozoic crustal thickness in the southern part of the Lachlan Orogen deduced from volcano and pluton-spacing geometry. *J. Geol Soc. Aust.* **28**, 19–32.

Ricketts, B. D., M. J. Ware and J. A. Donaldson 1982. Volcaniclastic rocks and volcaniclastic facies in the Middle Precambrian (Alphebian) Belcher Group, Northwest Territories. *Can. J. Earth Sci.* **19**, 1275–94.

Ridge, J. D. 1973. Volcanic exhalations and ore deposition in the vicinity of the sea floor. *Mineralium Deposita* **8**, 332–48.

Riehle, J. R. 1973. Calculated compaction profiles of rhyolitic ash-flow tuffs. *Geol Soc. Am. Bull.* **84**, 2193–216.

Rittmann, A. 1962. *Volcanoes and their activity*. New York: Wiley.

Roberts, B. and A. W. B. Siddans 1971. Fabric studies in the Llwyd Mawr ignimbrite, Caernarvonshire, North Wales. *Tectonophysics* **12**, 281–306.

Roberts, R. J. and D. W. Peterson 1961. *Suggested magmatic differences between welded 'ash' tuffs and welded crystal tuffs, Arizona and Nevada*, D73–9. US Geol Surv. Prof. Pap., no. 424–D.

Robson, G. R. 1967. Thickness of Etnean lavas. *Nature* **216**, 251–2.

Rodine, D. A. and A. M. Johnson 1976. The ability of debris, heavily freighted with coarse clastic materials, to flow on gentle slopes. *Sedimentology* **23**, 213–34.

Rodolfo, K. S. 1969. Bathymetry and marine geology of the Andaman Basin, and tectonic implications for Southeast Asia. *Geol Soc. Am. Bull.* **80**, 1203–30.

Rona, P. A. 1984. Hydrothermal mineralisation at seafloor spreading centers. *Earth Sci. Rev.* **20**, 1–104.

Roobol, M. J. 1976. Post-eruptive mechanical sorting of pyroclastic material — an example from Jamaica. *Geol Mag.* **113**, 429–40.

Roobol, M. J. and A. L. Smith 1976. Mount Pelée, Martinique: a pattern of alternating eruptive styles. *Geology* **4**, 521–4.

Roobol, M. J., J. V. Wright and A. L. Smith 1983. Gravity slides or calderas in the Lesser Antilles island arc? *J. Volcanol. Geotherm. Res.* **14**, 121–134.

Roobol, M. J., A. L. Smith and J. V. Wright 1985. Dispersal and characteristics of pyroclastic fall deposits from Mt. Misery volcano, West Indes. *Geol. Rundsch.* **74**, 321–335.

Rose, W. I., Jr 1972a. Santiaguito volcanic dome, Guatemala. *Geol Soc. Am. Bull.* **83**, 1413–34.

Rose, W. I., Jr 1972b. Notes on the 1902 eruption of Santá Maria volcano, Guatemala. *Bull. Volcanol.* **36**, 29–45.

Rose, W. I., Jr, S. Bonis, R. E. Stoiber, M. Keller and T. Bickford 1973. Studies of volcanic ash from two recent Central American eruptions. *Bull. Volcanol.* **37**, 338–64.

Rose, W. I., Jr, T. Pearson and S. Bonis 1977. Nuée ardente eruption from the foot of a dacite lava flow, Santiaguito volcano, Guatemala. *Bull. Volcanol.* **40**, 1–16.

Rose, W. I., Jr. A. T. Anderson, Jr, L. G. Woodruff and S. B. Bonis 1978. The October 1974 basaltic tephra from Fuego volcano: description and history of the magma body. *J. Volcanol. Geotherm. Res.* **4**, 3–53.

Rosi, M., A. Sbiana and C. Principe 1983. The Phlegrean Fields: structural evolution, volcanic history and eruptive mechanisms. *J. Volcanol. Geotherm. Res.* **17**, 273–88.

Rosner, D. R. and M. Epstein 1972. Effects of interface kinetics, capillarity and solute diffusion on bubble growth rates in highly supersaturated liquids. *Chem. Engng Sci.* **27**, 69–88.

Ross, G. S. and R. L. Smith 1961. *Ash-flow tuffs, their origin, geological relations and identification.* US Geol Surv. Prof. Pap., no. 366.

Rowley, P. D., M. A. Kuntz and N. S. MacLeod 1981. Pyroclastic flow deposits. In Lipman & Mullineaux (1981), 489–512.

Rundle, J. B. and J. C. Eichelberger 1983. Continental scientific drilling at Long Valley–Mono Craters. *EOS* **64**, 12–5.

Ryerson, F. J. and P. C. Hess 1980. The role of P_2O_5 in silicate melts. *Geochim. Cosmochim. Acta* **44**, 611–24.

Sadler, P. M. 1981. Sediment accumulation rates and the completeness of stratigraphic sections. *J. Geol.* **89**, 569–84.

Sample, J. C. and D. E. Karig 1982. A volcanic production rate for the Mariana island arc. *J. Volcanol. Geotherm. Res.* **13**, 73–82.

Sangster, D. F. 1972. *Precambrian volcanogenic massive sulphide deposits in Canada: a review.* Geol Surv. Can. Pap. 72–82.

Sarna–Wojcicki, A. M., S. Shipley, R. B. Waitt, Jr, D. Dzurisin and S. H. Wood 1981. Areal distribution thickness, mass, volume, and grain-size of air-fall ash from the six major eruptions of 1980. In Lipman & Mullineaux (1981), 577–600.

Sato, T. 1977. Kuroko deposits: their geology, geochemistry and origin. In *Volcanic processes in ore genesis*, 153–61. Geol Soc. Lond. Spec. Publn, no. 7.

Scandone, R. 1979. Effusion rate and energy balance of Paricutin eruption (1943–1952), Michoacan, Mexico. *J. Volcanol. Geotherm. Res.* **6**, 49–59.

Schmid, R. 1981. Descriptive nomenclature and classification of pyroclastic deposits and fragments: recommendations of the IUGS Subcommission on the Systematics of Igneous Rocks. *Geology* **9**, 41–3.

Schmincke, H.-U. 1967a. Fused tuff and peperites in south-central Washington. *Geol Soc. Am. Bull.* **78**, 319–330.

Schmincke, H.-U. 1967b. Graded lahars in the type sections of the Ellensburg Formation, south-central Washington. *J. Sed. Petrol.* **37**, 438–48.

Schmincke, H.-U. 1967c. Stratigraphy and petrography of four upper Yakima basalt flows in south-central Washington. *Geol Soc. Am. Bull.* **78**, 1385–1422.

Schmincke, H.-U. 1974. Volcanological aspects of peralkaline silicic welded ash-flows tuffs. *Bull. Volcanol.* **38**, 594–636.

Schmincke, H.-U. 1977. Phreatomagmatische Phasen in quartaren Vulkanen der Osteifel. *Geol. Jahrb. (A)* **39**, 3–45.

Schmincke, H.-U. and D. A. Swanson 1967. Laminar viscous flowage structures in ash-flow tuffs from Gran Canaria, Canary Islands. *J. Geol.* **75**, 641–64.

Schmincke, H.-U., R. V. Fisher and A. C. Waters 1973. Antidune and chute and pool structures in base surge deposits of the Laacher See area, Germany. *Sedimentology* **20**, 553–74.

Schmincke, H.-U. *et al.* 1979. Basaltic hyaloclàstites from Hole 396 B, DSDP Leg 46. In *Initial reports of the Deep-Sea Drilling Project* **46**, 341–56. Washington, DC: US Government Printing Office.

Scholz, C. H., M. Barazangi and M. Sbar 1971. Late Cenozoic evolution of the Great Basin, western United States, as an ensialic interarc basin. *Geol Soc. Am. Bull.* **82**, 2979–90.

Scholze, H. V. and H. O. Mulfinger 1959. Der Einbau des Wassers in Glasern, V. Die Diffusion des Wassers in Glasen bei hohen Temperaturen. *Glastech. Ber.* **3**, 381–6.

Sclater, J. G., J. W. Hawkins, J. Mammerickx and C. G. Chase 1972. Crustal extension between the Tonga and Lau Ridges: petrologic and geophysical evidence. *Geol Soc. Am. Bull.* **83**, 505–18.

Scott, R. B. 1971. Alkali exchange during devitrification and hydration of glasses in ignimbrite cooling units. *J. Geol.* **79**, 100–9.

Scrope, G. P. 1858. *The geology of extinct volcanoes of central France.* London: John Murray.

Searle, R. C. 1983. Submarine central volcanoes on the Nazca Plate — high resolution sonar observations. *Mar. Geol.* **53**, 77–102.

Searle, R. 1984. GLORIA survey of the East Pacific Rise near 3.5 S: tectonic and volcanic characteristics of a fast spreading mid-ocean rise. *Tectonophysics* **101**, 319–44.

Searle, R. C. and A. S. Laughton 1981. Fine-scale sonar study of tectonics and volcanism on the Reykjanes Ridge. In *Geology of oceans*, X. Le Pichon, J. Debyser and F. Vine (conveners), 5–13. Oceanologia Acta, 26th Int. Geol Congr., Symp. No. C4.

Secor, D. T. 1969. Mechanics of natural extension of fracturing at depth in the earth's crust. In *Research in tectonics*, A. J. Baer and D. K. Norris (eds), 3–48. Can. Geol Surv. Pap., 68–52.

Self, S. 1974. Explosive activity of Ngauruhoe, 27–30 March, 1974. *N.Z. J. Geol. Geophys.* **18**, 189–95.

Self, S. 1976. The recent volcanology of Terceira, Azores. *J. Geol Soc. Lond.* **132**, 645–66.

Self, S. 1983. Large-scale phreatomagmatic silicic volcanism: a case study from New Zealand. *J. Volcanol. Geotherm. Res.* 17, 433–69.

Self, S. and M. R. Rampino 1981. The 1883 eruption of Krakatau. *Nature* 294, 699–704.

Self, S. and R. S. J. Sparks 1978. Characteristics of widespread pyroclastic deposits formed by the interaction of silicic magma and water. *Bull. Volcanol.* 41, 196–212.

Self, S. and R. S. J. Sparks (eds) 1981. *Tephra studies.* Dordrecht: D. Reidel.

Self, S. and J. V. Wright 1983. Large wave-forms from the Fish Canyon Tuff, Colorado. *Geology* 11, 443–6.

Self, S., R. S. J. Sparks, B. Booth and G. P. L. Walker 1974. The 1973 Heimaey strombolian scoria deposit, Iceland. *Geol Mag.* 111, 539–48.

Self, S., L. Wilson and I. A. Nairn 1979. Vulcanian eruption mechanisms. *Nature* 277, 440–3.

Self, S., J. Kienle and J. P. Huot, 1980. Ukinrek maars, Alaska, II. Deposits and formation of the 1977 craters. *J. Volcanol. Geotherm. Res.* 7, 39–65.

Self, S., M. R. Rampino and J. J. Barbera 1981. The possible effects of large 19th and 20th century volcanic eruptions on zonal and hemispheric surface temperatures. *J. Volcanol. Geotherm. Res.* 11, 41–60.

Self, S., M. R. Rampino, M. S. Newton and J. A. Wolff 1984. Volcanological study of the great Tambora eruption of 1815. *Geology* 12, 659–63.

Selley, R. C. 1978. *Ancient sedimentary environments,* 2nd edn. London: Chapman & Hall.

Selley, R. C. 1982. *Introduction to sedimentology.* London: Academic Press.

Settle, M. 1978. Volcanic eruption clouds and the thermal output of explosive eruptions. *J. Volcanol. Geotherm. Res.* 3, 309–24.

Shand, S. J. 1947. *Eruptive rocks.* London: Allen & Unwin.

Shaw, H. R. 1963. Obsidian — H_2O viscosities at 1000 and 2000 bars in the temperature range 700° to 900°C. *J. Geophys. Res.* 68, 6337–43.

Shaw, H. R. 1969. Rheology of basalt in the melting range. *J. Petrol.* 10, 510–35.

Shaw, H. R. 1972. Viscosities of magmatic silicate liquids: An empirical method of prediction. *Am. J. Sci.* 272, 870–93.

Shaw, H. R. 1980. The fracture mechanisms of magma transport from the mantle to the surface. In Hargraves (1980), 201–54.

Shaw, H. R., D. L. Peck and A. R. Okamura 1968. The viscosity of basaltic magma: an analysis of field measurements in Makaopuhi lava lake, Hawaii. *Am. J. Sci.* 266, 225–64.

Sheffield, C. 1981. *Earth Watch: a survey of the world from space.* London: Sidgwick & Jackson.

Sheffield, C. 1983. *Man on Earth: The marks of man: a survey from space.* London: Sidgwick & Jackson.

Shepherd, J. B. and H. Sigurdsson 1982. Mechanism of the 1979 explosive eruption of Soufrière volcano, St. Vincent. *J. Volcanol. Geotherm. Res.* 13, 119–30.

Sheridan, M. F. 1970. Fumarolic mounds and ridges of the Bishop Tuff, California. *Geol Soc. Am. Bull.* 81, 851–68.

Sheridan, M. F. 1979. Emplacement of pyroclastic flows: a review. In Chapin & Elston (1979), 125–36.

Sheridan, M. F. 1980. Pyroclastic block flow from the September, 1976, eruption of La Soufrière volcano, Guadeloupe. *Bull. Volcanol.* 43, 397–402.

Sheridan, M. F. and J. R. Marshall 1983. Interpretation of pyroclast surface features using SEM images. *J. Volcanol. Geotherm. Res.* 16, 153–9.

Sheridan, M. F. and D. M. Ragan 1976. Compaction of ash-flow tuffs. In *Compaction of coarse-grained sediments,* G. V. Chilingarian and K. H. Wolf (eds), *Developments in sedimentology,* Vol. 18b, 677–713. Amsterdam: Elsevier.

Sheridan, M. F. and R. G. Updike 1975. Sugarloaf Mountain tephra — a Pleistocene rhyolitic deposit of base-surge origin in Northern Arizona. *Geol Soc. Am. Bull.* 86, 571–81.

Sheridan, M. F. and K. H. Wohletz 1981. Hydro-volcanic explosions: the systematics of water-pyroclast equilibration. *Science* 212, 1387–9.

Sheridan, M. F. and K. H. Wohletz 1983. Hydro-volcanism: basic considerations and review. *J. Volcanol. Geotherm. Res.* 17, 1–29.

Sheridan, M. F., F. Barberi, M. Rose and R. Santacroce 1981. A model for plinian eruptions of Vesuvius. *Nature* 289, 282–5.

Shiki, T. and Y. Misawa 1982. Forearc geological structure of the Japanese Islands. In Leggett (1982), 63–73.

Shreve, R. L. 1968. *The Blackhawk landslide.* Geol Soc. Am. Spec. Pap., no. 108.

Siebert, L. 1984. Large volcanic debris avalanches: characteristics of source areas, deposits and associated eruptions. *J. Volcanol. Geotherm Res.* 22, 163–97.

Sigurdsson, H. 1972. Partly-welded pyroclastic flow deposits in Dominica, Lesser Antilles. *Bull. Volcanol.* 36, 148–63.

Sigurdsson, H. 1981. *Geologic observations in the crater of Soufrière volcano, St. Vincent.* Univ. West Indies Seismic Res. Unit Spec. Publn, 1981/1.

Sigurdsson, H. 1982. Tephra from the 1979 Soufrière explosive eruption. *Science* 216, 1106–8.

Sigurdsson, H., R. S. J. Sparks, S. Carey and T. C. Huang 1980. Volcanogenic sedimentation in the Lesser Antilles arc. *J. Geol.* 88, 523–40.

Sillitoe, R. H. 1973. The tops and bottoms of porphyry copper deposits. *Econ. Geol.* 68, 799–815.

Sillitoe, R. H., E. M. Baker and W. A. Brook 1984. Gold deposits and hydrothermal eruption breccias associated with a maar volcano at Wau, Papua New Guinea. *Econ. Geol.* **79**, 638–55.

Simkin, T., L. Siebert, L. McClelland, D. Bridge, C. Newhall and J. H. Latter 1981. *Volcanoes of the World. A regional directory gazetteer and chronology of volcanism during the last 10,000 years.* Smithsonian Institution. Stroudsbourg, Pennsylvania: Hutchinson & Ross.

Simpson, J. E. 1972. Effects of the lower boundary on the head of a gravity current. *J. Fluid Mech.* **53**, 759–68.

Smith, A. L. and M. J. Roobol 1982. Andesitic pyroclastic flows. In *Orogenic andesites*, R. S. Thorpe (ed.), 415–33. New York: Wiley.

Smith, I. E. M. 1976. Peralkaline rhyolites from the D'Entrecastreux Islands, Papua New Guinea. In R. W. Johnson (1976), 275–85.

Smith, I. E. M. and R. W. Johnson 1981. Contrasting rhyolite suites in the late Cenozoic of Papua New Guinea. *J. Geophys. Res.* **86**, 10257–72.

Smith, R. B. 1978. Seismicity, crustal structure, and intraplate tectonics of the interior of the western Cordillera. In R. B. Smith and G. P. Eaton (1978), 111–44.

Smith, R. B. and G. P. Eaton (eds) 1978. *Cenozoic tectonics and regional geophysics of the western Cordillera.* Geol Soc. Am. Mem., no. 152.

Smith, R. L. 1960a. Ash-flows. *Geol Soc. Am. Bull.* **71**, 795–842.

Smith, R. L. 1960b. *Zones and zonal variations in welded ash-flows,* 149–59. US Geol Surv. Prof. Pap., no. 354–F.

Smith, R. L. 1979. Ash-flow magmatism. In Chapin & Elston (1979), 5–27.

Smith, R. L. and R. A. Bailey 1966. The Bandelier Tuff, a study of ash-flow eruption cycles from zoned magma chambers. *Bull. Volcanol.* **29**, 83–104.

Smith, R. L. and R. A. Bailey 1968. Resurgent cauldrons. In Coats *et al.* (1968), 153–210.

Sorem, R. K. 1982. Volcanic ash clusters: tephra rafts and scavengers. *J. Volcanol. Geotherm. Res.* **13**, 63–71.

Sourirajan, S. and G. C. Kennedy 1962. The system H_2O-NaCl at elevated temperatures and pressures. *Am. J. Sci.* **260**, 115–41.

Sparks, R. S. J. 1975. Stratigraphy and geology of the ignimbrites of Vulsini volcano, Central Italy. *Geol. Rundsch.* **64**, 497–523.

Sparks, R. S. J. 1976. Grain size variations in ignimbrites and implications for the transport of pyroclastic flows. *Sedimentology* **23**, 147–88.

Sparks, R. S. J. 1978a. The dynamics of bubble formation and growth in magmas: a review and analysis. *J. Volcanol. Geotherm. Res.* **3**, 1–37.

Sparks, R. S. J. 1978b. Gas release rates from pyroclastic flows: an assessment of the role of fluidisation

in their emplacement. *Bull. Volcanol.* **41**, 1–9.

Sparks, R. S. J. 1986. The dimensions and dynamics of volcanic eruption columns. *Bull. Volcanol.* **48**, 3–15.

Sparks, R. S. J. and S. Carey 1978. Subaqueous deposits and structures formed by the entry of pyroclastic flows into the sea in the Lesser Antilles arc. (abstr.). *Abstr. with Programme Joint Meeting GAC, MAC, GSA, Toronto.*

Sparks, R. S. J. and T. C. Huang 1980. The volcanological significance of deep-sea ash layers associated with ignimbrites. *Geol Mag.* **117**, 425–36.

Sparks, R. S. J. and H. Pinkerton 1978. Effects of degassing on rheology of basaltic magma. *Nature* **276**, 385–6.

Sparks, R. S. J. and G. P. L. Walker 1973. The ground surge deposit: a third type of pyroclastic rock. *Nature Phys. Sci.* **241**, 62–4.

Sparks, R. S. J. and G. P. L. Walker 1977. The significance of vitric-enriched air-fall ashes associated with crystal-enriched ignimbrites. *J. Volcanol. Geotherm. Res.* **2**, 329–41.

Sparks, R. S. J. and C. J. N. Wilson 1983. Flow-head deposits in ash turbidites. *Geology* **11**, 348–51.

Sparks, R. S. J. and L. Wilson 1976. A model for the formation of ignimbrite by gravitational column collapse. *J. Geol Soc. Lond.* **132**, 441–52.

Sparks, R. S. J. and L. Wilson 1982. Explosive volcanic eruptions — V. Observations of plume dynamics during the 1979 Soufrière · eruption, St. Vincent. *Geophys. J. R. Astr. Soc.* **69**, 551–70.

Sparks, R. S. J. and J. V. Wright 1979. Welded air-fall tuffs. In Chapin & Elston (1979), 155–66.

Sparks, R. S. J., J. G. Moore and C. J. Rice 1986 (*in press*). The initial giant umbrella cloud of the May 18th, 1980, explosive eruption of Mount St Helens. *J. Volcanol. Geotherm. Res.*

Sparks, R. S. J., H. Pinkerton and G. Hulme 1976. Classification and formation of lava levées on Mount Etna, Sicily. *Geology* **4**, 269–71.

Sparks, R. S. J., S. Self and G. P. L. Walker 1973. Products of ignimbrite eruptions. *Geology* **1**, 115–8.

Sparks, R. S. J., H. Sigurdsson and L. Wilson 1977. Magma mixing: a mechanism for triggering acid explosive eruptions. *Nature* **267**, 315–8.

Sparks, R. S. J., L. Wilson and G. Hulme 1978. Theoretical modelling of the generation, movement and emplacement of pyroclastic flows by column collapse. *J. Geophys. Res.* **83**, 1727–39.

Sparks, R. S. J., H. Sigurdsson and S. N. Carey 1980a. The entrance of pyroclastic flows into the sea, I. Oceanographic and geologic evidence from Dominica, Lesser Antilles. *J. Volcanol. Geotherm. Res.* **7**, 87–96.

Sparks, R. S. J., H. Sigurdsson and S. N. Carey 1980b. The entrance of pyroclastic flows into the sea, II. Theoretical considerations on subaqueous emplacement and welding. *J. Volcanol. Geotherm. Res.* **7**, 97–105.

Sparks, R. S. J., L. Wilson and H. Sigurdsson 1981. The pyroclastic deposits of the 1875 eruption of Askja, Iceland. *Phil Trans R. Soc. Lond. (A)* 299, 241–73.

Sparks, R. S. J., S. Brazier, T. C. Huang and D. Muerdter 1983. Sedimentology of the Minoan deep-sea tephra layer in the Aegean and eastern Mediterranean. *Mar. Geol.* 54, 131–67.

Spence, C. D. and A. F. de Rosen-Spence 1975. The place of sulphide mineralisation in the volcanic sequence at Noranda, Quebec. *Econ. Geol.* 70, 90–101.

Spencer, K. J. and D. H. Lindsley 1981. A solution model for coexisting iron-titanium oxides. *Am. Mineral.* 66, 1189–201.

Sporli, K. B. 1980. New Zealand and oblique-slip margins: tectonic development up to and during the Cainozoic. In Ballance & Reading (1980), 147–70.

Spry, A. 1962. The origin of columnar jointing particularly in basalt flows. *J. Geol. Soc. Aust.* 8, 191–216.

Stanton, W. I. 1960. The Lower Palaeozoic rocks of south-west Murrisk, Ireland. *Q. J. Geol Soc. Lond.* 116, 269–96.

Steven, T. A. and P. W. Lipman 1976. *Calderas of the San Juan volcanic field, south-western Colorado.* US Geol Surv. Prof. Pap., no. 958.

Stoiber, R. E. and W. I. Rose 1969. Recent volcanic and fumarolic activity at Santiaguito volcano, Guatemala. *Bull. Volcanol.* 33, 475–502.

Stolper, E. 1982. Water in silicate glasses: an infrared spectroscopic study. *Contrib. Mineral. Petrol.* 81, 1–17.

Stolper, E. and D. Walker 1980. Melt density and the average composition of basalt. *Contrib. Mineral. Petrol.* 74, 7–12.

Storey, B. C. and D. I. M. Macdonald 1984. Processes of formation and filling of a Mesozoic back arc basin on the island of South Georgia. In Kokelaar & Howells (1984), 207–17.

Stow, D. A. V. and J. P. B. Lovell 1979. Contourites: their recognition in modern and ancient sediments. *Earth-Sci. Rev.* 14, 251–91.

Strangway, D. W. (ed.) 1980. *The continental crust and its mineral deposits.* Geol Assoc. Can. Spec. Pap. no. 20.

Streckeisen, A. 1979. Classification and nomenclature of volcanic rocks, lamprophyres, carbonatites, and melilitic rocks: recommendations and suggestions of the IUGS Subcommission on the Systematics of Igneous Rocks. *Geology* 7, 331–5.

Subbarao, K. V. and R. N. Sukheswala (eds) 1981. *Deccan volcanism and related basalt provinces in other parts of the world.* Geol. Soc. India Mem. 3.

Sugimura, A. and S. Uyeda 1973. *Island arcs. Japan and its environs. Developments in geotectonics*, Vol. 3. Amsterdam: Elsevier.

Suthren, R. J. and H. Furnes 1980. Origin of some bedded welded tuffs. *Bull. Volcanol.* 43, 61–71.

Suzuki, K. and T. Ui 1982. Grain orientation and depositional ramps as flow direction indicators of a large-scale pyroclastic flow deposit in Japan. *Geology* 10, 429–33.

Suzuki, T., Y. Katsui and T. Nakamura 1973. Size distribution of the Tarumai Ta-b pumice-fall deposit. *Bull. Volcanol. Soc. Jap.* 18, 47–64.

Swanson, D. A. 1972. Magma supply rate of Kilauea volcano, 1952–1971. *Science* 175, 169–70.

Swanson, D. A. 1973. Pahoehoe flows from the 1969–71 Mauna Ulu eruption, Kilauea volcano, Hawaii. *Geol Soc. Am. Bull.* 84, 815–26.

Swanson, D. A. and T. L. Wright 1981. The regional approach to studying the Columbia River Basalt Group. In K. V. Subbarao and R. N. Sukheswala (1981), 58–80.

Swanson, D. A., T. H. Wright and R. T. Helz 1975. Linear vent systems and estimated rates of magma production and eruption for the Yakima basalt on the Columbia Plateau. *Am. J. Sci.* 275, 877–905.

Tamaki, K. 1985. Two modes of back arc spreading. *Geology* 13, 475–8.

Tanner, P. W. G., B. C. Storey and D. I. M. Macdonald 1981. Geology of an Upper Jurassic–Lower Cretaceous island-arc assemblage in Hauge Reef, the Pickersgill Islands and adjoining areas of South Georgia. *Br. Antarct. Surv. Bull.* 53, 77–117.

Tassé, N., J. Lajoie and E. Dimroth 1978. The anatomy and interpretation of an Archaean volcaniclastic sequence, Noranda region, Quebec. *Can. J. Earth Sci.* 15, 874–88.

Taylor, B. and G. D. Karner 1983. On the evolution of marginal basins. *Rev. Geophys. Space Phys.* 21, 1727–41.

Taylor, G. A. 1958. The 1951 eruption of Mount Lamington, Papua. *Aust. Bur. Miner. Resour. Geol. Geophys. Bull.* 38, 1–117.

Thompson, B. N., L. O. Kermode and A. Ewart (eds) 1965. *New Zealand volcanology. Central Volcanic Region.* N.Z. Dept Sci. Ind. Res. Info. Ser. 50.

Thompson, M. D. 1985. Evidence for a Late Pre-cambrian caldera in Boston, Massachusetts. *Geology* 13, 641–3.

Thorarinsson, S. 1953. The crater groups in Iceland. *Bull. Volcanol.* 14, 3–44.

Thorarinsson, S. 1954. *The eruption of Hekla 1947–1948. II, 3, The tephra-fall from Hekla on March 29th, 1947.* Visindafelag Islendinga. Reykjavik: Leiftur.

Thorarinsson, S. 1967. *Surtsey: The new island in the North Atlantic.* New York: The Viking Press.

Thorarinsson, S. 1968. On the rate of lava- and tephra

production and the upward migration of magmas in four Icelandic eruptions. *Geol. Rundsch.* 57, 705–18.

Thorarinsson, S. 1969. The Lakagigar eruption of 1783. *Bull. Volcanol.* 33, 910–27.

Thorarinsson, S. 1981. Tephra studies and tephrochronology: a historical review with special reference to Iceland. In Self & Sparks (1981), 1–12.

Thorarinsson, S. and G. Sigvaldason 1962. The eruption in Askja, 1961. A preliminary report. *Am. J. Sci.* 260, 641–51.

Thorarinsson, S., T. Einarsson, G. E. Sigvaldason and G. Ellison 1964. The submarine eruption off the Vestmann Islands. *Bull. Volcanol.* 27, 435–45.

Thorarinsson, S., S. Steinthorsson, T. Einarsson, H. Kristmannsdottir and N. Oskarsson 1973. The eruption of Heimaey, Iceland. *Nature* 241, 372–75.

Thorpe, R. S. (ed.) 1982. *Andesites: orogenic andesites and related rocks.* New York: John Wiley.

Thurston, P. C. 1980. Subaerial volcanism in the Archaean Uchi-Confederation volcanic belt. *Precamb. Res.* 12, 79–98.

Titley, S. R. 1982. The style and progress of mineralisation and alteration in porphyry copper systems: American Southwest. In *Advances in geology of the porphyry copper deposits: southwestern North America,* S. R. Titley (ed.), 93–116. Tucson: University of Arizona Press.

Turner, F. J. and J. Verhoogen 1960. *Igneous and metamorphic petrology.* New York: McGraw-Hill.

Turner, J. S. and L. B. Gustafson 1981. Fluid motions and compositional gradients produced by crystallisation or melting at vertical boundaries. *J. Volcanol. Geotherm. Res.* 11, 93–125.

Turner, J. S., H. E. Huppert and R. S. J. Sparks 1983. An experimental investigation of volatile exsolution in evolving magma chambers. *J. Volcanol. Geotherm. Res.* 16, 263–78.

Tuttle, O. F. and N. L. Bowen 1958. Origin of granite in the light of experimental studies in the system $NaAlSi_3O_8$–$KAlSi_3O_8$–SiO_2–H_2O. *Geol Soc. Am. Mem.* 74.

Ui, T. 1971. Genesis of magma and structure of magma chamber of several pyroclastic flows in Japan. *J. Fac. Sci. Tokyo Univ. Sec. II* 36, 135–46.

Ui, T. 1983. Volcanic dry avalanche deposits — identification and comparison with nonvolcanic debris stream deposits. *J. Volcan. Geotherm. Res.* 18, 135–50.

Urabe, T., S. D. Scott and K. Hattori 1983. A comparison of footwall-rock alteration and geothermal systems beneath some Japanese and Canadian volcanogenic massive sulfide deposits. In Ohmoto and Skinner (1983), 345–64.

Uyeda, S. 1983. Comparative subductology. *Episodes* 2, 19–24.

Uyeda, S. and H. Kanamori 1979. Back-arc opening and the mode of subduction. *J. Geophys. Res.* 84, 1049–61.

van Bemmelen, R. W. 1949. *The geology of Indonesia and adjacent archipelago.* The Hague: Government Printing Office.

van Bemmelen, R. W. 1961. Volcanology and geology of ignimbrites in Indonesia, North Italy and U.S.A. *Geol. Mijnbouw* 40, 399–411.

van Bemmelen, R. W. 1974. Driving forces of orogeny, with emphasis on blue-schist facies of metamorphism (Test-case III: Japan arc). *Tectonophysics* 22, 83–125.

van der Lingen, G. J. and J. R. Pettinga 1980. The Makara Basin: a Miocene slope-basin along the New Zealand sector of the Australian-Pacific obliquely convergent plate boundary. In Ballance & Reading (1980), 191–216.

van der Molen, I. and M. S. Paterson 1979. Experimental deformation of partially-melted granite. *Contrib. Miner. Petrol.* 70, 299–318.

Veevers, J. J. and D. Cotterill 1978. Western margin of Australia: evolution of a rifted arch system. *Geol Soc. Am. Bull.* 89, 337–55.

Verhoogen, J. 1951. Mechanics of ash formation. *Am. J. Sci.* 249, 729–39.

Vessell, R. K. and D. K. Davies 1981. Non-marine sedimentation in an active fore-arc basin. In *Recent and ancient non-marine depositional environments: models for exploration,* F. G. Ethridge and R. M. Flores (eds), 31–45. SEPM Spec. Publn 31.

Vogt, P. R. 1974. Volcano spacing, fractures and thickness of the lithosphere. *Earth Planet. Sci. Lett.* 21, 235–52.

Voight, B. (ed.) 1978. *Rock slides and avalanches, I. Natural phenomena.* Amsterdam: Elsevier.

Voight, B. 1981. Time scale for the first moments of the May 18 eruption. In Lipman & Mullineaux (1981), 69–80.

Voight, B., H. Glicken, R. J. Janda and P. M. Douglass 1981. Catastrophic rockslide avalanche of May 18. In Lipman & Mullineaux (1981), 347–77.

Von Huene, R. 1984. Tectonic processes along the front of modern convergent margins — research of the past decade. *A. Rev. Earth Planet. Sci.* 12, 359–81.

Von Huene, R. and M. A. Arthur 1982. Sedimentation across the Japan Trench off northern Honshu Island. In Leggett (1982), 27–48.

Vuagnat, M. 1975. Pillow lava flows: isolated sacks or connected tubes? *Bull. Volcanol.* 39, 581–9.

Wadge, G. 1978. Effusion rate and the shape of aa lava flow-fields on Mount Etna. *Geology* 6, 503–6.

Wadge, G. 1980. Output rate of magma from active central volcanoes. *Nature* 288, 253–5.

Wadge, G. 1982. Steady state volcanism: evidence from

eruption histories of polygenetic volcanoes. *J. Geophys. Res.* 87, 4035–49.

Wadge, G., G. P. L. Walker and J. E. Guest 1975. The output of the Etna volcano. *Nature* 213, 484–5.

Waitt, R. B. 1981. Devastating pyroclastic density flow and attendant air fall of May 18 — stratigraphy and sedimentology of deposits. In Lipman & Mullineaux (1981), 439–58.

Waitt, R. B. 1984. Comment on Walker & McBroome (1983). *Geology* 12, 693.

Walker, G. P. L. 1962. Tertiary welded tuffs in eastern Iceland. *Q. J. Geol Soc. Lond.* 118, 275–93.

Walker, G. P. L. 1963. The Breiddalur central volcano, Eastern Iceland. *Q. J. Geol Soc. Lond.* 119, 29–63.

Walker, G. P. L. 1970. Compound and simple lava flows and flood basalts. *Bull. Volcanol.* 35, 579–90.

Walker, G. P. L. 1971. Grainsize characteristics of pyroclastic deposits. *J. Geol.* 79, 696–714.

Walker, G. P. L. 1972. Crystal concentration in ignimbrites. *Contrib. Miner. Petrol.* 36, 135–46.

Walker, G. P. L. 1973a. Lengths of lava flows. *Phil Trans R. Soc. Lond.* 274, 107–18.

Walker, G. P. L. 1973b. Explosive volcanic eruptions — a new classification scheme. *Geol. Rundsch.* 62, 431–46.

Walker, G. P. L. 1979. A volcanic ash generated by explosions where ignimbrite entered the sea. *Nature* 281, 642–6.

Walker, G. P. L. 1980. The Taupo pumice: product of the most powerful known (ultraplinian) eruption? *J. Volcanol. Geotherm. Res.* 8, 69–94.

Walker, G. P. L. 1981a. Characteristics of two phreatoplinian ashes, and their water flushed origin. *J. Volcanol. Geotherm. Res.* 9, 395–407.

Walker, G. P. L. 1981b. Plinian eruptions and their products. *Bull. Volcanol.* 44, 223–40.

Walker, G. P. L. 1981c. The Waimihia and Hatepe plinian deposits from the rhyolitic Taupo volcanic centre. *N.Z. J. Geol. Geophys.* 24, 305–24.

Walker, G. P. L. 1981d. New Zealand case histories of pyroclastic studies. In Self & Sparks (1981), 317–30.

Walker, G. P. L. 1981e. Generation and dispersal of fine ash and dust by volcanic eruptions. *J. Volcanol. Geotherm. Res.* 11, 81–92.

Walker, G. P. L. 1983. Ignimbrite types and ignimbrite problems. *J. Volcanol. Geotherm. Res.* 17, 65–88.

Walker, G. P. L. 1984a. Downsag calderas, ring faults, caldera sizes, and incremental caldera growth. *J. Geophys. Res.* 89, 8407–16.

Walker, G. P. L. 1984b. Topographic evolution of eastern Iceland. *Jokull* 32, 13–20.

Walker, G. P. L. 1985. Origin of coarse lithic breccias near ignimbrite source vents. *J. Geotherm. Volcan. Res.* 25, 157–71.

Walker, G. P. L. and R. Croasdale 1971. Two Plinian-type eruptions in the Azores. *J. Geol Soc. Lond.* 127, 17–55.

Walker, G. P. L. and R. Croasdale 1972. Characteristics of some basaltic pyroclastics. *Bull. Volcanol.* 35, 303–17.

Walker, G. P. L. and L. A. McBroome 1983. Mount St. Helens 1980 and Mount Pelée 1902 — flow or surge? *Geology* 11, 571–4.

Walker, G. P. L. and L. Morgan 1984. Reply to Hoblitt & Miller (1984) and Waitt (1984). *Geology* 12, 693–5.

Walker, G. P. L. and C. J. N. Wilson 1983. Lateral variations in the Taupo ignimbrite. *J. Volcanol. Geotherm. Res.* 18, 117–23.

Walker, G. P. L., L. Wilson and E. L. G. Bowell 1971. Explosive volcanic eruptions — I. The rate of fall of pyroclasts. *Geophys. J. R. Astr. Soc.* 22, 377–83.

Walker, G. P. L., C. J. N. Wilson and P. C. Froggatt 1980a. Fines-depleted ignimbrite in New Zealand — the product of a turbulent pyroclastic flow. *Geology* 8, 245–9.

Walker, G. P. L., R. F. Heming and C. J. N. Wilson 1980b. Low-aspect ratio ignimbrites. *Nature* 283, 286–7.

Walker, G. P. L., S. Self and P. C. Froggatt 1981a. The ground layer of the Taupo Ignimbrite: a striking example of sedimentation from a pyroclastic flow. *J. Volcanol. Geotherm. Res.* 10, 1–11.

Walker, G. P. L., C. J. N. Wilson and P. C. Froggatt 1981b. An ignimbrite veneer deposit: the trail marker of a pyroclastic flow. *J. Volcanol. Geotherm. Res.* 9, 409–21.

Walker, G. P. L., R. F. Heming, T. J. Sprod and H. R. Walker 1981c. Last major eruptions of Rabaul volcano. In R. W. Johnson (1981), 181–93.

Walker, G. P. L., J. V. Wright, B. J. Clough and B. Booth 1981d. Pyroclastic geology of the rhyolitic volcano of La Primavera, Mexico. *Geol. Rundsch.* 70, 1100–18.

Walker, G. P. L., S. Self and L. Wilson 1984. Tarawera, 1886, New Zealand — a basaltic plinian fissure eruption. *J. Volcanol. Geotherm. Res.* 21, 61–78.

Walker, R. G. 1965. The origin and significance of the internal sedimentary structures of turbidites. *Yorkshire Geol Soc. Proc.* 35, 1–32.

Walker, R. G. 1975. Generalised facies models for resedimented conglomerates of turbidite association. *Geol Soc. Am. Bull.* 86, 737–48.

Walker, R. G. 1978. Deep-water sandstone facies and ancient submarine fans: models for exploration for stratigraphic traps. *Am. Assoc. Petrol. Geol. Bull.* 62, 932–66.

Walker, R. G. (ed.) 1984. *Facies models*, 2nd edn. Geol Assoc. Can., Geosci. Can. Repr. Ser. 1.

Waters, A. C. 1960. Determining direction of flow in basalts. *Am. J. Sci.* 258A, 350–66.

Waters, A. C. 1961. Stratigraphic and lithologic variations in the Columbia River basalt. *Am. J. Sci.* 259, 583–611.

Waters, A. C. and R. V. Fisher 1971. Base surges and their deposits: Capelinhos and Taal volcanoes. *J. Geophys. Res.* 76, 5596–614.

Watkins, N. D. and T. C. Huang 1977. Tephras in abyssal sediments east of the North Island, New Zealand: chronology, paleowind velocity, and paleo-explosivity. *N.Z. J. Geol. Geophys.* 20, 179–98.

Watkins, N. D., R. S. J. Sparks, H. Sigurdsson, T. C. Huang, A. Federman, S. Carey and D. Ninkovich 1978. Volume and extent of the Minoan tephra from Santorini volcano: new evidence from deep-sea cores. *Nature* 271, 122–26.

Wellman, P. 1974. Potassium–argon ages on the Cainozoic volcanic rocks of eastern Victoria, Australia. *J. Geol Soc. Aust.* 21, 359–76.

Wellman, P. and I. McDougall 1974. Cainozoic igneous activity in eastern Australia. *Tectonophysics* 23, 49–65.

Westgate, V. A. and M. P. Gorton 1981. Correlation techniques in tephra studies. In Self & Sparks (1981), 73–94.

White, W. S. 1960. The Keweenawan lavas of Lake Superior, an example of flood basalts. *Am. J. Sci.* 258A, 367–74.

Whitford, D. J., I. A. Nicholls and S. R. Taylor 1979. Spatial variations in the geochemistry of Quaternary lavas across the Sunda arc in Java and Bali. *Contrib. Miner. Petrol.* 70, 341–56.

Whitford-Stark, J. L. 1982. Factors influencing the morphology of volcanic landforms: an Earth-Moon comparison. *Earth Sci. Rev.* 18, 109–68.

Whitham, A. and R. S. J. Sparks *in press*. Pumice. *Bull. Volcanol.*

Williams, H. 1932. The history and characters of volcanic domes. *Univ. Calif. Publn Bull. Dept Geol Sci.* 21, 51–146.

Williams, H. 1941. Calderas and their origin. *Univ. Calif. Publn Bull. Dept Geol Sci. Bull.* 25, 239–346.

Williams, H. 1942. *The geology of Crater Lake National Park, Oregon, with a reconnaissance of the Cascade Range southward to Mt Shasta.* Carnegie Inst. Washington Publn 540.

Williams, H. and A. R. McBirney 1979. *Volcanology.* San Francisco: Freeman, Cooper.

Williams, H., F. J. Turner and C. M. Gilbert 1954. *Petrography. An introduction to the study of rocks in thin section.* San Francisco: W. H. Freeman.

Williams, S. N. 1983. Plinian airfall deposits of basaltic composition. *Geology* 11, 211–4.

Williams, S. N. and S. Self 1983. The October 1902 plinian eruptions of Santa Maria volcano, Guatemala. *J. Volcanol. Geotherm. Res.* 16, 33–56.

Wilson, C. J. N. 1980. The role of fluidisation in the emplacement of pyroclastic flows: An experimental approach. *J. Volcanol. Geotherm. Res.* 8, 231–49.

Wilson, C. J. N. 1981. *Studies on the origins and emplacement of pyroclastic flows.* Ph.D. thesis, Imperial College, University of London.

Wilson, C. J. N. 1984. The role of fluidisation in the emplacement of pyroclastic flows, 2: experimental results and their interpretation. *J. Volcanol. Geotherm. Res.* 20, 55–84.

Wilson, C. J. N. 1985. The Taupo eruption, New Zealand II. The Taupo ignimbrite. *Phil. Trans. R. Soc. Lond.* A.314, 229–310.

Wilson, C. J. N. and G. P. L. Walker 1981. Violence in pyroclastic flow eruptions. In Self & Sparks (1981), 441–8.

Wilson, C. J. N. and G. P. L. Walker 1982. Ignimbrite depositional facies: the anatomy of a pyroclastic flow. *J. Geol Soc. Lond.* 139, 581–92.

Wilson, C. J. N. and G. P. L. Walker 1985. The Taupo eruption, New Zealand I. General aspects. *Phil. Trans. R. Soc. Lond.* A.314, 199–228.

Wilson, J. T. 1963. Evidence from islands on the spreading of ocean floors. *Nature* 197, 536–8.

Wilson, L. 1972. Explosive volcanic eruptions — II. The atmospheric trajectories of pyroclasts. *Geophys. J. R. Astr. Soc.* 30, 381–92.

Wilson, L. 1976. Explosive volcanic eruptions — III. Plinian eruption columns. *Geophys. J. R. Astr. Soc.* 45, 543–56.

Wilson, L. 1978. Energetics of the Minoan eruption. In *Thera and the Aegean World.* Vol. I. In C. Doumas (1978), 221–28.

Wilson, L. 1980a. Relationships between pressure, volatile content and ejecta velocity in three types of volcanic explosion. *J. Volcanol. Geotherm. Res.* 8, 297–313.

Wilson, L. 1980b. Energetics of the Minoan eruption: some revisions. In Doumas (1980), 31–5.

Wilson, L. and J. W. Head 1981. Ascent and emplacement of basaltic magma on the Earth and Moon. *J. Geophys. Res.* 86, 2971–3001.

Wilson, L., R. S. J. Sparks, T. C. Huang and N. D. Watkins 1978. The control of eruption column heights by eruption energetics and dynamics. *J. Geophys. Res.* 83, 1829–36.

Wilson, L., R. S. J. Sparks and G. P. L. Walker 1980. Explosive volcanic eruptions — IV. The control of magma properties and conduit geometry on eruption column behaviour. *Geophys. J. R. Astr. Soc.* 63, 117–48.

Windley, B. F. and F. B. Davies 1978. Volcano spacings and lithospheric/crustal thickness in the Archaean. *Earth Planet. Sci. Lett.* 38, 291–7.

Wohletz, K. H. 1983. Mechanisms of hydrovolcanic pyroclast formation: grain-size, scanning electron microscopy, and experimental studies. *J. Volcanol. Geotherm. Res.* 17, 31–63.

Wohletz, K. H. and M. F. Sheridan 1979. A model of pyroclastic surge. In Chapin & Elston (1979), 177–94.

Wohletz, K. H. and M. F. Sheridan 1983. Hydro-volcanic explosions II. Evolution of basaltic tuff rings and tuff cones. *Am. J. Sci.* 283, 385–413.

Wolf, T. 1878. Der Cotopaxi und seinletzte Eruption am 26. Juni, 1877. *Neues Jahrb. Mineral. Geol. Palantol.* 113–67.

Wolfe, J. A. and S. Self 1982. Structural lineaments and Neogene volcanism in South-western Luzon. In *The tectonic and geologic evolution of the South-east Asia seas and islands*, 157–72. Geophys. Mon. Ser., no. 27.

Wolff, J. A. and J. V. Wright 1981. Rheomorphism of welded tuffs. *J. Volcanol. Geotherm. Res.* 10, 13–34.

Wolff, J. A. and J. V. Wright 1982. Formation of the Green Tuff, Pantelleria. *Bull. Volcanol.* 44, 681–90.

Wood, B. J. and S. Banno 1973. Garnet–orthopyroxene and orthopyroxene–climopyroxene relationships in simple and complex systems. *Contrib. Mineral. Petrol.* 42, 109–24.

Wood, C. A. 1977. Non-basaltic shield volcanoes. *Abstr. Planet. Geol. Field Conf. Snake River Plain, Idaho*, 35, NASA TM–78, 436.

Wood, C. A. 1978. Morphometric evolution of composite volcanoes. *Geophys. Res. Lett.* 5, 437–9.

Wood, C. A. 1979. Monogenetic volcanoes of the terrestrial planets. *Proc. 10th Lunar Planet. Sci. Conf.* Vol. 3, 2815–40.

Wood, C. A. 1980a. Morphometric evolution of cinder cones. *J. Volcanol. Geotherm. Res.* 7, 387–413.

Wood, C. A. 1980b. Morphometric analysis of cinder cone degradation. *J. Volcanol. Geotherm. Res.* 8, 137–60.

Wood, C. A. 1984. Continental rift jumps. *Tectonophysics* 94, 529–40.

Wood, C. A. *in press*. Maars. In *Encyclopedia of Volcanology*, J. Green (ed.).

Wood, C. P. 1980. Boninite at a continental margin. *Nature* 288, 692–4.

Worzel, J. L. 1959. Extensive deep-sea sub-bottom reflections identified as white ash. *Proc. Natn. Acad. Sci. USA* 45, 349–55.

Wright, A. E. and D. R. Bowes 1963. Classification of volcanic breccias: a discussion. *Geol Soc. Am. Bull.* 74, 79–86.

Wright, J. V. 1978. Remanent magnetism of poorly sorted deposits from the Minoan eruption of Santorini. *Bull. Volcanol.* 41, 131–5.

Wright, J. V. 1980. Stratigraphy and geology of the welded air-fall tuffs of Pantelleria, Italy. *Geol. Rundsch.* 69, 263–91.

Wright, J. V. 1981. The Rio Caliente ignimbrite: analysis of a compound intraplinian ignimbrite from a major late Quaternary Mexican eruption. *Bull. Volcanol.* 44, 189–212.

Wright, J. V. and M. P. Coward 1977. Rootless vents in welded ash-flow tuffs from Northern Snowdonia, North Wales. *Geol Mag.* 114, 133–40.

Wright, J. V. and E. Mutti 1981. The Dali Ash, Island of Rhodes, Greece: a problem in interpreting submarine volcanigenic sediments. *Bull. Volcanol.* 44, 153–67.

Wright, J. V. and G. P. L. Walker 1977. The ignimbrite source problem: significance of a co-ignimbrite lag-fall deposit. *Geology* 5, 729–32.

Wright, J. V. and G. P. L. Walker 1981. Eruption, transport and deposition of ignimbrite: a case study from Mexico. *J. Volcanol. Geotherm. Res.* 9, 111–31.

Wright, J. V., A. L. Smith and S. Self 1980. A working terminology of pyroclastic deposits. *J. Volcanol. Geotherm. Res.* 8, 315–36.

Wright, J. V., S. Self and R. V. Fisher 1981. Towards a facies model for ignimbrite-forming eruptions. In Self & Sparks (1981), 433–9.

Wright, J. V., M. J. Roobol, A. L. Smith, R. S. J. Sparks, S. A. Brazier, W. I. Rose, Jr and H. Sigurdsson 1984. Late Quaternary explosive silicic volcanism on St. Lucia, West Indies. *Geol Mag.* 121, 1–15.

Wyllie, P. J. 1977. Crustal anatexis: an experimental review. *Tectonophysics* 43, 41–71.

Yamada, E. 1973. Subaqueous pumice flow deposits in the Onikobe Caldera, Miyagi Prefecture, Japan. *J. Geol Soc. Jap.* 79, 585–97.

Yamada, E. 1984. Subaqueous pyroclastic flows: their development and their deposits. In Kokelaar & Howells (1984), 29–35.

Yamazaki, T., I. Kato, I. Muroi and M. Abe 1973. Textural analysis and flow mechanisms of the Donzurubo subaqueous pyroclastic flow deposits. *Bull. Volcanol.* 37, 231–44.

Yoder, H. S. and C. E. Tilley 1962. Origin of basaltic magmas: an experimental study of natural and synthetic rock systems. *J. Petrol.* 3, 342–532.

Yokoyama, S. 1974. Flow and emplacement mechanism of the Ito pyroclastic flow from Aira caldera, Japan. *Tokyo Kyoiku Daigaku Sci. Rep.*, Sec. C. 12, 17–62.

Young, A. 1969. Present rate of land erosion. *Nature* 224, 851–2.

Zeil, W. 1979. The Andes: a geological review. In *Beitrage zur Regionalen Geologie der Erde*, F. Bender, V. Jacobshage, J. D. de Jong & G. Luttig (eds), Vol. 13. Berlin: Gebrüder Bornträger.

Zoback, M. L., R. E. Anderson and G. A. Thompson 1981. Cainozoic evolution of the state of stress and style of tectonism of the Basin and Range province of the western United States. *Phil Trans R. Soc. Lond. (A)* 300, 407–34.

ACKNOWLEDGEMENTS

Many authors, organisations and publishers have generously consented to the use of their work. It is with great pleasure and gratitude that we acknowledge the following copyright holders:

Plate 2 reproduced from *Volcanism of the Eastern Snake River Plain, Idaho: a comparative planetary geology guidebook* (R. Greeley) by permission of the author and NASA; Figure 2.1 reproduced from A. Streckeisen, *Geology* 7, 331–5 by permission of the author and the Geological Society of America; Figures 2.2 and 2.6 reproduced from T. Murase and A. R. McBirney, *Geol. Soc. Am. Bull.* 84, 3563–92 by permission of A. R. McBirney and the Geological Society of America; Figures 2.3 and 6.46b reproduced from J. A. Wolff and J. V. Wright, *J. Volcanol. Geotherm. Res.* 10, 13–34 by permission of J. V. Wright and Elsevier Science Publishers; Figure 2.4 reproduced from *Physical processes of sedimentation – an introduction* (J. R. L. Allen) by permission of the author and Allen & Unwin; Figure 2.7 reproduced from I. Kushiro *et al.*, *J. Geophys. Res.* 81, 6351–6 by permission of I. Kushiro and the publisher, © 1976 by the American Geophysical Union; Figure 2.8 reproduced from T. Murase, *Hokkaido Univ. Fac. Sci. J.*, *Ser.* 7, 1 487–584 by permission of the author; Figures 2.12a and c reproduced from *Physical processes in geology* (A. R. Johnson) by permission of the author and Freeman, Cooper and Company; Plate 3 reproduced from *Volcano: ordeal by fire in Iceland's Westmann Islands* (A. Gunnarsson) by permission of S. Jónasson and the publisher.

Figures 3.1 and 3.2 reproduced from C. W. Burnham, *Earth Mineral Sci. (Penn. St. Univ.)* 41, 69–70 by permission of the author and the Editor; Figure 3.4 reproduced from R. S. J. Sparks, *J. Volcanol. Geotherm. Res.* 3, 1–37 by permission of the author and Elsevier Science Publishers; Figures 3.5 and 6.12 reproduced from L. Wilson, *J. Volcanol. Geotherm. Res.* 8, 297–313 by permission of the author and Elsevier Science Publishers; Figure 3.6 reproduced from B. P. Kokelaar, *J. Geol. Soc., Lond.* 139, 21–33 by permission of Blackwell Scientific Publications; Figure 3.7 reproduced from A. R. McBirney, *Bull. Volcanol.* 26, 455–69 by permission of the author and the publisher; Figure 3.8 reproduced from S. Sourirajan and G. C. Kennedy, *Am. J. Sci.* 260, 115–41 by permission of S. Sourirajan and the publisher; Figures 3.9 and 7.39 reproduced from K. H. Wohletz, *J. Volcanol. Geotherm. Res.* 17, 31–63 by permission of the author and Elsevier Science Publishers; Figures 3.20, 5.1, 6.2a and 6.25 reproduced from J. V. Wright *et al.*, *J. Volcanol. Geotherm. Res.* 8, 315–36 by permission of J. V. Wright and Elsevier Science Publishers; Figures 3.21 and 6.37 reproduced from R. S. J. Sparks *et al.*, *Phil Trans R. Soc.* A299, 241–73 by permission of R. S. J. Sparks and the Royal Society.

Figure 4.1a reproduced from S. Thorarinsson, *Bull. Volcanol.* 33, 910–27 by permission of the publisher; Figures 4.1b and 13.7 reproduced from D. A. Swanson *et al.*, *Am. J. Sci.* 275, 877–905 by permission of D. A. Swanson and the publisher; Figure 4.2 reproduced from G. P. L. Walker, *Phil Trans R. Soc.* A274, 107–18 by permission of the author and the Royal Society; Figures 4.3b and c reproduced from G. P. L. Walker, *Bull. Volcanol.* 35, 579–90 by permission of the author and the publisher; Figure 4.4 reproduced from J. P. Lockwood and P. W. Lipman, *Bull. Volcanol.* 43, 609–15 by permission of P. W. Lipman and the publisher; Figure 4.9 reproduced from R. S. J. Sparks *et al.*, *Geology* 4, 269–71 by permission of R. S. J. Sparks and the Geological Society of America; Figures 4.11 and 4.13 reproduced from R. Hargreaves and L. D. Ayres, *Can. J. Earth Sci.* 16, 1452–66 by permission of L. D. Ayres; Figure 4.15 reproduced from P. Lonsdale and R. Batiza, *Geol. Soc. Am. Bull.* 91, 545–54 by permission of P. Lonsdale and the Geological Society of America; Figures 4.16 and 4.17 reproduced from J. G. Jones and P. H. H. Nelson, *Geol. Mag.* 107, 13–21 by permission of J. G. Jones and Cambridge University Press; Figures 4.18c, 5.7b and 5.10b reproduced from R. L. Christiansen and D. W. Peterson, U.S. Geol. Survey Prof. Paper 1250, 17–30 by permission of M. Kraft, K. Kraft and the U.S. Geological Survey; Académie des Sciences d'Outre-Mer, Paris (4.18d, 5.10a); Figure 4.19 reproduced from H. Sigurdsson, Univ. West Indies Seismic Res. Spec. Publ. No. 1981/1 by permission of the author and the publisher; Figure 4.20 reproduced from T. Minakami *et al.*, *Bull. Volcanol.* 11, 45–160 by permission of T. Minakami and the publisher; Figures 4.21a, 4.22c and e reproduced by permission of B. Clough; Figure 4.29 reproduced from H. Pichler, *Bull. Volcanol.* 28, 293–310 by permission of the author and the publisher; Figure 4.30 reproduced from R. Cas, *Geol. Soc. Am. Bull.* 89, 1708–14 by permission of the author and the Geological Society of America; Figure 4.31 reproduced from C. H. Donaldson, in *Komatiites* (N. T. Arndt and E. G. Nisbet, eds) by permission of the author; Figure 4.32 reproduced from N. T. Arndt *et al.*, *J. Petrol.* 18, 319–69 (1977) by permission of N. T. Arndt and Oxford University Press; Table 4.2 reproduced from C. G. Newhall and W. G. Melson, *J. Volcanol. Geotherm. Res.* 17, 111–31 by permission of C. G. Melson and Elsevier Science Publishers.

Figures 5.3a and 7.13 reproduced from G. P. L. Walker, *J. Geol.* 79, 696–714 by permission of the author and the Editors, © 1971 by the University of Chicago, Figure 5.3a also from G. P. L. Walker *et al.*, *Geology* 8, 245–9 by permission of G. P. L. Walker and the Geological Society of America; Figures 5.3b and 5.14 reproduced from J. V. Wright, *Bull. Volcanol.* 44, 189–212 by permission of the author and the publisher; Figure 5.5 reproduced from D. K. Davies *et al.*, *Geol. Soc. Am. Bull.* 89, 369–84 by permission of the Geological Society of America; Figures 5.6a and 13.32 reproduced from R. K. Vessell and D. K. Davies, SEPM Spec. Publ. 31, 31–45 by permission of the Society of Economic Paleontologists and Mineralogists; Figure 5.7a repro-

duced from *Pictorial history of the Lassen volcano* (B. F. Loomis) by permission of the Loomis Museum Association; Figure 5.7b reproduced from R. L. Christiansen and D. W. Peterson, U.S. Geol. Survey Prof. Paper No. 1250, 17–30 by permission of J. W. Vallance and the U.S. Geological Survey; Figure 5.8 reproduced from A. M. Sarna-Wojcicki *et al.*, U.S. Geol. Survey Prof. Paper No. 1250, 577–600 by permission of A. M. Sarna-Wojcicki and the U.S. Geological Survey; Figure 5.9 reproduced from L. Wilson *et al.*, *J. Geophys. Res.* 83, 1829–36 by permission of L. Wilson and the publisher, © 1978 by the American Geophysical Union; Figure 5.10b reproduced from R. L. Christiansen and D. W. Peterson, U.S. Geol. Survey Prof. Paper No. 1250, 17–30 by permission of P. W. Lipman and the U.S. Geological Survey; Figure 5.15b reproduced from P. W. Francis *et al.*, *Geol. Rundschau* 63, 357–88 by permission of P. W. Francis and the publisher; Figures 5.15c and 5.20 reproduced from R. V. Fisher and G. Heiken, *J. Volcanol. Geotherm. Res.* 13, 339–71 by permission of R. V. Fisher and Elsevier Science Publishers, Figure 5.20 also from R. V. Fisher *et al.*, *Geology* 8, 472–6 by permission of R. V. Fisher and the Geological Society of America; Figure 5.17, 5.18a, 5.19a, b and d reproduced from J. G. Moore, *Bull. Volcanol.* 30, 337–63 by permission of the author and the publisher; Figure 5.18b reproduced from A. C. Waters and R. V. Fisher, *J. Geophys. Res.* 76, 5596–614 by permission of R. V. Fisher and the publisher, © 1971 by the American Geophysical Union; Figure 5.18c reproduced from J. Kienle *et al.*, *J. Volcanol. Geotherm. Res.* 7, 11–37 by permission of J. D. Faro and Elsevier Science Publishers.

Figure 6.4 reproduced from S. Self *et al.*, *Geol. Mag.* 111, 539–48 by permission of S. Self and Cambridge University Press; Figure 6.5 reproduced from B. F. Houghton and W. R. Hackett, *J. Volcanol. Geotherm. Res.* 21, 207–31 by permission of B. F. Houghton and Elsevier Science Publishers; Figures 6.7 and 6.31 reproduced from G. P. L. Walker and R. Croasdale, *Bull. Volcanol.* 35, 303–17 by permission of G. P. L. Walker and the publisher; Figures 6.9a, b, 6.11, 6.21 and 6.29 reproduced from S. Self, *J. Geol. Soc., Lond.* 132, 645–68 by permission of the author and Blackwell Scientific Publications, Figure 6.11 also from B. Booth *et al.*, *Phil Trans R. Soc.* A288, 271–319 by permission of G. P. L. Walker and the Royal Society, and G. P. L. Walker, *Geol. Rundschau* 62, 431–46 by permission of the author and the publisher; Figure 6.9c–e reproduced from B. Booth *et al.*, *Phil Trans R. Soc.* A288, 271–319 by permission of G. P. L. Walker and the Royal Society; Figures 6.13a and 6.15 reproduced from G. P. L.

Walker and R. Croasdale, *J. Geol. Soc., Lond.* 127, 17–55 by permission of G. P. L. Walker and Blackwell Scientific Publications; Figure 6.13d reproduced from K. Bloomfield *et al.*, *Geol. Rundschau* 66, 120–46 by permission of K. Bloomfield and the publisher; Figures 6.16a–d and 13.37–40 reproduced from G. P. L. Walker *et al.*, *Geol. Rundschau* 70, 1100–18 by permission of G. P. L. Walker and the publisher; Figures 6.17–19 reproduced from G. P. L. Walker, *J. Volcanol. Geotherm. Res.* 8, 69–94 by permission of the author and Elsevier Science Publishers, Figures 6.18 and 6.19 also from G. P. L. Walker, *Bull. Volcanol.* 44, 223–40 by permission of the author and the publisher; Figure 6.20 reproduced from L. Wilson, *Geophys. J. R. Astr. Soc.* 45, 543–56 by permission of the author and Blackwell Scientific Publications; Figure 6.22 reproduced from S. Self, *N.Z. J. Geol. Geophys.* 18, 189–95 by permission of the author and the publisher; Figure 6.23 reproduced from K. J. Murata *et al.*, *Bull. Volcanol.* 29, 765–96 by permission of the publisher; Figures 6.26 and 6.28 by S. Self *et al.*, reprinted from *Nature* Vol. 277, pp. 440–3 by permission of S. Self and the publisher, copyright © 1979 Macmillan Journals Limited; Figure 6.27 reproduced from I. A. Nairn and S. Self, *J. Volcanol. Geotherm. Res.* 3, 39–60 by permission of S. Self and Elsevier Science Publishers; Figures 6.30a–c reproduced from H. Sigurdsson, *Science* 216 (4 June 1982), 1106–8 by permission of the author and the publisher, © 1982 by the AAAS; Figures 6.33–36 reproduced from S. Self and R. S. J. Sparks, *Bull. Volcanol.* 41, 196–212 by permission of S. Self and the publisher; Figures 6.38 and 7.27 reproduced from G. P. L. Walker, in *Tephra studies* (S. Self and R. S. J. Sparks, eds), 317–30 by permission of the author and D. Reidel Publishing Company, Figure 6.38 also from S. N. Carey and H. Sigurdsson, *J. Geophys. Res.* 87, 7061–72 by permission of the author and the publisher, © 1982 by the American Geophysical Union; Figures 6.40 and 6.42–45 reproduced from R. S. J. Sparks and J. V. Wright, *Geol. Soc. Am. Spec. Paper* No. 180, 155–66 by permission of R. S. J. Sparks and the Geological Society of America; Figure 6.46a reproduced from J. V. Wright, *Geol. Rundschau* 69, 263–91 by permission of the author and the publisher.

Figure 7.1b reproduced from L. Wilson and J. W. Head, *U.S. Geol. Survey Bull.* No. 1250, 513–24 by permission of L. Wilson and the U.S. Geological Survey; Figures 7.2 and 7.3 reproduced from R. S. J. Sparks, *Sedimentology* 23, 147–88 by permission of the author, the Editor and Blackwell Scientific Publications; Figures 7.4, 7.5, 7.6b and 7.8 reproduced from J. N. Wilson, *J. Volcanol. Geotherm. Res.* 8, 231–49 by permission of the

author and Elsevier Science Publishers; Figure 7.7 by C. J. N. Wilson; Figures 7.17, 7.18b and 8.7a–c reproduced from R. S. J. Sparks, *Geol. Rundschau* 64, 497–523 by permission of the author and the publisher; Figures 7.18a and 8.3 reproduced from S. Yokoyama, *Tokyo Kyoiku Sci. Rep., Sect. C* 12, 17–62 by permission of the author; Figure 7.19 reproduced from J. V. Wright and G. P. L. Walker, *J. Volcanol. Geotherm. Res.* 9, 111–31 by permission of J. V. Wright and Elsevier Science Publishers; Figures 7.21–23 and 8.9 reproduced from R. S. J. Sparks *et al.*, *J. Geophys. Res.* 83, 1727–39 by permission of R. S. J. Sparks; Figures 7.25 and 7.29 reproduced from C. J. N. Wilson and G. P. L. Walker, *J. Geol. Soc., Lond.* 139, 581–92 by permission of Blackwell Scientific Publications; Figures 7.26a and 8.32a reproduced from G. P. L. Walker *et al.*, *J. Volcanol. Geotherm. Res.* 9 , 409–21 by permission of G. P. L. Walker and the publisher, © 1978 by the American Geophysical Union; Figures 7.36 and 7.48 reproduced from G. P. L. Walker and L. A. McBroome, *Geology* 1, 571–4 by permission of G. P. L. Walker and the Geological Society of America; Figures 7.38 and 7.43 reproduced from K. H. Wohletz and M. F. Sheridan, *Geol. Soc. Am. Spec. Paper* No. 180, 177–94 by permission of M. F. Sheridan and the Geological Society of America; Figure 7.44 reproduced from J. R. L. Allen, *Developments in sedimentology – 30B. Sedimentary structures* by permission of the author and Elsevier Science Publishers.

Figure 8.1 reproduced from T. A. Steven and P. W. Lipman, *U.S. Geol. Survey Prof. Paper* No. 958, 1–35 by permission of P. W. Lipman and the U.S. Geological Survey; Figure 8.2 by P. W. Francis *et al.*, reprinted from *Nature* Vol. 301, 51–3 by permission of the author and the publisher, copyright © 1983 Macmillan Journals Limited; Figure 8.5 reproduced from C. N. Fenner, *J. Geol.* 28, 569–606 by permission of the Editors, © 1920 by the University of Chicago, also from G. H. Curtis, *Geol. Soc. Am. Mem.* No. 116, 153–210 by permission of the author and the Geological Society of America; Figures 8.6 and 8.35 reproduced from P. D. Rowley *et al.*, *U.S. Geol. Survey Prof. Paper* No. 1250, 489–512 by permission of P. D. Rowley and the Geological Society of America; Figures 8.10 and 8.11a reproduced from L. Wilson *et al.*, *Geophys. J. R. Astr. Soc.* 63, 117–48 by permission of L. Wilson and Blackwell Scientific Publications; Figure 8.11b reproduced from S. N. Williams and S. Self, *J. Volcanol. Geotherm. Res.* 16, 33–56 by permission of S. Self and Elsevier Science Publishers; Figure 8.11c reproduced from L. Wilson, in *Thera and the Aegean World*, 31–5 by permission of the author and the publisher; Figures 8.12, 8.31, 8.33 and 8.47a reproduced from J. V. Wright, *Bull. Volcanol.* 44, 189–212 by permission of the author and the publisher; Figure 8.16 reproduced from C. H. Bacon, *J. Volcanol. Geotherm. Res.* 18, 57–115 by permission of the author and Elsevier Science Publishers; Figure 8.17 reproduced from S. Self and J. V. Wright, *Geology* 11, 443–6 by permission of J. V. Wright and the Geological Society of America; Figure 8.19 reproduced from J. V. Wright and G. P. L. Walker, *Geology* 5, 729–32 by permission of J. V. Wright and the Geological Society of America; Figures 8.20c, 8.21–23 and 13.27 reproduced from T. H. Druitt and R. S. J. Sparks, *J. Volcanol. Geotherm. Res.* 13, 147–71 by permission of R. S. J. Sparks and Elsevier Science Publishers; Figure 8.24 reproduced from D. Ninkovich *et al.*, *Bull. Volcanol.* 41, 286–98 by permission of D. Ninkovich and the publisher; Figure 8.25 reproduced from R. S. J. Sparks and T. C. Huang, *Geol. Mag.* 117, 425–36 by permission of R. S. J. Sparks and Cambridge University Press; Figure 8.30 reproduced from J. V. Wright *et al.*, in *Tephra studies* (S. Self and R. S. J. Sparks, eds), 433–9 by permission of J. V. Wright and D. Reidel Publishing Company; Figures 8.32b and 8.48 reproduced from G. P. L. Walker *et al.*, *J. Volcanol. Geotherm. Res.* 10, 1–11 by permission of G. P. L. Walker and Elsevier Science Publishers; Figure 8.32c reproduced from G. P. L. Walker *et al.*, *Geology* 8, 245–9 by permission of G. P. L. Walker and the Geological Society of America; Figure 8.37 reproduced from P. W. Lipman and R. L. Christiansen, U.S. Geol. Survey Prof Paper No. 501B, 74–8 by permission of P. W. Lipman and the U.S. Geological Survey; Figures 8.40 and 8.41 reproduced from D. H. Ragan and M. F. Sheridan, *Geol. Soc. Am. Bull.* 83, 95–106 by permission of M. F. Sheridan and the Geological Society of America; Figure 8.42 reproduced from R. L. Smith and R. A. Bailey, *Bull. Volcanol.* 29, 83–104 by permission of R. L. Smith and the publisher; Figure 8.45 reproduced from M. F. Sheridan, *Geol. Soc. Am. Bull.* 81, 851–68 by permission of the author and the Geological Society of America; Figures 8.51a and I.4 reproduced from G. P. L. Walker, *N.Z. J. Geol. Geophys.* 24, 304–24 by permission of the author and the publisher; Figures 8.51b, c and 8.53 reproduced from G. P. L. Walker, *J. Volcanol. Geotherm. Res.* 9, 395–407 by permission of the author and Elsevier Science Publishers; Figure 8.54 reproduced from G. P. L. Walker, *J. Volcanol. Geotherm. Res.* 8, 69–94 by permission of the author and Elsevier Science Publishers.

Figure 9.1a reproduced from R. S. Fiske, *Geol. Soc. Am. Bull.* 74, 391–406 by permission of the author and the Geological Society of America; Figures 9.1b and c reproduced from E. Yamada, *Journal of the Geological Society of Japan* 79, 585–97 by permission of the author and the publisher; Figures 9.1d and 9.2 reproduced from J. V. Wright and E. Mutti, *Bull. Volcanol.* 44, 153–67 by permission of J. V. Wright and the publisher; Figures 9.1e, 9.5 and 9.12 reproduced from R. S. Fiske and T. Matsuda, *Am. J. Sci.* 262, 76–106 by permission of R. S. Fiske and the publisher; Figure 9.3 reproduced from S. N. Carey and H. Sigurdsson, *J. Volcanol. Geotherm. Res.* 7, 67–86 by permission of H. Sigurdsson and Elsevier Science Publishers; Figure 9.4 reproduced from H. Sigurdsson *et al.*, *J. Geol.* 88, 523–40 by permission of the author and the Editors, © 1980 by the University of Chicago; Figure 9.6 reproduced from M. F. Howells *et al.*, Geol. Soc. Lond. Spec. Publ. No. 8, 611–8 by permission of M. F. Howells; Figures 9.7 and 9.8 reproduced from E. H. Francis and M. F. Howells, *J. Geol. Soc., Lond.* 129, 621–41 by permission of M. F. Howells and Blackwell Scientific Publications; Figure 9.9 reproduced from J. V. Wright and M. P. Coward, *Geol. Mag.* 114, 133–40 by permission of J. V. Wright and Cambridge University Press; Figure 9.10 by S. Self and M. R. Rampino, reprinted from *Nature* Vol. 294, pp. 699–704 by permission of S. Self and the publisher, copyright © 1981 Macmillan Journals Limited; Figure 9.14 by N. D. Watkins *et al.*, reprinted from *Nature* Vol. 271, pp. 122–6 by permission of R. S. J. Sparks and the publisher, copyright © 1978 Macmillan Journals Limited; Figure 9.15 reproduced from D. Ninkovich and N. J. Shackleton, *Earth Planet. Sci. Lett.* 27, 20–34 by permission of D. Ninkovich and Elsevier Science Publishers; Figures 9.16–18 reproduced from M. T. Ledbetter and R. S. J. Sparks, *Geology* 7, 240–4 by permission of R. S. J. Sparks and the Geological Society of America.

Figures 10.1, 13.33 and 13.34 reproduced from W. D. Kuenzi and O. H. Horst, *Geol. Soc. Am. Bull.* 90, 827–38 by permission of O. H. Horst and the Geological Society of America; Figure 10.4 reproduced from B. L. Jones *et al.*, *Geology* 12, 209–11 by permission of B. L. Jones and the Geological Society of America; Figure 10.6a reproduced from B. Voight, U.S. Geol. Survey Prof. Paper No. 1250, 69–80 by permission of G. Rosenquist and the U.S. Geological Survey; Figure 10.6b reproduced from P. W. Lipman and D. R. Mullineaux, U.S. Geol. Survey Prof. Paper No. 1250 by permission of the U.S. Geological Survey; Figure 10.8 reproduced from T. Ui, *J. Volcanol. Geotherm. Res.* 18, 135–50 by permission of the author and Elsevier Science Publishers; Figures 10.10 and 10.14a reproduced from

R. K. Fahnestock, in *Rockslides and avalanches, 1. Natural phenomena*, by permission of Elsevier Science Publishers, and from D. R. Crandell and R. K. Fahnestock, *U.S. Geol. Survey Bull.* 1221-A, 1–30 by permission of D. R. Crandell and the U.S. Geological Survey; Figure 10.14c by F. W. Williams; Figure 10.17 reproduced from J. C. Harms *et al.*, *Structures and sequences in clastic rocks*, SEPM short course No. 9 by permission of R. G. Walker and the Society of Economic Paleontologists and Mineralogists; Figures 10.26a and b reproduced from R. A. F. Cas, *J. Sediment. Petrol.* 49, 29–44 by permission of the author and the Society of Paleontologists and Mineralogists; Figure 10.27 reproduced from R. A. F. Cas, *J. Geol. Soc. Austral.* 24, 381–401 by permission of the author and the publisher; Figure 10.29a reproduced from G. V. Middleton and J. B. Southard, *Mechanics of sediment movement*, SEPM short course No. 3 by permission of G. V. Middleton and the Society of Economic Paleontologists and Mineralogists; Figure 10.29b reproduced from R. G. Walker, *Am. Assoc. Petrol. Geol. Bull.* 62, 932–66 by permission of the author; Figure 10.29c by R. A. F. Cas.

Figure 11.1 reproduced from R. Schmid, *Geology* 9, 41–3 by permission of the author and the Geological Society of America; Figure 11.4 reproduced from R. S. J. Sparks and G. P. L. Walker, *J. Volcanol. Geotherm. Res.* 2, 329–41 by permission of R. S. J. Sparks and Elsevier Science Publishers; Figure 11.5 by G. P. L. Walker, reprinted from *Nature* Vol. 281, pp. 642–6 by permission of the author and the publisher, © 1979 Macmillan Journals Limited.

Plates 13 and 15 reproduced from *Man on Earth – the marks of man: a survey from space* (C. Sheffield) by permission of Sidgwick & Jackson; Figure 13.1b reproduced from P. W. Lipman, *Bull. Volcanol.* 43, 703–25 by permission of the author and the publisher; Figure 13.4 reproduced from S. C. Porter, *Geol. Soc. Am. Bull.* 83, 3607–12 by permission of the author and the Geological Society of America; Figure 13.6 by P. Mohr, reprinted from *Nature* Vol. 303, pp. 577–84 by permission of the author and the publisher, copyright © 1983 Macmillan Journals Limited; Figures 13.10a and 13.12 reproduced from C. A. Wood, *J. Volcanol. Geotherm. Res.* 8, 137–60 by permission of the author and Elsevier Science Publishers; Figure 13.11 reproduced from G. Kieffer, in *Symp. J. Jung-géol., géomorph., struct. profonde du Massif Central français*, 479–510 by permission of the author; Figures 13.13 and 13.16 reproduced from E. B. Joyce, in *Quaternary studies* (R. P. Suggate and M. M. Creswell, eds), 169–76 by permission of the author and the Royal Society of New Zealand; Figure 13.18 reproduced from

C. A. Wood, in *Encyclopedia of volcanology* (J. Green, ed.) by permission of the author; Figure 13.19 reproduced from S. Self *et al.*, *J. Volcanol. Geotherm. Res.* 7, 39–65 by permission of S. Self and Elsevier Science Publishers; Figure 13.21 reproduced from K. H. Wohletz and M. F. Sheridan, *Am. J. Sci.* 283, 385–413 by permission of M. F. Sheridan and the publisher; Figure 13.27 reproduced from H. Pichler and S. Kussmaul, *Thera and the Aegean world* by permission of H. Pichler and the publisher; Figure 13.32 reproduced from R. K. Vessell and D. K. Davies, SEPM Spec. Publ. 31, 31–45 by permission of the Society of Economic Paleontologists and Mineralogists; Figure 13.35 reproduced from J. V. Wright *et al.*, *Geol. Mag.* 121, 1–15 by permission of J. V. Wright and Cambridge University Press; Figure 13.36 reproduced from J. A. Wolfe and S. Self, *Geophys. Mon. Ser.* 27, 157–72 by permission of S. Self and the publisher, © 1982 by the American Geophysical Union; Figure 13.39 reproduced from G. P. L. Walker *et al.*, *Geol. Rundschau* 70, 1100–18 by permission of G. P. L. Walker and the publisher, Figure 13.39 also from G. A. Mahood, *J. Volcanol. Geotherm. Res.* 8, 199–230 by permission of the author and Elsevier Science Publishers; Figure 13.42 reproduced from R. L. Smith and R. A. Bailey, Geol. Soc. Am. Mem. No. 116, 153–210 by permission of R. L. Smith and the Geological Society of America; Figures 13.45 and 13.46b by B. Clough *et al.*, reprinted from *Nature* Vol. 289, pp. 49–50 by permission of B. Clough and the publisher, copyright © 1981 Macmillan Journals Limited; Figure 13.48 reproduced from R. D. Ballard and T. H. van Andel, *Geol. Soc. Am. Bull.* 88, 523–30 by permission of R. D. Ballard and the Geological Society of America; Figure 13.49 by K. C. MacDonald, reproduced from the *Annual Review of Earth and Planetary Sciences* Vol. 10, 155–90 by permission of the author and the publisher, © 1982 by Annual Reviews Inc.; Figure 13.50 by J. G. Jones, reprinted from *Nature* Vol. 212, pp. 586–8 by permission of the author and the publisher, copyright © 1966 Macmillan Journals Limited; Figure 13.51 reproduced from G. Jones, *J. Geol. Soc., Lond.* 124, 197–211 by permission of the author and Blackwell Scientific Publications; Figure 13.52 reproduced from H. Furnes *et al.*, *J. Volcanol. Geotherm. Res.* 8, 95–110 by permission of H. Furnes and Elsevier Science Publishers.

Figure 14.12a reproduced from R. A. F. Cas, *J. Geol. Soc. Austral.* 24, 381–401 by permission of the author and the publisher; Figure 14.12b reproduced from R. A. F. Cas *et al.*, *J. Geol. Soc. Austral.* 28, 271–88 by permission of R. A. F. Cas and the publisher; Figure 14.13 reproduced from R. W. Hutchinson, *Econ. Geol.* 68, 1223–45

518 ACKNOWLEDGEMENTS

by permission of the Editor.

Figure 15.1a and b reproduced from E. M. Moores, *Rev. Geophys. Space Phys.* 20, 735–60 by permission of the author and the publisher, © 1982 by the American Geophysical Union; Figure 15.1c by R. N. Anderson *et al.*, reprinted from *Nature* Vol. 300, pp. 589–94 by permission of R. N. Anderson and the publisher, copyright © 1982 Macmillan Journals Limited; Figure 15.2 by D. E. Karig, reproduced from the *Annual Review of Earth and Planetary Sciences* Volume 2, 51–75 by permission of the author and the publisher, © 1974 by Annual Reviews Inc.; Figure 15.3 reproduced from G. Di Paola, *Bull. Volcanol.* 36, 517–60 by permission of the author and the publisher; Figure 15.4a reproduced from B. H. Baker *et al.*, Geol. Soc. Am. Spec. Paper No. 136, 1–67 by permission of B. H. Baker and the Geological Society of America; Figure 15.4b reproduced from P. Mohr and C. A. Wood, *Earth Planet. Sci. Lett.* 33, 126–44 by permission of P. Mohr and Elsevier Science Publishers; Figure 15.5a by G. P. Eaton, reproduced from the *Annual Review of Earth and Planetary Sciences* Volume 10, 409–40 by permission of the author and the publisher, © 1982 by Annual Reviews Inc.; Figure 15.5b reproduced from G. P. Eaton, *Tectonophysics* 102, 275–95 by permission of the author and Elsevier Science Publishers; Figure 15.6 reproduced from W. R. Dickinson, Pacific coast palaeogeographic symp. No. 3, 1–13 by permission of the author and the publisher; Figure 15.7 reproduced from D. E. Karig, *Geol. Soc. Am. Bull.* 82, 323–44 by permission of the author and the Geological Society of America; Figure 15.8 reproduced from K. B. Sporli, Int. Assoc. Sedimentologists Spec. Publ. No. 4, 147–70 by permission of the author, the Editor and Blackwell Scientific Publications; Figure 15.9 by J. C. Moore *et al.*, reprinted from *Nature* Vol. 281, pp. 638–42 by permission of J. C. Moore and the publisher, copyright © 1979 Macmillan Journals Limited; Figure 15.10 reproduced from A. Nur and Z. Ben-Avraham, *Tectonophysics* 99, 355–67 by permission of A. Nur and Elsevier Science Publishers; Figure 15.12 reproduced from D. E. Karig, *J. Geophys. Res.* 75, 239–54 by permission of the author and the publisher, © 1970 by the American Geophysical Union.

Figure I.2 reproduced from G. P. L. Walker *et al.*, *Geophys J. R. Astr. Soc.* 22, 377–83 by permission of G. P. L. Walker and Blackwell Scientific Publications; Figure I.3 reproduced from L. Wilson, *Geophys. J. R. Astr. Soc.* 30, 381–92 by permission of the author and Blackwell Scientific Publications. Fig. 1.4 reproduced from G. P. L. Walker, *N.Z.J. GEOL. Geophys.* 24, 305–24, by permission of the author and publisher.

INDEX

DATE DUE

DEMCO 38-297